Handbook of Experimental Pharmacology

Volume 172

Editor-in-Chief

K. Starke, Freiburg i. Br.

Editorial Board

G.V.R. Born, London
M. Eichelbaum, Stuttgart
D. Ganten, Berlin
F. Hofmann, München
W. Rosenthal, Berlin
G. Rubanyi, Richmond, CA

Molecular Chaperones in Health and Disease

Contributors

S. Alberti, I. J. Benjamin, M. Brunet, J. Buchner, E. S. Christians,
P. Csermely, C. Didelot, F. Edlich, J. G. Facciponte, G. Fischer,
M. Gaestel, C. Garrido, J. M. Harrell, J. Höhfeld, I. Judson,
H. H. Kampinga, R. J. Kaufman, M. Morange, Y. Morishima,
G. Multhoff, P. J. M. Murphy, L. Neckers, E. Papp, W. B. Pratt,
T. Scheibel, C. Soti, J. R. Subjeck, J. Tatzelt, R. Voellmy,
X.-Y. Wang, K. F. Winklhofer, P. Workman, K. Zhang

Editor
Matthias Gaestel

Professor
Matthias Gaestel
Medical School Hannover
Institute of Biochemistry
Carl-Neuberg-Str. 1
30625 Hannover
Germany
Gaestel.Matthias@mh-hannover.de

With 36 Figures and 20 Tables

ISSN 0171-2004

ISBN-10 3-540-25875-2 Springer Berlin Heidelberg New York

ISBN-13 978-3-540-25875-9 Springer Berlin Heidelberg New York

Library of Congress Control Number: 2005925882

This work is subject to copyright. All rights reserved, whether the whole or part of the material is concerned, specifically the rights of translation, reprinting, reuse of illustrations, recitation, broadcasting, reproduction on microfilm or in any other way, and storage in data banks. Duplication of this publication or parts thereof is permitted only under the provisions of the German Copyright Law of September 9, 1965, in its current version, and permission for use must always be obtained from Springer. Violations are liable for prosecution under the German Copyright Law.

Springer is a part of Springer Science + Business Media
springeronline.com

© Springer-Verlag Berlin Heidelberg 2006
Printed in Germany

The use of general descriptive names, registered names, trademarks, etc. in this publication does not imply, even in the absence of a specific statement, that such names are exempt from the relevant protective laws and regulations and therefore free for general use.

Product liability: The publishers cannot guarantee the accuracy of any information about dosage and application contained in this book. In every individual case the user must check such information by consulting the relevant literature.

Editor: S. Rallison
Editorial Assistant: S. Dathe
Cover design: *design&production* GmbH, Heidelberg, Germany
Typesetting and production: LE-TEX Jelonek, Schmidt & Vöckler GbR, Leipzig, Germany
Printed on acid-free paper   27/3151-YL - 5 4 3 2 1 0

Preface

More than 40 years after Ferruccio Ritossas's discovery of heat shock-induced formation of puffs in the polytene chromosomes of Drosophila, many different families of heat shock proteins and molecular chaperones have emerged and have been continuously studied in structure and function until today. As a result, it has now become clear that molecular chaperones are involved in a wide variety of essential processes in living cells. Molecular chaperones not only protect cells from stress damage by keeping cellular proteins in a folding competent state and preventing them from irreversible aggregation, but they are also expressed constitutively in the cell and participate in complex processes such as protein synthesis, intracellular protein transport, post-translational modification, and secretion of proteins as well as receptor signaling. Hence, it is not surprising that molecular chaperones are implicated in the pathogenesis of many relevant diseases and could be regarded as potential pharmacological targets.

Starting with the analysis of the mode of action of chaperones and the regulation of the heat shock response at the molecular, cellular, and organism level, this book then focuses on specific aspects, such as signal transduction, development, apoptosis, protein aggregation in vivo, oncogenic transformation, and immune response, where chaperones play a crucial role. Modulation of chaperone action and the use of so-called small-molecule chaperones is clearly of pharmacological interest and could be of therapeutic relevance for the treatment of diseases that result from deregulation of the above-mentioned cellular processes in which chaperones are involved. The chapters of the third and last part of this volume critically analyze this potential and illustrate significant progress in therapeutic application.

I am delighted that so many experts in the different areas of chaperone research contributed to this volume and I want to thank them for providing high-quality chapters in time to meet the publisher's deadline and ensure publication of this exciting volume in 2005. I want to acknowledge the very effective and professional support from Mrs. Susanne Dathe at Springer. I hope that this volume will give a balanced overview as well as specific and detailed information on most of the topics of interest for both specialists in the field and newcomers involved in experimental and therapeutic approaches of pharmacological modulation of heat shock response and chaperone action.

Hannover, May 2005   Matthias Gaestel

List of Contents

Chaperones in Preventing Protein Denaturation in Living Cells
and Protecting Against Cellular Stress 1
   H. H. Kampinga

Feedback Regulation of the Heat Shock Response 43
   R. Voellmy

Protein Folding in the Endoplasmic Reticulum
and the Unfolded Protein Response 69
   K. Zhang, R. J. Kaufman

Molecular Chaperones in Signal Transduction 93
   M. Gaestel

Chaperoning of Glucocorticoid Receptors 111
   W. B. Pratt, Y. Morishima, M. Murphy, M. Harrell

Heat Shock Response: Lessons from Mouse Knockouts 139
   E. S. Christians, I. J. Benjamin

HSFs in Development 153
   M. Morange

Heat Shock Proteins: Endogenous Modulators of Apoptotic Cell Death . 171
   C. Didelot, E. Schmitt, M. Brunet, L. Maingret, A. Parcellier, C. Garrido

Protein Aggregation as a Cause for Disease 199
   T. Scheibel, J. Buchner

The Role of Chaperones in Parkinson's Disease and Prion Diseases . . . 221
   K. F. Winklhofer, J. Tatzelt

Chaperoning Oncogenes: Hsp90 as a Target of Geldanamycin 259
   L. Neckers

Heat Shock Proteins in Immunity 279
   G. Multhoff

Molecular Chaperones and Cancer Immunotherapy 305
    X.-Y. Wang, J. G. Facciponte, J. R. Subjeck

Hsp90 Inhibitors in the Clinic 331
    S. Pacey, U. Banerji, I. Judson, P. Workman

Pharmacological Targeting of Catalyzed Protein Folding:
The Example of Peptide Bond *cis/trans* Isomerases 359
    F. Edlich, G. Fischer

Chemical Chaperones: Mechanisms of Action and Potential Use 405
    E. Papp, P. Csermely

Pharmacological Modulation of the Heat Shock Response 417
    C. Sőti, P. Csermely

Subject Index 437

List of Contributors

(Addresses stated at the beginning of respective chapters)

Banerji, U. , 331
Benjamin, I.J. , 139
Brunet, M. , 171
Buchner, J. , 199

Christians, E.S. , 139
Csermely, P. , 405, 417

Didelot, C. , 171

Edlich, F. , 359

Facciponte, J.G. , 305
Fischer, G. , 359

Gaestel, M. , 93
Garrido, C. , 171

Harrell, M. , 111

Judson, I. , 331

Kampinga, H.H. , 1
Kaufman, R.J. , 69

Maingret, L. , 171
Morange, M. , 153

Morishima, Y. , 111
Multhoff, G. , 279
Murphy, M. , 111

Neckers, L. , 259

Pacey, S. , 331
Papp, E. , 405
Parcellier, A. , 171
Pratt, W.B. , 111

Sőti, C. , 417
Scheibel, T. , 199
Schmitt, E. , 171
Subjeck, J.R. , 305

Tatzelt, Jörg , 221

Voellmy, R. , 43

Wang, X.-Y. , 305
Winklhofer, K.F. , 221
Workman, P. , 331

Zhang, K. , 69

Chaperones in Preventing Protein Denaturation in Living Cells and Protecting Against Cellular Stress

H. H. Kampinga

Department of Cell Biology, Section of Radiation and Stress Cell Biology, Faculty of Medical Sciences, University of Groningen, Ant. Deusinglaan 1, 9713 AV Groningen, The Netherlands
h.h.kampinga@med.rug.nl

1	The Molecular Chaperone Concept	2
2	Scope of This Chapter	3
3	The Hsp70 Gene Family	3
3.1	Regulation of the Mammalian Hsp70 ATP/ADP Cycle	7
3.1.1	The Hdj Family	7
3.1.2	Hip	9
3.1.3	The Bag Family	10
3.1.4	CHIP	11
3.1.5	Other Possible Regulators of the Hsp70 Chaperone Machine	11
3.2	Hsp70-Hsp90 Connections and Hop and Tpr-2 Co-factors	12
3.3	Hsp70-Hsp27 Connections	13
4	Internal and External Proteotoxic Stresses in Mammalian Cells	13
5	Thermotolerance and Hsp Reallocations in Mammalian Cells	14
6	Models Systems for Studying Hsp Chaperone-Like Activities in Stressed Living Mammalian Cells	16
6.1	Effects of Hsp Against Effects of Heat on Cellular Structures and Functions	16
6.2	Hsp Expression and Endogenous Reporter Proteins	17
6.3	Hsp Expression and Exogenous Reporter Proteins	17
6.3.1	Hsp70 and Hsp40	17
6.3.2	Hsp70 and Hip	20
6.3.3	Hsp70 and Bag-1	20
6.3.4	Hsp70 and CHIP	22
6.3.5	Chaperone Networks	22
7	Fate of Proteins: Folding or Degradation?	23
8	Proteotoxic Damage: Cellular Targets, Sorting, and Hsp-Mediated Protection	24
8.1	Processing of Damaged Proteins in the Cytosol	25
8.2	Processing of Damaged Proteins in the Nucleus	26
9	Cell Death Expression, Cell Transformation and Hsp	27
10	Concluding Remarks	30
	References	31

Abstract A variety of cellular internal and external stress conditions can be classified as proteotoxic stresses. Proteotoxic stresses can be defined as stresses that increase the fraction of proteins that are in an unfolded state, thereby enhancing the probability of the formation of intracellular aggregates. These aggregates, if not disposed, can lead to cell death. In response to the appearance of damaged proteins, cells induce the expression of heat shock proteins. These can function as molecular chaperones to prevent protein aggregation and to keep proteins in a state competent for either refolding or degradation. Most knowledge of the function and regulation (by co-factors) of individual heat shock proteins comes from cell free studies on refolding of heat- or chemically denatured, purified proteins. Unlike the experimental situation in a test tube, cells contain multiple chaperones and co-factors often moving in and out different subcompartments that contain a variety of protein substrates at different folding states. Also, within cells folding competes with the degradative machinery. In this chapter, an overview will be provided on how the main cytosolic/nuclear chaperone Hsp70 is regulated, what is known about its interaction with other main cytosolic/nuclear chaperone families (Hsp27, Hsp90, and Hsp110), and how it may function as a molecular chaperone in living mammalian cells to protect against proteotoxic stresses.

Keywords Chaperones · Mammalian cells · Hsp70 · Proteotoxic stress

1
The Molecular Chaperone Concept

The successful in vitro (cell-free system) folding of purified ribonuclease A by Anfinsen (Anfinsen 1973) led to the suggestion that all information necessary for a polypeptide to fold was an intrinsic feature of its primary structure and was independent of other factors (Ellis and Hemmingsen 1989). Most of these refolding experiments were performed by first denaturing a purified polypeptide with chemical agents and then removing the denaturant. The probability that such a polypeptide will fold correctly after removing the denaturant increases at low protein concentration (which limits interpolypeptide interactions) and low temperatures (which attenuates hydrophobic interactions). The high protein concentrations and temperatures within the cell lead to premature interactions of newly synthesized polypeptides, often accompanied by misfolding and aggregation (Jaenicke 1991). To assist polypeptide folding in vivo, a set of proteins, called molecular chaperones, exist whose function is to ensure that polypeptides will either fold or be transported properly. In biochemical terms, a molecular chaperone is defined as "a protein that prevents improper interactions between potentially complementary surfaces and disrupts any improper liaisons that may occur" (Ellis and Hemmingsen 1989). The proposed function of chaperones is to assist in self-assembly of proteins by inhibiting alternative assembly pathways that produce nonfunctional structures. It is important to note that a chaperone activity merely prevents aggregation and does not necessarily need to be associated with (re)folding of the bound substrate.

Most of our knowledge on the function and regulation of molecular chaperones and interactions between the main chaperone machines comes from cell-free experiments in which model proteins are either chemically or thermally denatured in the absence and presence of proposed chaperones (Buchner et al. 1991; Jakob et al. 1993; Knauf et al. 1994; Martin et al. 1993; Skowyra et al. 1990; Wiech et al. 1992). Aggregation of the polypeptides can be measured by light scattering and refolding to the native state by measuring the activity of the polypeptides. These studies have revealed that the mammalian heat shock proteins Hsp27, Hsp60, Hsp70, Hsp90, and Hsp110 can hold unfolded model substrates in a refoldable conformation during stress conditions. Subsequent refolding of the unfolded proteins requires the activities of the Hsp70 chaperone machine (Buchner 1996; Ehrnsperger et al. 1997; Freeman and Morimoto 1996; Jakob et al. 1993; Schneider et al. 1996; Wiech et al. 1992). This is why the Hsp70 chaperone machine is the most widely studied one in relation to cellular stresses and why this chapter will emphasize on this machine. However, some Hsp70 interactions with Hsp27, Hsp90, or Hsp110 will also be discussed because they are relevant for the protective action of Hsp70 against cellular stresses.

2
Scope of This Chapter

Unlike the experimental situation in a test tube, cells contain multiple chaperones in different (sub)compartments to which they often can move in and out dynamically and in which other macromolecules—protein substrates at different folding states, a degradative machinery, and a variety of co-factors— are present, which will influence the activities of the heat shock proteins. At the same time, when cells are exposed to external stresses such as heat shock, energy is absorbed throughout the cell, damaging nearly all classes of macromolecules and affecting many cellular structures and functions. This chapter provides an overview of the cytoprotective chaperone functions of heat shock proteins reflected in this complex environment of a living cell. To do so, first a short overview will be provided on the data obtained with cell-free systems regarding the regulation of the Hsp70 chaperone machine, as these have provided a basis and a toolbox for the work in living mammalian cells.

3
The Hsp70 Gene Family

Proteins encoded by the Hsp70 gene family are highly conserved and found in all organisms, from bacteria to humans. In prokaryotes, 3 Hsp70 family members—DnaK, Hsc66 and Hsc62—exist (Itoh et al. 1999). Most of our knowledge on the function and regulation of the Hsp70 family proteins comes from work on DnaK. DnaK (together with DnaJ and GrpE) was originally discovered as a protein involved in the replication of bacteriophage λ DNA (Friedman et al. 1989). DnaK is constitutively expressed, inducible by heat shock, and essential for cell growth at all temperatures (Lindquist and Craig 1988). Several studies reported on the interaction of DnaK with unfolded polypeptides and how it functions as a molecular chaperone (Hendrick and Hartl 1993; Schroder et al. 1993; Skowyra et al. 1990). Yet it is important to keep in mind that the function of DnaK (and likely also the mammalian family members) is not restricted to dealing with unfolded proteins alone.

In yeast, there are several Hsp70 family members with partially overlapping functions and localization (Craig et al. 1995). Yeast express a very special chaperone, Hsp104, that appears to be the critical component for thermotolerance (Glover and Lindquist 1998; Parsell et al. 1994). Expression of this yeast Hsp104 in mammalian cells can enhance the cellular chaperone capacity and inhibit heat-shock-induced loss of viability in cooperation with the endogenous human Hsc70 machine (Mosser et al. 2004). Yet, since no Hsp104 equivalent has been found in mammals so far, the yeast system may not be totally representative for what happens in higher eukaryotes and will not be further discussed here.

The nomenclature of the 70-kDa mammalian heat-shock proteins has been very confusing, but a few years ago a hitchhiker's guide was provided on all 70-kDa heat-shock proteins (Tavaria et al. 1997). In the database, 15 gene loci are found. Here, I will mostly concentrate on the two main cytosolic/nuclear isoforms, namely the stress-inducible Hsp70.1 (here referred to as Hsp70) and the constitutively expressed isoform Hsp70.8 (here referred to as Hsc70) (Table 1). Although these proteins share a high level of sequence identity, there is no evidence that they have completely overlapping functions and that they are equally responsive to co-factors. In this review, it will be cited whether Hsc70 or Hsp70 was used whenever possible, but caution is needed here because in many reports no distinction was made.

Several in vitro studies have demonstrated that mammalian Hsp70 and Hsc70 can function in the refolding of denatured proteins (Freeman and Morimoto 1996; Minami et al. 1996; Nimmesgern and Hartl 1993; Schumacher et al. 1994, 1996). Hsp70s consist of a conserved N-terminal 44-kDa ATPase domain (a.a. 1–392), a less well conserved 18-kDa peptide-binding domain (a.a. 393–536), and a C-terminal 10-kDa variable domain of unknown function (a.a. 537–638) (Fig. 1a).

Binding of substrates to the peptide binding domain is through consensus sequences in the client protein. These sequences are highly hydrophobic and are typically found within the interior of folded proteins (Rudiger et al. 1997). Statistically, such sites are present every 36 residues in all proteins. This would be consistent with the core function of Hsp70s, which is the binding and release of unfolded segments of a wide variety of polypeptides to prevent

Table 1 Summary of the subcellular localization and proposed functions of Hsp70-1 and Hsp70-8

Hsp70	Alternative names	Locus symbol	Subcellular localization	Functions	References
Hsp70-1	Hsp70 Hsp72 Hsp70i	HspA1A	Cytosol/ nucleus	Prevents aggregation of unfolded or misfolded proteins and assists in refolding	Freeman Myers et al. 1995b; Michels et al. 1997
				Binds AU-rich RNA	Shi et al. 1998
				Autoregulates heat shock response	Sturzbecher et al. 1987
				Binds (mutant) p53	Zylicz et al. 2001
			Extracellular/ membrane	Increases immunogenicity of tumor cells	Asea et al. 2000; Botzler et al. 1996; Multhoff et al. 1999
				Stimulates cytokine production	Melcher et al. 1998; Todryk et al. 1999
Hsp70-8	Hsc70 Hsp73	HspA8	Cytosol/ nucleus	Assists in folding of newly synthesized proteins	Frydman et al. 1994
				Assists in refolding of unfolded proteins	Ciavarra et al. 1994
				Involved in lysosomal protein degradation	Hayes and Dice 1996
				Required for proteasomal protein degradation	Bercovich et al. 1997
				Assists in polypeptide translocation	Terada et al. 1995
				Binds p53	Sturzbecher et al. 1987; Fourie et al. 1997b
				Involved in steroid receptor complex assembly	Hutchinson et al. 1996

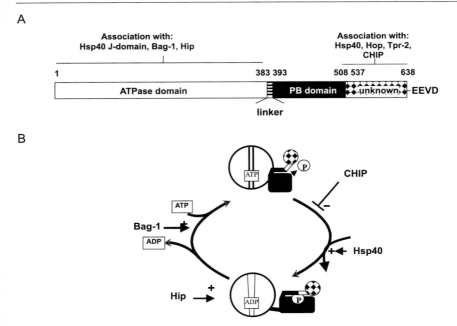

Fig. 1a Linear representation of Hsp70 domains. **b** Hsp70 ATPase cycle, its regulation by co-factors, and a model for domain coupling as derived from studies on DnaK (adapted from Rudiger et al. 1997). See text for further details

their aggregation (Rudiger et al. 1997). The conformation of the substrate-binding domain is under tight control of the ATPase domain and has at least two distinct conformations: an open conformation, which is formed when the ATPase domain binds ATP (Buchberger et al. 1994, 1995) and a closed conformation, which is formed upon ATP hydrolysis (Wall et al. 1994, 1995). The open conformation is characterized by a rapid substrate on/off rate and the closed form by a slow substrate on/off rate (Greene et al. 1995; Schmid et al. 1994; Takeda and Mckay 1996) (Fig. 1b). It is generally believed that ATP binding enables substrate binding, and that substrate binding is stabilized by subsequent ATP hydrolysis, locking the substrate to the binding domain (Beissinger and Buchner 1998; Bukau and Horwich 1998).

Central to the functioning of Hsp70s is the allosteric modulation of substrate affinity by ATP-induced conformational changes. ATP binding to the ATPase domain stimulates the release of a substrate from the peptide binding domain (Pellecchia et al. 2000; Zhu al. 1996) (Fig. 1b). The Hsp70 ATP binding domain is capable of changing the conformation of the substrate binding domain. Conversely, substrate binding stimulates ATP hydrolysis (Bukau and Horwich 1998; Ha et al. 1997) and thus alters the conformation of the ATPase domain.

The exact function of the 10-kDa domain still remains elusive. It has been proposed to be involved in self-association of Hsp70 into dimers and trimers (Benaroudj et al. 1997), suggesting that intermolecular association of mammalian Hsp70 proteins also may occur. The physiological relevance of oligomerization is not clear, but one might speculate that it is involved in regulation or storage of Hsp70 molecules. Interestingly, a bovine Hsc70 deleted of the entire 10-kDa C-terminal is almost as active as the wild type protein in the in vitro uncoating of coated vesicles (Ungewickell et al. 1997).

Only eukaryotic cytoplasmic Hsp70s terminate in the sequence GP(T/K)(V/I)EEVD (Boorstein et al. 1994). The interdomain communication is lost by deletion or mutation of this EEVD motif and the ATPase activity of such mutants is increased concomitantly. In apparent contrast with the entire deletion of the 10-kDa C-terminal domain (Ungewickell et al. 1997), deletion of the EEVD does reduce the ability to bind to and refold denatured substrates in vitro when compared to the full length Hsp70 (Freeman et al. 1995). EEVD deletion does completely abolish the regulation and binding of the cytoplasmic Hsp70 and Hsc70 by Hsp40/Hdj1 (Demand et al. 1998; Freeman et al. 1995). This suggests that at least for the cytoplasmic Hsp70 family members, part of their regulation is mediated by the extreme C-terminus.

3.1
Regulation of the Mammalian Hsp70 ATP/ADP Cycle

For the substrate-binding-and-release cycle, Hsp70 needs the assistance of co-chaperones. In the reaction cycle of the prokaryotic DnaK, the Hsp40 family member DnaJ delivers the substrate to DnaK and catalyses the hydrolysis of ATP. Subsequent exchange of ADP for ATP in order to release the substrate from DnaK requires the cooperation of GrpE (Liberek et al. 1991, 1995; Wawrzynow et al. 1995). The regulation of DnaK has long been held as a paradigm for Hsp70 regulation. However, the regulation of the mammalian Hsp70 machine is more complex: there are more co-factors regulating the substrate-binding-and-release cycle (Fig. 1b, Table 2) and many family members of at least some of these co-factors exist. Moreover, there are several factors that regulate the interaction with other major chaperones (Sects. 3.2 and 3.3). In general, Hsp40s and Hip enhance Hsp70-dependent refolding of unfolded model substrates, whereas Bag1, Chip, and HSBP1 inhibit Hsp70 chaperone activity in vitro (Table 2).

3.1.1
The Hdj Family

The *Escherichia coli* DnaJ contains four distinct domains: a J-domain, a glycine/phenylalanine (G/F)-rich domain, a cysteine-rich domain, and a C-terminal domain. Mammals have over 30 DnaJ homologs (Ohtsuka and Hata 2000) that

Table 2 Subcellular localization, interacting domains, and functions of Hsp/Hsc70 co-chaperones

Hsp/Hsc70 co-chaperone	Subcellular localization	Domain co-chaperone	Domain Hsp/Hsc70	Influence on Hsp/Hsc70 chaperone activity (in vitro)	References
Hdj-family	Cytosol/nucleus	J-domain	ATPase domain and C-terminal domain	Stimulates ATP hydrolysis: ↑ chaperone activity	Cheetham and Caplan 1998; Demand et al. 1998; Freeman et al. 1995; Greene et al. 1998; Minami et al. 1996; Suh et al. 1998
Hip	Cytosol	TPR domains + charged α-helical domain	ATPase domain	Stabilizes ADP-bound conformation: ↑ chaperone activity	Hohfeld et al. 1995; Irmer and Hohfeld 1997; Prapapanich et al. 1996
Bag-1 family	Cytosol/nucleus	Bag domain	ATPase domain	Accelerates nucleotide exchange: ↓ chaperone activity	Bimston et al. 1998; Takayama et al. 1997; Zeiner et al. 1997
CHIP	Cytosol	TPR1 domain	C-terminus	Inhibits ATP hydrolysis: ↓ chaperone activity	Ballinger et al. 1999
HspBP$_1$	Unknown	Unknown	ATPase domain	Prevents ATP binding: ↓ chaperone activity	Raynes and Guerriero 1998

can be classified into three types (Cheetham and Caplan 1998): type I proteins with full domain conservation with DnaJ, type II proteins containing the N-terminal J- and the G/F-rich domains, and type III proteins containing only a J-domain, positioned anywhere in the protein.

Type I DnaJ-like proteins have chaperone activity on their own: they can bind some (unfolded) substrates, thereby preventing their aggregation, and present them to DnaK/Hsp70 (Bukau and Horwich 1998; Cheetham and Caplan 1998). This capacity seems absent in type II proteins, but this is still being debated. The most studied human Hdj family members are Hsp40/Hdj1 (Hata et al. 1996) and Hsj-1 (Cheetham et al. 1992), type II proteins. The most conserved action of nearly all Hdj family proteins is their ability to enhance the ATPase activity of Hsp70 (Cheetham and Caplan 1998) by their J-domain. This J-domain (together with the G/F-rich domain) is required for interaction with the C-terminal ATPase domain of Hsp70 and may also be responsible for recruitment of Hsp70 family members to specific cellular processes, thereby increasing the local concentration of Hsp70 (Kelley 1998). Since N-terminal EEVD deletions in Hsp70 also abolish interactions with Hdj proteins (see above), Hsp70 and Hdj-1 proteins likely interact in a bipartite manner.

Cooperation with Hsp40/Hdj1 seems essential for Hsp70 to function in the in vitro refolding of denatured luciferase or β-galactosidase (Freeman et al. 1995; Freeman and Morimoto 1996). The refolding reaction is in some assays dependent on addition of reticulocyte lysate (Minami et al. 1996), suggesting that additional factors are required. A mutation in the conserved HDP motif of the J-domain abolishes the ability of Hdj-proteins to interact with Hsp70, leading to a loss in refolding capacity of the Hsp70 machine (Kelley 1998). A substoichiometric concentration of Hsp40/Hdj1 seems sufficient for stimulation of the ATPase activity of Hsc70 (Nagata et al. 1998).

3.1.2
Hip

The Hsc70-interacting protein (Hip), for which no homolog is present in *E. coli* binds to Hsc70/Hsp70, but only when the Hsp70 ATPase activity has been stimulated by Hsp40/Hdj1 (Höhfeld et al. 1995). Hip belongs to the tetratricopeptide repeat (TPR) protein family and contains three repeats of a degenerated 34 amino acids motif (reviewed by Smith 1998), which are required for interaction with the ATPase domain of Hsp70 (Prapapanich et al. 1996). Furthermore, the N-terminus of Hip is involved in its oligomerization, probably into homotetramers (Höhfeld et al. 1995). This oligomerization is dispensable for Hsc70 association (Prapapanich et al. 1996; Irmer and Höhfeld 1997).

The presence of Hip in an in vitro mixture of Hsc70, Hsp40/Hdj1, and ATP results in stabilization of the Hsc70 ADP state (Höhfeld et al. 1995; Höhfeld and Jentsch 1998). Hip can enhance the refolding of chemically and thermally denatured luciferase (Höhfeld et al. 1995; Lüders et al. 1998), albeit not under conditions where Hsp40 levels are high (Gebauer et al. 1997).

It has been suggested that Hip also has a chaperone activity on its own, as it can bind denatured proteins and prevent in vitro protein aggregation (Höhfeld et al. 1995; Bruce and Churchich 1997). However, prevention from aggregation by Hip alone is not sufficient for enhanced protein refolding. In fact, the presence of Hip alone inhibits the recovery of chemically denatured malate dehydrogenase and alkaline phosphatase (Bruce and Churchich 1997). This would be consistent with binding and trapping of denatured proteins by Hip and indicates that cooperation with other chaperones is essential to allow protein refolding.

3.1.3
The Bag Family

Bag-1 was first identified as an anti-apoptotic protein, associating with Bcl2, a protein that blocks programmed cell death (Takayama et al. 1995). The same proteins was independently cloned from human liver cells as a protein associating with the glucocorticoid receptor and referred to as Rap46 (receptor associating protein 46) (Zeiner and Gehring 1995). Subsequently, Bag-1 was identified as an Hsp70 co-factor (Takayama et al. 1997; Zeiner et al. 1997; Gebauer et al. 1997, 1998b: Bimston et al. 1998). Four different isoforms of Bag-1 are expressed in human cells, which are encoded by the same mRNA and translated from different translation initiation sites through a leaky scanning mechanism. These isoforms are 50, 46, 33, and 29 kDa in size, have identical C-termini, but differ in their N-terminal extensions and subcellular localization (Nollen et al. 2000; Yang et al. 1998). All these isoforms interact with the N-terminal ATPase domain of Hsp70 via the C-terminal domain (Takayama et al. 1997). In human cells, so far five different Bag family members (now termed Bag-1 to Bag-5) have been identified of which, besides Bag-1, also Bag-2 and Bag-3 bind to Hsp70 (Takayama et al. 1999). Both human and mouse Bag-1 contain an ubiquitin-like domain (Hohfeld 1998) that may serve as an attachment site for the covalent binding of other ubiquitin molecules that could serve as a degradation-independent targeting signal. Alternatively, this domain might regulate the turnover of Bag-1 (Takayama et al. 1995) or even enable the degradation in trans of a Bag-1-associated substrate (Hohfeld 1998; Zeiner et al. 1997) together with CHIP (see Sect. 3.1.4). The ubiquitin-like domain of Bag-1 is not essential for association with Hsc70, nor for modulation of the Hsc70 functionality in the refolding of denatured β-galactosidase (Takayama et al. 1997).

In cell-free systems, Bag-1 as well as Bag-2 and Bag-3 inhibit Hsp70-dependent refolding of chemically or heat-denatured substrates (Bimston et al. 1998; Takayama et al. 1997, 1999; Zeiner et al. 1997). Initially, there was a debate on how this inhibition came about. On one hand, it was proposed that Bag-1, by binding to the ATPase domain, would prevent ATP-dependent substrate release, thereby maintaining substrates in a ternary complex with Hsp70 (Bimston et al. 1998). An alternative was that Bag-1, by stimulating the exchange of ADP for ATP similar to GrpE, caused a premature release of the Hsp70-bound unfolded substrate, thereby destabilizing the complex between Hsp70 and the substrate (Hohfeld and Jentsch 1997). The subsequent finding that the crystal structure of the Bag/Hsc70 complex closely resembled that of the GrpE/DnaK complex (Sondermann et al. 2001), despite the lack of sequence homology between Bag-1 and GrpE, now favors the model that Bag-1 has GrpE-like functions. It is noteworthy that binding of human Bag-1 to Hsp70 results in a displacement of the Hsp70 co-factors Hsp40, Hip, and Hop (Gebauer et al. 1997, 1998a, 1998b; Hohfeld and Jentsch 1997; Zeiner et al. 1997). This strongly suggests that Bag-1 acts on Hsp70 in the absence of co-factors and, moreover, acts dominantly over other Hsp70 regulators.

3.1.4
CHIP

The 35-kDa cytoplasmic protein CHIP (carboxyl terminus of Hsc70-interacting protein) is yet another mammalian co-factor involved in the regulation of the Hsp70 chaperone machine. CHIP interacts with the C-terminus of Hsc70 via its N-terminal residues 1–197, containing a tetratricopeptide domain and an adjacent charged domain (Ballinger et al. 1999). CHIP also contains a C-terminal U-box that is not required for interaction with Hsc70 but that provides CHIP with an E3 ubiquitin ligase activity, with Hsc70 itself as a major target (Jiang et al. 2001).

Two different groups have shown that CHIP, similar to Hsp40, acts on Hsc70 in the ATP-bound state. However, in contrast to Hsp40, CHIP inhibits the Hsp40-stimulated ATPase activity of Hsc70 (Ballinger et al. 1999; Jiang et al. 2001). In parallel, CHIP inhibits the refolding of thermally denatured firefly luciferase in vitro in the presence of Hsc70 and Hsp40 at a molar CHIP to Hsp40 to Hsp70 ratio of 2:2:1 (Jiang et al. 2001). CHIP also interacts with the TPR acceptor site of Hsp90, which leads to remodeling of the Hsp90 complex (Connell. et al. 2001).

3.1.5
Other Possible Regulators of the Hsp70 Chaperone Machine

There are still a number of other, less investigated factors that have been shown to affect the Hsp70-mediated refolding. First, the pro-apoptotic factor Scythe can inhibit Hsp70-mediated refolding in vitro: a Bag-like domain appears responsible for this action. Reaper, a *Drosophila* protein that acts upstream of Scythe in the apoptotic program, but for which no human homolog has been discovered yet, can reverse the Scythe-mediated inhibition, probably by displacing Scythe from Hsp70 through inducing conformational changes in Scythe and via competitive binding to Hsp70 (Thress et al. 2001). Another factor, HspBP1, prevents ATP binding (Raynes and Guerriero 1998) and was later demonstrated to be a nucleotide exchange factor of Hsp70 (Kabani et al. 2002). Although mechanistically different from Bag-1 (Shomura et al. 2005), like Bag-1 this factor can stimulate folding at very low HspBP1:Hsc/p70 ratios (Shomura et al. 2005), but folding is inhibited at higher ratios (Raynes and Guerriero 1998; Kabani et al. 2002), likely due to premature substrate release.

3.2
Hsp70-Hsp90 Connections and Hop and Tpr-2 Co-factors

Although Hsp90 seems a more specialized chaperone, dealing with maturation of hormone receptors and many other growth factors and signaling molecules, it also can function under stress conditions to capture heat-denatured proteins. But, in vitro Hsp90 alone does not stimulate refolding of non-native β-galactosidase, nor does it enhance Hsp70-dependent refolding. Yet Hsp90 can maintain the non-native substrate in a folding-competent state which, upon addition of Hsp70, Hdj-1, and nucleotide, leads to refolding (Freeman and Morimoto 1996).

A number of proteins are involved in the connection between Hsc70 and Hsp90. The first co-factor of Hsp90 and Hcp70, STI1 in yeast (Nicolet and Craig 1989) or Hop (Hsc70/Hsp90 organizing protein) in mammals, is a 63-kDa protein with several TPR motifs (Pratt and Dittmar 1998). The N-terminal TPR regions are mainly responsible for interaction with the C-terminus of Hsc70, where Hsp40/Hdj1 also binds. However, Hop and Hsp40 do not compete for interaction with Hsc70. This may allow Hsp40 to cooperate with Hsc70 even when the chaperone interacts with Hop. As for Hsp40, deletion of the last four amino acids of Hsc70, the EEVD sequence, abolishes binding of Hop (Demand et al. 1998). Hop does not bind to Hsc70 in the ATP but only in the ADP-bound form (Johnson et al. 1998; Prapapanich et al. 1998). The central TPR domains in Hop are involved in binding to the C-terminus of Hsp90 (Gebauer et al. 1998a; Lassle et al. 1997). Interestingly, Hsp90 also terminates with EEVD, but here, deletion does not affect its ability to stabilize misfolded proteins in an in vitro assay (Smith 1998). Yet mutation of EEVD into AAVD abolishes binding of Hop to Hsp90 (Chen et al. 1998). Hop has no chaperone activity on its own (Bose et al. 1996; Freeman et al. 1996), has no effect on the ATP cycle of Hsc70, and does not affect the Hsc70 chaperone activity (Chen and Smith 1998; Gebauer et al. 1998a, 1998b). Yet Hop can have a positive effect on luciferase refolding assays in the presence of Hsc70 and Hsp90, probably by serving as a physical link between Hsc70 and Hsp90 (Johnson et al. 1998; Prapapanich et al. 1998) and thus facilitating the cooperation between the two chaperone systems.

Recently a second linker, Tpr-2, was described to regulate substrate shuttling between Hsc70 and Hsp90 (Brychzy et al. 2003). Tpr-2 contains two TPR domains and a J domain. It recognizes both Hsc70 and Hsp90 through its TPR domains. The J domain seems responsible for transfer of substrates from Hsp90 onto Hsc70. Tpr-2 induces ATP-independent dissociation of Hsp90 but not of Hsc70 from chaperone–substrate complexes.

3.3
Hsp70-Hsp27 Connections

The small heat shock proteins have been mostly known for their stabilizing effects of cytoskeletal elements (Bryantsev et al. 2002; Lavoie et al. 1995), but also do have chaperone-like activities in vitro (Horwitz 1992; Jakob et al. 1993; Knauf et al. 1994; Merck et al. 1993). These ATP-independent chaperones can keep unfolded proteins in a refolding-competent form, but seem to depend on the action of the Hsp70 machine for their refolding to the active state (Ehrnsperger et al. 1997). Little or nothing is known about how substrates are transferred from these small Hsp to Hsp70. Also, there is an ongoing debate about what determines the vitro chaperone activity of Hsp27: these debates concern especially the role of the phosphorylation status and oligomeric size. These issues, however, go beyond the scope of this overview.

4
Internal and External Proteotoxic Stresses in Mammalian Cells

During a number of fundamental cellular processes [protein synthesis, protein transport (to organelles), protein functioning (e.g., in subunit–subunit interactions, and organelle biosynthesis)] interactive protein surfaces are transiently exposed to the intracellular environment. This is an unavoidable internal form of stress, as improper interactions may occur against which a cell has to defend itself. When cells age or when they become genetically unstable, genetic alterations may accumulate and result in the formation of aberrant gene products; these may never fold properly during and after translation, resulting in a different and augmented form of internal stress. Finally, heritable protein folding diseases such as cystic fibrosis (CF), Huntington's disease (HD), spinal and bulbar muscular atrophy (SBMA), and spinocerebellar ataxia (SCA) cause an accumulation of damaged proteins in cells (Bailey et al. 2002; Cummings al. 1998; Kopito 1999; Wyttenbach et al. 2001) that requires the lifelong enhanced activity of chaperones, but again in a different manner, as these proteins are nonfoldable.

In this chapter, I will mainly focus on the response of cells to external forms of stress that affect the stability of proteins. The prototype of such a stress is heat shock, which can cause denaturation of proteins. Direct evidence for protein denaturation and its relevance to heat-induced lethality in mammalian cells has been, for example, provided by studies using differential scanning calorimetry

(DSC) (Lepock al. 1990, 1993) and electron spin resonance (Burgman and Konings 1992). Subsequent to this protein denaturation, proteins will aggregate, and a tight relation between heat-induced protein aggregation and the cell biological consequences of heat has been found (Kampinga, 1993). However, many other forms of stress such as arsenite treatment, treatment with amino acid analogs, or oxidative stress can induce (different types of) toxic protein damage. It is important to note that all forms of stress that cause some kind of protein damage also induce the heat shock response, i.e., the activation of the heat shock factor-1 that transactivates the heat-inducible chaperones (Morimoto 1998). Moreover, the induction of chaperones by one proteotoxic stress generally not only results in a transient resistance against the subsequent similar stress, but also results in cross-resistance to many, if not all, other proteotoxic stresses (Hahn and Li, 1982). In general the induced synthesis of Hsp upon stress can be viewed as an amplification of their basic chaperone function (Ellis and van der Vies 1991) and thus heat shock seems a valid model to study chaperone function and regulation in mammalian cells. Besides direct thermal effects on protein folding, heat shock also causes oxidative (protein) damage (Freeman et al. 1990), may be related to enhanced metabolic rates, which, in part, may require specialized chaperones.

It is important to note that purely genotoxic stresses such as ionizing radiation do not lead to the elevated expression of heat shock proteins (Anderson et al. 1988) and inversely, that the transiently elevated expression of heat shock proteins by, for example, heat stress does not affect the sensitivity to ionizing radiation, which is particularly relevant to the discussion on the role of Hsp in the process of apoptosis (Sect. 9).

5
Thermotolerance and Hsp Reallocations in Mammalian Cells

When cells are exposed to a nonlethal heat shock, they can develop a transient resistance against a subsequent heat treatment (Gerner and Schneider 1975). In these so-called thermotolerant cells, damage to cellular structures and functions is either reduced or repaired more rapidly (Laszlo 1992b). Although thermotolerance can develop in the absence of synthesis of Hsp (Kampinga 1993), several observations have indicated that heat-shock proteins play an important role. Firstly, there is a good concordance between the kinetics of the transient heat-induced increase in the expression of Hsp and the kinetics of the induction and decay of thermotolerance (Li and Werb 1982). Also, as mentioned above, agents that induce heat shock proteins, such as heavy metals and ethanol, induce thermotolerance. Conversely, heat causes an increased resistance to these agents (Hahn and Li 1982). Secondly, heat-resistant cell lines (e.g., generated by repeated heating and selection of survivors) have elevated expression of one or more HSPs (Anderson et al. 1989; Laszlo and

Li 1985). Thirdly, constitutive overexpression of some of the single members of the various HSP families has been shown to result in a permanent thermoresistant state (Angelidis et al. 1991; Landry et al. 1989; Li et al. 1991). With regards to the latter, it must be stated that it can not be excluded that the observed resistant phenotypes of Hsp70 overexpressing stable clones are the consequence of secondary changes in the cell since especially Hsp70 family proteins have such important functions under nonstress conditions. Indeed, cell lines with (tetracycline) regulated expression of Hsp70 show only moderate heat resistance compared to cells with chronic Hsp70 overexpression (Nollen et al. 1999). Also, in these transient settings, the elevation of levels of Hsp70 alone to a level similar to that in thermotolerant cells does not result in the same level of heat resistance, certainly not for heat damage in the nuclear compartment (Nollen et al. 1999), indicating the need for the coordinated induction of "all" heat-inducible chaperones for full heat resistance in all cellular compartments.

Heat is aspecific in depositing its energy in cells and thus causes damage in all cellular compartments. It is therefore difficult to determine critical targets for heat-induced cell death. Moreover, they may be strongly cell-type-dependent, partially determined by the level and intracellular localization of the endogenously expressed heat shock proteins. Yet there are indications that there could be two more critical compartments. The first evidence comes from DSC studies from Lepock's group, showing that the fraction of protein denaturation in cells is the highest in the nuclear compartment (Lepock et al. 2001). Our data that luciferase is more rapidly inactivated when residing in the nucleus than when in the cytosol confirm the relative high heat sensitivity of this compartment (Michels et al. 1995). Also the dynamic translocation of several major cytosolic chaperones, including Hsp70/Hsp40 and Hsp25/27 to the nucleus (Arrigo et al. 1988; Bryantsev et al. 2002; Hattori et al. 1993; Michels et al. 1997, 1999; Nollen et al. 1999, 2001b; van de Klundert et al. 1998; Welch and Suhan 1985) suggests that there is a high need for these chaperones to repair (critical) damage in the nucleus. Nuclear protein aggregation will interfere with many nuclear processes that are all known to be very heat-sensitive, and indeed a tight correlation was found between the extent of heat-induced nuclear protein aggregation and thermal killing under many different conditions (Kampinga 1993; Kampinga et al. 1989; Laszlo 1992a).

The second most critical target may be the cytoskeleton or/and the microtubule organizing center (MTOC). The cytoskeletal collapse is one of the most obvious features that can be observed after cellular heating (Welch and Suhan 1985) and the findings that some Hsp, in particular Hsp27, move to the cytoskeletal elements (Bryantsev et al. 2002; Lavoie et al. 1993) suggest that this also is a very critical, heat sensitive element in mammalian cells. However, the cytoskeleton is highly dynamic and damage is likely to be very reversible unless the MTOC is nonfunctional. A nonfunctional MTOC will not only lead to lack of restoration of microtubules, but also prohibit cell division and thus affect the proliferative capacity of cells. Indeed, the MTOC has been shown to accumulate large amounts of protein aggregates (Vidair et al. 1995, 1996) and these were related to severe mitotic delays and abnormalities (Hut et al. 2005; Nakahata et al. 2002). Furthermore, Hsp70 also relocates to the MTOC upon heating in interphase or mitosis and this correlates with the regain of MTOC function (Brown et al. 1996).

6
Models Systems for Studying Hsp Chaperone-Like Activities in Stressed Living Mammalian Cells

From the above discussion, it is evident that, for studying Hsp chaperone-like activities in living mammalian cells after stress, one requires a system that allows examination of HSP functions in different cellular compartments in relation to thermotolerance. This system must be thermoresponsive, which means heat-sensitive, and it must allow recovery after heat shock.

6.1
Effects of Hsp Against Effects of Heat on Cellular Structures and Functions

One can assess the effects of modulation of Hsp expression on heat damage to various cellular functions (mRNA synthesis, protein synthesis, etc.) or structures (cytoskeleton stability, insolubilization of proteins to nuclear or centrosomal structures) in relation to Hsp (re)localization. Thermotolerant (TT) cells (expressing all heat-inducible proteins) usually are protected against this heat-induced loss of functions or recover better (Laszlo 1992a, 1992b; Kampinga 1993). However, such experiments do not provide insight on whether and how this may relate to the chaperone activity of Hsp.

The amount of TX-100 protein insolubilization after heat shock has often been used to study total cellular or nuclear protein aggregation. This protein aggregation is reduced in TT cells and/or the rate of resolubilization of the aggregates is enhanced (Laszlo 1992a, 1992b; Kampinga 1993). In cells overexpressing Hsp70 constitutively, the amount of aggregation was reduced and in parallel disaggregation rates were enhanced, whereas in cells stably overexpressing Hsp27, the rate of disaggregation was enhanced but no effect on the initial amount of heat-induced aggregates was found (Kampinga et al. 1994; Stege et al. 1994a, 1994b). However, a precise mechanism of action (e.g., folding, degradation, or indirect effects) cannot be deduced from such experiments. Moreover, these studies require the generation of stable cell lines overexpressing single or multiple Hsp, and indirect effects due to adaptation to the high Hsp levels can never be excluded.

6.2
Hsp Expression and Endogenous Reporter Proteins

In heat-stressed cells, endogenous proteins such as DNA polymerases (Kampinga et al. 1985; Spiro et al. 1982), p68 kinase (Dubois et al. 1991), topoiomerase I, (Ciavarra et al. 1994), the proto-oncogene c-myc, and p53 (Evan and Hancock 1985) become detergent insoluble. Insolubilization and inactivation of these endogenous proteins is reversible and usually enhanced in TT cells, but defined studies on the role of individual Hsp as chaperones in these processes are lacking, mainly because such studies require stable cell lines expressing the Hsp of interest.

6.3
Hsp Expression and Exogenous Reporter Proteins

Using heat-sensitive exogenous reporter proteins may have several advantages over endogenous reporters. Rapid studies on HSP activities are possible using transient co-transfections of the reporter with the HSP and co-factors, allowing examination of just the cells that express the HSP, as they will be the only cells that express the reporter. Another advantage of exogenous reporters is that some of them catalyze an enzymatic reaction with absolutely no cellular enzyme equivalent. Thus, no cellular background reaction will disturb the measurements. Moreover, exogenous reporters can be manipulated, e.g., targeted into specific cellular compartments, thus allowing compartment-specific investigations with the same reporter enzyme. A disadvantage of exogenous reporters is that a foreign protein is used and effects on such reporters might differ from effects on essential cellular proteins and the real physiological situation.

The first report on heat inactivation and insolubilization of exogenous reporters in mammalian cells is from the group of Bensaude (Nguyen et al. 1989). They demonstrated that wild type luciferase from the firefly *Photinus pyralis* and β-galactosidase expressed in mammalian or *Drosophila* cells rapidly inactivate during a heat shock at 42 °C. This loss in activity is linked to a loss in Triton-X100 solubility, or in other words, an increase in protein aggregation. The reporters are protected from heat inactivation when cells are made thermotolerant by prior exposure to heat (Nguyen et al. 1989), suggesting that they may be in vivo substrates for Hsp.

6.3.1
Hsp70 and Hsp40

In collaboration with Bensaude, our group next generated cytosolic and nuclear-targeted variants of firefly luciferase and found that the same protein was inactivated and insolubilized more rapidly in the nucleus of hamster O23 cells

(Michels et al. 1995), confirming that the nucleus may provide an environment in which protein aggregation rates by heat shock are faster than in the cytosol (Sect. 5). Subsequently, transient co-expression of Hsp70 was shown to retard the rate of inactivation and to enhance the rate and yield of reactivation after heat shock (Michels et al. 1997; Fig. 2). Also, heat resistant Rif-1 variants overexpressing Hsp70 (Anderson et al. 1989) showed this effect (Kampinga et al. 1997). Interestingly, the amount of protection was always lower for the nuclear than for the cytoplasmic luciferase (Michels et al. 1997). Physiological upregulation of Hsp70 alone comparable to levels seen in thermotolerant cells enhanced the rate and extent of cytosolic luciferase to the same level as in the thermotolerant cells; however, the yield of recovery of heat-inactivated nuclear luciferase was only marginally increased at this elevated level of Hsp70 alone (Nollen et al. 1999).

The latter means that other heat-inducible factors are needed for full nuclear chaperone activity. Such a factor could be one of the Hsp70 co-chaperones, Hsp40 in particular, as this protein is known to co-localize with Hsp70 to the nucleus and the nucleoli upon heat shock (Welch and Suhan 1985; Hattori et al. 1993). Indeed, Hsp40 and Hsp70 co-expression in cells enhanced luciferase refolding (Michels at al. 1997; Fig. 2). By co-expressing a mutant Hsp40 (Hsp40-H/Q) that cannot interact with Hsp70, we demonstrated that the effects of Hsp40 and Hsp70 on luciferase reactivation must occur via a direct interaction (Fig. 2). Furthermore, we found that the cooperation between Hsp40 and Hsp70 was lost by deletion of the C-terminal EEVD sequence of Hsp70 (Michels et al. 1999; Fig. 2). This is consistent with in vitro data (Freeman et al. 1995) that suggest that this sequence is essential for the cooperation between Hsp70 and Hsp40.[1] In addition, these data revealed that overexpression of Hsp40 together with Hsp70 increased the refolding at all temperatures for nuclear luciferase, while for the cytoplasmic luciferase this effect was only seen at higher temperatures (Michels et al. 1997). Given that a deletion construct expressing only the J-domain of Hsp40 acted as a dominant negative for luciferase refolding (Michels et al. 1999), it is concluded that levels of endogenous Hsp40 expressed in the hamster cells are sufficient to assist Hsp70 in the cytoplasm (at least at the lower stress levels), but that for nuclear refolding processes it is readily a limiting factor.

Whereas the activity of the Hsp70 mutant lacking the EEVD could not be enhanced by Hsp40 co-expression, its overexpression alone strongly enhanced refolding rates (Michels et al. 1999; Fig. 2). Therefore, although Hsp70-EEVD

---

[1] It is important to state that mutants are crucial in the study for co-chaperone interactions since immunoprecipitation (IP) experiments are noninformative due to two potential artifacts. First, Hsp70 is so sticky that it will likely co-immunoprecipitate any protein under all conditions (including at 37 °C), since there will always be a fraction of any protein in a partially unfolded form. Secondly, after heat shock (when interactions become even more important), many proteins including the (co)chaperones will end up in the pellet of the cell lysates that cannot be included in the IP protocol.

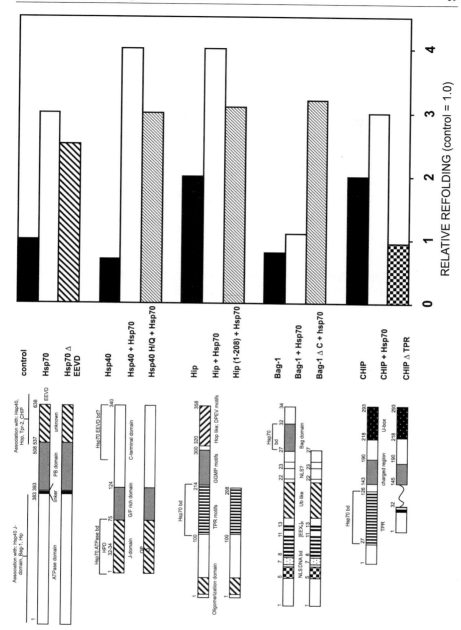

Fig. 2 Effect of Hsp70 and its co-factors on refolding of heat-denatured luciferase. Linear representations of Hsp70 and co-factors and the mutants used are given. The effect of their overexpression on the yield of active luciferase 1 h after heating relative to that seen in nontransfected hamster O23 cells cells (=1.0) is provided. See text for references and further details

mutants may have lost most of their refolding stimulating activity when used as isolated proteins in vitro (Sect. 3), they may still be quite functional in the context of cells with all other (co)chaperones present. Hence, caution is required to translate data from in vitro directly to the situation in living cells and the findings with use of Hsp70-EEVD mutants in cells cannot be interpreted as chaperone-independent actions of Hsp70.

Overexpression of Hsp40 alone in hamster cells led to an unexpected reduction in luciferase refolding (Michels et al. 1997; Fig. 2). This effect was not seen for the Hsp40-H/Q mutant, indicating that they are mediated via Hsc70, constitutively present in the cells. These data reveal that high Hsp40:Hsp/c70 ratios are unfavorable for refolding in the cell, which is supported by in vitro findings (Schumacher et al. 1996; Szabo et al. 1996).

6.3.2
Hsp70 and Hip

Overexpression of Hip alone in hamster cells clearly leads to an enhanced yield of cytoplasmic luciferase reactivation; Hip also further enhanced the refolding in the presence of overexpressed Hsp70 (Nollen et al. 2001; Fig. 2). Analysis with a variety of Hip mutants indicated that the last 110 carboxyl-terminal amino acids (amino acids 258–368) are not essential for the effects of Hip and that the effects of Hip depend on the interaction with Hsp70 or Hsc70 because mutants with deletions of the Hsp70 binding domain resulted in loss of its effects on refolding (Nollen et al. 2001a: Fig. 2).

Consistent with its exclusive presence in the cytoplasm both before and after heat shock, Hip did not affect the refolding of nuclear luciferase (unpublished data). Also, Hip (unlike Hsp70 and Hsp40) is not reported to be (very) stress-inducible, implying that its main function may be to assist cytosolic Hsp70 (and Hsp90) under physiological conditions. Indeed, Hip has been shown to be involved in GR receptor maturation; here it seems not to be a component of the assembly machinery but rather it could play a regulatory role by competing with Bag-1 for binding to the ATPase domain of Hsp70 (Zeiner et al. 1997). To my knowledge, no other data exist showing an effect of Hip on stress resistance of mammalian cells.

6.3.3
Hsp70 and Bag-1

Bag-1 is probably the most provocative co-chaperone of Hsp70, as it reduces the Hsp70-mediated refolding of heat-denatured luciferase in cells (Nollen et al. 2000; Fig. 2). The localization of the effect of Bag-1 on the activity of Hsp70 depends on the localization of the Bag-1 isoform that is expressed. Expression of the longest 50-K isoform of Bag-1 that is expressed only in the nucleus only influences the Hsp70 activities in the nucleus. Iversety, the shorter versions mainly affect refolding of the cytoplasmic luciferase (Nollen et al. 2000). Consistent with in vitro data, we also find that Bag-1 competes with Hip for effects on luciferase refolding and the action of Bag-1 seems dominant over Hip (Nollen et al. 2001a).

There are two apparent controversies in these findings. First, is this a realistic physiologically relevant finding considering that Bag-1 is likely the functional mammalian GrpE homolog that accelerates the ADP:ATP exchange (Sect. 3) and should Bag-1 not act in vivo as an enhancer of folding? Indeed, one could envision that transient transfections would lead to artificially high Bag-1:Hsp70 ratios and these would result in nonphysiologically relevant inhibition of refolding. However, using a tet-regulatable system, we found that only a twofold increase in the basal expression level of Bag-1 already leads to inhibitory effects on Hsp70-mediated refolding (Nollen et al. 2000). This change in ratio is within the range than can be found in a variety of mammalian cell lines. Therefore, indeed one must conclude that Bag-1 usually inhibits refolding of stress-denatured proteins. However, Bag-1 is not stress-inducible: so, when Hsp70 levels increase after a stress, the Bag-1:Hsp70 ratio will decline, providing a physiologically relevant increase in the functionality of the Hsp70 machine.

The second contradiction-in-terms is why one cytoprotective protein (Bag-1) would inhibit another cytoprotective protein (Hsp70). One possibility is that Bag-1 may enhance degradation of unfolded proteins in an Hsc70-dependent manner (Luders et al. 1998). However, this was not found to be the case for heat-denature luciferase (Nollen et al. 2000). Another hypothesis could be that the interactions between Hsp70 and Bag-1 may not enhance the Hsp70 chaperone function but rather enable Bag-1 to function in growth regulation. In addition to Bcl-2 and Hsp70, Bag proteins also interact with a large set of other important regulatory proteins, including steroid hormone receptors, the plasma membrane bound growth factor receptors HGF and PDGF, and the kinase Raf-1 (Bardelli et al. 1996; Takayama et al. 1997; Wang et al. 1996; Zeiner et al. 1997; Zeiner and Gehring 1995). In response to stress, Bag-1 may coordinate signals for cells by downregulating the activity of Raf-1 kinase. When levels of Hsp70 are elevated, Raf-1 is displaced from Bag1 by Hsp70, and DNA synthesis is arrested (Song et al. 2001). Although attractive, the caveat in this model is that inhibition of DNA synthesis and S-phase progression occurs almost instantly upon heating (Warters and Stone 1983) far before Hsp70 levels are elevated.

Here, I would like to speculate that Bag-1 may be required for temporal storage of denatured substrates during heating and, at the correct stoichiometry, this will assure that the cells are not too rapidly depleted from all their Hsp70/Hsc70 under stress conditions (Sect. 8).

6.3.4
Hsp70 and CHIP

As for Bag-1, the effects of CHIP on the Hsp70 machine in cells are still puzzling. Whereas in cell-free systems, CHIP is a clear inhibitor of Hsc70-mediated refolding activities (Sect. 3.1), we demonstrated that its overexpression in cells with low levels of Hsc/Hsp70 resulted in increased yields of active cytoplasmic luciferase after heat shock, whereas little or no effect was seen when Hsp70 was co-overexpressed. This action required the TRP domain of CHIP (Kampinga et al. 2003; Fig. 2) that is capable of binding to both Hsp90 and Hsc/Hsp70. However, under conditions of inhibiting Hsp90, the protective effect of CHIP could still be seen. Moreover, coexpression of Hsp40 acted competitively with the effect of CHIP on the rate of luciferase inactivation, and dominant-negative constructs of Hsp40 (Michels et al. 1999) completely abolished the effects of CHIP on protein folding, indicating that this activity of CHIP is Hsc70/Hsp70-dependent (Kampinga et al. 2003).

The opposing effects of overexpressing CHIP alone or Hsp40 alone are still consistent with their opposing effects on the Hsc70/Hsp70 ATPase activity. As stated, excess Hsp40 inhibits refolding (Fig. 2). This may be caused by reduced Hsp70 recycling because substrates will remain bound to Hsp70 in the ADP-bound state for a longer time and no new substrates can be captured. Consequently, the pool of Hsp70 that can bind substrates is more rapidly depleted. CHIP may increase Hsp70 folding activity through a functional interference with the interaction between Hsp40 and Hsp70, thus ameliorating the consequences of the functional access of Hsp40 to Hsp70. Alternatively, overexpression of CHIP may enhance the likelihood of substrate binding by Hsp70 in cells by increasing the fraction of Hsp70 in the ATP-bound form, which has a higher substrate on-rate (Fig. 1). Although the off-rate of substrates from Hsp70 is even higher when it is in the ATP-bound state (i.e., lower substrate affinity), attenuating forward cycling at this step of hydrolysis by CHIP could lead to a better loading of denaturing substrates to Hsp70.

Interestingly, the U-box domain is dispensable for the actions of CHIP on refolding heat denatured substrates and luciferase degradation is not accelerated by CHIP (Kampinga et al. 2003), uncoupling its effect on luciferase refolding from the E3-ubiquitin ligase activity of CHIP (Jiang et al. 2001) and suggesting that the latter activity may be less relevant to the stress protective action of the Hsp70 machine.

6.3.5
Chaperone Networks

Little is known about how the Hsp70 machine may interact with Hsp27 or Hsp90 in stress protection in mammalian cells. Clearly, ansamycins known to inhibit Hsp90 also inhibit luciferase refolding in cells (Kampinga et al. 2003;

Schneider et al. 1996) and seem to enhance the degradation of the unfolded substrate (Schneider et al. 1996). Considering that Hsp90 in vitro is merely a "holder" (Sect. 3.2), one can envision that here also substrate delivery to the Hsp70 machine is normally required for refolding. But so far no evidence for this has been provided.

Similarly, little or no data have been published on possible interactions between Hsp70 and Hsp27 in cells. In fact, although Hsp27 has been shown to be an chaperone in vitro, no data are available showing that Hsp27 can enhance refolding of denatured proteins in mammalian cells. Rather, some indirect evidence suggests that the small Hsp as well as the αB-crystallins may be involved in protein degradation (Aki et al. 2003; den Engelsman et al. 2003; Parcellier et al. 2003). Whether this requires collaboration with the Hsp70 machine remains unknown.

Finally, there seems to be an intricate interaction between Hsp70 and its large Hsp105/110 family members (HspH1, /APG-1, and APG-2) (Lee Yoon et al. 1995; Yasuda et al. 1995), here referred to as Hsp110. Hsp110 forms heteromeric complexes in cells with Hsc70 (Hatayama and Yasuda 1998) and overexpression of Hsp110 can confer substantial heat resistance to both Rat-1 and HeLa cells (Oh et al. 1997). In vitro, Hsp110 can maintain unfolded substrates in a soluble, folding-competent state and does so significantly more efficient than Hsp70. For efficient refolding of denatured substrates, it requires the activity of Hsp70, Hsp40, and reticulocyte lysate (Oh et al. 1997). In fact, if both Hsp70 and Hsp110 are present together during denaturation in vitro, Hsp110 inhibits refolding by inhibiting the Hsp70 ATPase activity (Yamagishi et al. 2004). Interestingly, it was found that Hsp110 is phosphorylated by protein kinase CK2 at Ser(509) and that the inhibition of Hsp70-mediated refolding was phosphorylation-dependent. Phosphorylated Hsp110 had no effect on Hsp70-mediated refolding of heat-denatured luciferase, whereas a non-phosphorylatable mutant of Hsp110 suppressed the Hsp70-mediated refolding of heat-denatured luciferase in mammalian cells (Ishihara et al. 2003). This regulation of Hsp110 by phosphorylation may play an important role in the regulation of protein maintenance under stress (Sect. 8).

7
Fate of Proteins: Folding or Degradation?

It has been suggested that the fate of Hsp70 client proteins in terms of folding or degradation may be determined by its co-chaperones. Whereas Hsp40 and Hip as "positive" regulators of the Hsp70 ATPase cycle should lead to refolding, Bag-1 and CHIP as "negative" regulators may convert Hsc70 (and also Hsp90) from a folding machinery into a degradation machine (Demand et al. 2001). Since Bag-1 has a ubiquitin domain and CHIP is a E3-ubiquitin ligase, this hypothesis appears quite appealing and indeed in vitro Bag and CHIP were found to enhance protein ubiquitination (Demand et al. 2001). Moreover, overexpression of CHIP in mammalian cells has been shown to enhance the degradation of the cystic fibrosis transmembrane-conductance regulator (CFTR) in an Hsc70-related manner (Meacham et al. 2001). However, it must be noted that the effect of CHIP as an enhancer of (Hsc70-dependent) protein degradation is substrate specific, as maturation and degradation of proteins such as the transferrin receptor and apolipoprotein B48 are unaffected by CHIP overexpression (Meacham et al. 2001) and as CHIP can stimulate the refolding of heat-unfolded luciferase rather than enhancing its degradation (Kampinga et al. 2003). Inversely, Hsp40 not always is related to protein folding in cells: e.g., overexpression of Hsp40 (alone or with Hsp70) in cases of protein misfolding diseases such as SBMA was shown to lead to enhanced degradation of the (mutant) client protein (Bailey et al. 2002). Even for the same client protein, we found that whereas Hsp70 and Hsp40 overexpression enhanced refolding of firefly luciferase (Michels et al. 1997: Fig. 2), it also reduced the half-life of the activity of the same luciferase upon prolonged incubation at 37 °C. This effect was not seen when Hsp70 was overexpressed alone (Michels, Kanon, and Kampinga, unpublished observations). Consequently, the triage between folding, degradation, and aggregation is not (solely) determined by the (co)chaperones. Rather, a selection mechanism could be based on differences in the rate of refolding and degradation: when the rate of degradation is lower than the rate of refolding, proteins that constantly reappear in the pool of unfolded proteins after binding and release by chaperone machines because they cannot be refolded have a greater chance of being degraded. In other words, the foldability of a protein may be the most determining factor for its ultimate fate, rather that the co-chaperones. Yet if folding occurs at low speed due to inhibition of the chaperones, client proteins that are good substrates for the proteasome will show accelerated degradation. Also, effects of CHIP on degradation may more likely go via Hsp90 than via Hsc70 (Connell et al. 2001).

8
Proteotoxic Damage: Cellular Targets, Sorting, and Hsp-Mediated Protection

In cells, active refolding of unfolded proteins during heat shock may not be very productive because when the substrate is released from the chaperone, it will immediately unfold again. In addition, Hsp70 may not be recycled and cells will be rapidly depleted from Hsp70 chaperone activity. Also, given the complexity of compartmentalization in mammalian cells and the movement of heat shock proteins in and out of different compartments, questions on the spatial organization arise: does the cells utilize special (sub)compartments for (re)folding and/or degradation?

8.1
Processing of Damaged Proteins in the Cytosol

Heat-denatured cytosolic proteins such as p68, AUF, and aggregates of unidentified proteins have been found to accumulate in the peri-nuclear region and in the proximity of the centrosome of heat-stressed cells (Brown et al. 1996; Dubois et al. 1996). Originally these data were interpreted to suggest that centrosome proteins are labile, readily unfold during heat shock, and can rapidly form aggregates with other soluble or insoluble proteins. An alternative explanation may be proposed based on recent observations that under normal growth conditions, overexpressed misfolded proteins are actively transported to the centrosome (Garcia-Mata et al. 1999; Johnston et al. 1998). It was proposed that these so-called aggresomes may enable cells to coordinately regulate and dispose protein damage when the capacity of heat shock proteins to refold and the proteasome to degrade damaged proteins is exceeded. Hereby random protein aggregation and cellular dysfunction can be prevented. Although the reversibility of the aggresomes seen in protein-folding diseases may be limited (Holmberg et al. 2004) and cells may only dispose them via autophagy (Ravikumar et al. 2002), disappearance of heat-unfolded proteins accumulated at centrosomes was observed after a heat shock. In fact, it was found that Hsp70 also accumulates the MTOC upon heat shock and that thermotolerant cells display a more rapid recovery from heat-induced damage to the centrosome structure and centrosome function (assayed by regrowth of microtubules), suggesting a role for Hsp in assisting repair of centrosomes after heat shock (Brown et al. 1996; Hut et al. 2005; Vidair et al. 1995). More specifically, microinjection of Hsc70 antibodies retarded recovery of interphase centrosome structure and microtubule regrowth abilities after heat shock, whereas injection of purified Hsc70 prior to heat shock enhanced these processes (Brown et al. 1996).

The finding that a major holder chaperone, Hsp90, is a structural component of the centrosome (Lange et al. 2000) further makes this organelle and excellent place to serve as a storage place. Hsp90 could assist in keeping the accumulated damaged proteins in a conformation that will allow their refolding by components of the Hsp70 chaperone machine or their degradation by the proteasome when normal growth conditions are restored. It is interesting to note that the in vitro refolding activity of the Hsp70 machine is reversibly inhibited above 41 °C, whereas holding by Hsp90 is still effective (Freeman and Morimoto 1996). Thus, storage at high temperatures by Hsp90 at the centrosome also will allow recycling of Hsp70 and prevent its depletion during heat shock. Whether the proteins accumulated at or in aggresomes are refolded or degraded remains to be elucidated.

8.2
Processing of Damaged Proteins in the Nucleus

Also in the nucleus, heat-unfolded proteins seem to undergo intranuclear sorting during heat shock. Studies on nuclear firefly luciferase fused to the heat-stable green fluorescent protein (N-luc-EGFP) have shown that heat inactivation of the luciferase part of N-luc-EGFP results in its translocation to intranuclear granules and the nucleolus.

As mentioned, especially Hsp70 and Hsp40 are known to translocate to the nucleolus after a heat shock. As for the centrosome, this was originally interpreted to suggest that the nucleolus is especially sensitive to heat. Nucleoli become swollen and nucleolar processes, such as ribosomal RNA synthesis and assembly of ribosomal precursor particles, are inhibited by a heat shock. The movement of Hsp70 in and out of the nucleolus was therefore interpreted as repairing heat-induced nucleolar protein damage (Pelham 1984; Welch and Suhan 1986). However, alternatively the translocation of Hsp70 to and from the nucleolus may be associated with a temporal storage of nonnucleolar unfolded proteins during heating for subsequent repair at folding-permissive temperatures. Consistent with our observations that denatured luciferase is stored in the nucleolus for later processing (Nollen et al. 2001b), a variety of other proteins have been reported to be sequestered in the nucleolus during stress (Dundr et al. 2000). The model of the nucleolus as a storage site for Hsp70-mediated refolding was further supported by the finding that nucleolar storage and refolding of luciferase were not observed in cells expressing a chaperone-defective Hsp70(1–543) (Nollen et al. 2001b). Intriguingly, when the nuclear isoform of the Hsp70-inhibitory protein Bag1 was coexpressed with Hsp70, the N-luc-EGFP as well as Hsp70 and Bag-1 remained nucleolar (Nollen, Brunsting, and Kampinga, unpublished observations). Together these data indicate that nucleolar transport during heating is dependent on the peptide binding ability of Hsp70 but is not sensitive to the effects of Bag-1. However, refolding is inhibited by Bag-1 in parallel with inhibition of the exit of the denatured substrate from the nucleolus, strongly supporting that folding at permissive temperatures is subsequent to nucleolar storage. During heat, physiological levels of nucleolar Bag-1 may assure that the substrate is released from Hsp70 in the nucleus, allowing Hsp70 to shuttle dynamically to and from the nucleolus (Kim et al. 2002) in order to capture other unfolded proteins in the nucleosol for transport to the nucleolus.

For this type of storage mechanism, again a "holder" chaperone seems required that would bind unfolded proteins in the nucleolus during heat shock and release them to the Hsp70 machine after return to normal temperatures. Although purely speculative at this time, candidates for such activities could be the large Hsp70 family members. As mentioned (Sect. 6.3), Hsp110 is an extremely good holder and can inhibit refolding via Hsp70 in a manner that depends on their phosphorylation status (Ishihara et al. 2003). The kinase involved in Hsp110 phosphorylation (CK2) shows similar translocation to and from the nucleolus during and after heat shock as N-luc-EGFP (Gerber et al. 2000). Combined with the early observations that Hsp110 localizes to nucleolus (Shyy et al. 1983) provokes the speculation that Hsp110 is used in a phosphorylation-dependent manner for storage of unfolded proteins during heat shock that can be transferred to Hsp70 after heat shock for refolding.

Although Hsp70 and Hsp40 are found to concentrate in the nucleolus after heat stress, Hsp25/27 appears to accumulate mainly in non-nucleolar nuclear foci after a variety of cellular stresses in various cell cultures (Arrigo et al. 1988; Bryantsev et al. 2002; Loktionova et al. 1996). Recent observations revealed that the non-nucleolar N-luc-EGFP distribution after heat shock overlaps with that of these Hsp25 foci (Bryantsev et al. 2002). Although these non-nucleolar foci of heat-denatured N-luc-EGFP were not readily reversible after a severe heat shock (Nollen et al. 2001b), reversibility was seen after more mild heat treatments (Bryantsev et al., unpublished observations). It has been proposed (Bryantsev et al. 2002) that by binding heat-denatured nucleosolic proteins to the Hsp25, these granular structures may serve to maintain these proteins in a state competent for processing by the proteasome, but this requires further testing.

In summary, rapid storage of damaged proteins at specific sites in the cell as opposed to random aggregation may have several advantages. It may clear the cell from damaged proteins and in this way prevent indirect damage to other cellular proteins, protein structures, and processes. In addition, no energy is wasted on the refolding of proteins, which may immediately become denatured again because of a continued exposure to stress. Moreover, chaperones such as Hsp70 may be recycled during stress. Finally, delivery of unfolded proteins to the centrosome or the nucleolus might prevent irreversible damage to a cell population because both cell division and new protein synthesis are hindered until the protein-damaging conditions have disappeared and damaged proteins have been refolded or degraded.

9
Cell Death Expression, Cell Transformation and Hsp

After proteotoxic stress, interphase (differentiated) cells may either initiate apoptosis (if this program is available in the cells) or they may undergo necrosis. Hsp, by their chaperone activity, may inhibit or repair the damage and thus take away the trigger for activating cell death. Similarly, storage of poly-Q-diseased proteins may prevent (mini)aggregates from disturbing cellular functions; thus cells maintain their biochemical activity and stay alive (Arrasate et al. 2004). In how far Hsp play a role in this storage mechanism remains to be elucidated, but, for example, aggresome formation clearly requires an intact cytoskeleton (Johnston et al. 2002) and thus stabilization of the cytoskeletal by, for instance,

Hsp27 may certainly be important for storage under conditions of heat stress. For dividing cells, life or death ultimately depends on the ability to move through the cell cycle and divide. If checkpoints are not functional and/or apoptosis is not initiated, cells may attempt to progress through S-phase, but nuclear protein damage may obstruct DNA synthesis, leading to replication errors and lethal chromatid aberrations (Dewey et al. 1990) and subsequent death by apoptosis or necrosis. When cells enter mitosis with unresolved protein damage (or when heated in mitosis), cells may undergo a mitotic catastrophe or aberrant divisions (Hut et al. 2005; Nakahata et al. 2002) that will mostly result in secondary apoptosis or necrosis of the daughter cells (Fig. 3). There are clear indications that the accumulation of heat-unfolded proteins at the centrosome are linked to these aberrant mitoses (Hut et al. 2005; Nakahata et al. 2002), implying that storage of heat damage at centrosomes must be restored before proper cell divisions can occur.

Besides being involved in heat damage repair, Hsp70 (as well as Hsp27 and Hsp90) have been implicated in the process of apoptosis execution. Hsp70 was suggested to inhibit JNK activation after heat shock (Gabai et al. 1997), but these effects may not be uncoupled from the Hsp70 chaperone activity either upstream or at the level of JNK (Mosser et al. 2000). However, based on in

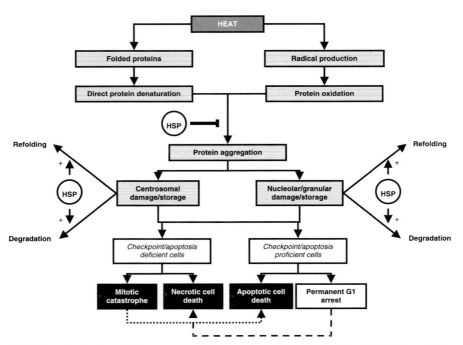

Fig. 3 Hypothetical model for causes and expression of heat-induced cell lethality and the protective effects of Hsp. See text for further explanation

vitro observations, Hsp70 was also suggested to be able to interfere with the formation of the apoptosome (Beere et al. 2000; Saleh et al. 2000), one of the cell death execution caspase complexes that is induced after the leakage of cytochrome-c from the mitochondria. This would suggest that Hsp70 could act downstream in the apoptosis pathway and thus could be cytoprotective, irrespective of the type of cell death trigger. This has not yet been confirmed in living cells and recent in vitro data suggest that the presumed Hsp70 effects on apoptosome formation may have been due to salt artifacts (Steel et al. 2004). Moreover, if true, these or other presumed specific interferences of Hsp70 with the apoptosis pathway should also have an impact when nonproteotoxic stresses such as FAS-ligand or radiation are used.

When investigating FAS-mediated apoptosis, Jaattela's group had already repeatedly shown that Hsp70 did not prevent apoptosome formation in cells, and in this case Hsp70 acts even further downstream in the apoptotic cascade (Jaattela et al. 1998; Nylandsted et al. 2004). In fact, their studies are among the few that do indicate that Hsp70 has anti-cell death functions that may be separated from its chaperone activity acting upstream in the apoptotic pathway to prevent proteotoxic damage. Nevertheless, here also a link with protein quality control is suggested by their findings that Hsp70-mediated protection against FAS-induced cell death seems to occur at the level of lysosomes (Nylandsted et al. 2004). Also, it was recently shown that human Fas associated factor 1 (hFAF1) inhibits Hsp70 chaperone activity in mammalian cells (Kim et al. 2004), suggesting that stimulating the FAS pathway may inhibit physiological Hsc70/Hsp70 functions (including those involved in lysosomal control) and thus explain why overexpression of Hsp70 is protective in the case of FAS-induced apoptosis.

Maybe the most compelling data to show that Hsp70 (and perhaps other Hsp) may have no general inhibitory function in the pathway of apoptosis execution comes from studies on the interaction between hyperthermia and radiation, long before specific studies on apoptosis became popular. These data showed that the temporal and physiological upregulation of the cohort of all heat-inducible proteins sufficient to provide cells with a thermotolerant state does not alter the cellular radiosensitivity (Dikomey and Jung 1992; Hartson-Eaton et al. 1984; Haveman et al. 1987; Jorritsma et al. 1986; Kampinga et al. 1997; Mivechi and Li 1987; Raaphorst and Azzam 1983, and many more). Also, when Hsp70 or Hsp27 are transiently upregulated using inducible systems, no effect of radiosensitivity was found (unpublished data). Inversely, RNAi-mediated transient downregulation of Hsp70 was shown not to affect radiation-induced apoptosis (Ekedahl et al. 2003). Finally, although some stable cell lines overexpressing individual Hsp may become radiation resistant (Gehrmann et al. 2005; Lee et al. 2001), in other cell systems, Hsp overexpression does not protect (Fortin et al. 2000; Kampinga et al. 1997; Stege et al. 1995) or even sensitizes for radiation-induced cell death (Liu et al. 2003). As mentioned before, clonal overexpression of Hsp may require adaptation of cellular gene and protein expression profiles. Depending on cell type-specific adaptations, indirect effects rather than functions of Hsp per se may be causally related to cellular phenotypes such as an altered radiation sensitivity.

The presumed function of Hsp70 in apoptosis has often been used to explain why it is repeatedly found to be associated with tumorigenesis. For example, elevated Hsp70 levels have been found in almost all oncogene-transformed cell lines and tumor cells (Li et al. 1995). Hsp70 is highly expressed in many tumors (Jaattela, 1999) and the oncogenic potential of Hsp70 has been demonstrated (Jaattela 1995; Volloch and Sherman 1999). If not through direct interference with the apoptotic pathway, how can one explain these associations? First, cellular transformation may require the Hsp70 machine, as suggested by findings that several of the transforming viral T antigens contain regions of significant homology with the conserved J domain of the DnaJ co-chaperones (Kelley and Georgopoulos 1997). Alternatively, Hsp70 can regulate the stability and function of the tumor suppressor p53 (King et al. 2001; Zylicz et al. 2001). A third possibility, attractive but even more speculative, is that Hsp70 may act as a buffer for the negative consequences of aberrant protein expression due to the genomic instability in tumor cells. As such, its elevated expression is required for tumor cell survival, without Hsp70 being in the transformation process as such.

10
Concluding Remarks

The most conclusive data on the function of heat shock proteins, in particular Hsp70, in mammalian cells is that they indeed act as chaperones to prevent irreversible aggregates and assist in either the folding or degradation of their client proteins. Whereas most concepts for the mechanisms of regulation of the Hsp70 chaperone machine derived from cell free studies can be extrapolated to the situation in living mammalian cells, the outcome and impact on the cells' ability to deal with protein damage may not always be straightforward (e.g., effects of Hsp70-ΔEEVD or CHIP overexpression). Besides stoichiometry issues and expression of endogenous (co)chaperones aside from the ones investigated by ectopic overexpression, factors such as intracellular distribution and storage sites for unfolded proteins (centrosomes, nucleolus, Hsp27 nuclear granules) require careful consideration before one can draw firm conclusions. The impact on chaperones and their co-regulators on the decision between folding and degradation is far from clear and may be more dependent on the foldability of the protein itself and/or its transport to intracellular storage sites.

Cell death after proteotoxic stresses is multifactorial (Fig. 3): for heat shock, the most sensitive targets may be those that function as storage sites for unfolded proteins at the same time (i.e., centrosome, nucleolus/nuclear granules). Hsp can prevent damage induction or repair proteins after stress and for repair they may possibly utilize a mechanism of temporal storage of the unfolded proteins during heat stress. These actions may result in less apoptosis or, depending on the cell type, other forms of cell death, without Hsp being directly involved in later steps of the cell death execution program.

Acknowledgements I would like to thank my PhD students Gerard Stege, Annemieke Michels, and Ellen Nollen for their excellent work in my lab. Their theses provided me with a great deal of background material for this chapter. Also, I would like to thank my current PhD students Jurre Hageman, Maria Rujano, Michel Vos, Rita Setroikroma for their critical feedback on this manuscript.

References

Aki T, Yoshida K, Mizukami Y (2003) The mechanism of alphaB-crystallin gene expression by proteasome inhibition. Biochem Biophys Res Commun 311:162–167

Anderson RL, Shiu E, Fisher GA, Hahn GM (1988) DNA damage does not appear to be a trigger for thermotolerance in mammalian cells. Int J Radiat Biol 54:285–298

Anderson RL, Van Kersteren I, Kraft PE, Hahn GM (1989) Biochemical analysis of heat-resistant mouse tumor cell strains: a new member of the HSP70 family. Mol Cell Biol 9:3509–3516

Anfinsen CB (1973) Principles that govern folding of protein chains. Science 181:223–230

Angelidis CE, Lazaridis I, Pagoulatos GN (1991) Constitutive expression of heat-shock protein 70 in mammalian cells confers thermoresistance. Eur J Biochem 199:35–39

Arrasate M, Mitra S, Schweitzer ES, Segal MR, Finkbeiner S (2004) Inclusion body formation reduces levels of mutant huntingtin and the risk of neuronal death. Nature 431:805–810

Arrigo AP, Suhan JP, Welch WJ (1988) Dynamic changes in the structure and intracellular locale of the mammalian low-molecular-weight heat shock protein. Mol Cell Biol 8:5059–5071

Asea A, Kraeft SK, Kurt-Jones EA, Stevenson MA, Chen LB, Finberg RW, Koo GC, Calderwood SK (2000) HSP70 stimulates cytokine production through a CD14-dependant pathway, demonstrating its dual role as a chaperone and cytokine. Nat Med 6:435–442

Bailey CK, Andriola IF, Kampinga HH, Merry DE (2002) Molecular chaperones enhance the degradation of expanded polyglutamine repeat androgen receptor in a cellular model of spinal and bulbar muscular atrophy. Hum Mol Genet 11:515–523

Ballinger CA, Connell P, Wu Y, Hu Z, Thompson LJ, Yin LY, Patterson C (1999) Identification of CHIP, a novel tetratricopeptide repeat-containing protein that interacts with heat shock proteins and negatively regulates chaperone functions. Mol Cell Biol 19:4535–4545

Bardelli A, Longati P, Albero D, Goruppi S, Schneider C, Ponzetto C, Comoglio PM (1996) HGF receptor associates with the anti-apoptotic protein BAG-1 and prevents cell death. EMBO J 15:6205–6212

Beere HM, Wolf BB, Cain K, Mosser DD, Mahboubi A, Kuwana T, Tailor P, Morimoto RI, Cohen GM, Green DR (2000) Heat-shock protein 70 inhibits apoptosis by preventing recruitment of procaspase-9 to the Apaf-1 apoptosome. Nat Cell Biol 2:469–475

Beissinger M, Buchner J (1998) How chaperones fold proteins. Biol Chem 379:245–259

Benaroudj N, Fouchaq B, Ladjimi MM (1997) The COOH-terminal peptide binding domain is essential for self-association of the molecular chaperone HSC70. J Biol Chem 272:8744–8751

Bercovich B, Stancovski I, Mayer A, Blumenfeld N, Laszlo A, Schwartz AL, Ciechanover A (1997) Ubiquitin-dependent degradation of certain protein substrates in vitro requires the molecular chaperone Hsc70. J Biol Chem 272:9002–9010

Bimston D, Song JH, Winchester D, Takayama S, Reed JC, Morimoto RI (1998) BAG-1, a negative regulator of Hsp70 chaperone activity, uncouples nucleotide hydrolysis from substrate release. EMBO J 17:6871–6878

Boorstein WR, Ziegelhoffer T, Craig EA (1994) Molecular evolution of the Hsp70 multigene family. J Mol Evol 38:1–17

Bose S, Weikl T, Bugl H, Buchner J (1996) Chaperone function of Hsp90-associated proteins. Science 274:1715–1717

Botzler C, Issels R, Multhoff G (1996) Heat-shock protein 72 cell-surface expression on human lung carcinoma cells is associated with an increased sensitivity to lysis mediated by adherent natural killer cells. Cancer Immunol Immunother 43:226–230

Brown CR, Hong-Brown LQ, Doxsey SJ, Welch WJ (1996) Molecular chaperones and the centrosome. A role for HSP 73 in centrosomal repair following heat shock treatment. J Biol Chem 271:833–840

Bryantsev AL, Loktionova SA, Ilyinskaya OP, Tararak EM, Kampinga HH, Kabakov AE (2002) Distribution, phosphorylation, and activities of Hsp25 in heat-stressed H9c2 myoblasts: a functional link to cytoprotection. Cell Stress Chaperones. 7:146–155

Brychzy A, Rein T, Winklhofer KF, Hartl FU, Young JC, Obermann WM (2003) Cofactor Tpr2 combines two TPR domains and a J domain to regulate the Hsp70/Hsp90 chaperone system. EMBO J 22:3613–3623

Buchberger A, Valencia A, Mcmacken R, Sander C, Bukau B (1994) The chaperone function of DnaK requires the coupling of ATPase activity with substrate binding through residue E171. EMBO J 13:1687–1695

Buchberger A, Theyssen H, Schroder H, Mccarty JS, Virgallita G, Milkereit P, Reinstein J, Bukau B (1995) Nucleotide-induced conformational changes in the ATPase and substrate binding domains of the DnaK chaperone provide evidence for interdomain communication. J Biol Chem 270:16903–16910

Buchner J (1996) Supervising the fold: functional principles of molecular chaperones. FASEB J 10:10–19

Buchner J, Schmidt M, Fuchs M, Jaenicke R, Rudolph R, Schmid FX, Kiefhaber T (1991) GroE facilitates refolding of citrate synthase by suppressing aggregation. Biochemistry 30:1586–1591

Bukau B, Horwich AL (1998) The Hsp70 and Hsp60 chaperone machines. Cell 92:351–366

Burgman PWJJ, Konings AWT (1992) Heat induced protein denaturation in the particulate fraction of HeLa S3 cells: effects of thermotolerance. J Cell Physiol 153:88–94

Cheetham ME, Brion JP, Anderton BH (1992) Human homologues of the bacterial heat-shock protein DnaJ are preferentially expressed in neurons. Biochem J 284:469–476

Cheetham ME, Caplan AJ (1998) Structure, function and evolution of DnaJ: conservation and adaptation of chaperone function. Cell Stress Chaperones. 3:28–36

Chen SY, Smith DF (1998) Hop as an adaptor in the heat shock protein 70 (Hsp70) and Hsp90 chaperone machinery. J Biol Chem 273:35194–35200

Chen SY, Sullivan WP, Toft DO, Smith DF (1998) Differential interactions of p23 and the TPR-containing proteins Hop, Cyp40, FKBP52 and FKBP51 with Hsp90 mutants. Cell Stress Chaperones 3:118–129

Ciavarra RP, Goldman C, Wen KK, Tedeschi B, Castora FJ (1994) Heat stress induces Hsc70/nuclear topoisomerase I complex formation in vivo: evidence for Hsc70-mediated, ATP-independent reactivation in vitro. Proc Natl Acad Sci U S A 91:1751–1755

Connell P, Ballinger CA, Jiang J, Wu Y, Thompson LJ, Hohfeld J, Patterson C (2001) The co-chaperone CHIP regulates protein triage decisions mediated by heat-shock proteins. Nat Cell Biol 3:93–96

Craig E, Ziegelhoffer T, Nelson J, Laloraya S, Halladay J (1995) Complex multigene family of functionally distinct Hsp70s of yeast. Cold Spring Harb Symp Quant Biol 60:441–449

Cummings CJ, Mancini MA, Antalffy B, Defranco DB, Orr HT, Zoghbi HY (1998) Chaperone suppression of aggregation and altered subcellular proteasome localization imply protein misfolding in SCA1. Nat Gen 19:148–154

Demand J, Luders J, Hohfeld J (1998) The carboxy-terminal domain of Hsc70 provides binding sites for a distinct set of chaperone cofactors. Mol Cell Biol 18:2023–2028

Demand J, Alberti S, Patterson C, Hohfeld J (2001) Cooperation of a ubiquitin domain protein and an E3 ubiquitin ligase during chaperone/proteasome coupling. Curr Biol 11:1569–1577

Den Engelsman J, Keijsers V, de Jong WW, Boelens WC (2003) The small heat-shock protein alpha B-crystallin promotes FBX4-dependent ubiquitination. J Biol Chem 278:4699–4704

Dewey WC, Li X, Wong RSL (1990) Cell killing, chromosomal aberrations, and division delay as thermal sensitivity is modified during the cell cycle. Radiat Res 122:268–274

Dikomey E, Jung H (1992) Effect of thermotolerance and step-down heating on thermal radiosensitization in CHO cells. Int J Radiat Biol 61:235–242

Dubois MF, Hovanessian AG, Bensaude O (1991) Heat-shock-induced denaturation of proteins. Characterization of the insolubilization of the interferon-induced p68 kinase. J Biol Chem 266:9707–9711

Dundr M, Misteli T, Olson MO (2000) The dynamics of postmitotic reassembly of the nucleolus. J Cell Biol 150:433–446

Ehrnsperger M, Graber S, Gaestel M, Buchner J (1997) Binding of non-native protein to Hsp25 during heat shock creates a reservoir of folding intermediates for reactivation. EMBO J 16:221–229

Ekedahl J, Joseph B, Marchetti P, Fauvel H, Formstecher P, Lewensohn R, Zhivotovsky B (2003) Heat shock protein 72 does not modulate ionizing radiation-induced apoptosis in U1810 non-small cell lung carcinoma cells. Cancer Biol Ther 2:663–669

Ellis RJ, Hemmingsen SM (1989) Molecular chaperones: proteins essential for the biogenesis of some macromolecular structures. TIBS 14:339–342

Ellis RJ, Van der Vies SM (1991) Molecular chaperones. Annu Rev Biochem 60:321–347

Evan GI, Hancock DC (1985) Studies on the interaction of the human c-myc protein with cell nuclei: p62c-myc as a member of a discrete subset of nuclear proteins. Cell 43:253–261

Fortin A, Raybaud-Diogene H, Tetu B, Deschenes R, Huot J, Landry J (2000) Overexpression of the 27 kDa heat shock protein is associated with thermoresistance and chemoresistance but not with radioresistance. Int J Radiat Oncol Biol Phys 46:1259–1266

Fourie AM, Hupp TR, Lane DP, Sang BC, Barbosa MS, Sambrook JF, Gething MJH (1997) HSP70 binding sites in the tumor suppressor protein p53. J Biol Chem 272:19471–19479

Freeman BC, Morimoto RI (1996) The human cytosolic molecular chaperones hsp90, Hsp70 (hsc70) and hdj-1 have distinct roles in recognition of a non-native protein and protein refolding. EMBO J 15:2969–2979

Freeman BC, Myers MP, Schumacher R, Morimoto RI (1995) Identification of a regulatory motif in Hsp70 that affects ATPase activity, substrate binding and interaction with HDJ-1. EMBO J 14:2281–2292

Freeman BC, Toft DO, Morimoto RI (1996) Molecular chaperone machines: chaperone activities of the cyclophilin Cyp-40 and the steroid aporeceptor-associated protein p23. Science 274:1718–1720

Freeman ML, Spitz DR, Meredith MJ (1990) Does heat shock enhance oxidative stress? Studies with ferrous and ferric iron. Radiat Res 124:288–293

Friedman DI, Olson ER, Georgopoulos C, Tilly K, Herskowitz I, Banuett F (1984) Interactions of bacteriophage and host macromolecules in the growth of bacteriophage-lambda. Microbiol Rev 48:299–325

Frydman J, Nimmesgern E, Ohtsuka K, Hartl FU (1994) Folding of nascent polypeptide chains in a high molecular mass assembly with molecular chaperones. Nature 370:111–117

Gabai VL, Meriin AB, Mosser DD, Caron AW, Rits S, Shifrin VI, Sherman MY (1997) Hsp70 prevents activation of stress kinases—a novel pathway of cellular thermotolerance. J Biol Chem 272:18033–18037

Garcia-Mata R, Bebok Z, Sorscher EJ, Sztul ES (1999) Characterization and dynamics of aggresome formation by a cytosolic GFP-chimera. J Cell Biol 146:1239–1254

Gebauer M, Zeiner M, Gehring U (1997) Proteins interacting with the molecular chaperone hsp70/hsc70: physical associations and effects on refolding activity. FEBS Lett 417:109–113

Gebauer M, Melki R, Gehring U (1998a) The chaperone cofactor Hop/p60 interacts with the cytosolic chaperonin-containing TCP-1 and affects its nucleotide exchange and protein folding activities. J Biol Chem 273:29475–29480

Gebauer M, Zeiner M, Gehring U (1998b) Interference between proteins Hap46 and Hop/p60, which bind to different domains of the molecular chaperone hsp70/hsc70. Mol. Cell Biol 18:6238–6244

Gehrmann M, Marienhagen J, Eichholtz-Wirth H, Fritz E, Ellwart J, Jaattela M, Zilch T, Multhoff G (2005) Dual function of membrane-bound heat shock protein 70 (Hsp70), Bag-4, and Hsp40: protection against radiation-induced effects and target structure for natural killer cells. Cell Death Differ 12:38–51

Gerber DA, Souquere-Besse S, Puvion F, Dubois MF, Bensaude O, Cochet C (2000) Heat-induced relocalization of protein kinase CK2. Implication of CK2 in the context of cellular stress. J Biol Chem 275:23919–23926

Gerner EW, Schneider MJ (1975) Induced thermal resistance in HeLa cells. Nature 256:500–502

Glover JR, Lindquist S (1998) Hsp104, Hsp70, and Hsp40: a novel chaperone system that rescues previously aggregated proteins. Cell 94:73–82

Greene LE, Zinner R, Naficy S, Eisenberg E (1995) Effect of nucleotide on the binding of peptides to 70-kDa heat shock protein. J Biol Chem 270:2967–2973

Greene MK, Maskos K, Landry SJ (1998) Role of the J-domain in the cooperation of Hsp40 with Hsp70. Proc Natl Acad Sci U S A 95:6108–6113

Ha JH, Hellman U, Johnson ER, Li LS, Mckay DB, Sousa MC, Takeda S, Wernstedt C, Wilbanks SM (1997) Destabilization of peptide binding and interdomain communication by an E543K mutation in the bovine 70-kDa heat shock cognate protein, a molecular chaperone. J Biol Chem 272:27796–27803

Hahn GM, Li GC (1982) Thermotolerance and heat shock proteins in mammalian cells. Radiat Res 92:452–457

Hartson-Eaton M, Malcolm AW, Hahn GM (1984) Radiosensitivity and thermosensitization of thermotolerant Chinese hamster cells and RIF-1 tumors. Radiat Res 99:175–184

Hata M, Okumura K, Seto M, Ohtsuka K (1996) Genomic cloning of a human heat shock protein 40 (Hsp40) gene (HSPF1) and its chromosomal localization to 19p13.2. Genomics 38:446–449

Hatayama T, Yasuda K (1998) Association of HSP105 with HSC70 in high molecular mass complexes in mouse FM3A cells. Biochem Biophys Res Commun 248:395–401

Hattori H, Kaneda T, Lokeshwar B, Laszlo A, Ohtsuka K (1993) A stress-inducible 40 kDa protein (hsp40): purification by modified two-dimensional gel electrophoresis and co-localization with hsc70 (p73) in heat-shocked HeLa cells. J Cell Sci 104:629–638

Haveman J, Hart AAM, Wondergem J (1987) Thermal radiosensitization and thermotolerance in cultured cells from a murine mammary carcinoma. Int J Radiat Biol 51:71–80

Hayes SA, Dice JF (1996) Roles of molecular chaperones in protein degradation. J Cell Biol 132:255–258

Hendrick JP, Hartl F-U (1993) Molecular chaperone functions of heat-shock proteins. Annu Rev Biochem 62:49–384

Hohfeld J (1998) Regulation of the heat shock cognate Hsc70 in the mammalian cell: the characterization of the anti-apoptotic protein BAG-1 provides novel insights. Biol Chem 379:269–274

Hohfeld J, Jentsch S (1997) GrpE-like regulation of the Hsc70 chaperone by the anti-apoptotic protein BAG-1. EMBO J 16:6209–6216

Hohfeld J, Minami Y, Hartl FU (1995) Hip, a novel cochaperone involved in the eukaryotic Hsc70/Hsp40 reaction cycle. Cell 83:589–598

Holmberg CI, Staniszewski KE, Mensah KN, Matouschek A, Morimoto RI (2004) Inefficient degradation of truncated polyglutamine proteins by the proteasome. EMBO J 23:4307–4318

Horwitz J (1992) α-Crystallin can function as a molecular chaperone. Proc Natl Acad Sci U S A 89:10449–10453

Hut HMJ, Kampinga HH, Sibon OCM (2005) Hsp70 protects mitotic cells against heat-induced centrosome damage and division abnormalities. Mol Biol Cell (in press)

Hutchison KA, Dittmar KD, Stancato LF, Pratt WB (1996) Ability of various members of the hsp70 family of chaperones to promote assembly of the glucocorticoid receptor into a functional heterocomplex with hsp90. J. Steroid Biochem Mol Biol 58:251–258

Irmer H, Hohfeld J (1997) Characterization of functional domains of the eukaryotic co-chaperone Hip. J Biol Chem 272:2230–2235

Ishihara K, Yamagishi N, Hatayama T (2003) Protein kinase CK2 phosphorylates Hsp105 alpha at Ser509 and modulates its function. Biochem J 371:917–925

Itoh T, Matsuda H, Mori H (1999) Phylogenetic analysis of the third Hsp70 homolog in Escherichia coli: a novel member of the Hsc66 subfamily and its possible co-chaperone. DNA Res 6:299–305

Jaattela M (1995) Over-expression of hsp70 confers tumorigenicity to mouse fibrosarcoma cells. Int J Cancer 60:689–693

Jaattela M (1999) Escaping cell death: survival proteins in cancer. Exp Cell Res 248:30–43

Jaattela M, Wissing D, Kokholm K, Kallunki T, Egeblad M (1998) Hsp70 exerts its anti-apoptotic function downstream of caspase-3-like proteases. EMBO J 17:6124–6134

Jaenicke R (1991) Protein stability and protein folding. Ciba Found Symp 161:206–216

Jakob U, Gaestel M, Engel K, Buchner J (1993) Small heat shock proteins are molecular chaperones. J Biol Chem 268:1517–1520

Jiang J, Ballinger CA, Wu Y, Dai Q, Cyr DM, Hohfeld J, Patterson C (2001) CHIP is a U-box-dependent E3 ubiquitin ligase: identification of Hsc70 as a target for ubiquitylation. J Bio Chem. 276:42938–42944

Johnson BD, Schumacher RJ, Ross ED, Toft DO (1998) Hop modulates hsp70/hsp90 interactions in protein folding. J Biol Chem 273:3679–3686

Johnston JA, Ward CL, Kopito RR (1998) Aggresomes: a cellular response to misfolded proteins. J Cell Biol 143:1883–1898

Johnston JA, Illing ME, Kopito RR (2002) Cytoplasmic dynein/dynactin mediates the assembly of aggresomes. Cell Motil Cytoskel 53:26–38

Jorritsma JBM, Burgman P, Kampinga HH, Konings AWT (1986) DNA polymerase activity in heat killing and hyperthermic radiosensitization of mammalian cells as observed after fractionated heat treatments. Radiat Res 105:307–319

Kampinga HH (1993) Thermotolerance in mammalian cells. Protein denaturation and aggregation, and stress proteins. J Cell Sci 104:11–17

Kampinga HH, Jorritsma JBM, Konings AWT (1985) Heat-induced alterations in DNA polymerase activity of HeLa cells and of isolated nuclei. Relation to cell survival. Int J Radiat Biol 47:29–40

Kampinga HH, Turkel-Uygur N, Roti Roti JL, Konings AW (1989) The relationship of increased nuclear protein content induced by hyperthermia to killing of HeLa S3 cells. Radiat Res 117:511–522

Kampinga HH, Brunsting JF, Stege GJJ, Konings AWT, Landry J (1994) Cells overexpressing Hsp27 show accelerated recovery from heat-induced nuclear protein aggregation. Biochem Biophys Res Comm 204:1170–1177

Kampinga HH, Konings AW, Evers AJ, Brunsting JF, Misfud N, Anderson RL (1997) Resistance to heat radiosensitization and protein damage in thermotolerant and thermoresistant cells. Int J Radiat Biol 71:315–326

Kampinga HH, Kanon B, Salomons FA, Kabakov AE, Patterson C (2003) Overexpression of the cochaperone CHIP enhances Hsp70-dependent folding activity in mammalian cells. Mol Cell Biol 23:4948–4958

Kabani M, McLellan C, Raynes DA, Guerriero V, Brodsky JL (2002) HspBP1, a homologue of the yeast Fes1 and Sls1 proteins, is an Hsc70 nucleotide exchange factor. FEBS Lett 531:339–342

Kelley WL (1998) The J-domain family and the recruitment of chaperone power. Trends Biochem Sci 23:222–227

Kelley WL, Georgopoulos C (1997) The T/t common exon of simian virus 40, JC, and BK polyomavirus T antigens can functionally replace the J-domain of the Escherichia coli DnaJ molecular chaperone. Proc Natl Aad Sci U S A 94:3679–3684

Kim HJ, Song EJ, Lee YS, Kim E, Lee KJ (2005) Human Fas associated factor 1 interacts with heat shock protein 70 and negatively regulates chaperone activity. J Biol Chem 280:8125–8133

Kim S, Nollen EA, Kitagawa K, Bindokas VP, Morimoto RI (2002) Polyglutamine protein aggregates are dynamic. Nature Cell Biol 4:826–831

King FW, Wawrzynow A, Hohfeld J, Zylicz M (2001) Co-chaperones Bag-1, Hop and Hsp40 regulate Hsc70 and Hsp90 interactions with wild-type or mutant p53. EMBO J 20:6297–6305

Knauf U, Jakob U, Engel K, Buchner J, Gaestel M (1994) Stress- and mitogen-induced phosphorylation of the small heat shock protein Hsp25 by MAPKAP kinase 2 is not essential for chaperone properties and cellular thermoresistance. EMBO J 13:54–60

Kopito RR (1999) Biosynthesis and degradation of CFTR. Physiol Rev 79:S167–S173

Landry J, Chretien P, Lambert H, Hickey E, Weber LA (1989) Heat shock resistance conferred by expression of the human hsp27 gene in rodent cells. J Cell Biol 109:7–15

Lange BM, Bachi A, Wilm M, Gonzalez C (2000) Hsp90 is a core centrosomal component and is required at different stages of the centrosome cycle in Drosophila and vertebrates. EMBO J 19:1252–1262

Lassle M, Blatch GL, Kundra V, Takatori T, Zetter BR (1997) Stress-inducible, murine protein mSTI1—characterization of binding domains for heat shock proteins and in vitro phosphorylation by different kinases. J Biol Chem 272:1876–1884

Laszlo A (1992a) The effects of hyperthermia on mammalian cell structure and function. Cell Prolif 25:59–87

Laszlo A (1992b) The thermoresistant state: protection from initial damage or better repair? Exp Cell Res 202:519–531

Laszlo A, Li GC (1985) Heat-resistant variants of Chinese hamster fibroblasts altered in expression of heat shock protein. Proc Natl Acad Sci U S A 82:8029–8033

Lavoie JN, Hickey E, Weber LA, Landry J (1993) Modulation of actin microfilament dynamics and fluid phase pinocytosis by phosphorylation of heat shock protein 27. J Biol Chem 268:24210–24214

Lavoie JN, Lambert H, Hickey E, Weber LA, Landry J (1995) Modulation of cellular thermoresistance and actin filament stability accompanies phosphorylation-induced changes in the oligomeric structure of heat shock protein 27. Mol Cell Biol 15:505–516

Lee SJ, Choi SA, Lee KH, Chung HY, Kim TH, Cho CK, Lee YS (2001) Role of inducible heat shock protein 70 in radiation-induced cell death. Cell Stress Chaperones. 6:273–281

Leeyoon D, Easton D, Murawski M, Burd R, Subjeck JR (1995) Identification of a major subfamily of large hsp70-like proteins through the cloning of the mammalian 110-kDa heat shock protein. J Biol Chem 270:15725–15733

Lepock JR, Frey HE, Inniss WE (1990) Thermal analysis of bacteria by differential scanning calorimetry: relationship of protein denaturation in situ to maximum growth temperature. Biochim Biophys Acta 1055:19–26

Lepock JR, Frey HE, Ritchie KP (1993) Protein denaturation in intact hepatocytes and isolated cellular organelles during heat shock. J Cell Biol 122:1267–1276

Lepock JR, Frey HE, Heynen ML, Senisterra GA, Warters RL (2001) The nuclear matrix is a thermolabile cellular structure. Cell Stress Chaperones. 6:136–147

Li GC, Werb Z (1982) Correlation between synthesis of heat shock proteins and development of thermotolerance in Chinese hamster fibroblasts. Proc Natl Acad Sci U S A 79:3218–3222

Li GC, Li L, Liu Y-K, Mak JY, Chen L, Lee WMF (1991) Thermal response of rat fibroblasts stably transfected with the human 70-kDa heat shock protein-encoding gene. Proc Natl Acad Sci U S A 88:1681–1685

Li GC, Mivechi NF, Weitzel G (1995) Heat shock proteins, thermotolerance, and their relevance to clinical hyperthermia. Int J Hyperthermia 11:459–488

Liberek K, Marszalek J, Ang D, Georgopoulos C, Zylicz M (1991) Escherichia coli DnaJ and GrpE heat shock proteins jointly stimulate ATPase activity of DnaK. Proc Natl Acad Sci U S A 88:2874–2878

Liberek K, Wall D, Georgopoulos C (1995) The DnaJ chaperone catalytically activates the DnaK chaperone to preferentially bind the sigma(32) heat shock transcriptional regulator. Proc Natl Acad Sci U S A 92:6224–6228

Lindquist S, Craig EA (1988) The heat-shock proteins. Annual Rev Genet 22:631–677

Liu QL, Kishi H, Ohtsuka K, Muraguchi A (2003) Heat shock protein 70 binds caspase-activated DNase and enhances its activity in TCR-stimulated T cells. Blood 102:1788–1796

Loktionova SA, Ilyinskaya OP, Gabai VL, Kabakov AE (1996) Distinct effects of heat shock and ATP depletion on distribution and isoform patterns of human Hsp27 in endothelial cells. FEBS Lett 392:100–104

Luders J, Demand J, Schonfelder S, Frien M, Zimmermann R, Hohfeld J (1998) Cofactor-induced modulation of the functional specificity of the molecular chaperone Hsc70s. Biol Chem 379:1217–1226

Martin H, Bruecher J, Claude R, Hoelzer D (1993) Cumulative chemotherapy increases mafosfamide toxicity for normal progenitor cells in aml patients—rationale for cryopreserving adapted-dose purged marrow early in first complete remission. Bone Marrow Transplant 12:495–499

Meacham GC, Patterson C, Zhang W, Younger JM, Cyr DM (2001) The Hsc70 co-chaperone CHIP targets immature CFTR for proteasomal degradation. Nat Cell Biol 3:100–105

Melcher A, Todryk S, Hardwick N, Ford M, Jacobson M, Vile RG (1998) Tumor immunogenicity is determined by the mechanism of cell death via induction of heat shock protein expression. Nat Med 4:581–587

Merck KB, Groenen PJTA, Voorter CEM, de Haard-Hoekman WA, Horwitz J, Bloemendal H, De Jong WW (1993) Structural and functional similarities of bovine alpha-crystallin and mouse small heat-shock protein. J Biol Chem 268:1046–1052

Michels AA, Nguyen VT, Konings AW, Kampinga HH, Bensaude O (1995) Thermostability of a nuclear-targeted luciferase expressed in mammalian cells—destabilizing influence of the intranuclear microenvironment. Eur J Biochem 234:382–389

Michels AA, Kanon B, Konings AW, Ohtsuka K, Bensaude O, Kampinga HH (1997) Hsp70 and Hsp40 chaperone activities in the cytoplasm and the nucleus of mammalian cells. J Biol Chem 272:33283–33289

Michels AA, Kanon B, Bensaude O, Kampinga HH (1999) Heat shock protein (Hsp) 40 mutants inhibit Hsp70 in mammalian cells. J Biol Chem 274:36757–36763

Minami Y, Hohfeld J, Ohtsuka K, Hartl FU (1996) Regulation of the heat-shock protein 70 reaction cycle by the mammalian DnaJ homolog, Hsp40. J Biol Chem 271:19617–19624

Mivechi NF, Li GC (1987) Lack of effect of thermotolerance on radiation response and thermal radiosensitization of murine bone marrow progenitors. Cancer Res 47:1538–1541

Morimoto RI (1998) Regulation of the heat shock transcriptional response: cross talk between a family of heat shock factors, molecular chaperones, and negative regulators. Genes Dev 12:3788–3796

Mosser DD, Caron AW, Bourget L, Meriin AB, Sherman MY, Morimoto RI, Massie B (2000) The chaperone function of hsp70 is required for protection against stress-induced apoptosis. Mol Cell Biol 20:7146–7159

Mosser DD, Ho S, Glover JR (2004) Saccharomyces cerevisiae Hsp104 enhances the chaperone capacity of human cells and inhibits heat stress-induced proapoptotic signaling. Biochemistry 43:8107–8115

Multhoff G, Mizzen L, Winchester CC, Milner CM, Wenk S, Eissner G, Kampinga HH, Laumbacher B, Johnson J (1999) Heat shock protein 70 (Hsp70) stimulates proliferation and cytolytic activity of natural killer cells. Exp Hematol 27:1627–1636

Nakahata K, Miyakoda M, Suzuki K, Kodama S, Watanabe M (2002) Heat shock induces centrosomal dysfunction, and causes non-apoptotic mitotic catastrophe in human tumour cells. Int J Hyperthermia 18:332–343

Nguyen VT, Morange M, Bensaude O (1989) Protein denaturation during heat shock and related stress. Escherichia coli b-galactosidase and photinus pyralis luciferase inactivation in mouse cells. J Biol Chem 264:10487–10492

Nicolet CM, Craig EA (1989) Isolation and characterization of STI1, a stress-inducible gene from Saccharomyces cerevisiae. Mol Cell Biol 9:3638–3646

Nimmesgern E, Hartl F-U (1993) ATP-dependent protein refolding activity in reticulocyte lysate. Evidence for the participation of different chaperone components. FEBS 331:25–30

Nollen EA, Brunsting JF, Roelofsen H, Weber LA, Kampinga HH (1999) In vivo chaperone activity of heat shock protein 70 and thermotolerance. Mol Cell Biol 19:2069–2079

Nollen EA, Brunsting JF, Song J, Kampinga HH, Morimoto RI (2000) Bag1 functions in vivo as a negative regulator of Hsp70 chaperone activity. Mol Cell Biol 20:1083–1088

Nollen EA, Kabakov AE, Brunsting JF, Kanon B, Hohfeld J, Kampinga HH (2001a) Modulation of in vivo HSP70 chaperone activity by Hip and Bag-1. J Biol Chem 276:4677–4682

Nollen EA, Salomons FA, Brunsting JF, Want JJ, Sibon OC, Kampinga HH (2001b) Dynamic changes in the localization of thermally unfolded nuclear proteins associated with chaperone-dependent protection. Proc Natl Acad Sci U S A 98:12038–12043

Nylandsted J, Gyrd-Hansen M, Danielewicz A, Fehrenbacher N, Lademann U, Hoyer-Hansen M, Weber E, Multhoff G, Rohde M, Jaattela M (2004) Heat shock protein 70 promotes cell survival by inhibiting lysosomal membrane permeabilization. J Exp Med 200:425–435

Oh HJ, Chen X, Subjeck JR (1997) Hsp110 protects heat-denatured proteins and confers cellular thermoresistance. J Biol Chem 272:31636–31640

Ohtsuka K, Hata M (2000) Mammalian HSP40/DNAJ homologs: cloning of novel cDNAs and a proposal for their classification and nomenclature. Cell Stress Chaperones 5:98–112

Parcellier A, Schmitt E, Gurbuxani S, Seigneurin-Berny D, Pance A, Chantome A, Plenchette S, Khochbin S, Solary E, Garrido C (2003) HSP27 is a ubiquitin-binding protein involved in I-kappaBalpha proteasomal degradation. Mol Cell Biol 23:5790–5802

Parsell DA, Kowal AS, Lindquist S (1994) Saccharomyces cerevisiae hsp104 protein—purification and characterization of ATP-induced structural changes. J Biol Chem 269:4480–4487

Pelham HRB (1984) Hsp70 accelerates the recovery of nucleolar morphology after heat shock. EMBO J 3:3095–3100

Pellecchia M, Montgomery DL, Stevens SY, Vander Kooi CW, Feng HP, Gierasch LM, Zuiderweg ERP (2000) Structural insights into substrate binding by the molecular chaperone DnaK. Nat Struct Biol. 7:298–303

Prapapanich V, Chen SY, Toran EJ, Rimerman RA, Smith DF (1996) Mutational analysis of the hsp70-interacting protein hip. Mol Cell Biol 16:6200–6207

Prapapanich V, Chen SY, Smith DF (1998) Mutation of hip's carboxy-terminal region inhibits a transitional stage of progesterone receptor assembly. Mol Cell Biol 18:944–952

Pratt WB, Dittmar KD (1998) Studies with purified chaperones advance the understanding of the mechanism of glucocorticoid receptor hsp90 heterocomplex assembly. Trends Endocrinol Metab 9:244–252

Raaphorst GP, Azzam EI (1983) Thermal radiosensitization in Chinese hamster (V79) and mouse C3H 10T 1/2 cells. The thermotolerance effect. Br J Cancer 48:45–54

Ravikumar B, Duden R, Rubinsztein DC (2002) Aggregate-prone proteins with polyglutamine and polyalanine expansions are degraded by autophagy. Hum Mol Genet 11:1107–1117

Raynes DA, Guerriero V (1998) Inhibition of Hsp70 ATPase activity and protein renaturation by a novel Hsp70-binding protein. J Biol Chem 273:32883–32888

Rudiger S, Buchberger A, Bukau B (1997a) Interaction of Hsp70 chaperones with substrates. Nat Struct Biol 4:342–349

Rudiger S, Germeroth L, Schneidermergener J, Bukau B (1997b) Substrate specificity of the DnaK chaperone determined by screening cellulose-bound peptide libraries. EMBO J 16:1501–1507

Saleh A, Srinivasula SM, Balkir L, Robbins PD, Alnemri ES (2000) Negative regulation of the Apaf-1 apoptosome by Hsp70. Nat Cell Biol 2:476–483

Schmid D, Baici A, Gehring H, Christen P (1994) Kinetics of molecular chaperone action. Science 263:971–973

Schneider C, SeppLorenzino L, Nimmesgern E, Ouerfelli O, Danishefsky S, Rosen N, Hartl FU (1996) Pharmacologic shifting of a balance between protein refolding and degradation mediated by Hsp90. Proc Natl Acad Sci U S A 93:14536–14541

Schroder H, Langer T, Hartl FU, Bukau B (1993) DnaK, DnaJ and GrpE form a cellular chaperone machinery capable of repairing heat-induced protein damage. EMBO J 12:4137–4144

Schumacher RJ, Hurst R, Sullivan WP, Mcmahon NJ, Toft DO, Matts RL (1994) ATP-dependent chaperoning activity of reticulocyte lysate. J Biol Chem 269:9493–9499

Schumacher RJ, Hansen WJ, Freeman BC, Alnemri E, Litwack G, Toft DO (1996) Cooperative action of Hsp70, Hsp90, and DnaJ proteins in protein renaturation. Biochemistry 35:14889–14898

Shi Y, Mosser DD, Morimoto RI (1998) Molecular chaperones as HSF1-specific transcriptional repressors. Genes Dev 12:654–666

Shomura Y, Dragovic Z, Chang HC, Tzvetkov N, Young JC, Brodsky JL, Guerriero V, Hartl FU, Bracher A (2005) Regulation of Hsp70 function by HspBP1: structural analysis reveals an alternate mechanism for Hsp70 nucleotide exchange. Mol Cell 17:367–379

Shyy T, Subjeck JR, Johnson RJ (1983) The effects of heat shock and growth state on the nucleolar localization of hsp 110. J Cell Biol 97:1389–1395

Skowyra D, Georgopoulos C, Zylicz M (1990) The E. coli dnaK gene product, the hsp70 homolog, can reactivate heat-inactivated RNA polymerase in an ATP hydrolysis-dependent manner. Cell 62:939–944

Smith DF (1998) Sequence motifs shared between chaperone components participating in the assembly of progesterone receptor complexes. Biol Chem 379:283–288

Sondermann H, Scheufler C, Schneider C, Hohfeld J, Hartl FU, Moarefi I (2001) Structure of a Bag/Hsc70 complex: convergent functional evolution of Hsp70 nucleotide exchange factors. Science 291:1553–1557

Song J, Takeda M, Morimoto RI (2001) Bag1-Hsp70 mediates a physiological stress signalling pathway that regulates Raf-1/ERK and cell growth. Nat Cell Biol 3:276–282

Spiro IJ, Denman DL, Dewey WC (1982) Effect of hyperthermia on CHO DNA polymerases α and β. Radiat Res 89:134–149

Steel R, Doherty JP, Buzzard K, Clemons N, Hawkins CJ, Anderson RL (2004) Hsp72 inhibits apoptosis upstream of the mitochondria and not through interactions with Apaf-1. J Biol Chem 279:51490–51499

Stege GJ, Li GC, Li L, Kampinga HH, Konings AW (1994a) On the role of hsp72 in heat-induced intranuclear protein aggregation. Int J Hyperthermia 10:659–674

Stege GJ, Li L, Kampinga HH, Konings AW, Li GC (1994b) Importance of the ATP-binding domain and nucleolar localization domain of HSP72 in the protection of nuclear proteins against heat-induced aggregation. Exp Cell Res 214:279–284

Stege GJ, Kampinga HH, Konings AW (1995) Heat-induced intranuclear protein aggregation and thermal radiosensitization. Int J Radiat Biol 67:203–209

Sturzbecher HW, Chumakov P, Welch WJ, Jenkins JR (1987) Mutant P53 proteins bind Hsp 72/73 cellular heat shock-related proteins in Sv40-transformed monkey cells. Oncogene 1:201–211

Suh WC, Burkholder WF, Lu CZ, Zhao X, Gottesman ME, Gross CA (1998) Interaction of the Hsp70 molecular chaperone, DnaK, with its cochaperone DnaJ. Proc Natl Acad Sci U S A 95:15223–15228

Szabo A, Korszun R, Hartl FU, Flanagan J (1996) A zinc finger-like domain of the molecular chaperone DnaJ is involved in binding to denatured protein substrates. EMBO J 15:408–417

Takayama S, Sato T, Krajewski S, Kochel K, Irie S, Millan JA, Reed JC (1995) Cloning and functional analysis of BAG-1: a novel Bcl-2-binding protein with anti-cell death activity. Cell 80:279–284

Takayama S, Bimston DN, Matsuzawa S, Freeman BC, Aimesempe C, Xie ZH, Morimoto RI, Reed JC (1997) BAG-1 modulates the chaperone activity of Hsp70/Hsc70. EMBO J 16:4887–4896

Takayama S, Xie ZH, Reed JC (1999) An evolutionarily conserved family of Hsp70/Hsc70 molecular chaperone regulators. J Biol Chem 274:781–786

Takeda S, Mckay DB (1996) Kinetics of peptide binding to the bovine 70 kDa heat shock cognate protein, a molecular chaperone. Biochemistry 35:4636–4644

Tavaria M, Gabriele T, Kola I, Anderson RL (1996) A hitchhiker's guide to the human Hsp70 family. Cell Stress Chaperones. 1:23–28

Terada K, Ohtsuka K, Imamoto N, Yoneda Y, Mori M (1995) Role of heat shock cognate 70 protein in import of ornithine transcarbamylase precursor into mammalian mitochondria. Mol Cell Biol 15:3708–3713

Thress K, Song J, Morimoto RI, Kornbluth S (2001) Reversible inhibition of Hsp70 chaperone function by Scythe and Reaper. EMBO J 20:1033–1041

Todryk S, Melcher AA, Hardwick N, Linardakis E, Bateman A, Colombo MP, Stoppacciaro A, Vile RG (1999) Heat shock protein 70 induced during tumor cell killing induces Th1 cytokines and targets immature dendritic cell precursors to enhance antigen uptake. J Immunol 163:1398–1408

Ungewickell E, Ungewickell H, Holstein SEH (1997) Functional interaction of the auxilin J domain with the nucleotide- and substrate-binding modules of Hsc70. Biol. Chem 272:19594–19600

Van de Klundert FAJM, Gijsen MLJ, Van den IJssel PRLA, Snoeckx LHEH, De Jong WW (1998) alpha B-crystallin and hsp25 in neonatal cardiac cells—differences in cellular localization under stress conditions. Eur J Cell Biol 75:38–45

Vidair CA, Doxsey SJ, Dewey WC (1995) Thermotolerant cells possess an enhanced capacity to repair heat-induced alterations to centrosome structure and function. J Cell Physiol 163:194–203

Vidair CA, Huang RN, Doxsey SJ (1996) Heat shock causes protein aggregation and reduced protein solubility at the centrosome and other cytoplasmic locations. Int J Hyperthermia. 12:681–695

Volloch VZ, Sherman MY (1999) Oncogenic potential of Hsp72. Oncogene 18:3648–3651

Wall D, Zylicz M, Georgopoulos C (1994) The NH2-terminal 108 amino acids of the Escherichia Coli DnaJ protein stimulate the ATPase activity of Dnak and are sufficient for lambda replication. J Biol Chem 269:5446–5451

Wall D, Zylicz M, Georgopoulos C (1995) The conserved G/F motif of the DnaJ chaperone is necessary for the activation of the substrate binding properties of the DnaK chaperone. J Biol Chem 270:2139–2144

Wang HG, Takayama S, Rapp UR, Reed JC (1996) Bcl-2 interacting protein, BAG-1, binds to and activates the kinase Raf-1. Proc Natl Acad Sci U S A 93:7063–7068

Warters RL, Stone OL (1983) The effects of hyperthermia on DNA replication in HeLa cells. Radiat Res 93:71–84

Wawrzynow A, Banecki B, Wall D, Liberek K, Georgopoulos C, Zylicz M (1995) ATP hydrolysis is required for the DnaJ-dependent activation of DnaK chaperone for binding to both native and denatured protein substrates. J Biol Chem 270:19307–19311

Welch WJ, Suhan JP (1985) Morphological study of the mammalian stress response: characterization of changes in cytoplasmic organelles, cytoskeleton, and nucleoli, and appearance of intranuclear actin filaments in rat fibroblasts after heat-shock treatment. J Cell Biol 101:1198–1211

Welch WJ, Suhan JP (1986) Cellular and biochemical events in mammalian cells during and after recovery from physiological stress. J Cell Biol 103:2035–2052

Wiech H, Buchner J, Zimmermann R, Jakob U (1992) Hsp90 chaperones protein folding in vitro. Nature 358:169–170

Wyttenbach A, Swartz J, Kita H, Thykjaer T, Carmichael J, Bradley J, Brown R, Maxwell M, Schapira A, Orntoft TF, Kato K, Rubinsztein DC (2001) Polyglutamine expansions cause decreased CRE-mediated transcription and early gene expression changes prior to cell death in an inducible cell model of Huntington's disease. Hum Mol Genet 10:1829–1845

Yamagishi N, Ishihara K, Hatayama T (2004) Hsp105alpha suppresses Hsc70 chaperone activity by inhibiting Hsc70 ATPase activity. J Biol Chem 279:41727–41733

Yang XL, Chernenko G, Hao YW, Ding ZH, Pater MM, Pater A, Tang SC (1998) Human BAG-1/RAP46 protein is generated as four isoforms by alternative translation initiation and overexpressed in cancer cells. Oncogene 17:981–989

Yasuda K, Nakai A, Hatayama T, Nagata K (1995) Cloning and expression of murine high molecular mass heat shock proteins, HSP105. J Biol Chem 270:29718–29723

Zeiner M, Gebauer M, Gehring U (1997) Mammalian protein RAP46: an interaction partner and modulator of 70 kDa heat shock proteins. EMBO J 16:5483–5490

Zeiner M, Gehring U (1995) A protein that interacts with members of the nuclear hormone receptor family: identification and cDNA cloning. Proc Natl Acad Sci U S A 92:11465–11469

Zhu XT, Zhao X, Burkholder WF, Gragerov A, Ogata CM, Gottesman ME, Hendrickson WA (1996) Structural analysis of substrate binding by the molecular chaperone DnaK. Science 272:1606–1614

Zylicz M, Ang D, Liberek K, Georgopoulos C (1989) Initiation of lambda-Dna replication with purified host-encoded and bacteriophage-encoded proteins—the role of the Dnak, Dnaj and Grpe heat-shock proteins. EMBO J 8:1601–1608

Zylicz M, King FW, Wawrzynow A (2001) Hsp70 interactions with the p53 tumour suppressor protein. EMBO J 20:4634–4638

Feedback Regulation of the Heat Shock Response

R. Voellmy

Department of Biochemistry and Molecular Biology, University of Miami,
Miller School of Medicine, Miami FL, 33136, USA
rvoellmy@med.miami.edu

1	Introduction	43
2	Aspects of Feedback Regulation of HSF1 Activity and Function	45
2.1	Repression of Trimerization of HSF1 by Hsps and Co-chaperones	45
2.2	Repression of HSF1 Transcriptional Activity by Hsps and Co-chaperones . . .	49
2.3	Repression and Co-activation of HSF1 Transcriptional Activity by Phosphorylation	53
2.4	Stress Granules: Sequestration of HSF1 and Splicing Factors	57
3	Post-Transcriptional Aspects of Feedback Regulation of the Heat Shock Response	58
3.1	Disposal of Stress-Unfolded Proteins by Refolding or Proteasome-Mediated Degradation	58
3.2	Other Post-Transcriptional Effects: Hsp mRNA Stability and Translatability .	59
4	Synopsis	60
References	. . . .	62

Abstract The heat shock response is triggered primarily by nonnative proteins accumulating in a stressed cell and results in increased expression of heat shock proteins (Hsps), i.e., of chaperones capable of participating in the refolding or elimination of nonnative proteins. Best known is the transcriptional part of this response that is mediated predominantly by heat shock factor 1 (HSF1). HSF1 activity is regulated at different levels by Hsps and co-chaperones and is modulated further by a number of mechanisms involving other stress-regulated aspects of cell metabolism.

Keywords HSF1 · Feedback regulation · Transcription · Stress · Heat shock

1
Introduction

For the purposes of this review, the heat shock response will be understood as the induction of heat shock or stress protein (Hsp) synthesis that occurs in eukaryotic cells subsequent to heat treatment or exposure to another proteotoxic stress. In their early studies of cultured *Drosophila* cells, Lindquist and colleagues observed that, under conditions of moderately severe heat stress, the heat shock response was self-regulated, i.e., Hsp expression increased immediately after initiation of the heat treatment, persisted for a period of time and then returned to a level approximately corresponding to the prestress level (Lindquist 1980; DiDomenico et al. 1982). This phenomenon, referred to as attenuation, was subsequently also observed and extensively characterized in mammalian cells (Abravaya et al. 1991). Slowing heat-induced transcription or translation using appropriate inhibitors prolonged the period of continued Hsp expression during stress recovery, suggesting that one or more Hsps feedback-regulate the heat shock response (DiDomenico et al. 1982). It was also recognized that the heat shock response is regulated both transcriptionally and post-transcriptionally.

The transcriptional response to heat and other proteotoxic stress is mediated by so-called heat shock transcription factors or, shortened, heat shock factors (HSF). Originally, these factors were defined as proteins that are capable of specifically binding heat shock elements (HSEs). HSEs are short sequence elements that are present in the promoters of hsp genes and are required for heat/stress induction of their expression. Vertebrate animals and plants have several different but related HSFs (reviewed, e.g., in Pirkkala et al. 2001). Among the different vertebrate HSFs, the factor named HSF1 is absolutely required for stress regulation of Hsp expression (McMillan et al. 1998; Zhang et al. 2002). This review is primarily concerned with the regulation of HSF1 and the single HSF species in other organisms such as yeast and *Drosophila* (also referred to herein as HSF1 for convenience) that perform the same function as vertebrate HSF1. Functionally relevant regulation of HSF1 largely occurs at the level of HSF1 activity and not of synthesis/degradation of the transcription factor (Zimarino and Wu 1987; Rabindran et al. 1991; Kline and Morimoto 1997; information from mutagenesis studies, e.g., Zuo et al. 1995, may also be considered pertinent).

Although this review portrays feedback regulation of the heat shock response as occurring predominantly at the level of activity and function of HSF1, it is noted that other HSFs may contribute and, in some organisms, play a critical role in this regulation. He et al. (2003) reported evidence that mammalian HSF2 can enhance heat-induced HSF1 activity. Avian cells lacking avian-specific HSF3 activity are defective in stress-induced Hsp expression (Tanabe et al. 1998). In avian organisms, both HSF1 and HSF3 contribute to stress tolerance (Inouye et al. 2003). In Arabidopsis, only inactivation of both hsf1 and hsf3 genes results in significant impairment of the heat shock response (Lohmann et al. 2004).

2
Aspects of Feedback Regulation of HSF1 Activity and Function

2.1
Repression of Trimerization of HSF1 by Hsps and Co-chaperones

An early step in activation of HSF1 is the homotrimerization of the transcription factor. This oligomerization is accompanied by the acquisition of HSE DNA-binding activity. Several studies, including extensive mutagenesis experiments (Zuo et al. 1994), provided strong evidence for a causal relationship between HSF1 oligomeric conformation and HSE DNA-binding activity. Originally discovered by Westwood and colleagues (1991) studying *Drosophila* HSF1, homotrimerization was soon discovered to be a general feature of eukaryotic HSF1 regulation (see, e.g., Sarge et al. 1993 and Baler et al. 1993). Even for yeast HSF1, which was long thought to be constitutively trimeric, careful analysis revealed that, although the factor has a high basal level, HSE DNA-binding activity can still be stimulated 10- to 15-fold by heat shock (Giardina and Lis 1995).

Although oligomerization of purified HSF1 can be induced by heat or hydrogen peroxide exposure in vitro (Goodson and Sarge 1995; Larson et al. 1995; Zhong et al. 1998), suggesting an inherent ability of HSF1 to sense stress, several observations strongly argue for the existence of a cellular mechanism (or mechanisms) that keeps HSF1 in a nonhomo-oligomeric state in the absence of a stress. Overexpression of HSF1 in homologous or heterologous systems resulted in accumulation of trimeric factor, suggesting that titration of a cellular repression mechanism had occurred (see, e.g., Zuo et al. 1995). Hsp expression can be induced in *Drosophila* and *Xenopus laevis* at temperatures (heat shock temperatures) that are at or below the normal growth temperature of mammalian cells. When human HSF1 was introduced into *Drosophila* or *Xenopus laevis* cells, the factor initially accumulated in a nonhomo-oligomeric form but was induced to trimerize at the low heat shock temperatures of the latter cells, i.e., at temperatures at which it remains nonhomo-oligomeric in human cells (Clos et al. 1993; Zuo et al. 1995). These results implied that the temperature at which homotrimerization of HSF1 is induced is determined by cellular factors, not by intrinsic properties of HSF1.

A possible mechanism for repression of HSF1 homotrimerization was suggested by what had been learned about how the heat shock response is mediated. It was recognized early on that most conditions and chemicals that induce Hsp expression have the potential of causing denaturation of cellular protein or, in the case of amino acid analogs, of resulting in the synthesis of nonnative proteins (Kelley and Schlesinger 1978; Hightower 1980). Subsequent work supported this concept and provided evidence that most Hsp inducers including heat cause oxidation of nonprotein and protein thiols, resulting in glutathione-adducted and cross-linked proteins (Freeman et al. 1995; Liu et al.

1996; McDuffee et al. 1997; Senisterra et al. 1997; Zou et al. 1998a). Hence, accumulation of nonnative proteins in the cell appeared to be a reasonable initiating event that leads to induced Hsp synthesis and, since induced Hsp synthesis is mediated mainly by HSF1, to activation of this factor. That nonnative proteins are capable of inducing Hsp expression was demonstrated directly by experiments, in which injection of chemically denatured proteins but not of the corresponding native proteins in *Xenopus* oocytes, or cytoplasmic misexpression of a normally secreted protein requiring disulfide bond formation for correct folding, were shown to induce expression of a reporter gene controlled by an hsp gene promoter (Ananthan et al. 1986; Guo et al. 2001; see Goff et al. 1985 for analogous observations in bacteria). Chaperone proteins, i.e., Hsps and co-chaperones, offered themselves as obvious potential links between protein denaturation and derepression of HSF1 activity. The most basic mechanism of regulation of HSF1 oligomeric status that could be imagined involved an Hsp or other chaperone protein that acted by binding to HSF1 and repressing its trimerization. A heat stress capable of inducing Hsp expression can result in the unfolding of as much as one to several percent of cellular protein (Lepock et al. 1993; Lepock 2005). Such a large amount of nonnative protein can be expected to compete effectively for Hsps and co-chaperones with their normal cellular targets. Thus, in the above hypothetical mechanism, HSF1 molecules would be stripped effectively of their repressor Hsp or co-chaperone in a heat-stressed cell and be free to assemble into homotrimers.

Initial searches for a repressor of HSF1 homotrimerization concentrated on Hsp70. Hsp70 appeared to be an attractive candidate because of the strong stress inducibility of its expression, which inducibility was thought to provide for a highly effective mechanism of feedback regulation. Obviously, this "Hsp70 hypothesis" was based on a narrow view of feedback regulation of the stress protein response. The broader hypothesis discussed herein postulates that feedback regulation is provided both by new Hsp synthesis and through the disposal of stress-unfolded proteins, either by refolding or proteolytic degradation, resulting in a gradual re-association of Hsps and co-chaperones with their original cellular targets, including HSF1. More recently, the focus changed to Hsp90 and Hsp90-containing multichaperone complexes. This new interest was prompted by observations that benzoquinone ansamycins such as herbimycin A and geldanamycin were capable of inducing Hsp expression (Murakami et al. 1991; Hedge et al. 1995; Conde et al. 1997; Zou et al. 1998a, 1998b). The compounds were shown to specifically bind in the ATP-binding domain of Hsp90 and modulate Hsp90 conformation and function (Whitesell et al. 1994; Grenert et al. 1997; Prodromou et al. 1997). In addition, in vitro experiments suggested that HSF1 has an affinity for Hsp90 and that Hsp90-containing multichaperone complexes resembling those found in association with steroid receptors can assemble on HSF1 (Nadeau et al. 1993; Nair et al. 1996).

Several subsequent studies provided strong evidence that HSF1 homotrimerization is repressed by Hsp90-containing multichaperone complexes. A first such study described a cell-free system capable of reproducing important aspects of in vivo regulation (Zou et al. 1998b). HSF1 oligomerization and acquisition of HSE DNA-binding activity could be induced in this system by heat, geldanamycin, and addition of chemically denatured proteins. Immunodepletion of Hsp90 induced HSE DNA-binding activity. Back addition of purified Hsp90 after immunodepletion of Hsp90 prevented this induction of DNA-binding activity. Ali et al. (1998) demonstrated that microinjection of anti-Hsp90 antibody but not control antibody into *Xenopus* oocytes resulted in induction of HSF1 DNA-binding activity. This effect of anti-Hsp90 antibody injection could be neutralized by subsequent injection of purified Hsp90. Genetic evidence for an analogous repressive role of Hsp90 in yeast was provided by Duina et al. (1998).

Because of its abundance relative to that of other chaperones, an important fraction of Hsp90 may not be complexed with other chaperones. However, when associated with target proteins such as steroid receptors or protein kinases, Hsp90 is typically found in the form of an Hsp90-containing multichaperone complex (Pratt and Toft 1997). As mentioned before, Nair et al. (1996) had shown that Hsp90-containing multichaperone complexes could be assembled on recombinantly produced HSF1. Hence, it appeared likely that HSF1 would also associate with such complexes in vivo. Indeed, when testing antibodies against a large number of chaperones and co-chaperones in the *Xenopus* oocyte system, Bharadwaj et al. (1999) found that in vivo depletion of p23 also led to induction of HSF1 DNA-binding activity in the absence of a stress. In addition, results obtained by Duina et al. (1998) provided evidence that an immunophilin, a Cyp40-like cyclophilin, participated in repression of yeast HSF1 activity. While direct evidence for an in vivo interaction of HSF1 and Hsp90 also emerged from other studies (Nadeau et al. 1993; Ali et al. 1998, Bharadwaj et al. 1999), more definitive experiments made use of in situ cross-linking techniques to ensure that complexes observed had in fact formed in vivo. Cross-linked human HSF1–Hsp90 complexes could be immunoprecipitated under stringent conditions from unstressed cells (Zou et al. 1998b; Guo et al. 2001). The majority of these complexes disappeared in heat-treated cells with a rate comparable to that of HSF1 oligomerization. Taken together, these findings strongly support the hypothesis that HSF1 homotrimerization is repressed by a multichaperone complex comprising Hsp90, p23, and an immunophilin. This complex is associated with HSF1 in unstressed cells and is released upon stress, allowing HSF1 to self-associate to form homotrimers. Further evidence was provided by experiments with rat cardiac myocytes, showing that geldanamycin reduced the HSF1-Hsp90 interaction (Knowlton and Sun 2001). Moreover, Zhao et al. (2002) found that over-expression of Hsp90 led to reduced heat-induced HSF1 DNA-binding activity in mouse NIH3T3 cells.

The Hsp90-containing complex (or complexes) that apparently associates with nonhomotrimeric HSF1 is similar to the multichaperone complexes that are part of mature steroid aporeceptor complexes (Pratt and Toft 1997). The latter associations are known to be dynamic end products of a pathway of interactions between chaperones and between chaperones and target protein. What was learned about these interactions from work on steroid receptors is presented in Fig. 1 in the context of a model of repression/derepression of HSF1 trimerization. Briefly, the known steps involve initial binding of target protein by Hsp70 and Hsp40, introduction of Hsp90 by Hsp90- and Hsp70-binding protein Hop, and substitution of the first Hsp90-containing complex with a final "mature" Hsp90–p23–immunophilin complex. Observations made by Marchler and Wu (2001) can readily be interpreted as in vivo evidence that such assembly reactions in fact occur on HSF1. These authors found that depletion by specific siRNA of Hsp70, Hsp40 (DroJ1) or Hsp90 increased *Drosophila*

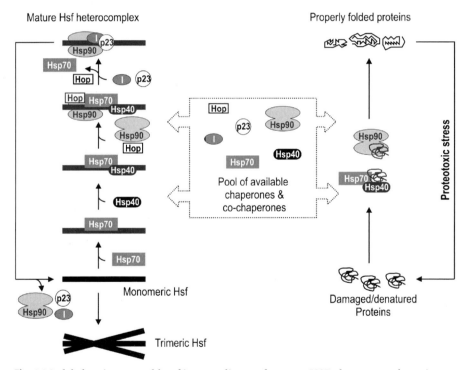

Fig. 1 Model showing assembly of intermediate and mature HSF1 heterocomplexes in unstressed cells. It is believed that a majority of HSF1 molecules in unstressed cells are associated with Hsp90-p23-immunophilin complexes (mature complexes). When a cell is stressed, an important fraction of cellular proteins unfold. Stress-unfolded proteins compete with HSF1 for Hsps and co-chaperones. Because HSF1 heterocomplexes are dynamic assemblies, this competition results in rapid disappearance of the complexes and inhibition of complex assembly. Unbound HSF1 monomers self-associate to form homotrimers. *I* = immunophilin. [Reproduced from Voellmy 2004 with permission from Cell Stress Society International (CSSI)]

HSF1 DNA-binding activity. Co-depletion of Hsp70 and Hsp40 or of Hsp90 and Hsp40 had synergistic inducing effects. It is noted that the more recent finding by Dai et al. (2003) that CHIP acts a co-regulator of HSF1 can also be explained in the context of the latter interactions. CHIP is a co-chaperone that can bind to Hsp90 and Hsp70. It is known to attenuate the ATPase activity of Hsp70 and to have the ability to remodel Hsp90-containing chaperone complexes, resulting in the removal of p23 (Ballinger et al. 1999; Connell et al. 2001; McDonough and Patterson 2003). The role of CHIP in HSF1 regulation may be related to the maintenance of a proper rate of multichaperone complex assembly on nonhomotrimeric (and, as will become clear from the discussion below, homotrimeric) HSF1.

Based on the above-discussed model of repression/derepression of HSF1 homotrimerization, it can be expected that feedback regulation will be provided by release of Hsps and co-chaperones from stress-unfolded proteins as they are successfully refolded or are disposed via the ubiquitin-proteasome pathway and by increased synthesis of Hsps. As a consequence, Hsps and co-chaperones will become increasingly available for interacting with HSF1 during recovery of cells from a stress or during an extended exposure to a mild stress. These interactions will initially occur on homotrimeric HSF1 and are likely to be part of an active mechanism that returns HSF1 to the nonhomotrimeric state.

It is noted that there is no direct evidence that nonhomotrimeric HSF1 in unstressed cells is quantitatively associated with Hsp90-containing multichaperone complex. Based on the observed effects of geldanamycin, one can only extrapolate that homotrimerization of an important fraction of HSF1 must be repressed by such a complex (Zou et al. 1998a, 1998b; see also the experiments of Duina et al. 1998). Therefore, it remains possible that a fraction of HSF1 may be regulated by an entirely different, unknown mechanism.

2.2
Repression of HSF1 Transcriptional Activity by Hsps and Co-chaperones

HSF1 oligomeric status and transcriptional activity are independently regulated. A so-called regulatory domain was mapped in mammalian HSF1 that represses the transactivation competence of the factor (Green et al. 1995; see also Zuo et al. 1995 and Shi et al. 1995). This domain encompasses residues 201–330 of the 529-residue human HSF1 sequence (see the schematic map in Fig. 2).

Guo et al. (2001) tested the hypothesis that transcriptional competence was also repressed by an Hsp90-containing multichaperone complex (or complexes), which was expected to bind to the regulatory domain of trimerized HSF1 molecules. Observations by Nair et al. (1996) provided an initial motivation for the study. As introduced before, the latter authors found that Hsp90-containing multichaperone complexes could be assembled on recombinant HSF1. Because recombinant forms of HSF1 tend to be trimeric (unless

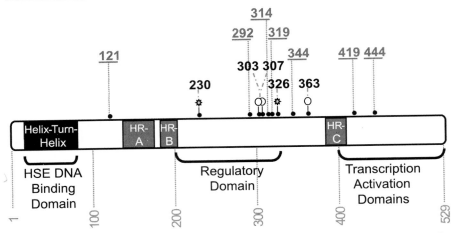

Fig. 2 Sequence and functional properties of human HSF1. *HR-A*, *HR-B*, and *HR-C* refer to characteristic hydrophobic repeat sequences (leucine zippers). *Numbers* indicate amino acid positions. The relative positions of the 12 known sites of phosphorylation of HSF1 are indicted *above* the sequence bar. *Stars* identify phosphorylation sites involved in factor activation (Ser^{230}, Ser^{326}), and *open circles* sites associated with factor deactivation (Ser^{303}, Ser^{307}, Ser^{363})

specific precautions are taken to prevent oligomerization), the findings suggested that the multichaperone complexes were formed on trimeric HSF1. To confirm that Nair and colleagues had indeed observed assembly of complexes on trimeric forms of HSF1, reconstitution experiments were repeated using as target recombinant HSF1 bound to HSE DNA-containing beads instead of tagged HSF1 captured on anti-tag antibody-containing beads employed in the earlier study. Because only trimeric HSF1 has HSE DNA-binding activity, these experiments exclusively monitored chaperone complex assembly on trimeric HSF1. Results confirmed that geldanamycin- and apyrase-sensitive Hsp90–p23–immunophilin complexes could be assembled on trimeric HSF1.

In vivo evidence for interactions between homotrimeric HSF1 and multichaperone complexes that were distinguishable from the complexes associating with nonhomotrimeric HSF1 was obtained by immunoprecipitation, using various anti-chaperone antibodies, of in situ cross-linked complexes before and at different times during a moderately severe heat treatment of human cells (43 °C for 1–15 min) (Guo et al. 2001). The experiments showed that the HSF1–Hsp90 interaction was maximal prior to the heat treatment and decreased to a low level during heat treatment. Conversely, an interaction between HSF1 and FKBP52 was observed that could not be detected in cells that were not heat-treated. Already after 1 min of heat treatment, a cross-linked HSF1–FKBP52 complex was readily detectable. The apparent abundance of this complex increased over the course of the 15-min heat treatment. It is noted that virtually all HSF1 was in a trimeric form after 15 min of heat treatment. Similar observations were made when cells were stressed by sodium arsenite or azetidine carboxylate instead of by heat. FKBP52 was known to bind target proteins as part of Hsp90-containing multichaperone complexes. The findings were interpreted to reflect dissociation of the initial interaction between nonhomotrimeric HSF1 and Hsp90–p23–immunophilin complex and trimerization of unbound HSF1 monomer according to the model described in the previous section. These events were followed by rapid assembly, on a fraction of homotrimerized HSF1, of Hsp90-containing complexes that also included FKBP52. Although other explanations remained possible, it appeared that the complexes forming on nonhomotrimeric and homotrimeric HSF1 differed in the type of immunophilin they contained. The homotrimeric but not (or less) the nonhomotrimeric HSF1 association comprised FKBP52.

Experiments aimed at obtaining evidence for a role of multichaperone complexes in repression of HSF1 transcriptional activity exploited previous observations that a substantial fraction of HSF1 overexpressed from transduced genes accumulates in a trimeric and DNA-binding but transcriptionally incompetent form (Zuo et al. 1994, 1995). Overexpressed mutants of HSF1 lacking the regulatory domain are transcriptionally active. Hence, in the overexpression situation, regulation of the HSF1 oligomeric state is bypassed, allowing for direct examination of HSF1 transcriptional regulation. To distinguish transcriptional activities of endogenous and exogenous HSF1 forms, the experiments overexpressed LEXA-HSF1 rather than HSF1. LEXA-HSF1 was identical with HSF1, except that the HSE DNA-binding domain was substituted with the DNA-binding domain of bacterial repressor LEXA. Because of the abundance of overexpressed LEXA-HSF1 relative to endogenous HSF1, the experiments also allowed for direct comparison of LEXA-HSF1 transcriptional activity and assembly of chaperone complexes. The HSF1–FKBP52 interaction was used to specifically assess assembly of chaperone complexes on homotrimeric HSF1 forms. The study produced two important results (Guo et al. 2001):

- Three different maneuvers that could be expected to reduce the availability of Hsps and co-chaperones for binding to HSF1, i.e., overexpression of a steroid receptor, exposure of cells to azetidine carboxylate, forcing synthesis of nonnative proteins, and overexpression of a protein incapable of reaching a native conformation intracellularly, caused transcriptional activation of LEXA-HSF1 as well as readily detectable reduction in the abundance of multichaperone complexes associated with LEXA-HSF1.
- Multichaperone complexes formed substantially less well on mutants of LEXA-HSF1 lacking portions of the regulatory domain than on prototype LEXA-HSF1.

Together, these findings strongly suggested that transcriptional activity of trimeric HSF1 is repressed by a multichaperone complex (or complexes) that associates with the HSF1 regulatory domain. Furthermore, the data showed that a reduction in the availability of Hsps and co-chaperones for binding to trimeric HSF1 such as occurs during induction of the heat shock response results in release of HSF1 from transcriptional repression.

A different mechanism that represses HSF1 transcriptional activity was described by Shi et al. (1998). These authors demonstrated using in vitro reconstitution assays that Hsp70 and Hsp40 (Hdj1) interact with the transcription activation domain region of mouse HSF1. Mapping experiments revealed that both full-length Hsp70 and an isolated substrate-binding domain but not mutant Hsp70 AAAA lacking chaperoning activity bound the HSF1 activation domains. Mapping of the Hsp70 binding site in HSF1 led to the definition of an interaction domain located between mouse HSF1 residues 425 and 439. Overexpression in mammalian cells of either Hsp70 or Hsp40 (Hdj1) inhibited the transcriptional activity of a co-expressed GAL4-HSF1 activation domain chimera. The transcriptional competence of a control GAL4-VP16 fusion protein remained unaffected. Regarding the relative importance of the two different mechanisms of transcriptional repression, available data suggest that the first mechanism that is mediated by multichaperone complex interaction with the HSF1 regulatory domain is more sensitive than the second mechanism that involves direct inhibition of transcription activation domains by Hsp70 and/or Hsp40: even when expressed at relatively modest levels, HSF1 mutants lacking part or all of the regulatory domain were active transcription factors in the absence of a stress (Zuo et al. 1995; Xia et al. 1999; Hall and Voellmy, unpublished observations).

As discussed in Sect. 2.1, an early step in the HSF1 activation process is the conversion of nonhomotrimeric, non-DNA-binding factor to homotrimeric, DNA-binding factor. An effective repression mechanism controls this conversion. Why then is there an apparent need for controlling activation at a subsequent step? The answer may lie in observations suggesting that the homotrimer may be the thermodynamically favored form of HSF1, at least in mammalian cells (but see below). Hence, homotrimerization of HSF1 may be an essentially irreversible process. As a consequence, even a modest decrease in the availability of Hsps and co-chaperones for binding to HSF1 polypeptide may result in nearly complete oligomerization of the factor. Hence, if regulation of the oligomeric state were the only mechanism for regulating HSF1 activity, it may not be possible to achieve activation of the factor that is proportional to the intensity of stress experienced by a cell. Repression mechanisms at a second level may provide for this proportionality. Unbound and chaperone-bound, trimeric HSF1 may readily equilibrate as a function of chaperone availability. To illustrate the effects of such regulation at two successive levels, at an intermediate intensity of stress the availability of Hsps and co-chaperones for binding HSF1 will only be moderately reduced. This reduction may be sufficient to cause near quantitative trimerization of the factor. However, because of the proportionality of chaperone binding to trimeric HSF1, only a fraction of (trimeric) HSF1 would remain unbound and proceed to become transcriptionally active. The remaining fraction would be bound by Hsps and/or multichaperone complex and remain transcriptionally inert. The above-discussed in situ cross-linking and immunoprecipitation experiments appeared to lend support to this concept (Guo et al. 2001). In cells exposed to a moderately severe heat stress, virtually all HSF1 was converted to the trimeric form, but a fraction of (trimeric) HSF1 became bound by an Hsp90- and FKBP52-containing chaperone complex. It is noted that the ratio of unbound to multichaperone complex-bound trimeric HSF1 may also be influenced by DAXX, a co-activator of HSF1, whose availability for binding to trimeric HSF1 is stress-regulated (see Sect. 2.3 for a more extensive description).

The mechanisms for inhibiting transcriptional competence described in this section may be equally important for feedback regulation of HSF1 activity during prolonged exposure to a moderate stress or subsequent to a stress. Because the trimeric conformation of HSF1 appears to be the thermodynamically favored conformation, the existence of an active mechanism (or mechanisms) that disassembles HSF1 trimers must be postulated. As long as the concentration of nonnative proteins remains sufficiently high to engage a sizable fraction of Hsps and co-chaperones, a futile cycle of assembly and disassembly of trimeric HSF1 can be expected to occur. Hence, mechanisms that repress the transcriptional competence of trimeric HSF1 may be the only mechanisms capable of reducing HSF1 activity progressively during stress recovery. Only late in the recovery process will the amounts of Hsps and co-chaperones available for binding to HSF1 polypeptides reach levels that are effective in curtailing oligomerization. It is further noted that disassembly of HSF1 trimers is likely to be catalyzed by a chaperone-mediated process. Therefore, the associations of Hsps and chaperone complexes with trimeric HSF1 that inhibit transcriptional activity may also represent initial steps in chaperone-mediated disassembly of trimeric HSF1. Because the abundance of these associations may be proportional to the amount of Hsps and co-chaperones available for binding, disassembly may occur at a rate that is proportional to the progress of recovery, which is determined by the rates of disposal of unfolded proteins and induced synthesis of Hsps.

2.3
Repression and Co-activation of HSF1 Transcriptional Activity by Phosphorylation

Phosphorylation would appear to be an ideal means for controlling a rapid physiological response such as the heat shock response. Not surprisingly, investigations relating to phosphorylation and dephosphorylation of heat shock factors began as soon as they became feasible. Results suggested that HSF1 activity indeed was influenced by phosphorylation/dephosphorylation events. It appeared that this regulation primarily concerned HSF1 transcriptional competence: several protein kinase C inhibitors were found to reduce heat-induced (hsp promoter-directed) reporter gene activity in mammalian cells, without affecting HSF1 DNA-binding activity (Erdos and Lee 1994; Xia and Voellmy 1997). Analogous observations were reported for *Drosophila* HSF1 (Fritsch and Wu 1999). Exposure of certain cell types to salicylate, menadione or hydrogen peroxide induced HSF1 HSE DNA-binding activity but not transcriptional activity (Jurivich et al. 1992; Bruce et al. 1993). HSF1 was only weakly phosphorylated under these conditions (Jurivich et al. 1995; Cotto et al. 1996; Xia and Voellmy 1997). Co-exposure to Ser/Thr phosphatase inhibitor calyculin A resulted in increased phosphorylation of HSF1 and derepression of transcriptional activity (Xia and Voellmy 1997). As discussed before, overexpressed HSF1 is HSE DNA-binding but transcriptionally inactive in the absence of a stress. Exposure to calyculin A or phorbol ester resulted in activation of HSF1 in cells overexpressing the factor (Xia and Voellmy 1997).

In spite of this ample indirect evidence for a role of phosphorylation in the HSF1 activation process, specific information only emerged relatively recently. Holmberg et al. (2001) reported that a Ser^{230}-to-Ala substitution reduced the heat-induced activity of human HSF1 by about twofold as measured by the ability of the mutant factor to mediate Hsp70 synthesis in mouse hsf1$^{-/-}$ MEF cells. Experiments that used a phosphospecific antibody revealed that phosphorylation of Ser^{230} was stress-inducible. Ser^{230} lies within a CaMKII consensus site. Overexpression of CaMKII resulted in enhanced heat-induced hsp70 promoter-directed reporter gene expression and phosphorylation at Ser^{230}. Conversely, kinase inhibitor KN-62 reduced reporter activity. Guettouche et al. (2005) recently conducted a comprehensive physical analysis of phosphorylation in heat-stressed human cells and found that HSF1 was multiply phosphorylated on Ser^{121}, Ser^{230}, Ser^{292}, Ser^{303}, Ser^{307}, Ser^{314}, Ser^{319}, Ser^{326}, Ser^{344}, Ser^{363}, Ser^{419}, and Ser^{444}. In agreement with phosphoamino acid analyses reported earlier, no Thr or Tyr residues were found phosphorylated. The same authors also carried out a systematic alanine scan of human HSF1. Every Ser, Thr, or Tyr residue was mutated, and consequences for HSF1 transcriptional competence were tested. The study uncovered a single residue, Ser^{326}, whose substitution resulted in a several-fold reduced transcriptional activity when assayed in the hsf1$^{-/-}$ MEF cell system. Like that of Ser^{230}, phosphorylation of Ser^{326} was found to be inducible by stress. When directly compared with the effect of Ser^{326} phosphorylation on stress-induced transcriptional activity, the effect of Ser^{230} phosphorylation appeared to be much smaller. The two studies revealed that inducible phosphorylation of at least two HSF1 Ser residues contributes to transcriptional activity. Clearly, however, stress activation of HSF1, albeit to a reduced level, also occurs in the absence of this phosphorylation. Therefore, the presently available evidence does not support models of regulation of the heat shock response that are based on activation of the response by inducible phosphorylation of HSF1 (Rieger et al. 2005).

HSF1 activity is also controlled negatively by phosphorylation. Evidence for this mode of regulation was obtained for yeast as well as mammalian cells (Hoi and Jacobsen 1994; Knauf et al. 1996; Kline and Morimoto 1997; see also Kim et al. 1997). After realizing that the regulatory domain of human HSF1 included six proline-directed serine motifs, Knauf et al. (1996) tested whether elimination of the serines in these motifs affected heat-inducible transactivation by GAL4-HSF1 chimeras. Heat inducibility of transactivation was substantially reduced for mutants lacking all six serines as well as for mutants lacking only Ser^{303} and Ser^{307}. Immune complex kinase assays revealed that the HSF1 regulatory domain (including Ser^{303} and Ser^{307}) could be phosphorylated by ERK1/ERK2 and p38 MAP kinases (Knauf et al. 1996; Kim et al. 1997). Kline and Morimoto (1997) first demonstrated that HSF1 was in fact phosphorylated on Ser^{303} and Ser^{307} in vivo. Studies by Chu et al. (1996, 1998) provided evidence that Ser^{303} is phosphorylated by glycogen synthase kinase 3 (GSK3) subsequent to phosphorylation of Ser^{307} by a MAP kinase. By now, there appears to be better, albeit not universal, support for a role of GSK3 in repressive phosphorylation of HSF1 than for any other protein kinase (Xavier et al. 2000; but see also Hietakangas et al. 2003). Several studies, using different assay systems, confirmed that Ser^{303}/Ser^{307} double substitutions have deregulated transcriptional activity (Knauf et al. 1996; Chu et al. 1996; Kline and Morimoto 1997; Xia et al. 1998; Guettouche et al. 2005). Similarly, in vivo phosphorylation of the two residues was observed by several laboratories (Kline and Morimoto 1997; Xia et al. 1998; Hietakangas et al. 2003; Guettouche et al. 2005). It can therefore be considered as established that HSF1 transcriptional activity is repressed by phosphorylation at Ser^{303} and Ser^{307}. Two independent studies provided evidence that phosphorylation of human HSF1 residue Ser^{363} also has a repressive effect on HSF1 transcriptional activity and/or an enhancing effect on HSF1 deactivation subsequent to a stress (Chu et al. 1998; Dai et al. 2000). It was more recently reported that phosphorylation of Ser^{303} and Ser^{307} also inhibits HSF1 function via a mechanism that is not directly related to transcriptional competence (Wang et al. 2003, 2004): HSF1 phosphorylated on the latter residues interacts with a 14-3-3 protein. In the absence of a stress, binding of 14-3-3 to HSF1 leads to sequestration of the transcription factor in the cytoplasm. This mechanism is believed to play a role during HSF1 deactivation subsequent to a stress.

While phosphorylation of Ser^{230} and Ser^{326} enhances, and phosphorylation of Ser^{303}, Ser^{307}, and Ser^{363} represses heat stress-induced transcriptional competence, human HSF1 is also phosphorylated in heat-treated cells on residues Ser^{121}, Ser^{292}, Ser^{314}, Ser^{319}, Ser^{344}, Ser^{419}, and Ser^{444} (Guettouche et al. 2005). The hypothesis that these phosphorylation events may be gratuitous is difficult to accept, considering that they concern a majority of the sites on which HSF1 is phosphorylated in heat-stressed cells. It seems more reasonable to believe that the transactivation experiments carried out to date were inappropriate for discovering the roles of the latter phosphorylation events. What was learned recently about an unexpected consequence of phosphorylation of HSF1 residue Ser^{419} may illustrate this difficulty. Kim et al. (2005) found that Polo-like kinase 1 is capable of phosphorylating HSF1 on Ser^{419}. Phosphorylation of cyclin B1 by the latter kinase is known to be essential for the nuclear translocation of the cyclin that takes place in prophase. A Ser^{419}-to-Ala substitution failed to exhibit the typical exclusive nuclear localization of HSF1 in heat-treated cells. This observation suggested that phosphorylation of Ser^{419} is important for stress-induced nuclear concentration of HSF1. Apparently, a Ser^{419}-to-Ala substitution remained without a detectable activity phenotype because the nuclear concentration of the mutant failed to become limiting under the conditions of the transactivation assays.

Phosphorylation of human HSF1 at Ser^{230} and Ser^{326} is stress-induced (Holmberg et al. 2001; Guettouche et al. 2005). As was persuasively shown by Hietakangas et al. (2003), phosphorylation of residues Ser^{303} and Ser^{307} also occurs during stress and is followed by gradual dephosphorylation during recovery from the stress. GSK3 (Ser^{303}) and JNK1 (Ser^{363}) kinase activities were also found to be heat stress-induced (He et al. 1998; Dai et al. 2000; Xavier et al. 2000). Hence, exposure of human cells to heat and, presumably, also other stressors appear to trigger a sequence of phosphorylation events beginning with transcription-enhancing phosphorylation at HSF1 residues Ser^{230} and Ser^{326} and ending with repressive phosphorylation at Ser^{303}, Ser^{307}, and Ser^{363}. It is not known how the sequence is initiated, and whether it represents a fixed program or is susceptible to modulation. Heat-induced phosphorylation of HSF1 is known to be feedback-regulated: when mammalian cells are exposed to a continuous mild stress, the level of HSF1 phosphorylation, estimated from SDS-PAGE mobility shift information, is initially increased but then eventually returns to the prestress level (Kline and Morimoto 1997). The mechanism of this feedback regulation is not known but is expected to comprise several components. As discussed previously, several of the protein kinases putatively involved in stress-induced phosphorylation of HSF1 are known, and the activity of at least two of them is heat stress-induced (GSK3, JNK1). Stress activation of these kinases may be susceptible to feedback regulation. Presumably, the target for transcription-modulating phosphorylation is the trimeric form of HSF1. Trimerization of HSF1 is regulated by the availability of Hsps and co-chaperones for binding to nonhomotrimeric factor. Presumably, the mechanism that returns trimeric HSF1 to the nonhomotrimeric form is also dependent on Hsps and co-chaperones. Finally, trimeric HSF1 associated with multichaperone complex may not be recognized as a target by the relevant protein kinases. HSF1 co-activator DAXX may function to increase the accessibility of trimeric HSF1 for activity-enhancing phosphorylation (Boellmann et al. 2004).

DAXX was isolated as a protein that interacts with trimeric HSF1 (Boellmann et al. 2004). Although not absolutely required, DAXX plays an important role in stress activation of HSF1, as evidenced by the substantial negative effects of deletion of DAXX, RNA interference depletion, and expression of dominant-negative mutants. As discussed previously, overexpression of HSF1 results in accumulation of trimeric factor that is repressed by interacting multichaperone complex and only exhibits minimal transcriptional activity in the absence of a stress. Co-expression of DAXX has a large synergistic effect on HSF1 activity. This effect is accompanied by increased phosphorylation of HSF1 (Boellmann and Voellmy, unpublished data). These observations suggest that DAXX binding to HSF1 may displace repressing multichaperone complex and prime the transcription factor for phosphorylation.

DAXX is a largely nuclear protein that predominantly associates with so-called promyelocytic leukemia oncogenic domains (PODs) or nuclear domains 10 (Ishov et al. 1999; Torii et al. 1999; Zhong et al. 2000). These subnuclear domains are organized by promyelocytic leukemia protein (PML). DAXX is released from these stores into the nucleoplasm during different types of stress (Maul et al. 1995; Nefkens et al. 2003). Subsequent to this release, DAXX becomes available for binding to trimeric HSF1 and promoting its transcriptional activation. Hence, HSF1 activity is also modulated indirectly by factors that control the release of DAXX from PODs. This release has been found to correlate with desumoylation of PML protein, which may be triggered by phosphorylation of the protein (Muller et al. 1998; Everett et al. 1999; Nefkens et al. 2003). Interaction of HIPK1 with DAXX also results in release of DAXX from PODs (Ecsedy et al. 2003). Furthermore, the biochemical properties of DAXX may also be modulated by its phosphorylation by HIPK1. Finally, DAXX is known to interact with several protein kinases and to function as an adapter for apoptosis signal-regulating kinase 1 that phosphorylates JNK (Chang et al. 1998; Rochat-Steiner et al. 2000; Ecsedy et al. 2003). Hence, there is a precedent for the possibility that DAXX may direct a protein kinase to HSF1 that could phosphorylate the transcription factor or an associated protein.

2.4
Stress Granules: Sequestration of HSF1 and Splicing Factors

In primate cells exposed to a stress, a large fraction of HSF1 was observed to rapidly redistribute from the nucleoplasm (and cytoplasm) to a small number of subnuclear structures referred to as stress granules (Cotto et al. 1997). Although it soon became clear that stress granules are not associated with genes for major hsp genes (Jolly et al. 1997), it only recently could be shown that the structures form on a particular heterochromatic region of (human) chromosome 9 (9q12) to which HSF1 binds (Jolly et al. 2002). Overexpression of Hsp70 prevents the concentration of HSF1 in stress granules (Jolly et al. 2002). Within these structures, HSF1 activates transcription of satellite III repeats by RNA polymerase II (Jolly et al. 2004; Rizzi et al. 2004). The transcripts formed remain stably associated with the 9q12 region. Stress granules are also sites of sequestration of specific splicing factors such as hSF2/ASF and hSRp30c, which is dependent on HSF1 and satellite III transcripts (Metz et al. 2004, and references cited therein). Although this remains speculative, it may be hypothesized that sequestration of the majority of stress-activated HSF1 molecules in stress granules will alter the dynamics of hsp gene expression. In terms of feedback regulation, the degree of competition by nonnative proteins for Hsp70 during stress recovery may influence HSF1 sequestration in stress granules, which in turn may affect expression from hsp genes. The expression of proteins from non-hsp genes may also be impacted owing to the sequestration of certain splicing factors in stress granules.

3
Post-Transcriptional Aspects of Feedback Regulation of the Heat Shock Response

3.1
Disposal of Stress-Unfolded Proteins by Refolding or Proteasome-Mediated Degradation

Under normal physiological conditions, a cell's ability to fold or refold nonnative polypeptides and, if necessary, to discard them is perfectly adapted to its needs. Probably, the largest workload for a chaperone system under these conditions is generated by new protein synthesis (Baler et al. 1992). Only when the cell is exposed to a proteotoxic stress, e.g., heat, osmotic stress, or exposure to a wide variety of chemicals, denaturation of proteins temporarily exceeds the cell's capacity, the level of unfolded proteins rises, and Hsps and co-chaperones are monopolized by the unfolded proteins, resulting in activation of HSF1 and induction of the heat shock response. The fine balance between protein unfolding, folding/refolding, and degradation via the ubiquitin-proteasome pathway is illustrated by the observation that inhibition of proteasome function by small-molecule inhibitors such as MG132 or lactacystin results in induction of Hsp synthesis, i.e., in a heat shock response (Bush et al. 1997; Lee and Goldberg 1998). The compulsory function of HSF1 in the heat shock response that is triggered by an elevation of the concentration of unfolded proteins is confirmed by the finding that proteasome inhibitors failed to induce Hsp synthesis in hsf1$^{-/-}$ cells and that the response could be restored by back addition of HSF1 (Pirkkala et al. 2000). It is now clear that both folding/refolding of proteins and disposal of nonrepairable proteins via the ubiquitin-proteasome pathway are chaperone-mediated processes. Unfolded proteins associate with chaperones and chaperone complexes. The decision of whether an unfolded protein is to be degraded involves association with and modulation of the latter chaperone complexes by co-chaperones such as CHIP (reviewed by Esser et al. 2004). As discussed before, CHIP is capable of binding to and affecting the activities of Hsp70 and Hsp90. In addition, CHIP also possesses E3 ubiquitin ligase activity. Hence, in addition to downregulation of HSF1 activity by direct feedback mechanisms, the increased levels of Hsps resulting from stress induction of the heat shock response are expected to accelerate both refolding of stress-unfolded proteins as well as their disposal by the ubiquitin-proteasome pathway (Nollen et al. 1999; Fisher et al. 1997; Gusarova et al. 2001; Zhang et al. 2001). These effects should further increase the rate of attenuation of the heat shock response during a moderate stress or in the wake of stress exposure. It is noted that CHIP has additional functions that are not linked to protein degradation. Its role as a co-regulator of HSF1 discussed earlier in this review is an example of such a function.

3.2
Other Post-Transcriptional Effects: Hsp mRNA Stability and Translatability

Feedback regulation of the heat shock response may also be mediated through post-transcriptional mechanisms that affect the synthesis of Hsps. During a stress, cells efficiently produce Hsp mRNAs, which are relatively stable and are preferentially translated into Hsps. Primary hsp gene transcripts typically do not contain introns and, therefore, escape inhibition of splicing that occurs under conditions of severe stress (Yost and Lindquist 1986). Preferential translation of Hsp mRNAs appears to be a reflection of a profound inhibition of translation of non-Hsp mRNAs. Once formed, Hsp mRNAs appear to be translated similarly under stress and nonstress conditions (Theodorakis and Morimoto 1987 and references cited therein). Hence, translation of Hsp mRNAs does not appear to be actively regulated. In contrast, hsp70 mRNA stability is dramatically greater during heat or arsenite stress than during recovery from stress or in unstressed cells (DiDomenico et al. 1982; Theodorakis and Morimoto 1987). There appears to exist no direct evidence that Hsp70 or another Hsp destabilizes Hsp70 mRNA, although 3′ UTR sequences were found to be responsible for heat stress-induced stability, and Hsp70 and Hsp110 were found to interact with AU-rich sequences in 3′ UTR sequences (Petersen and Lindquist 1989; Moseley et al. 1993; Henics et al. 1999). Xing et al. (2004) recently reported an interaction between HSF1 and symplekin (and CstF-64) in heat-stressed but not unstressed human cells. Symplekin is known to interact with polyadenylation factors CstF and CPSF and may function as a scaffold to assemble the latter proteins into a complex. Overexpression of an HSF1 mutant containing an inactivating point mutation in the HSE DNA-binding domain resulted in interference with polyadenylation but not synthesis of hsp70 gene transcripts in heat-stressed cells and in reduced synthesis of Hsp70, suggesting that HSF1 is instrumental in bringing polyadenylation factors to the site at which Hsp70 pre-mRNA is synthesized and polyadenylated. This mechanism may be an important aspect of efficient Hsp70 expression in heat-stressed cells.

4
Synopsis

The heat shock response is triggered when cells are exposed to heat or another proteotoxic stress. The immediate effect of this stress exposure is denaturation of a significant fraction of cellular proteins, and the end result is increased levels of Hsp chaperones that accelerate refolding of stress-unfolded proteins and degradation by the ubiquitin-proteasome pathway. Stress-unfolded proteins also appear to be the primary signal for activating the key transcription factor, HSF1, which mediates stress-induced expression of hsp genes. HSF1 activation is repressed at the levels of homotrimerization and acquisition of transcriptional competence by Hsps and multichaperone complexes (Fig. 3). It is believed that most HSF1 molecules are dynamically associated with an Hsp90–p23–immunophilin complex in an unstressed cell. Stress-unfolded proteins accumulating in a stressed cell compete with HSF1 for Hsps and co-chaperones. As a result, HSF1 is released from the multichaperone complex and self-associates to form homotrimers. A fraction (whose importance is related to stress intensity) of trimeric HSF1 escapes capture by a repressive Hsp90- and FKBP52-containing multichaperone complex and other repressive interactions with Hsps, acquires transcriptional competence, and begins to transactivate hsp genes. Consequently, levels of Hsps rise. During recovery from a stress or during prolonged exposure to a mild to moderately severe stress, stress-unfolded proteins are either refolded or eliminated by proteolytic degradation. Levels of Hsps and co-chaperones available for interaction with HSF1 increase again. This is expected to result in formation of repressive interactions between HSF1 and Hsps and/or multichaperone complexes, accelerated chaperone-mediated disassembly of HSF1 trimers, and the eventual re-association of HSF1 polypeptide with Hsp90-p23-immunophilin complex. This direct control of HSF1 activity by Hsps and co-chaperones is modulated by several interactions/reactions that link HSF1 regulation to a wider set of stress-regulated aspects of cell metabolism and may serve to integrate multiple stress signals. These reactions include stress-induced activating and repressive phosphorylation of HSF1, which is, in part, mediated by ERK1, GSK3, JNK1, and CaMKII kinases. The activities of some of these kinases were increased in the stressed cell. Transcriptional activation of HSF1 is also enhanced by DAXX, a protein that only becomes available for interaction with HSF1 subsequent to stress-induced disintegration of PODs. Finally, HSF1-mediated hsp gene expression may be indirectly influenced by stress-induced and Hsp70-repressed sequestration of HSF1 in stress granules.

The author hopes that the present review has not only informed on what is known about the regulation of HSF1 and the heat shock response, but has also led the reader to recognize the large lacunae that still exist in our understanding. Clearly, one of the least understood aspects is the pathway of deactivation of HSF1. The mechanism (or mechanisms) by which HSF1

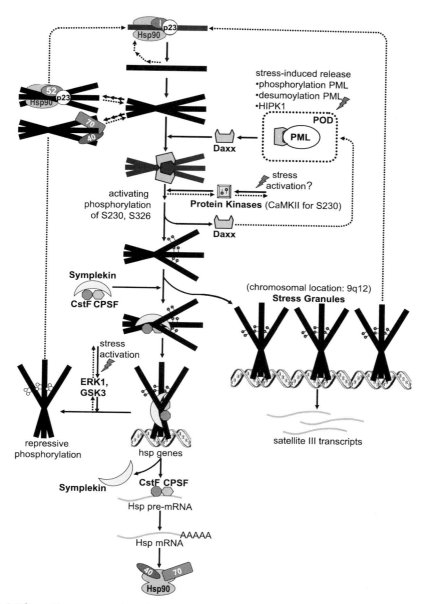

Fig. 3 Schematic representation of various aspects and mechanisms involved in the regulation of HSF1 activity and the heat shock response. *70*, *52*, and *40* refer to Hsp70, FKBP52, and Hsp40, respectively

homotrimers are disassembled as well as the relationship between this mechanism and repressive phosphorylation and repressive chaperone interactions with trimeric HSF1 remain entirely unknown. Furthermore, the present understanding of the regulation of the heat shock response is largely based on results from cell culture studies. Continued examination of the response in complex organisms such as plants or metazoan animals, or their organs and tissues, will likely result in significant adjustments in our present thinking, in particular relating to interactions between HSF1 and other HSFs, and the role of different HSFs in specific differentiated cell types. Finally, it is noted that, to avoid duplication, a discussion of interactions between HSF1 (and the heat shock response) and other signaling pathways including apoptotic pathways and steroid receptor-mediated regulation was omitted.

Acknowledgements I thank Alexis Hall for the artwork and HSF Pharmaceuticals S.A. for supporting this project.

References

Abravaya K, Philips B, Morimoto RI (1991) Attenuation of the heat shock response in HeLa cells is mediated by the release of bound heat shock transcription factor and is modulated by changes in growth and in heat shock temperatures. Genes Dev 5:2117–2127

Ali A, Bharadwaj S, O'Carroll R, Ovsenek N (1998) Hsp90 interacts with and regulates the activity of heat shock factor 1 in *Xenopus* oocytes. Mol Cell Biol 18:4949–4960

Ananthan J, Goldberg AL, Voellmy R (1986) Abnormal proteins serve as eukaryotic stress signals and trigger the activation of heat shock genes. Science 232:522–524

Baler R, Welch WJ, Voellmy R (1992) Heat shock gene regulation by nascent polypeptides and denatured proteins: Hsp70 as a potential autoregulatory factor. J Cell Biol 117:1151–1159

Baler R, Dahl G, Voellmy R (1993) Activation of human heat shock genes is accompanied by oligomerization, modification, and rapid translocation of heat shock transcription factor Hsf1. Mol Cell Biol 13:2486–2496

Ballinger CA, Connell P, Wu Y, Hu Z, Thompson LJ, Yin LY, Patterson C (1999) Identification of Chip, a novel tetratricopeptide repeat-containing protein that interacts with heat shock proteins and negatively regulates chaperone functions. Mol Cell Biol 19:4535–4545

Bharadwaj S, Ali A, Ovsenek N (1999) Multiple components of the Hsp90 chaperone complex function in regulation of heat shock factor 1 in vivo. Mol Cell Biol 19:8033–8041

Boellmann F, Guettouche T, Guo Y, Fenna M, Mnayer L, Voellmy R (2004) DAXX interacts with heat shock factor 1 during stress activation and enhances its transcriptional activity. Proc Natl Acad Sci U S A 101:4100–4105

Bruce JL, Price BD, Coleman CN, Calderwood SK (1993) Oxidative injury rapidly activates the heat shock transcription factor but fails to increase levels of heat shock proteins. Cancer Res 53:12–15

Bush KT, Goldberg AL, Nigam SK (1997) Proteasome inhibition leads to a heat shock response, induction of endoplasmic reticulum chaperones, and thermotolerance. J Biol Chem 272:9086–9092

Chang HY, Nishitoh H, Yang X, Ichijo H, Baltimore D (1998) Activation of apoptosis signal-regulating kinase 1 (ASK 1) by the adapter protein Daxx. Science 281:1860–1863

Chu B, Soncin F, Price BD, Stevenson, MA, Calderwood SK (1996) Sequential phosphorylation by mitogen-activated protein kinase and glycogen synthase kinase 3 represses transcriptional activation by heat shock factor-1. J Biol Chem 271:30847–30857

Chu B, Zhong R, Soncin F, Stevenson MA, Calderwood SK (1998) Transcriptional activity of heat shock factor 1 at 37 degrees C is repressed through phosphorylation on two distinct serine residues by glycogen synthase kinase 3 and protein kinases Calpha and Czeta. J Biol Chem 273:18640–18646

Clos J, Rabindran S, Wisniewski J, Wu C (1993) Induction temperature of human heat shock factor is reprogrammed in a *Drosophila* cell environment. Nature 364:252–255

Conde AG, Lau SS, Dillmann WH, Mestril R (1997) Induction of heat shock proteins by tyrosine kinase inhibitors in rat cardiomyocytes and myogenic cells confers protection against simulated ischemia. J Mol Cell Cardiol 29:1927–1938

Connell P, Ballinger CA, Jiang J, Wu Y, Thompson LJ, Hoehfeld J, Patterson C (2001) Regulation of heat shock protein-mediated protein triage decisions by the co-chaperone Chip. Nat Cell Biol 3:93–96

Cotto JJ, Kline M, Morimoto RI (1996) Activation of heat shock factor 1 DNA binding precedes stress-induced serine phosphorylation. Evidence for a multistep pathway of regulation. J Biol Chem 271:3355–3358

Cotto J, Fox S, Morimoto R (1997) HSF1 granules: a novel stress-induced nuclear component of human cells, J Cell Sci 110:2925–2934

Dai R, Frejtag W, He B, Zhang Y, Mivechi NF (2000) c-Jun NH2-terminal kinase targeting and phosphorylation of heat shock factor-1 suppress its transcriptional activity. J Biol Chem 275:18210–18218

Dai Q, Zhang C, Wu Y, McDonough H, Whaley RA, Godfrey V, Li HH, Madamanchi N, Xu W, Neckers L, Cyr D, Patterson C (2003) Chip activates Hsf1 and confers protection against apoptosis and cellular stress. EMBO J 22:5446–5458

DiDomenico BJ, Bugaisky GE, Lindquist S (1982). The heat shock response is self-regulated at both the transcriptional and post-transcriptional levels. Cell 31:593–603

Duina AA, Kalton HM, Gaber RF (1998) Requirement for Hsp90 and a Cyp40-type cyclophilin in negative regulation of the heat shock response. J Biol Chem 273:18974–18978

Ecsedy JA, Michaelson JS, Leder P (2003) Homeodomain-interacting protein kinase 1 modulates Daxx localization, phosphorylation, and transcriptional activity. Mol Cell Biol 23:950–960

Erdos G, Lee YI (1994) Effect of staurosporine on the transcription of *hsp70* heat shock gene in HT-29 cells. Biochem Biophys Res Commun 202:476–483

Esser C, Alberti S, Hoehfeld J (2004) Cooperation of molecular chaperones with the ubiquitin/proteasome system. Biochim Biophys Acta 1695:171–188

Everett RD, Lomonte P, Sternsdorf T, van Driel R, Orr A (1999) Cell cycle regulation of PML modification and ND10 composition. J Cell Sci 112:4581–4588

Fisher EA, Zhou M, Mitchell DM, Wu X, Omura S, Wang H, Goldberg AL, Ginsberg HN (1997) The degradation of apolipoprotein B100 is mediated by the ubiquitin-proteasome pathway and involves heat shock protein 70. J Biol Chem 272:20427–20434

Freeman ML, Borrelli MJ, Syed K, Senisterra G, Stafford DM, Lepock JR (1995) Characterization of a signal generated by oxidation of protein thiols activates the heat shock transcription factor. J Cell Physiol 164:356–366

Fritsch M, Wu C (1999) Phosphorylation of Drosophila heat shock transcription factor. Cell Stress Chaperones 4:102–117

Giardina C, Lis JT (1995) Dynamic protein-DNA architecture of a yeast heat shock promoter. Mol Cell Biol 15:2737–2744

Green M, Schuetz TJ, Sullivan EK, Kingston RE (1995) A heat-shock-responsive domain of human Hsf1 that regulates transcription activation domain function. Mol Cell Biol 15:3354–3362

Grenert JP, Sullivan WP, Fadden P, Haystead TA, Clark J, Mimnaugh E, Krutzsch H, Ochel HJ, Schulte TW, Sausville E, Neckers LM, Toft DO (1997) The amino-terminal domain of heat shock protein 90 (HSP90) that binds geldanamycin is an ATP/ADP switch domain that regulates HSP90 conformation. J Biol Chem 272:23843–23850

Goff SA, Goldberg AL (1985) Production of abnormal proteins in E. coli stimulates transcription of lon and other heat shock genes. Cell 41:587–595

Goodson ML, Sarge KD (1995) Heat-inducible DNA binding of purified heat shock transcription factor 1. J Biol Chem 270:2447–2450

Guettouche T, Boellmann F, Lane WS, Voellmy R (2005) Analysis of phosphorylation of human heat shock factor 1 in cells experiencing a stress. BMC Biochemistry 6:4

Guo Y, Guettouche T, Fenna M, Boellmann F, Pratt WB, Toft DO, Smith DF, Voellmy R (2001) Evidence for a mechanism of repression of heat shock factor 1 transcriptional activity by a multichaperone complex. J Biol Chem 276:45791–45799

Gusarova V, Caplan AJ, Brodsky JL, Fisher EA (2001) Apolipoprotein B degradation is promoter by the molecular chaperones hsp90 and hsp70. J Biol Chem 276:24891–24900

He B, Meng YH, Mivechi NF (1998) Glycogen synthase kinase 3beta and extracellular signal-regulated kinase inactivate heat shock transcription factor 1 by facilitating the disappearance of transcriptionally active granules after heat shock. Mol Cell Biol 18:6624–6633

He H, Soncin F, Grammatikakis N, Li Y, Siganou A, Gong J, Brown SA, Kingston RE, Calderwood SK (2003) Elevated expression of heat shock factor (HSF) 2A stimulates HSF1-induced transcription during stress. J Biol Chem 278:35465–35475

Hedge RS, Zuo J, Voellmy R, Welch WJ (1995) Short circuiting stress protein expression via a tyrosine kinase inhibitor, herbimycin A. J Cell Physiol 165:186–200

Henics T, Nagy E, Oh HJ, Csermely P, von Gabai A, Subjeck JR (1999) Mammalian Hsp70 and Hsp110 proteins bind to RNA motifs involved in mRNA stability. J Biol Chem 274:17318–17324

Hietakangas V, Ahlskog JK, Jakobsson AM, Hellesuo M, Sahlberg NM, Holmberg CI, Mikhailov A, Palvimo JJ, Pirkkala L, Sistonen L (2003) Phosphorylation of serine 303 is a prerequisite for the stress-inducible SUMO modification of heat shock factor 1. Mol Cell Biol 23:2953–2968

Hightower LE (1980) Cultured animal cells exposed to amino acid analogues or puromycin rapidly synthesize several polypeptides. J Cell Physiol 102:407–427

Hoj A, Jacobsen BK (1994) A short element required for turning off heat shock transcription factor: evidence that phosphorylation enhances deactivation. EMBO J 13:2617–2624

Holmberg CI, Hietakangas V, Mikhailov A, Rantanen JO, Kallio M, Meinander A, Hellman J, Morrice N, MacKintosh C, Morimoto RI, Eriksson JE, Sistonen L (2001) Phosphorylation of serine 230 promotes inducible transcriptional activity of heat shock factor 1. EMBO J 20:3800–3810

Inouye S, Katsuki K, Izu H, Fujimoto M, Sugahara K, Yamada S-I, Shinkai Y, Oka Y, Katoh Y, Nakai A (2003) Activation of heat shock genes is not necessary for protection by heat shock transcription factor 1 against cell death due to a single exposure to high temperatures. Mol Cell Biol 23:5882–5895

Ishov AM, Sotnikov AG, Negorev D, Vladimirova OV, Neff N, Kamitani T, Yeh ETH, Strauss III JF, Maul GG (1999) PML is critical for ND10 formation and recruits the PML-interacting protein DAXX to this nuclear structure when modified by SUMO-1. J Cell Biol 147:221–233

Jolly C, Morimoto RI, Robert-Nicoud M, Vourc'h C (1997) HSF1 transcription factor concentrates in nuclear foci during heat shock: relationship with transcription sites. J Cell Sci 110:2935–2941

Jolly C, Konecny L, Grady DL, Kurskova YA, Cotto JJ, Morimoto RI, Vourc'h C (2002) In vivo binding of active heat shock transcription factor 1 to human chromosome 9 heterochromatin during stress. J Cell Biol 156:775–781

Jolly C, Metz A, Govin J, Vigneron M, Turner BM, Khochbin S, Vourc'h C (2004) Stress-induced transcription of satellite III repeats. J Cell Biol 164:25–33

Jurivich DA, Sistonen L, Kroes RA, Morimoto RI (1992) Effect of sodium salicylate on the human heat shock response. Science 255:1243–1245

Jurivich DA, Pachetti C, Qiu L, Welk JF (1995) Salicylate triggers heat shock factor differently than heat. J Biol Chem 270:24489–24495

Kelley PM, Schlesinger MJ (1978) The effect of amino acid analogues and heat shock on gene expression in chicken embryo fibroblasts. Cell 15:1277–1286

Kim J, Nueda A, Meng YH, Dynan WS, Mivechi NF (1997) Analysis of the phosphorylation of human heat shock transcription factor-1 by MAP kinase family members. J Cell Biochem 67:43–54

Kim S-A, Yoon J-H, Lee S-H, Ahn S-G (2005) Polo-like kinase 1 phosphorylates HSF1 and mediates its nuclear translocation during heat stress. J Biol Chem 280:12653–12657; e-pub Jan 20

Kline MP, Morimoto RI (1997) Repression of the heat shock factor 1 transcriptional activation domain is modulated by constitutive phosphorylation. Mol Cell Biol 17:2107–2115

Knauf U, Newton EM, Kyriakis J, Kingston RE (1996) Repression of human heat shock factor 1 activity at control temperature by phosphorylation. Genes Dev 10:2782–2793

Knowlton AA, Sun L (2001) Heat shock factor-1, steroid hormones, and the regulation of heat-shock protein expression in the heart. Am J Physiol Heart Circ Physiol 280:H455–H464

Larson JS, Schuetz TJ, Kingston RE (1995) In vitro activation of purified human heat shock factor by heat. Biochemistry 34:1902–1911

Lee DH, Goldberg AL (1998) Proteasome inhibitors cause induction of heat shock proteins and trehalose, which together confer thermotolerance in Saccharomyces cerevisiae. Mol Cell Biol 18:30–38

Lepock JR (2005) Measurement of protein stability and protein denaturation in cells using differential scanning microcalorimetry. Methods 35:117–125

Lepock JR, Frey HE, Ritchie KP (1993) Protein denaturation in intact hepatocytes and isolated cellular organelles during heat shock. J Cell Biol 122:1267–1276

Lindquist S (1980) Varying patterns of protein synthesis in Drosophila during heat shock: implications for regulation. Dev Biol 77:463–479

Liu H, Lightfoot R, Stevens JL (1996) Activation of heat shock factor by alkylating agents is triggered by glutathione depletion and oxidation of protein thiols. J Biol Chem 271:4805–4812

Lohmann C, Eggers-Schumacher G, Wunderlich M, Schoeffl F (2004) Two different heat shock transcription factors regulate immediate early expression of stress genes in Aradopsis. Mol Genet Genomics 271:11–21

Marchler G, Wu C (2001) Modulation of *Drosophila* heat shock transcription factor activity by the molecular chaperone DroJ1. EMBO J 20:499–509

Maul GG, Yu E, Ishov AM, Epstein AL (1995) Nuclear domain 10 (ND10) associated proteins are also present in nuclear bodies and redistribute to hundreds of nuclear sites after stress. J Cell Biochem 59:498–513

McDonough H, Patterson C (2003) Chip: a link between the chaperone and proteasome systems. Cell Stress Chaperones 8:303–308

McDuffee AT, Senisterra G, Huntley S, Lepock JR, Sekhar KR, Meredith MJ, Borrelli MJ, Morrow JD, Freeman ML (1997) Proteins containing non-native disulfide bonds generated by oxidative stress can act as signals for the induction of the heat shock response. J Cell Physiol 171:143–151

McMillan DR, Xiao X, Shao L, Graves K, Benjamin IJ (1998) Targeted disruption of heat shock transcription factor 1 abolishes thermotolerance and protection against heat-inducible apoptosis. J Biol Chem 273:7523–7528

Metz A, Soret J, Vourc'h C, Tazi J, Jolly C (2004) A key role for stress-induced satellite III transcripts in the relocalization of splicing factors into nuclear stress granules. J Cell Sci 177:4551–4558

Moseley PL, Wallen ES, McCafferty JD, Flanagan S, Kern JA (1993) Heat stress regulates the human 70-kDa heat-shock gene through the $3'$-untranslated region. Am J Physiol 264:L533–L537

Muller S, Matunis MJ, Dejean A (1998) Conjugation with the ubiquitin-related modifier SUMO-1 regulates the partitioning of PML within the nucleus. EMBO J 17:61–70

Murakami Y, Uehara Y, Yamamoto C, Fukazawa H, Mizuno S (1991) Induction of HSP72/73 by herbimycin A, an inhibitor of transformation by tyrosine kinase oncogenes. Exp Cell Res 195:338–344

Nadeau K, Das A, Walsh CT (1993) Hsp90 cochaperonins possess ATPase activity and bind heat shock transcription factors and peptidyl prolyl isomerases. J Biol Chem 268:1479–1487

Nair SC, Toran EJ, Rimerman RA, Hjermstad S, Smithgall TE, Smith DF (1996) A pathway of multi-chaperone interactions common to diverse regulatory proteins: estrogen receptor, FES tyrosine kinase, heat shock transcription factor 1, and the aryl hydrocarbon receptor. Cell Stress Chaperones 1:237–250

Nefkens I, Negorev DG, Ishov AM, Michaelson JS, Yeh ETH, Tanguay RM, Mueller WEG, Maul GG (2003) Heat shock and Cd^{2+} exposure regulate PML and DAXX release from ND10 by independent mechanisms that modify the induction of heat-shock proteins 70 and 25 differently. J Cell Sci 116:513–524

Nollen EA, Brunsting JF, Roelofsen H, Weber LA, Kampinga HH (1999) In vivo chaperone activity of heat shock protein 70 and thermotolerance. Mol Cell Biol 19:2069–2079

Petersen RB, Lindquist S (1989) Regulation of Hsp70 synthesis by messenger RNA degradation. Cell Regulation 1:135–149

Pirkkala L, Alastalo T, Zuo X, Benjamin IJ (2000) Disruption of heat shock factor 1 reveals an essential role in the ubiquitin proteolytic pathway. Mol Cell Biol 20:2670–2675

Pirkkala L, Nykanen P, Sistonen L (2001) Roles of the heat shock transcription factors in regulation of the heat shock response and beyond. FASEB J 15:1118–1131

Pratt WB, Toft DO (1997) Steroid receptor interactions with heat shock protein and immunophilin chaperones. Endocr Rev 18:306–360

Prodromou C, Roe SM, O'Brien R, Ladbury JE, Piper PW, Pearl LH (1997) Identification and structural characterization of the ATP/ADP-binding site in the HSP90 molecular chaperone. Cell 90:65–75

Rabindran SK, Giorgi G, Clos J, Wu C (1991) Molecular cloning and expression of a human heat shock factor, HSF1. Proc Natl Acad Sci U S A 88:6906–6910

Rieger TR, Morimoto RI, Hatzimanikatis V (2005) Mathematical modeling of the eukaryotic heat shock response: dynamics of the hsp70 promoter. Biophys J 88:1646–1658

Rizzi N, Denegri M, Chiodi L, Corioni M, Valgardsdottir R, Cobianchi F, Riva S, Biamonti G (2004) Transcriptional activation of a constitutive heterochromatic domain of the human genome in response to heat shock. Mol Biol Cell 15:543–551

Rochat-Steiner V, Becker K, Micheau O, Schneider P, Burns K, Tschopp J (2000) FIST/HIPK3: a Fas/FADD-interacting serine/threonine kinase that induces FADD phosphorylation and inhibits fas-mediated Jun NH(2)-terminal kinase activation. J Exp Med 192:1165–1174

Sarge KD, Murphy SP, Morimoto RI (1993) Activation of heat shock gene transcription by heat shock factor 1 involves oligomerization, acquisition of DNA-binding activity, and nuclear translocation and can occur in the absence of stress. Mol Cell Biol 13:1392–1407

Senisterra GA, Huntley SA, Escaravage M, Sekhar KR, Freeman ML, Borrelli M, Lepock JR (1997) Destabilization of the Ca^{2+}-ATPase of sarcoplasmic reticulum by thiol-specific, heat shock inducers results in thermal denaturation at 37 degrees C. Biochemistry 36:11002–11011

Shi Y, Kroeger PE, Morimoto RI (1995) The carboxyl-terminal transactivation domain of heat shock factor 1 is negatively regulated and stress responsive. Mol Cell Biol 15:4309–4318

Shi Y, Mosser DD, Morimoto RI (1998). Molecular chaperones as Hsf1-specific transcriptional repressors. Genes Dev 12:654–666

Tanabe M, Kawazoe Y, Takeda S, Morimoto RI, Nagata K, Nakai A (1998) Disruption of the hsf3 gene results in the severe reduction of heat shock gene expression and loss of thermotolerance. EMBO J 17:1750–1758

Theodorakis NG, Morimoto RI (1987) Posttranscriptional regulation of Hsp70 expression in human cells: effects of heat shock, inhibition of protein synthesis, and adenovirus infection on translation and mRNA stability. Mol Cell Biol 7:4357–4368

Torii S, Egan DA, Evans RA, Reed JC (1999) Human DAXX regulates FAS-induced apoptosis from nuclear PML oncogenic domains (PODs). EMBO J 18:6037–6049

Voellmy R (2004) On mechanisms that control heat shock transcription factor activity in metazoan cells. Cell Stress Chaperones 9:122–133

Wang X, Grammatikakis N, Sikanou A, Calderwood SK (2003) Regulation of molecular chaperone gene transcription involves the serine phosphorylation, 14-3-3ε binding, and cytoplasmic sequestration of heat shock factor 1. Mol Cell Biol 23:6013–6026

Wang X, Grammatikakis N, Sikanou A, Stevenson MA, Calderwood SK (2004) Interaction between extracellular signal-regulated protein kinase 1, 14-3-3ε, and heat shock factor 1 during stress. J Biol Chem 279:49460–49469

Westwood JT, Clos J, Wu C (1991) Stress-induced oligomerization and chromosomal relocalization of heat-shock factor. Nature 353:822–827

Whitesell L, Mimnaugh EG, De Costa B, Myers CE, Neckers LM (1994) Inhibition of heat shock protein HSP90-PP60V-SRC heteroprotein complex formation by benzoquinone ansamycins: essential role for stress proteins in oncogenic transformation. Proc Natl Acad Sci. U S A 91:8324–8328

Xavier IJ, Mercier PA, McLoughlin CM, Ali A, Woodgett N JR, Ovsenek, N (2000) Glycogen synthase kinase 3beta negatively regulates both DNA-binding and transcriptional activities of heat shock factor 1. J Biol Chem 275:29147–29152

Xia W, Voellmy R (1997) Hyperphosphorylation of heat shock transcription factor 1 is correlated with transcriptional competence and slow dissociation of active factor trimers. J Biol Chem 272:4094–4102

Xia W, Guo Y, Vilaboa N, Zuo J, Voellmy, R (1998) Transcriptional activation of heat shock factor HSF1 probed by phosphopeptide analysis of factor ^{32}P-labeled in vivo. J Biol Chem 273:8749–8755

Xia W, Vilaboa N, Martin J, Mestril R, Guo Y, Voellmy R (1999) Modulation of tolerance by mutant heat shock transcription factors. Cell Stress Chaperones 4:8–18

Xing H, Mayhew CN, Cullen KE, Park-Sarge O-K, Sarge KD (2004) HSF1 modulation of Hsp70 mRNA polyadenylation via interaction with symplekin. J Biol Chem 279:10551–10555

Yost HJ, Lindquist S (1986) RNA splicing is interrupted by heat shock and is rescued by heat shock protein synthesis. Cell 45:185–193

Zhang Y, Nijbroek G, Sullivan ML, McCracken AA, Watkins SC, Michaelis S, Brodsky JL (2001) Hsp70 molecular chaperone facilitates endoplasmic reticulum-associated protein degradation of cystic fibrosis transmembrane conductance regulator in yeast. Mol Biol Cell 12:1303–1314

Zhang Y, Huang L, Zhang J, Moskophidis D, Mivechi NF (2002) Targeted disruption of hsf1 leads to lack of thermotolerance and defines tissue-specific regulation for stress-inducible HSP molecular chaperones. J Cell Biochem 86:376–393

Zhao C, Hashiguchi K, Kondoh W, Du W, Hata J, Yamada T (2002) Exogenous expression of heat shock protein 90 kDa retards the cell cycle and impairs the heat shock response. Exp Cell Res 275:200–214

Zhong M, Orosz A, Wu C (1998) Direct sensing of heat and oxidation by Drosophila heat shock transcription factor. Mol Cell 2:101–108

Zhong S, Salomoni P, Ronchetti S, Guo A, Ruggero D, Pandolfi PP (2000) Promyelocytic leukemia protein (PML) and DAXX participate in a novel nuclear pathway for apoptosis. J Exp Med 191:631–640

Zimarino V, Wu C (1987) Induction of sequence-specific binding of Drosophila heat shock activator protein without protein synthesis. Nature 327:727–730

Zou J, Salminen WF, Roberts SM, Voellmy R (1998a) Correlation between glutathione oxidation and trimerization of heat shock factor 1, an early step in stress induction of the Hsp response. Cell Stress Chaperones 3:130–141

Zou J, Guo Y, Guettouche T, Smith DF, Voellmy R (1998b) Repression of heat shock transcription factor Hsf1 activation by Hsp90 (Hsp90 complex) that forms a stress-sensitive complex with HSF1. Cell 94:471–480

Zuo J, Baler R, Dahl G, Voellmy R (1994) Activation of the DNA-binding ability of human heat shock transcription factor 1 may involve the transition from an intramolecular to an intermolecular triple-stranded coiled-coil structure. Mol Cell Biol 14:7557–7568

Zuo J, Rungger D, Voellmy R (1995) Multiple layers of regulation of human heat shock transcription factor 1. Mol Cell Biol 15:4319–4330

Protein Folding in the Endoplasmic Reticulum and the Unfolded Protein Response

K. Zhang · R. J. Kaufman (✉)

Howard Hughes Medical Institute, Department of Biological Chemistry,
University of Michigan Medical Center, 1150 W. Medical Center Dr., Ann Arbor MI, 48109, USA
kaufmanr@umich.edu

1	Polypeptide Modification, Folding, and Assembly in the Endoplasmic Reticulum Lumen	70
2	Historical Perspective of the Unfolded Protein Response	71
3	The Unfolded Protein Response Sensors in Higher Eukaryotes	72
4	Activation of Unfolded Protein Response Sensors	75
5	The Transcriptional Response to Endoplasmic Reticulum Stress	76
6	The Translational Response to Endoplasmic Reticulum Stress	77
7	Endoplasmic Reticulum-Associated Protein Degradation	78
8	Unfolded Protein Response-Induced Apoptosis	79
9	The Physiological Roles of the Unfolded Protein Response	80
9.1	The Unfolded Protein Response in B Cell Differentiation	81
9.2	The Unfolded Protein Response in Glucose Homeostasis and Diabetes	81
9.3	The Unfolded Protein Response in Organelle Expansion	82
9.4	The Unfolded Protein Response in Neurological Diseases	83
10	Perspectives	84
	References	85

Abstract In all eukaryotic cells, the endoplasmic reticulum (ER) is an intracellular organelle where folding and assembly occurs for proteins destined to the extracellular space, plasma membrane, and the exo/endocytic compartments (Kaufman 1999). As a protein-folding compartment, the ER is exquisitely sensitive to alterations in homeostasis, and provides stringent quality control systems to ensure that only correctly folded proteins transit to the Golgi and unfolded or misfolded proteins are retained and ultimately degraded. A number of biochemical and physiological stimuli, such as perturbation in calcium homeostasis or redox status, elevated secretory protein synthesis, expression of misfolded proteins, sugar/glucose deprivation, altered glycosylation, and overloading of cholesterol can disrupt ER homeostasis, impose stress to the ER, and subsequently lead to accumulation of unfolded or misfolded proteins in the ER lumen. The ER has evolved highly specific signaling pathways called the unfolded protein response (UPR) to cope with the accumulation of unfolded or misfolded proteins. Elucidation of the molecular mechanisms by which accumulation of unfolded proteins in the ER transmits a signal to the cytoplasm and nucleus has led to major new insights into the diverse cellular and physiological processes that are regulated by the UPR. This chapter summarizes how cells respond to the accumulation of unfolded proteins in the cell and the relevance of these signaling pathways to human physiology and disease.

Keywords Endoplasmic reticulum · Unfolded protein response · Translational control · ERAD · Apoptosis

Abbreviations

UPR	Unfolded protein response
ER	Endoplasmic reticulum
GRP	Glucose-regulated protein
IRE1	Inositol requiring 1
ATF6	Activating transcription factor 6
PERK	PKR-like ER kinase
XBP1	X-box binding protein 1
UPRE	Unfolded protein response element
ERSE	ER stress response element
ERAD	ER-associated protein degradion

1
Polypeptide Modification, Folding, and Assembly in the Endoplasmic Reticulum Lumen

The endoplasmic reticulum (ER) is the site of biosynthesis for sterols, lipids, and membrane and secreted proteins. Approximately one-third of all cellular protein synthesis occurs on the membrane of the rough ER. For some specialized cells that function to secrete proteins, such as plasma cells, hepatocytes, pancreatic acinar and islet cells, over 90% of the translated polypeptides are directed into the ER lumen, the entrance site into the secretory pathway (Kaufman 2004). Since the protein concentration in the ER lumen is approximately 100 mg/ml, it is essential that protein chaperones facilitate protein folding by preventing aggregation of protein folding intermediates and by correcting misfolded proteins that are caught in kinetic low-energy traps. These energy-requiring processes ensure high-fidelity protein folding in the lumen of the ER. For example, the most abundant ER chaperone BiP/GRP78 uses the energy from ATP hydrolysis to promote folding and prevent aggregation of proteins within the ER. In addition, the oxidizing environment of the ER creates a constant demand for cellular protein disulfide isomerases to catalyze and monitor disulfide bond formation in a regulated and ordered manner. Only those polypeptides that are properly folded and assembled in the ER can transit to the Golgi compartment, a process called quality control. Proteins that are misfolded in the ER are retained and eventually translocated back through into the cytosol for degradation by the 26S proteasome in a process called ER-associated degradation (ERAD) (Tsai et al. 2002).

The recognition and modification of oligosaccharide structures in the lumen of the ER is intimately coupled to polypeptide folding (Helenius 1994). As the growing nascent chain is translocated into the lumen of the ER, a 14-oligosaccharide core, N-acetylglucosamine$_2$-mannose$_9$-glucose$_3$(GlcNAc$_2$-Man$_9$Glc$_3$), is added to consensus asparagine residues (Asn-X-Ser/Thr; where X is any amino acid except Pro). Immediately after the addition of this core, the three terminal glucose residues are cleaved by the sequential action of glucosidases I and II to yield a GlcNAc$_2$Man$_9$ structure. If the polypeptide is not folded properly, a UDP-glucose:glycoprotein glucosyltransferase recognizes the unfolded nature of the glycoprotein and reglucosylates the core structure to re-establish the glucose-α(1,3)-mannose glycosidic linkage (Ritter and Helenius 2000). Monoglucosylated oligosaccharides containing this bond bind to the ER-resident protein chaperones calnexin and calreticulin. Glycoprotein interaction with calnexin and calreticulin promotes interaction with the oxidoreductase ERp57 that promotes proper disulfide bond rearrangement. This quality control process ensures that unfolded glycoproteins do not exit the ER.

The ER has evolved highly specific signaling pathways to ensure that its protein-folding capacity is not overwhelmed. Upon accumulation of unfolded proteins in the ER lumen, several adaptive pathways are activated to reduce the amount of new protein translocated into the ER lumen, to increase the retrotranslocation and degradative potential of ER-localized malfolded proteins, and to increase the protein-folding capacity of the ER.

2
Historical Perspective of the Unfolded Protein Response

In the mid-1970s, it was observed that transformation of fibroblasts with Rous sarcoma virus induces expression of a set of genes (Pouyssegur et al. 1977). The same set of genes was found to be upregulated upon glucose deprivation, hence the products of these genes were termed glucose-regulated proteins (GRPs, i.e., GRP78 and GRP94) (Lee et al. 1984, 1983). Independently, a protein was identified that binds unassembled immunoglobulin heavy chains in the ER of pre-B cells and prevents their secretion until the immunoglobulin light chains are expressed. This protein was named the immunoglobulin binding protein (BiP) (Haas and Wabl 1983; Lee 1987) and is identical to GRP78. Subsequently, it was demonstrated that overexpression of an unfolded mutant of the influenza hemagglutinin protein was sufficient to induce expression of BiP and GRP94 (Gething and Sambrook 1992), leading to the designation of this signaling pathway as the unfolded protein response (UPR). In addition, overexpression of a wild-type protein, coagulation factor VIII, was also shown to induce expression of BiP and GRP94 (Dorner et al. 1987, 1988).

The UPR is conserved in all eukaryotic cells. Analysis of the UPR pathway in the budding yeast *Saccharomyces cerevisiae* identified a 22-bp *cis*-acting UPR element (UPRE) that was necessary and sufficient for ER stress induction of a reporter gene upon accumulation of unfolded proteins in the ER (Mori et al. 1992). A genetic screen was used to isolate mutants in this pathway. The first mutant identified independently by two groups was defective in a gene encoding an ER transmembrane serine/threonine protein kinase (Ire1p/Ern1p). Ire1p was characterized as a proximal UPR transducer required for transcriptional induction of UPR genes and for survival upon ER stress (Cox et al. 1993; Mori et al. 1993). Ire1p has an N-terminal lumenal domain that senses the ER stress signal and a C-terminal cytoplasmic domain that has a serine/threonine kinase activity. Subsequently, it was discovered that the C-terminus has homology with RNaseL, a nonspecific endoribonuclease that signals one arm of the interferon response in mammalian cells. Indeed, it was demonstrated that Ire1p has a site-specific endoribonuclease (RNase) activity required for activating KAR2/BiP transcription (Mori et al. 1993; Shamu and Walter 1996; Welihinda and Kaufman 1996). The presence of unfolded proteins in the ER lumen promotes dimerization and *trans*-autophosphorylation of Ire1p, activating its RNase activity to cleave at two sites within the mRNA encoding HAC1, a transcription factor that potently activates KAR2/BiP transcription. The Ire1p-dependent cleavage initiates *HAC1* mRNA splicing through an unconventional reaction (Kawahara et al. 1998; Sidrauski et al. 1996; Sidrauski and Walter 1997). The 5' and 3' ends of *HAC1* mRNA are subsequently ligated by transfer RNA ligase. The unconventional *HAC1*-mRNA processing reaction removes a 252-nucleotide intron, and the result is replacement of the carboxy-terminal ten amino acids in Hac1p (Hac1up) with a new 19-amino-acid segment (Hac1ip). This splicing reaction regulates UPR transcriptional activation in two ways. First, the new carboxyl terminus on Hac1ip converts Hac1p into a tenfold more potent transcriptional activator. Second, removal of the intron increases the translational efficiency of *HAC1* mRNA.

Spliced *HAC1* mRNA encodes a basic leucine zipper (b-ZIP) transcription factor that binds to a DNA sequence motif, termed the unfolded protein response element (UPRE; consensus CAGCGTG), as a dimer in the promoters of many UPR responsive genes. It was reported that Hac1p activates transcription of approximately 381 UPR target genes in yeast (Travers et al. 2000).

3
The Unfolded Protein Response Sensors in Higher Eukaryotes

All metazoan cells have conserved the essential and unique properties of the UPR in yeast, but have also evolved additional sensors and a greater number of downstream targets to generate a diversity of responses (Fig. 1). In mammals, the counterpart of yeast Ire1p has two isoforms: IRE1α and IRE1β. While IRE1a is expressed in most cells and tissues, with high-level expression in the pancreas and placenta, IRE1β expression is primarily restricted to intestinal epithelial cells (Tirasophon et al. 1998; Wang et al. 1998). Both molecules respond to the accumulation of unfolded proteins in the ER to activate their kinase and subsequent RNase activities. In vitro cleavage reactions using yeast *HAC1*-mRNA substrate showed that the cleavage specificities of IRE1α and IRE1β are similar, indicating that they did not evolve to recognize different sets of substrates, but rather to generate temporal- and tissue-specific expression. All cells in multicellular organisms also constitutively express two additional stress sensors that respond to the accumulation of unfolded proteins in the ER lumen and coordinate either adaptive or cell-death responses: PERK and ATF6.

PERK has a carboxy-terminal protein kinase domain homologous to the double-stranded RNA-activated protein kinase PKR that phosphorylates the alpha subunit of eukaryotic translation initiation factor 2 (eIF2α) (Fig. 1). However, where PKR is a soluble cytosolic protein, PERK contains a transmembrane domain and an amino-terminal domain that resides in the ER lumen and responds to the accumulation of unfolded proteins in the ER lumen (Harding et al. 1999; Liu et al. 2000; Shi et al. 1998). The luminal domain of PERK shares a low degree of homology with the IRE1 luminal domain. Surprisingly, the human PERK luminal domain can functionally substitute for the yeast Ire1p luminal domain, although the yeast genome does not have a PERK

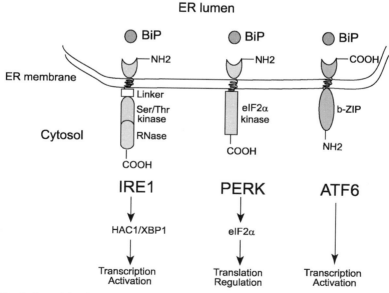

Fig. 1 Depiction of the three UPR transducers. The functional domains of IRE1, PERK, and ATF6 are shown

homolog. This supports the idea that the fundamental mechanism for sensing the accumulation of unfolded proteins in the ER is conserved between IRE1 and PERK and between yeast and humans (Harding et al. 1999; Liu et al. 2000; Shi et al. 1998). In an evolutionary sense, it is significant that the eIF2α kinase PKR and the endoribonuclease RNaseL are both components of the interferon antiviral response that respond to the presence of double-stranded RNA produced during viral replication and prevent virus production. PERK and IRE1 share a common ancestry with PKR and RNAse L, although they have evolved to respond to accumulation of unfolded proteins in the ER. As the effector domains of PERK and PKR and of IRE1 and RNAse L share functions, it is interesting to speculate that PERK and IRE1 may also provide a protective function in infectious disease.

ATF6 is transcription factor with a b-ZIP domain in the cytosol that also contains a large ER luminal domain to sense ER stress (Haze et al. 1999). There are two isoforms of ATF6, ATF6α (90 kDa) and ATF6β (110 kDa, also known

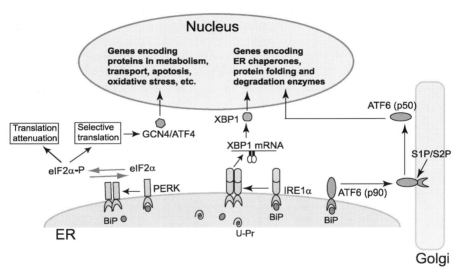

Fig. 2 Translational and transcriptional regulation upon ER stress. Upon accumulation of unfolded proteins in the ER lumen, PERK is released from BiP, thus permitting its dimerization and activation. Activated PERK phosphorylates eIF2α to reduce the frequency of the mRNA translation initiation in general. However, selective mRNAs, such as *GCN4* and *ATF4* mRNA, are preferentially translated in the presence of phosphorylated eIF2α. Upon accumulation of unfolded protein in the ER lumen, BiP release from IRE1 permits dimerization to activate its kinase and RNase activities to initiate *XBP1* mRNA splicing. Spliced *XBP1* mRNA encodes a potent transcription factor that binds to UPRE and ERSE sequences of many UPR target genes. BiP release from ATF6 permits ATF6 transport to the Golgi compartment where full-length ATF6 (90 kDa) is cleaved by S1P and S2P proteases to yield a cytosolic fragment (50 kDa) that migrates to the nucleus to activate transcription of UPR responsive genes. U-Pr unfolded protein

as CREB-RP). On activation of the UPR, ATF6α and ATF6β transit to the Golgi where they are cleaved by site-1 protease (S1P) and site-2 protease (S2P) to generate 50-kDa cytosolic b-ZIP-containing fragments that migrate to the nucleus to activate transcription of UPR target genes (Shen et al. 2002; Ye et al. 2000) (Fig. 2). Notably, ER stress-induced cleavage of ATF6 is processed by the same proteases S1P and S2P that cleave the ER-associated transmembrane sterol-response element binding protein (SREBP), which is a transcription factor required for induction of sterol biosynthetic genes (Ye et al. 2000). Regulated intramembrane proteolysis of ATF6 and SREBP is controlled at the step of trafficking of these transcription factors from the ER to the Golgi compartment. Whereas only unfolded protein accumulation in the ER promotes ATF6 transit to the Golgi for cleavage, cholesterol deprivation induces trafficking of SREBP to the Golgi.

4
Activation of Unfolded Protein Response Sensors

UPR signaling is an adaptive mechanism for cells to survive accumulation of unfolded proteins in the ER lumen. The UPR reduces the amount of new protein translocated into the ER lumen, increases retrotranslocation and degradation of misfolded ER-localized proteins, and bolsters the protein-folding capacity of the ER. The UPR is orchestrated by transcriptional activation of multiple genes mediated by IRE1 and ATF6, and a general decrease in translation initiation and a selective translation of several specific mRNAs mediated by PERK.

The most proximal UPR event is the activation of the ER stress sensors by a common stimulus, the accumulation of unfolded proteins in the ER lumen. Current studies support the idea that BiP serves as a master UPR regulator that plays a central role in activating all three transducers IRE1, PERK, and ATF6 in response to ER stress (Bertolotti et al. 2000; Dorner et al. 1992; Shen et al. 2002). BiP is a member of the heat shock protein family of 70 kDa that has a peptide-dependent ATPase activity. BiP binds to exposed hydrophobic patches on unfolded proteins and requires ATP to promote release. BiP interaction with unfolded proteins is part of the cellular quality control mechanism that only permits trafficking of properly folded proteins to the Golgi compartment. Reduction in the level of free BiP leads to UPR activation, whereas overexpression of BiP inhibits UPR activation (Dorner et al. 1988, 1992; Morris et al. 1997). Under nonstressed conditions, BiP binds to IRE1, PERK, and ATF6 to prevent their signaling. As the ER is overloaded by newly synthesized unfolded proteins or is "stressed" by agents that cause protein misfolding, the pool of free BiP in the ER lumen is depleted. As the free pool of BiP drops, IRE1 and PERK are released to permit homodimerization and autophosphorylation, leading to their activation (Bertolotti et al. 2000; Liu et al. 2003). Concomitantly, release of ATF6 from BiP permits ATF6 transport to the Golgi compartment, where it is cleaved to generate the cytosolic activated form of ATF6 (Shen et al. 2002). Thus, this BiP-regulated activation provides a direct mechanism for the three UPR transducers to sense the "stress" in the ER. However, in certain cells, different stresses or physiologic conditions can selectively activate only one or two of the ER stress sensors. For example, in B cell differentiation, the IRE1α-mediated UPR subpathway is activated and indispensable while the PERK-mediated UPR subpathway through phosphorylation of eIF2α is not required for the B cell differentiation process (Gass et al. 2002; Zhang et al. 2005). In contrast, in pancreatic β cells, glucose limitation appears to activate PERK prior to activation of IRE1 (Scheuner and Kaufman, unpublished observation). It will be important to elucidate how a general BiP repression mechanism permits the selective activation of individual components of the UPR that mediate various downstream effects.

5
The Transcriptional Response to Endoplasmic Reticulum Stress

To cope with accumulation of unfolded or misfolded protein in the ER lumen, the UPR is activated to alter transcriptional programs through IRE1 and ATF6 (Fig. 2). In mammals, the promoter regions of many UPR-inducible genes, such as BiP, GRP94, and calreticulin, contain a mammalian ER stress response element (ERSE, minimal motif: CCAAT(N_9)CCACG) that is necessary and sufficient for ER stress-induced gene transcription (Yoshida et al. 1998). Using ERSE as a probe in a yeast one-hybrid screen, researchers isolated two UPR-specific b-ZIP transcription factors, the X-box DNA binding protein 1 (XBP1) and ATF6 (Yoshida et al. 1998). XBP1 was identified as a homolog of yeast Hac1p that is a substrate for mammalian IRE1 RNase activity (Calfon et al. 2002; Shen et al. 2001; Yoshida et al. 2001). On activation of the UPR, IRE1 RNase cleaves *XBP1* mRNA to remove a 26-nucleotide intron, generating a translational frame-shift. Spliced *XBP1* mRNA encodes a protein with a novel carboxy-terminus that acts as a potent transcriptional activator for many UPR target genes. ATF6 is a UPR transducer that can bind ERSE motifs in the promoter regions of UPR responsive genes (Yoshida et al. 1998). Both ATF6α and ATF6β, in the presence of the CCAAT-binding factor (CBF; also called NF-Y), bind to the 3' half ERSE sequence (CCACG) in the promoter regions of UPR-responsive genes to activate transcription. Whereas CBF is a factor that constitutively binds the CCAAT motif, ATF6 is the inducible factor that binds the CCACG motif (Haze et al. 1999; Li et al. 2000; Yoshida et al. 2000).

The two bZIP transcription factors of ATF/CREB family, XBP1 and ATF6, serve as key regulators of transcriptional control in response to ER stress. ATF6 regulates a group of genes encoding ER-resident molecular chaperones and folding enzymes, while XBP1 regulates a subset of ER-resident chaperone genes that are essential for protein folding, maturation, and degradation in the ER (Lee et al. 2003; Okada et al. 2002). It was previously proposed that *XBP1* mRNA is induced by ATF6 in response to ER stress to generate more substrate *XBP1* mRNA for IRE1-mediated splicing (Lee et al. 2002; Yoshida et al. 2000, 2001). However, UPR induction of *XBP1* transcripts and proteins was not altered in the cells having defective or reduced ATF6 cleavage (Lee et al. 2003; Lee et al. 2002). Induction of *ATF6* mRNA upon ER stress was partially compromised in the absence of XBP1; therefore it was proposed that ATF6 lies downstream of XBP1 in some cases (Lee et al. 2003). These results suggest that XBP1 and ATF6 are situated largely in parallel pathways and may interact with each other upon ER stress.

6
The Translational Response to Endoplasmic Reticulum Stress

An immediate response to the accumulation of unfolded proteins in the ER of metazoan cells, is activation of PERK to inhibit protein biosynthesis through phosphorylation of eIF2α (Kaufman 2004). When eIF2α is phosphorylated, the formation of the ternary translation initiation complex eIF2/GTP/ tRNA$_i^{Met}$ is prevented, leading to reduced efficiency of AUG initiation codon recognition and general translational attenuation to reduce the workload of the ER (Harding et al. 2000, 2001; Scheuner 2001) (Fig. 2). Murine cells deleted in PERK, or mutated at Ser51 in eIF2α to prevent phosphorylation, did not attenuate protein synthesis upon ER stress. As a consequence, these cells were not able to survive ER stress (Scheuner 2001). Whereas phosphorylation of eIF2α by PERK leads to attenuation of global mRNA translation, phosphorylated eIF2α selectively stimulates translation of a specific subset of mRNAs in response to stress (Fig. 2).

In yeast, the Gcn2p-mediated phosphorylation of eIF2α upon amino acid starvation promotes translation of *GCN4* mRNA that encodes a b-ZIP transcription factor required for induction of genes encoding amino-acid biosynthetic functions (Hinnebusch 2000). *GCN4* mRNA contains four upstream open reading frames (uORFs) in its 5′ UTR, which ordinarily inhibit the ability of the ribosome to scan through the 5′ end of the mRNA and reach the correct AUG initiation codon. Phosphorylation of eIF2α limits 60S-ribosomal-subunit joining to allow the 40S ribosomal subunit to scan through the ORFs and initiate polypeptide-chain synthesis at the authentic GCN4 ORF, thus allowing translation of *GCN4* (Kaufman 2004). This control mechanism is also utilized in mammalian cells to regulate translation in response to ER stress and amino acid starvation. For example, upon ER stress, phosphorylated eIF2α selectively promotes translation of activating transcription factor 4 (ATF4) mRNA (Harding et al. 2000; Scheuner 2001). ATF4 subsequently activates transcription of genes involved in amino acid metabolism and transport, oxidation-reduction reactions, and ER stress-induced apoptosis (Harding et al. 2003).

7
Endoplasmic Reticulum-Associated Protein Degradation

Protein folding in the oxidizing environment of the ER is an energy-requiring process (Braakman et al. 1992; Dorner et al. 1990). Under nonstressed conditions, newly synthesized proteins exist as unfolded intermediates along the protein-folding pathway. Once ER stress is imposed, for example, by depletion of energy, many folding intermediates become irreversibly trapped in low-energy states and accumulate. These unfolded proteins are retained in the ER through interactions with BiP, calnexin, and calreticulin. Eventually, unfolded or misfolded proteins in the ER lumen are retrotranslocated to the cytoplasm, where they are ubiquitinated and degraded by the proteasome (Werner et al. 1996). This process is called ER-associated degradation (ERAD) and is regulated by the UPR.

Many specific components of the ERAD pathway in yeast, such as *DER1*, *HRD1/DER3*, *HRD3*, and *UBC7*, are induced by the UPR (Travers et al. 2000). Hrd1p is an ER type I-transmembrane protein having E3 ubiquitin ligase activity (Bays et al. 2001). The interaction between the ER lumenal domains of Hrd3p and Hrd1p stabilizes the cytosolic RING-H2 motif of Hrd1p, which is required for its ubiquitin ligase activity (Gardner et al. 2000, 2001). Hrd1p prefers a misfolded protein as a ubiquitination substrate and uses only Ubc7p or Ubc1p, E2 ubiquitin-conjugating enzymes, to specifically mediate ubiquitination of ERAD substrates. Although UBC1 mRNA is unaffected by dithiothreitol treatment, induction of *UBC7* and *HRD1* upon ER stress is completely dependent on *HAC1* and *IRE1*, indicating that the UPR may upregulate components of the ERAD system in yeast (Friedlander et al. 2000). Yeast cells unable to perform ERAD are constantly susceptible to folding stress, as indicated by a constitutive activation of the UPR and a requirement for the UPR for normal growth and survival even under conditions of mild stress (Friedlander et al. 2000). For example, absence of both Ubc1p and Ubc7p or absence of Hrd1p results in marked stabilization of an ERAD substrate and induction of the UPR. Moreover, the ER-associated AAA-ATPase Cdc48p–Ufd1p–Npl4p complex that is required for ERAD functions as a cytosolic chaperone complex to extract ER degradation substrates from the ER lumen (Ye et al. 2001). Efficient dislocation of ERAD substrates ensures their subsequent proteolysis by the 26S proteasome. Although the UPR does not upregulate the expression of *CDC48*, mutations in either of *CDC48*, *UFD1*, or *NPL4* cause accumulation of ERAD substrates in the ER and activate the UPR (Jarosch et al. 2002). Rather than being individually dispensable, the UPR and ERAD are intimately coordinated, complementary mechanisms that prevent unfolded protein accumulation and mitigate its toxic consequences.

Regulation of ERAD by the UPR is further suggested by identification of EDEM in mammalian cells (Yoshida et al. 2003). EDEM is a type-II transmembrane protein localized to the ER, and its lumenal domain shows significant homology to a 1,2-mannosidase but lacks such enzymatic activity. IRE1α-deficient cells were defective in degradation of a mutant α1-antitrypsin, an ERAD substrate, and this defect was completely restored by expression of EDEM, supporting that EDEM-mediated ERAD is solely dependent on the IRE1-mediated UPR pathway. (Hosokawa et al. 2001; Molinari et al. 2003; Oda et al. 2003; Yoshida et al. 2003). On the other hand, if the overload of unfolded or misfolded proteins in the ER is not resolved, prolonged UPR activation will lead to programmed cell death.

8
Unfolded Protein Response-Induced Apoptosis

Three known pro-apoptotic pathways emanating from the ER are mediated by IRE1α, caspase-12, and PERK/CHOP, respectively (Fig. 3). Under ER stress, activated IRE1α can bind c-Jun-N-terminal inhibitory kinase (JIK) and recruit cytosolic adapter TRAF2 to the ER membrane (Urano et al. 2000; Yoneda et al. 2001). TRAF2 activates the apoptosis-signaling kinase 1 (ASK1), a mitogen-activated protein kinase kinase kinase (MAPKKK) (Nishitoh et al. 2002). Activated ASK1 leads to activation of the JNK protein kinase and mitochondria/Apaf1-dependent caspase activation (Leppa and Bohmann 1999; Nishitoh et al. 2002; Urano et al. 2000). Caspase-12 is an ER-associated proximal effector of the caspase activation cascade, and cells defective in this enzyme are partially resistant to ER stress-induced apoptosis (Nakagawa et al. 2000). ER stress induces TRAF2 release from procaspase 12, allowing it to bind activated IRE1. This also permits clustering of procaspase-12 at the ER membrane, thus leading to procaspase-12 activation (Yoneda et al. 2001). Caspase-12 can activate caspase-9, which in turn activates caspase-3 (Morishima et al. 2002). Procaspase-12 can also be activated by m-calpain in response to calcium release from the ER, although the physiological significance of this pathway is not known (Nakagawa and Yuan 2000). In addition, upon ER stress, procaspase-7 is activated and recruited to the ER membrane (Rao et al. 2001). Finally, a second death-signaling pathway activated by ER stress is mediated by transcriptional activation of genes encoding pro-apoptotic functions. Activation of the UPR transducer PERK leads to translation of the transcription factor ATF4, which subsequently activates transcription of CHOP/GADD 153, a b-ZIP transcription factor that potentates apoptosis, possibly through repressing expression of the apoptotic repressor BCL2 (Harding et al. 2000; Ma et al. 2002). In addition to its transcriptional induction by PERK/ATF4, CHOP is also regulated at the post-translational level by phosphorylation mediated by p38 MAP kinase (Wang et al. 1996; Wang and Ron 1996). CHOP forms stable heterodimers with C/EBP family members and controls expression of a set of stress-induced genes, which may be involved in apoptosis or organ regeneration. CHOP-deficient

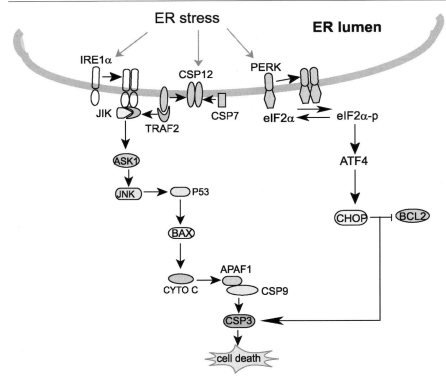

Fig. 3 Apoptosis mediated by UPR signaling. Upon ER stress, activated IRE1α can recruit JIK and TRAF2 to activate ASK1 and JNK, leading to activation of mitochondria/Apaf1-dependent caspases. Upon activation of the UPR, c-Jun-N-terminal inhibitory kinase (*JIK*) release from procaspase-12 permits clustering and activation of procaspase-12. Caspase-12 activates procaspase-9 to activate procaspase-3, the executioner of cell death. Activated PERK phosphorylates eIF2α that enhances translation of ATF4 mRNA. ATF4 induces transcription of the pro-apoptotic factor CHOP, which can inhibit expression of apoptotic suppressor BCL2. *CSP,* caspase; *CYTO C,* cytochrome C

mice have reduced apoptosis in renal epithelium in response to tunicamycin, a reagent that induces ER stress by blocking protein glycosylation (Zinszner et al. 1998).

9
The Physiological Roles of the Unfolded Protein Response

During cell growth, differentiation, and environmental stimuli, there are different levels of protein-folding load imposed upon the ER. Cells have evolved the ability to augment their folding capacity and remodel their secretory pathway in response to developmental demands and physiological changes. Accumulating evidence suggests that the UPR plays important roles in differentiation and function of specialized cells. Moreover, pathological conditions that interfere with ER homeostasis produce prolonged activation of the UPR that may contribute to the pathogenesis of many diseases.

9.1
The Unfolded Protein Response in B Cell Differentiation

On terminal differentiation of B lymphoid cells to plasma cells, the ER compartment expands approximately fivefold to accommodate the large increase in immunoglobulin (Ig) synthesis (Wiest et al. 1990). The UPR transcriptional activator XBP1 is required for plasma cell differentiation (Reimold et al. 2001). XBP1-deficient B lymphoid cells express Ig genes and undergo isotype switching, but are defective in plasma cell differentiation and do not secrete high levels of Igs. Expression of the spliced form of XBP1 efficiently restores production of secreted Igs in XBP1-deficient B cells, suggesting a physiological role for the UPR in high-rate production of secreted antibodies (Iwakoshi et al. 2003). During plasma cell differentiation, IRE1α-mediated splicing of *XBP1* mRNA was found to depend on increased translation of Ig chains (Gass et al. 2002; Iwakoshi et al. 2003; van Anken et al. 2003). These observations support the hypothesis that increased synthesis of Ig produces greater amounts of nascent, unfolded, and unassembled subunits that bind and sequester BiP, leading to UPR activation. Indeed, BiP is the most abundantly expressed UPR-dependent gene and was first identified as encoding a protein that binds Ig heavy chains in the absence of light chains (Haas and Wabl 1983). In addition, the UPR transducer ATF6 may be involved in the process of terminal differentiation of B cells by regulating secretion of Igs (Gass et al. 2002; Gunn et al. 2004).

It is possible that the UPR may signal a B cell differentiation program that occurs prior to increased antibody synthesis. Recently, our group demonstrated that mouse IRE1α is required at two distinct steps during B cell lymphopoiesis (Zhang et al. 2005). IRE1α plays essential roles in both early and late stages of B cell development. In the very early stage, IRE1α regulates transcription of the VDJ recombination-activating genes *rag1*, *rag2*, and *TdT*, which is required for initiation of VDJ rearrangement and B cell receptor formation. In the late stage of B cell differentiation, IRE1α is required to splice the *XBP1* mRNA for terminal differentiation of mature B cells into antibody-secreting plasma cells (Zhang et al. 2005).

9.2
The Unfolded Protein Response in Glucose Homeostasis and Diabetes

The metabolism of glucose is tightly controlled at the levels of synthesis and utilization through hormonal regulation. Glucose not only promotes the secretion of insulin but also stimulates insulin transcription and translation (Itoh and

Okamoto 1980; Lang 1999; Permutt 1974). UPR signaling is essential to maintain glucose homeostasis. It is noteworthy that the UPR was first characterized as transcriptional activation of a set of genes, encoding glucose-regulated proteins, in response to glucose/energy deprivation (Pouyssegur et al. 1977). We now know that pancreatic β cells uniquely require the UPR for survival during intermittent fluctuations in blood glucose (Harding et al. 2001; Scheuner 2001). Humans and mice with deletions in PERK have a profound pancreatic β cell dysfunction and develop infancy-onset diabetes (Delepine et al. 2000; Harding et al. 2001). Mice with a homozygous Ser51Ala mutation at the PERK phosphorylation site in eIF2α display a β cell loss in utero, suggesting that translational control through PERK-mediated phosphorylation of eIF2 is required to maintain β cell survival (Scheuner 2001). Pancreatic β cells are exquisitely sensitive to physiological fluctuations in blood glucose, because they lack hexokinase, an enzyme with a high affinity for glucose as a substrate. We propose that blood glucose levels influence the protein-folding status in the ER. As glucose levels decline, the energy supply decreases, so protein folding becomes less efficient and PERK is activated. The UPR regulates transcriptional induction of glucose-regulated proteins that might provide a protective function by increasing the cellular capacity for the uptake and use of glucose. Conversely, as blood glucose levels rise, eIF2α would be dephosphorylated so that translation would accelerate to increase proinsulin synthesis (Scheuner 2001). This would allow entry of new preproinsulin into the ER, and is consistent with the glucose-stimulated increase in total protein and proinsulin synthesis observed in isolated β cell preparations. Eventually, after prolonged proinsulin translation, the UPR would be turned on to inhibit further protein synthesis and prevent overload of the ER folding capacity. In this manner, a balance between glucose level and PERK-eIF2α UPR signaling is essential for the glucose-regulated periodic fluctuations in proinsulin translation, β cell function, and survival.

The UPR may also play an important role in the regulation of cellular responses to insulin. A recent study showed that ER stress serves as a central feature of peripheral insulin resistance and type 2 diabetes and that the IRE1α-XBP1 UPR pathway is critical for this process (Ozcan et al. 2004). Mice deficient in XBP1 develop insulin resistance. ER stress in obese mice leads to suppression of insulin receptor signaling through hyperactivation of c-Jun N-terminal kinase (JNK) and subsequent serine phosphorylation of insulin receptor substrate-1.

9.3
The Unfolded Protein Response in Organelle Expansion

When the protein-folding load exceeds the capacity of the ER to fold proteins, the UPR maintains ER homeostasis by inhibiting protein synthesis and enhancing transcription of resident ER proteins that facilitate protein maturation, secretion, and degradation. The UPR is required for ER expansion that occurs upon differentiation of highly specialized secretory cells (Kaufman 2002). During differentiation of certain secretory cells, such as those in the pancreas or liver, membrane expansion is accompanied by a dramatic increase in protein secretion and UPR activation. Recent evidence support that one role of UPR activation is to expand the quantity of the ER in order to promote more productive protein folding and secretion. In mature B cells, ectopic expression of XBP1 induced a wide spectrum of secretory pathway genes and physically expanded the ER (Shaffer et al. 2004). Overexpression of spliced XBP1 increased cell size, lysosome content, mitochondrial mass and function, ribosome number, and total protein synthesis. Thus, XBP1 coordinates diverse changes in cellular structure and function resulting in the characteristic phenotype of professional secretory cells. Furthermore, another study showed that spliced XBP1 could induce membrane biosynthesis and ER proliferation in a cell type different from B lymphocytes (Sriburi et al. 2004). Overexpression of spliced XBP1 in NIH-3T3 cells was sufficient to induce synthesis of phosphatidylcholine, the primary phospholipid of the ER membrane. Cells overexpressing spliced XBP1 exhibit elevated levels of membrane phospholipids, increased surface area and volume of rough ER, and enhanced activity of the cytidine diphosphocholine pathway of phosphatidylcholine biosynthesis.

9.4
The Unfolded Protein Response in Neurological Diseases

Neurological disease caused by expansion of polyglutamine repeats and neurodegenerative diseases, such as Alzheimer's and Parkinson's disease, are associated with accumulation of abnormal protein and dysfunction of the ER. Analysis of the polyglutamine repeat associated with the spinocerebrocellular atrophy protein (SCA3) in Machado-Joseph disease suggests that cytoplasmic accumulation of the SCA3 aggregate can inhibit proteasome function, thereby interfering with ERAD to induce the UPR and elicit caspase-12 activation (Nishitoh et al. 2002). Alzheimer's disease is a progressive neurodegenerative disorder that is characterized clinically by progressive loss of memory and cognitive impairment, and pathologically by the extracellular deposition of senile plaques. Mutations of genes that encode amyloid precursor protein, presenilin-1 (PS1) and presenilin-2 (PS2) were found to cause familial Alzheimer's disease (Goate et al. 1991; Levy-Lahad et al. 1995; Rogaev et al. 1995; Sherrington et al. 1995). Interestingly, it was observed that PS1 bound directly to IRE1α on the ER membrane and that the autophosphorylation of IRE1α in response to ER stress was diminished in cells expressing mutant PS1 compared with cells expressing wild-type PS1 (Katayama et al. 1999, 2001). Mutant PS1 was also found to suppress the activation of the other two UPR transducers, ATF6 and PERK, so the global ER response to stress seems to be reduced by mutant PS1. Indeed, cells expressing mutant PS1 show increased vulnerability to ER stress (Guo et al. 1999; Katayama et al. 1999, 2001). The mechanisms by which mutant PS1 affects the ER stress response are attributed to the inhibited activation of the ER stress transducers IRE1, PERK, and ATF6. However, in sporadic Alzheimer's disease, the spliced isoform of PS2 induces expression of high-mobility group A1a protein (HMGA1a), and also downregulates the UPR signaling pathway in a manner similar to that of PS1 mutant in familial Alzheimer disease (Sato et al. 2001). It was suggested that caspase-4, the human homolog of murine caspase-12, plays critical roles in ER stress-induced neuronal cell death in Alzheimer disease (Katayama et al. 2004).

In addition, autosomal recessive juvenile parkinsonism (AR-JP) results from defects in the Parkin gene (Kitada et al. 1998), encoding a ubiquitin protein ligase (E3) that functions with ubiquitin-conjugating enzyme UbcH7 or UbcH8 to tag proteins for degradation. Overexpression of Parkin suppresses cell death associated with ER stress (Imai et al. 2000). PAEL-R is a putative transmembrane receptor protein that is detected in an insoluble form in the brains of AR-JP patients (Imai et al. 2001). Inherited Parkinson's disease is associated with the accumulation of PAEL-R in the ER of dopaminergic neurons. The accumulation of PAEL-R results from defective Parkin that does not maintain the proteasome-degrading activity necessary to maintain ER function (Imai et al. 2002).

Finally, translational inhibition through UPR activation was observed in cerebral ischemia. PERK is the only eIF2α kinase that is known to be activated after cerebral ischemia (Kumar et al. 2001) and *XBP1* mRNA splicing was detected after transient cerebral ischemia (Paschen et al. 2003). Together these findings indicate that the etiology of many neurological diseases is significantly related to impaired ER homeostasis and activation of the UPR.

10
Perspectives

Over the past 10 years, tremendous progress has been made in identifying the components that regulate the UPR upon accumulation of unfolded protein in the ER lumen. However, little is known regarding the physiological roles of the different UPR pathways in maintaining cell homeostasis and in disease pathogenesis. Additional studies are required to elucidate the mechanisms by which selective UPR transducers are activated under different physiological conditions. Research efforts to understand the upstream events in the UPR pathway promise to expand our knowledge of UPR regulation and its physiological functions. It is known that a variety of environmental insults and genetic defects result in accumulation of unfolded or misfolded proteins in the ER that contribute to the pathogenesis of different disease states. As new animal models with defects in different signaling components of the UPR are generated, we will gain a more precise knowledge of how these pathways cause or are a consequence of different pathological conditions. Elucidating which components of the UPR that are beneficial versus those that are detrimental under different conditions of stress represents a major avenue for future research. As we gain a greater understanding of UPR signaling and the physiological roles of the UPR in health and disease, it should be possible to design novel therapeutic strategies to activate or inhibit UPR signaling in order to intervene in diseases associated with abnormal accumulation of unfolded or misfolded proteins in the ER.

Acknowledgements This work is supported by NIH grant DK42394 to R.J. Kaufman.

References

Bays NW, Gardner RG, Seelig LP, Joazeiro CA, Hampton RY (2001) Hrd1p/Der3p is a membrane-anchored ubiquitin ligase required for ER- associated degradation. Nat Cell Biol 3:24–29

Bertolotti A, Zhang Y, Hendershot LM, Harding HP, Ron D (2000) Dynamic interaction of BiP and ER stress transducers in the unfolded- protein response. Nat Cell Biol 2:326–332

Braakman I, Helenius J, Helenius A (1992) Role of ATP and disulphide bonds during protein folding in the endoplasmic reticulum. Nature 356:260–262

Calfon M, Zeng H, Urano F, Till JH, Hubbard SR, Harding HP, Clark SG, Ron D (2002) IRE1 couples endoplasmic reticulum load to secretory capacity by processing the XBP-1 mRNA. Nature 415:92–96

Cox JS, Shamu CE, Walter P (1993) Transcriptional induction of genes encoding endoplasmic reticulum resident proteins requires a transmembrane protein kinase. Cell 73:1197–1206

Delepine M, Nicolino M, Barrett T, Golamaully M, Mark Lathrop G, Julier C (2000) EIF2AK3, encoding translation initiation factor 2-alpha kinase 3, is mutated in patients with Wolcott-Rallison syndrome [In Process Citation]. Nat Genet 25:406–409

Dorner AJ, Bole DG, Kaufman RJ (1987) The relationship of N-linked glycosylation and heavy chain-binding protein association with the secretion of glycoproteins. J Cell Biol 105:2665–2674

Dorner AJ, Krane MG, Kaufman RJ (1988) Reduction of endogenous GRP78 levels improves secretion of a heterologous protein in CHO cells. Mol Cell Biol 8:4063–4070

Dorner AJ, Wasley LC, Kaufman RJ (1990) Protein dissociation from GRP78 and secretion are blocked by depletion of cellular ATP levels. Proc Natl Acad Sci U S A 87:7429–7432

Dorner AJ, Wasley LC, Kaufman RJ (1992) Overexpression of GRP78 mitigates stress induction of glucose regulated proteins and blocks secretion of selective proteins in Chinese hamster ovary cells. EMBO J 11:1563–1571

Friedlander R, Jarosch E, Urban J, Volkwein C, Sommer T (2000) A regulatory link between ER-associated protein degradation and the unfolded-protein response. Nat Cell Biol 2:379–384

Gardner RG, Swarbrick GM, Bays NW, Cronin SR, Wilhovsky S, Seelig L, Kim C, Hampton RY (2000) Endoplasmic reticulum degradation requires lumen to cytosol signaling. Transmembrane control of Hrd1p by Hrd3p. J Cell Biol 151:69–82

Gardner RG, Shearer AG, Hampton RY (2001) In vivo action of the HRD ubiquitin ligase complex: mechanisms of endoplasmic reticulum quality control and sterol regulation. Mol Cell Biol 21:4276–4291

Gass JN, Gifford NM, Brewer JW (2002) Activation of an unfolded protein response during differentiation of antibody-secreting B cells. J Biol Chem 277:49047–49054

Gething MJ, Sambrook J (1992) Protein folding in the cell. Nature 355:33–45

Goate A, Chartier-Harlin MC, Mullan M, Brown J, Crawford F, Fidani L, Giuffra L, Haynes A, Irving N, James L et al (1991) Segregation of a missense mutation in the amyloid precursor protein gene with familial Alzheimer's disease. Nature 349:704–706

Gunn KE, Gifford NM, Mori K, Brewer JW (2004) A role for the unfolded protein response in optimizing antibody secretion. Mol Immunol 41:919–927

Guo Q, Sebastian L, Sopher BL, Miller MW, Ware CB, Martin GM, Mattson MP (1999) Increased vulnerability of hippocampal neurons from presenilin-1 mutant knock-in mice to amyloid beta-peptide toxicity: central roles of superoxide production and caspase activation. J Neurochem 72:1019–1029

Haas IG, Wabl M (1983) Immunoglobulin heavy chain binding protein. Nature 306:387–389

Harding HP, Zhang Y, Ron D (1999) Protein translation and folding are coupled by an endoplasmic-reticulum-resident kinase [In Process Citation]. Nature 397:271–274

Harding HP, Zhang Y, Bertolotti A, Zeng H, Ron D (2000) Perk is essential for translational regulation and cell survival during the unfolded protein response. Mol Cell 5:897–904

Harding HP, Zeng H, Zhang Y, Jungries R, Chung P, Plesken H, Sabatini DD, Ron D (2001) Diabetes mellitus and exocrine pancreatic dysfunction in perk-/- mice reveals a role for translational control in secretory cell survival. Mol Cell 7:1153–1163

Harding HP, Zhang Y, Zeng H, Novoa I, Lu PD, Calfon M, Sadri N, Yun C, Popko B, Paules R, Stojdl DF, Bell JC, Hettmann T, Leiden JM, Ron D (2003) An integrated stress response regulates amino acid metabolism and resistance to oxidative stress. Mol Cell 11:619–633

Haze K, Yoshida H, Yanagi H, Yura T, Mori K (1999) Mammalian transcription factor ATF6 is synthesized as a transmembrane protein and activated by proteolysis in response to endoplasmic reticulum stress. Mol Biol Cell 10:3787–3799

Helenius A (1994) How N-linked oligosaccharides affect glycoprotein folding in the endoplasmic reticulum. Mol Biol Cell 5:253–265

Hinnebusch A (2000) Mechanism and regulation of initiator methionyl-tRNA binding ribosomes. Cold Spring Harbor Press, Cold Spring Harbor, NY

Hosokawa N, Wada I, Hasegawa K, Yorihuzi T, Tremblay LO, Herscovics A, Nagata K (2001) A novel ER alpha-mannosidase-like protein accelerates ER-associated degradation. EMBO Rep 2:415–422

Imai Y, Soda M, Takahashi R (2000) Parkin suppresses unfolded protein stress-induced cell death through its E3 ubiquitin-protein ligase activity. J Biol Chem 275:35661–5664

Imai Y, Soda M, Inoue H, Hattori N, Mizuno Y, Takahashi R (2001) An unfolded putative transmembrane polypeptide, which can lead to endoplasmic reticulum stress, is a substrate of Parkin. Cell 105:891–902

Imai Y, Soda M, Hatakeyama S, Akagi T, Hashikawa T, Nakayama KI, Takahashi R (2002) CHIP is associated with Parkin, a gene responsible for familial Parkinson's Disease, and enhances its ubiquitin ligase activity. Mol Cell 10:55–67

Itoh N, Okamoto H (1980) Translational control of proinsulin synthesis by glucose. Nature 283:100–102

Iwakoshi NN, Lee AH, Vallabhajosyula P, Otipoby KL, Rajewsky K, Glimcher LH (2003) Plasma cell differentiation and the unfolded protein response intersect at the transcription factor XBP-1. Nat Immunol 4:321–329

Jarosch E, Taxis C, Volkwein C, Bordallo J, Finley D, Wolf DH, Sommer T (2002) Protein dislocation from the ER requires polyubiquitination and the AAA-ATPase Cdc48. Nat Cell Biol 4:134–139

Katayama T, Imaizumi K, Sato N, Miyoshi K, Kudo T, Hitomi J, Morihara T, Yoneda T, Gomi F, Mori Y, Nakano Y, Takeda J, Tsuda T, Itoyama Y, Murayama O, Takashima A, St George-Hyslop P, Takeda M, Tohyama M (1999) Presenilin-1 mutations downregulate the signalling pathway of the unfolded-protein response. Nat Cell Biol 1:479–485

Katayama T, Imaizumi K, Honda A, Yoneda T, Kudo T, Takeda M, Mori K, Rozmahel R, Fraser P, George-Hyslop PS, Tohyama M (2001) Disturbed activation of endoplasmic reticulum stress transducers by familial Alzheimer's disease-linked presenilin-1 mutations. J Biol Chem 276:43446–43454

Katayama T, Imaizumi K, Manabe T, Hitomi J, Kudo T, Tohyama M (2004) Induction of neuronal death by ER stress in Alzheimer's disease. J Chem Neuroanat 28:67–78

Kaufman RJ (1999) Stress signaling from the lumen of the endoplasmic reticulum: coordination of gene transcriptional and translational controls. Genes Dev 13:1211–1233

Kaufman RJ (2002) Orchestrating the unfolded protein response in health and disease. J Clin Invest 110:1389–1398

Kaufman RJ (2004) Regulation of mRNA translation by protein folding in the endoplasmic reticulum. Trends Biochem Sci 29:152–158

Kawahara T, Yanagi H, Yura T, Mori K (1998) Unconventional splicing of HAC1/ERN4 mRNA required for the unfolded protein response. Sequence-specific and non-sequential cleavage of the splice sites [In Process Citation]. J Biol Chem 273:1802–1807

Kitada T, Asakawa S, Hattori N, Matsumine H, Yamamura Y, Minoshima S, Yokochi M, Mizuno Y, Shimizu N (1998) Mutations in the parkin gene cause autosomal recessive juvenile parkinsonism. Nature 392:605–608

Kumar R, Azam S, Sullivan JM, Owen C, Cavener DR, Zhang P, Ron D, Harding HP, Chen JJ, Han A, White BC, Krause GS, DeGracia DJ (2001) Brain ischemia and reperfusion activates the eukaryotic initiation factor 2alpha kinase, PERK. J Neurochem 77:1418–1421

Lang J (1999) Molecular mechanisms and regulation of insulin exocytosis as a paradigm of endocrine secretion. Eur J Biochem 259:3–17

Lee AH, Iwakoshi NN, Glimcher LH (2003) XBP-1 regulates a subset of endoplasmic reticulum resident chaperone genes in the unfolded protein response. Mol Cell Biol 23:7448–7459

Lee AS (1987) Coordinated regulation of a set of genes by glucose and calcium ionophore in mammalian cells. Trends Biochem Sci 12:20–24

Lee AS, Delegeane AM, Baker V, Chow PC (1983) Transcriptional regulation of two genes specifically induced by glucose starvation in a hamster mutant fibroblast cell line. J Biol Chem 258:597–603

Lee AS, Bell J, Ting J (1984) Biochemical characterization of the 94- and 78-kilodalton glucose-regulated proteins in hamster fibroblasts. J Biol Chem 259:4616–4621

Lee K, Tirasophon W, Shen X, Michalak M, Prywes R, Okada T, Yoshida H, Mori K, Kaufman RJ (2002) IRE1-mediated unconventional mRNA splicing and S2P-mediated ATF6 cleavage merge to regulate XBP1 in signaling the unfolded protein response. Genes Dev 16:452–466

Leppa S, Bohmann D (1999) Diverse functions of JNK signaling and c-Jun in stress response and apoptosis. Oncogene 18:6158–6162

Levy-Lahad E, Wijsman EM, Nemens E, Anderson L, Goddard KA, Weber JL, Bird TD, Schellenberg GD (1995) A familial Alzheimer's disease locus on chromosome 1. Science 269:970–973

Li M, Baumeister P, Roy B, Phan T, Foti D, Luo S, Lee AS (2000) ATF6 as a transcription activator of the endoplasmic reticulum stress element: thapsigargin stress-induced changes and synergistic interactions with NF-Y and YY1. Mol Cell Biol 20:5096–5106

Liu CY, Schroder M, Kaufman RJ (2000) Ligand-independent dimerization activates the stress response kinases IRE1 and PERK in the lumen of the endoplasmic reticulum. J Biol Chem 275:24881–24885

Liu CY, Xu Z, Kaufman RJ (2003) Structure and intermolecular interactions of the luminal dimerization domain of human IRE1alpha. J Biol Chem 278:17680–17687

Ma Y, Brewer JW, Diehl JA, Hendershot LM (2002) Two distinct stress signaling pathways converge upon the CHOP promoter during the mammalian unfolded protein response. J Mol Biol 318:1351–1365

Molinari M, Calanca V, Galli C, Lucca P, Paganetti P (2003) Role of EDEM in the release of misfolded glycoproteins from the calnexin cycle. Science 299:1397–1400

Mori K, Sant A, Kohno K, Normington K, Gething MJ, Sambrook JF (1992) A 22 bp cis-acting element is necessary and sufficient for the induction of the yeast KAR2 (BiP) gene by unfolded proteins. EMBO J 11:2583–2593

Mori K, Ma W, Gething M-J, Sambrook J (1993) A transmembrane protein with a cdc2$^+$/CDC28-related kinase activity is required for signalling from the ER to the nucleus. Cell 74:743–756

Morishima N, Nakanishi K, Takenouchi H, Shibata T, Yasuhiko Y (2002) An ER stress-specific caspase cascade in apoptosis: cytochrome c-independent activation of caspase-9 by caspase-12. J Biol Chem 3:3

Morris JA, Dorner AJ, Edwards CA, Hendershot LM, Kaufman RJ (1997) Immunoglobulin binding protein (BiP) function is required to protect cells from endoplasmic reticulum stress but is not required for the secretion of selective proteins. J Biol Chem 272:4327–4334

Nakagawa T, Yuan J (2000) Cross-talk between two cysteine protease families. Activation of caspase-12 by calpain in apoptosis. J Cell Biol 150:887–894

Nakagawa T, Zhu H, Morishima N, Li E, Xu J, Yankner BA, Yuan J (2000) Caspase-12 mediates endoplasmic-reticulum-specific apoptosis and cytotoxicity by amyloid-beta. Nature 403:98–103

Nishitoh H, Matsuzawa A, Tobiume K, Saegusa K, Takeda K, Inoue K, Hori S, Kakizuka A, Ichijo H (2002) ASK1 is essential for endoplasmic reticulum stress-induced neuronal cell death triggered by expanded polyglutamine repeats. Genes Dev 16:1345–1355

Oda Y, Hosokawa N, Wada I, Nagata K (2003) EDEM as an acceptor of terminally misfolded glycoproteins released from calnexin. Science 299:1394–1397

Okada T, Yoshida H, Akazawa R, Negishi M, Mori K (2002) Distinct roles of ATF6 and PERK in transcription during the mammalian unfolded protein response. Biochem J 366:585–594

Ozcan U, Cao Q, Yilmaz E, Lee AH, Iwakoshi NN, Ozdelen E, Tuncman G, Gorgun C, Glimcher LH, Hotamisligil GS (2004) Endoplasmic reticulum stress links obesity, insulin action, and type 2 diabetes. Science 306:457–461

Paschen W, Aufenberg C, Hotop S, Mengesdorf T (2003) Transient cerebral ischemia activates processing of xbp1 messenger RNA indicative of endoplasmic reticulum stress. J Cereb Blood Flow Metab 23:449–461

Permutt MA (1974) Effect of glucose on initiation and elongation rates in isolated rat pancreatic islets. J Biol Chem 249:2738–2742

Pouyssegur J, Shiu R, Pastan I (1977) Induction of two transformation-sensitive membrane polypeptides in normal fibroblasts by a block in glycoprotein synthesis or glucose deprivation. Cell 11:941–947

Rao RV, Hermel E, Castro-Obregon S, del Rio G, Ellerby LM, Ellerby HM, Bredesen DE (2001) Coupling endoplasmic reticulum stress to the cell death program. Mechanism of caspase activation. J Biol Chem 276:33869–33874

Reimold AM, Iwakoshi NN, Manis J, Vallabhajosyula P, Szomolanyi-Tsuda E, Gravallese EM, Friend D, Grusby MJ, Alt F, Glimcher LH (2001) Plasma cell differentiation requires the transcription factor XBP-1. Nature 412:300–307

Ritter C, Helenius A (2000) Recognition of local glycoprotein misfolding by the ER folding sensor UDP-glucose:glycoprotein glucosyltransferase. Nat Struct Biol 7:278–280

Rogaev EI, Sherrington R, Rogaeva EA, Levesque G, Ikeda M, Liang Y, Chi H, Lin C, Holman K, Tsuda T et al (1995) Familial Alzheimer's disease in kindreds with missense mutations in a gene on chromosome 1 related to the Alzheimer's disease type 3 gene. Nature 376:775–778

Sato N, Imaizumi K, Manabe T, Taniguchi M, Hitomi J, Katayama T, Yoneda T, Morihara T, Yasuda Y, Takagi T, Kudo T, Tsuda T, Itoyama Y, Makifuchi T, Fraser PE, St George-Hyslop P, Tohyama M (2001) Increased production of beta-amyloid and vulnerability to endoplasmic reticulum stress by an aberrant spliced form of presenilin 2. J Biol Chem 276:2108–2114

Scheuner D, Song B, McEwen E, Lui C, Laybutt R, Gillespie P, Saunders T, Bonner-Weir S, Kaufman RJ (2001) Translational control is required for the unfolded protein response and in vivo glucose homeostasis. Mol Cell 7:1165–1176

Shaffer AL, Shapiro-Shelef M, Iwakoshi NN, Lee AH, Qian SB, Zhao H, Yu X, Yang L, Tan BK, Rosenwald A, Hurt EM, Petroulakis E, Sonenberg N, Yewdell JW, Calame K, Glimcher LH, Staudt LM (2004) XBP1, downstream of Blimp-1, expands the secretory apparatus and other organelles, and increases protein synthesis in plasma cell differentiation. Immunity 21:81–93

Shamu CE, Walter P (1996) Oligomerization and phosphorylation of the Ire1p kinase during intracellular signaling from the encoplasmic reticulum to the nucleus. EMBO J 15:3028–3039

Shen J, Chen X, Hendershot L, Prywes R (2002) ER stress regulation of ATF6 localization by dissociation of BiP/GRP78 binding and unmasking of Golgi localization signals. Dev Cell 3:99–111

Shen X, Ellis RE, Lee K, Liu CY, Yang K, Solomon A, Yoshida H, Morimoto R, Kurnit DM, Mori K, Kaufman RJ (2001) Complementary signaling pathways regulate the unfolded protein response and are required for C. elegans development. Cell 107:893–903

Sherrington R, Rogaev EI, Liang Y, Rogaeva EA, Levesque G, Ikeda M, Chi H, Lin C, Li G, Holman K et al (1995) Cloning of a gene bearing missense mutations in early-onset familial Alzheimer's disease. Nature 375:754–760

Shi Y, Vattem KM, Sood R, An J, Liang J, Stramm L, Wek RC (1998) Identification and characterization of pancreatic eukaryotic initiation factor 2 alpha-subunit kinase, PEK, involved in translational control. Mol Cell Biol 18:7499–7509

Sidrauski C, Cox JS, Walter P (1996) tRNA ligase is required for regulated mRNA splicing in the unfolded protein response [see comments]. Cell 87:405–413

Sidrauski C, Walter P (1997) The transmembrane kinase Ire1p is a site-specific endonuclease that initiates mRNA splicing in the unfolded protein response. Cell 90:1031–1039

Sriburi R, Jackowski S, Mori K, Brewer JW (2004) XBP1: a link between the unfolded protein response, lipid biosynthesis, and biogenesis of the endoplasmic reticulum. J Cell Biol 167:35–41

Tirasophon W, Welihinda AA, Kaufman RJ (1998) A stress response pathway from the endoplasmic reticulum to the nucleus requires a novel bifunctional protein kinase/endoribonuclease (Ire1p) in mammalian cells. Genes Dev 12:1812–1824

Travers KJ, Patil CK, Wodicka L, Lockhart DJ, Weissman JS, Walter P (2000) Functional and genomic analyses reveal an essential coordination between the unfolded protein response and ER-associated degradation. Cell 101:249–258

Tsai B, Ye Y, Rapoport TA (2002) Retro-translocation of proteins from the endoplasmic reticulum into the cytosol. Nat Rev Mol Cell Biol 3:246–255

Urano F, Wang X, Bertolotti A, Zhang Y, Chung P, Harding HP, Ron D (2000) Coupling of stress in the ER to activation of JNK protein kinases by transmembrane protein kinase IRE1. Science 287:664–666

Van Anken E, Romijn EP, Maggioni C, Mezghrani A, Sitia R, Braakman I, Heck AJ (2003) Sequential waves of functionally related proteins are expressed when B cells prepare for antibody secretion. Immunity 18:243–253

Wang XZ, Ron D (1996) Stress-induced phosphorylation and activation of the transcription factor CHOP (GADD153) by p38 MAP Kinase. Science 272:1347–1349

Wang XZ, Lawson B, Brewer JW, Zinszner H, Sanjay A, Mi LJ, Boorstein R, Kreibich G, Hendershot LM, Ron D (1996) Signals from the stressed endoplasmic reticulum induce C/EBP-homologous protein (CHOP/GADD153). Mol Cell Biol 16:4273–4280

Wang XZ, Harding HP, Zhang Y, Jolicoeur EM, Kuroda M, Ron D (1998) Cloning of mammalian Ire1 reveals diversity in the ER stress responses. EMBO J 17:5708–5717

Welihinda AA, Kaufman RJ (1996) The unfolded protein response pathway in Saccharomyces cerevisiae. Oligomerization and trans-phosphorylation of Ire1p (Ern1p) are required for kinase activation. J Biol Chem 271:18181–18187

Werner ED, Brodsky JL, McCracken AA (1996) Proteasome-dependent endoplasmic reticulum-associated protein degradation: an unconventional route to a familiar fate. Proc Natl Acad Sci U S A 93:13797–13801

Wiest DL, Burkhardt JK, Hester S, Hortsch M, Meyer DI, Argon Y (1990) Membrane biogenesis during B cell differentiation: most endoplasmic reticulum proteins are expressed coordinately. J Cell Biol 110:1501–1511

Ye J, Rawson RB, Komuro R, Chen X, Dave UP, Prywes R, Brown MS, Goldstein JL (2000) ER stress induces cleavage of membrane-bound ATF6 by the same proteases that process SREBPs. Mol Cell 6:1355–1364

Ye Y, Meyer HH, Rapoport TA (2001) The AAA ATPase Cdc48/p97 and its partners transport proteins from the ER into the cytosol. Nature 414:652–656

Yoneda T, Imaizumi K, Oono K, Yui D, Gomi F, Katayama T, Tohyama M (2001) Activation of caspase-12, an endoplastic reticulum (ER) resident caspase, through tumor necrosis factor receptor-associated Factor 2-dependent mechanism in response to the ER stress. J Biol Chem 276:13935–13940

Yoshida H, Haze K, Yanagi H, Yura T, Mori K (1998) Identification of the cis-acting endoplasmic reticulum stress response element responsible for transcriptional induction of mammalian glucose-regulated proteins. Involvement of basic leucine zipper transcription factors. J Biol Chem 273:33741–33749

Yoshida H, Okada T, Haze K, Yanagi H, Yura T, Negishi M, Mori K (2000) ATF6 activated by Proteolysis Binds in the Presence of NF-Y (CBF) Directly to the cis-acting element responsible for the mammalian unfolded protein response. Mol Cell Biol 20:6755–6767

Yoshida H, Matsui T, Yamamoto A, Okada T, Mori K (2001) XBP1 mRNA is induced by ATF6 and spliced by IRE1 in response to ER stress to produce a highly active transcription factor. Cell 107:881–891

Yoshida H, Matsui T, Hosokawa N, Kaufman RJ, Nagata K, Mori K (2003) A time-dependent phase shift in the mammalian unfolded protein response. Dev Cell 4:265–271

Zhang K, Wong HN, Song B, Miller CN, Scheuner D, Kaufman RJ (2005) The unfolded protein response sensor IRE1alpha is required at 2 distinct steps in B cell lymphopoiesis. J Clin Invest 115:268–281

Zinszner H, Kuroda M, Wang X, Batchvarova N, Lightfoot RT, Remotti H, Stevens JL, Ron D (1998) CHOP is implicated in programmed cell death in response to impaired function of the endoplasmic reticulum. Genes Dev 12:982–995

Molecular Chaperones in Signal Transduction

M. Gaestel

Institute of Biochemistry, Medical School Hannover, Carl-Neuberg-Str. 1, 30625 Hannover, Germany
gaestel.matthias@mh-hannover.de

1	Chaperones and Stress-Activated Signaling Pathways	94
1.1	Heat Shock Proteins and HSF1 Activation	94
1.1.1	Hsp70	94
1.1.2	Hsp90	94
1.2	Unfolded Protein Response	95
1.3	Redox Signaling	95
1.4	Stress-Activated Protein Kinase Cascades	96
2	Chaperones and Signaling by Nuclear Hormone Receptors	97
3	Regulation of Intrinsic Properties of Other Signaling Molecules by Chaperones	99
3.1	Growth Factor Receptors	99
3.2	Protein Kinases	99
3.3	Protein Phosphatases	101
4	Chaperones as Direct and Specific Signaling Molecules?	101
5	New Classes of Chaperones Potentially Involved in Signaling	102
5.1	Histone Chaperones	102
5.2	RNA Chaperones	103
6	Conclusions	104
	References	104

Abstract Many cellular signaling molecules exist in different conformations corresponding to active and inactive states. Transition between these states is regulated by reversible modifications, such as phosphorylation, or by binding of nucleotide triphosphates, their regulated hydrolysis to diphosphates, and their exchange against fresh triphosphates. Specificity and efficiency of cellular signaling is further maintained by regulated subcellular localization of signaling molecules as well as regulated protein–protein interaction. Hence, it is not surprising that molecular chaperones—proteins that are able to specifically interact with distinct conformations of other proteins—could *per se* interfere with cellular signaling. Hence, it is not surprising that chaperones have co-evolved as integral components of signaling networks where they can function in the maturation as well as in regulating the transition between active and inactive state of signaling molecules, such as receptors, transcriptional regulators and protein kinases. Furthermore, new classes of specific chaperones are emerging and their role in histone-mediated chromatin remodeling and RNA folding are under investigation.

Keywords Cell signaling · Receptor · Protein kinase · Transcription factor · Chaperone

1
Chaperones and Stress-Activated Signaling Pathways

1.1
Heat Shock Proteins and HSF1 Activation

1.1.1
Hsp70

Trimerization and activation of heat shock factor 1 (HSF1) is necessary for transcriptional activation of the heat shock response (see also the chapter by R. Voellmy, this volume). Targeted disruption of the HSF1 gene in mice embryonic fibroblasts leads to complete loss of thermotolerance (McMillan et al. 1998), indicating that loss HSF1 is not compensated by other pathways in response to heat shock (see also the chapter by E. Christians and I. Benjamin and the chapter by M. Morange, this volume). While multiple phosphorylations at HSF1 regulate its activity in different ways (Knauf et al. 1996; Holmberg et al. 2001), there is also a clear role of chaperones such as Hsp70 in feedback regulation of HSF1 activation. Preliminary evidence came from biochemical experiments that demonstrate that complexes containing Hsp70 and HSF1 exist in eukaryotic cells (Baler et al. 1992). The subsequent finding that Hsp70–HSF1 interaction is preferentially detected in cells with increased Hsp70 levels after heat shock response suggested Hsp70 as a negative regulator of HSF1 (Abravaya et al. 1992). It was assumed that heat shock gene stimulation by nascent polypeptides and denatured proteins withdraws Hsp70 from its complex with HSF1 and releases HSF1 for further activation. Increased Hsp70 levels as a result of expression of HSF1-dependent genes could then bind and inactivate HSF1 again and provide feedback control of the heat shock response (Morimoto 1993). Interestingly, Hsp70 and its co-chaperone Hdj1 bind to the transactivation domain of HSF1 and directly inhibits transcriptional activation (Shi et al. 1998), providing a plausible molecular mechanism for feedback control of the heat shock response by Hsp70.

At least in the yeast *Saccharomyces cerevisiae*, some observations challenge the involvement of specific Hsp70-like chaperones in HSF1 regulation in this organism (Hjorth-Sorensen et al. 2001).

1.1.2
Hsp90

It turned out that Hsp70 and its co-chaperones are not the only molecules involved in HSF1 regulation. In an *in* vitro system where human HSF1 could be activated by non-native protein or heat treatment, addition of Hsp90 was demonstrated to inhibit HSF1 activation (Zou et al. 1998). Inversely, reduction of the level of Hsp90 dramatically activates HSF1. Most interestingly, geldanamycin, a potent inhibitor of Hsp90, activates HSF1 under conditions in which it acts as an Hsp90-specific reagent (Zou et al. 1998). Since a Hsp90-containing HSF1 complex could be detected in nonstressed cells and its dissociation during stress was demonstrated, it was supposed that Hsp90 is a major repressor of HSF1 (Ali et al. 1998; Zou et al. 1998). Since multiple components of the HSP90 chaperone complex including p23 and immunophilins were demonstrated to bind to HSF1 in vivo (Bharadwaj et al. 1999), it is suggested that HSF1 oligomerization is regulated by a foldosome-type mechanism similar to steroid receptor pathways (see below and the chapter by W.B. Pratt et al., this volume). Further elegant evidence for the physiological role of Hsp90 in regulation of the heat shock response comes from the finding that in scrapie-infected mouse neuroblastoma cells, which fail to induce the expression of Hsp72 and Hsp28 after various stress conditions, geldamycin restores defective heat shock response in vivo (Winklhofer et al. 2001).

Here, it should be mentioned that other HSF1-interacting proteins such as CHIP (C-terminus of Hsp70-interacting protein) (Dai et al. 2003), DAXX (Boellmann et al. 2004), and HSBP1 (heat shock factor binding protein 1) (Satyal et al. 1998) could also positively or negatively affect HSF1 activity. For DAXX, a protein described to stimulate HSF1-mediated transactivation (Boellmann et al. 2004), a specific interaction with the phosphorylated isoforms of the small heat shock protein Hsp27 has been detected (Charette et al. 2000). This indirectly also links small Hsps to HSF1 regulation.

1.2
Unfolded Protein Response

While HSF1 activation can be triggered by misfolded proteins located in the cytoplasm, misfolding of proteins to be secreted in the lumen of the endoplasmic reticulum (ER) stimulates another cellular signaling pathway, namely the unfolded protein response (UPR). In regulation of UPR, the ER homolog of Hsp70, BiP, is at least indirectly involved (see the chapter by K. Zhang and R.J. Kaufman, this volume).

1.3
Redox Signaling

Activation of the heat shock response by diverse extracellular stresses can be explained by the fact that all these conditions lead to misfolding of intracellular proteins and to activation of HSF1 by the feedback mechanism discussed in Sect. 1.1. However, there is evidence that stress stimuli such as hypoxia or sodium arsenite treatment require specific cellular pathways for complete activation of the heat shock response. In this regard, it was shown that the small GTP-ase rac1 and rac1-regulated reactive oxygen species (ROS) are necessary for redox-dependent HSF1 activation (Ozaki et al. 2000). Interestingly, mouse heat shock transcription factor 1 deficiency alters cardiac redox homeostasis and increases mitochondrial oxidative damage. This is mainly due to a lower intracellular glutathione (GSH)/glutathione disulfate (GSSG) ratio and demonstrates that HSF1 is also necessary to maintain redox homeostasis and antioxidative defenses at physiological temperatures (Yan et al. 2002).

In bacteria, Hsp33 is a cytoplasmic protein sensing redox conditions by highly reactive cysteins, which bind a Zn^{2+} ion in the reduced state. Oxidizing conditions, such as H_2O_2 treatment, cause intramolecular disulfide formation, Zn^{2+} release, and activation of the chaperone function of Hsp33 (Jakob et al. 1999; Hoffmann et al. 2004). So far, it is not completely clear whether eukaryotic Hsps can be regulated by redox signals and undergo cysteine oxidation (Sitia and Molteni 2004). Although the molecular mechanism is not understood so far, the fact that Hsp72 displays enhanced binding to DHFR upon oxidative exposure and protects this enzyme against irreversible modification (Musch et al. 2004) could support this idea.

1.4
Stress-Activated Protein Kinase Cascades

In yeast and mammalian cells, parallel protein kinase cascades exist that connect extracellular signals to various specific intracellular responses. Central elements of these cascades are the mitogen-activated protein kinases/extracellular-regulated kinases, MAPKs or ERKs, and the stress-activated protein kinases, SAPKs, which can be subdivided into two groups, the jun-N-terminal kinases (JNKs) and the 38-kDa SAPKs (p38). A major substrate of the p38 kinase cascade, which is activated by heat shock, changes in osmolarity, UV treatment, various chemicals such as arsenite and anisomycin as well as by bacterial lipopolysaccharide and chemotactic peptides, is the small heat shock protein Hsp25 (mouse)/Hsp27 (human). Hsp25/27 is directly phosphorylated by the p38-activated protein kinase MK2 (MAPK-activated protein kinase 2, MAPKAPK 2) (Stokoe et al. 1992). Phosphorylation of Hsp25/27 changes its intracellular oligomerization and stimulates its ATP-independent chaperone properties in vitro (Lambert et al. 1999; Rogalla et al. 1999). Furthermore, Hsp25 may contribute in a phosphorylation-dependent and, hence, signal-dependent manner to stress-dependent actin remodeling (Guay et al. 1997), to proteasome-targeting of specific signaling proteins such as IkB (Parcellier et al. 2003) to activation of the IKK complex (Park et al. 2003) and to recovery of mRNA translation following cellular stress by association with the eIF4F complex (Cowan and Morley 2004). The role of small heat shock protein and its phosphorylation in apoptotic signaling is discussed in the chapter by C. Didelot et al.

αB-crystallin, another member of the small heat shock protein family, is a substrate targeted by different kinase cascades. While phosphorylation by the classical MAPK/ERK cascade is responsible for increased phosphorylation of serine residue 45, serine 59 in αB-crystallin is phosphorylated by the p38 cascade (Kato et al. 1998). As for Hsp25/27, phosphorylation leads to changes in oligomerization of αB-crystallin and alters chaperone properties (Ito et al. 2001). A phosphorylation-dependent function of αB-crystallin in protein degradation has been proposed (den Engelsman et al. 2004).

It should be mentioned that besides the chaperone-kinase cascade connection, several cross-talks between the transcriptional-regulated heat shock response and different protein kinase cascades has been described (Kim et al. 1997). Downregulation of HSF1 activity at control temperatures proceeds by constitutive phosphorylation at two serine residues (S303, 307) and this phosphorylation is further increased by stimulation of the MAPK/ERK kinase cascade together with glycogen synthase kinase 3 (GSK3), providing a mechanism to maintain the physiological balance between heat shock response and cell proliferation (Chu et al. 1996; Knauf et al. 1996; Kline and Morimoto 1997). A similar HSF1-activity-suppressing effect is also described for phosphorylation by the SAPK JNK (Dai et al. 2000). Physical association of the MAPK ERK1 with HSF1 promotes kinase activity of ERK in heat-shocked cells and, together with 14-3-3 binding to HSF1, enhances cytoplasmic localization of HSF1 (Wang et al. 2004). This could contribute to HSF1 deactivation in cells recovering from stress.

In contrast to the role of ERK1/GSK3, calcium/calmodulin-dependent kinase, casein kinase 2 and polo-like kinase phosphorylate HSF1 at S230, T142, and T419, respectively, and stimulate HSF1 activation and nuclear translocation (Holmberg et al. 2001; Soncin et al. 2003; Kim et al. 2005b).

2
Chaperones and Signaling by Nuclear Hormone Receptors

Similar to their function in feedback control of the heat shock response by chaperoning HSF1, Hsp70, Hsp90, and its co-chaperones are also involved in chaperoning and regulating nuclear hormone receptors. Association of the heat shock protein hsp90 with steroid hormone receptor products was described two decades ago (Catelli et al. 1985; Sanchez et al. 1985). Since then, a complex scenario of different steps in hormone receptor maturation, activation, and deactivation by chaperones has emerged (see the chapter by W.B. Pratt et al., this volume). Maturation of hormone receptor starts with binding of Hsp70 and Hsp40 to the nascent aporeceptor at the ribosome and the subsequent transfer to Hsp90 through the adapter protein Hop (Kosano et al.

1998). Replacement of Hop by p23 and binding of immunophilins then leads to the complete aporeceptor complex containing a dimer of Hsp90 (Young and Hartl 2002). Upon hormone binding, the aporeceptor complex falls apart and the hormone-bound DNA-binding and transactivating receptor is released and translocated to the respective hormone response elements at the DNA. PPIase activity of interacting immunophilins such as FKBP51 contributes to transactivation properties of the protein–hormone complex (Riggs et al. 2003). This demonstrates a role of prolyl-*cis-trans*-isomerization in associated proteins for transactivation. In addition, the PPIase domain of receptor complex bound immunophilins and immunophilin-like tetratricopeptide repeat domain proteins such as protein phosphatase 5 (Silverstein et al. 1997) can specifically interact with the cytoplasmic transport protein dynein, which could be responsible for retrograde transport of the receptor complex to the nuclear pore (Pratt et al. 2004).

Since the physiologically relevant hormone response in most cases should be transient, downregulation often proceeds within minutes and involves p23 and Hsp90 action. The binding of p23 and Hsp90 to the DNA-bound hormone receptor leads to disassembly of hormone receptor complex and release from DNA (Freeman et al. 2000). In addition, a transcriptional repressing activity in *cis* at the hormone responsive and other promoters could be measured for p23 (Freeman and Yamamoto 2002; Morimoto 2002), suggesting that this co-chaperone could be involved in transcription factor complexes and broadly contribute to transcriptional regulation under certain conditions.

An isoform of the protein BAG-1 (Bcl2-associated anhanogene-1; Takayama et al. 1995), BAG-1M, was initially identified as a glucocorticoid receptor (GR) binding protein, which translocates to the nucleus and preferentially interacts with the activated receptor (Zeiner and Gehring 1995; Zeiner et al. 1999). It turned out that BAG-1 is a ADP-exchange factor for Hsp70 and Hsc70 that probably acts similarly to the GrpE-protein for DnaK, the prokaryotic homolog of Hsp70 (Hohfeld and Jentsch 1997; Takayama et al. 1997). Interestingly, Hsp70 together with the BAG-1M or a shorter isoform, BAG-1L, also contribute to hormone receptor regulation: BAG-1M,L interact with the hinge region of the GR and inhibit dexamethasone-induced receptor-mediated transcription (Schneikert et al. 1999). Possibly, BAG-1–Hsp70 complexes are involved in downregulation of the activated GR (Nollen and Morimoto 2002). In contrast, BAG-1L, probably together with Hsp/Hsc70, interacts with the androgen receptor and vitamin D_3 receptor and stimulates transcriptional activation of respective genes (Froesch et al. 1998; Guzey et al. 2000).

3
Regulation of Intrinsic Properties of Other Signaling Molecules by Chaperones

3.1
Growth Factor Receptors

Preliminary evidence for a connection between specific transmembrane receptors with components of the chaperone machinery comes from the observation that BAG-1, an ADP exchange factor for Hsp70 and Hsc70, interacts with the intracellular domain of the hepatocyte growth factor (HGF) receptor and cooperates in HGF-induced protection from apoptosis (Bardelli et al. 1996). Another co-chaperone of Hsp70, Tid1, containing a DnaJ domain interacts with the cytoplasmic domain of the receptor tyrosine kinase ErbB-2 and modulates the uncontrolled proliferation of ErbB-2-overexpressing carcinoma cells by reducing ErbB-2 expression and tumor progression. Since a functional DnaJ domain of Tid1 is required for the effects, the co-chaperone function of Tid1 on HSP70 most likely plays an essential role (Kim et al. 2004b).

The ErbB-2 growth factor receptor is not only a target for Hsp70 co-chaperones but also for Hsp90. Hsp90 binds to a specific loop within the kinase domain of the receptor and stabilizes ErbB-2. Inversely, the Hsp90 inhibitor geldanamycin disrupts ErbB-2/Hsp90 association and stimulates ErbB2 proteasomal degradation (Xu et al. 2001). In addition to stabilization, there is also an inhibitory effect of Hsp90 on catalytic function and ligand-induced receptor heterodimerization of ErbB-2 (Citri et al. 2004). In contrast, Hsp90 binding seems to be necessary for effective signal transduction of the VEGF-receptor 2 to focal adhesion kinase (FAK), since inhibition of HSP90 binding to a region in the last 130 amino acids of VEGFR2 blocks VEGF-dependent phosphorylation of FAK (Le Boeuf et al. 2004). Obviously, the effect of Hsp90-binding on receptor signaling strongly depends on the receptor region(s) involved.

3.2
Protein Kinases

A first specific interaction between a protein kinase, the viral oncogenic tyrosine kinase v-src, and Hsp90 has been described by Brugge et al. (1981) and Oppermann et al. (1981). Meanwhile, more than 100 client proteins, mainly transcription factors and protein kinases, have been identified to be regulated by Hsp90 containing protein complexes (Pratt and Toft 2003). While Hsp90–p23–immunophilin complexes seem specific for transcription factors (see above), protein kinases are often bound by a Hsp90–p23–p50cdc37 complex, making p50cdc37 a kinase-targeting "subunit" of Hsp90 (Kimura et al. 1997). Examples of protein kinase regulation by Hsp90 are the oncogene raf-1, several tyrosine kinases, CK2, PKR, Ste11, and IKKs.

Raf-1 acts as a central element in the mitogen-activated signaling pathway, which was demonstrated to be compromised by deletion of Hsp90 and cdc37 in *Drosophila* (Cutforth and Rubin 1994), and Ste11, a yeast equivalent of Raf, forms complexes with wild-type Hsp90 (Louvion et al. 1998). In mammalian cells, a Hsp90–p23–p50cdc37 complex is necessary for raf-1 maturation and subsequent activation after release from the complex (Grammatikakis et al. 1999). Interestingly, binding of the co-chaperone BAG-1 stabilizes the active conformation of raf-1 (Wang et al. 1996) and competitive binding of BAG-1 to Hsp70 is involved in raf-1 inactivation (Song et al. 2001; Nollen and Morimoto 2002).

Recently, Hsp90 and p50cdc37 have been isolated as binding partners of the protein kinase IRAK1-1 (De Nardo et al. 2005), a component downstream from the interleukin-1 receptor and also involved in toll-like receptor (TLR) signaling via TRAF6. Pharmacologic inhibition of Hsp90 leads to destabilization of IRAK-1 in macrophages (De Nardo et al. 2005). Hence, Hsp90 can be involved in regulation of TLR-signaling and immunotolerance.

PKR is a protein kinase activated by intracellular occurrence of double-stranded (ds) RNA as a result of virus infection. By phosphorylation of the eukaryotic initiation factor eIF-2α, PKR inhibits protein biosynthesis of the virus-infected cell and contributes to limiting virus infection. Hsp90 and p23 bind to PKR through its N-terminal dsRNA binding region as well as through its kinase domain and stabilize the inactive conformation of PKR. Both dsRNA and geldanamycin induce the rapid dissociation of Hsp90 and p23 from mature PKR and activate PKR both in vivo and in vitro within minutes (Donze et al. 2001).

In contrast to PKR, Hsp90-binding is also able to activate protein kinases. Casein kinase 2 (CK2), an enzyme implicated in critical cellular processes such as proliferation, apoptosis, differentiation, and transformation, is protected from self-aggregation by Hsp90-binding and its kinase activity is stimulated (Miyata and Yahara 1992). Interestingly, CK2 can phosphorylate the co-chaperone p50cdc37 at a specific serine residue (S13). This phosphorylation enhances binding of p50cdc37 and Hsp90 to other protein kinases and, hence, CK2 may act as a master regulator for Hsp90-mediated kinase chaperoning (Miyata and Nishida 2004). Recent findings indicate that Hsp90-association regulates many relevant processes such as completion of mitosis by Polo-like kinase 1 (de Carcer 2004) and inflammatory response via I-kappa-B kinases (Broemer et al. 2004).

In addition to reversible modification of signaling molecules under certain cellular situations where sustained activation or inactivation of a pathway is required, for example, in specific phases of the cell cycle, regulated stability and degradation of signaling components such as protein kinases or transcriptional regulators is observed. Hsp90 contributes to stabilization not only of receptors such as ErbB-2, but also of protein kinases. Different isoforms of the cyclin-dependent kinase 11, CDK11p46 and CDK11p110, are stabilized by Hsp90-binding, since treatment of cells with geldanamycin leads to ubiquitination and enhanced degradation of both CDK11p110 and CDK11p46 through proteasome-dependent pathways (Mikolajczyk and Nelson 2004). Nucleophosmin-anaplastic lymphoma kinase (NPM-ALK), a constitutively active fusion tyrosine kinase involved in lymphomagenesis of human anaplastic large cell lymphomas, is also stabilized by Hsp90. When Hsp90 is inhibited by geldanamycin, a rapid Hsp70-assisted ubiquitin-dependent proteasomal degradation of this kinase is observed (Bonvini et al. 2004).

Besides Hsp90, other chaperones may also contribute to regulation of protein kinase activity: Hsc70 associates with newly synthesized cyclin D1 and is an essential component of the mature, active cyclin D1–CDK4 holoenzyme complex (Diehl et al. 2003). Hsp27 associates with the I-kappa-B kinase (IKK) complex, and p38-dependent phosphorylation of Hsp27 enhanced its association with IKKβ and resulted in decreased IKK activity (Park et al. 2003). Another small Hsp, HspB2, has been described to be specifically associated with myotonic dystrophy protein kinase (DMPK) in muscle cells and, hence, is also designated myotonic dystrophy protein kinase-binding protein, MKBP (Suzuki et al. 1998). Interestingly, MKBP enhances the kinase activity of DMPK and protects it from heat-induced inactivation in vitro, suggesting that MKBP also participates in a novel stress-responsive muscle cell-specific regulation in vivo.

3.3
Protein Phosphatases

Protein phosphatase 5 (PP5) is a serine/threonine protein phosphatase proposed to participate in several signaling pathways of mammalian cells. Interestingly, PP5 binds to Hsp90 via its tetratricopeptide repeat (TPR) domain and is co-purified with the glucocorticoid receptor chaperone complex (Silverstein et al. 1997). Furthermore, loss of interaction of Hsp90 and Hsp70 with PP5 leads to an increase of PP5 activity paralleled by proteolytic cleavage of the N- and C-termini, with the subsequent appearance of high-molecular-mass species. This indicates that PP5 is activated by proteolysis on dissociation from Hsps, and is destroyed via the proteasome after ubiquitination (Zeke et al. 2005).

4
Chaperones as Direct and Specific Signaling Molecules?

An interesting example for a folding catalyst that acts in a signal-specific manner is the protein-prolyl-isomerase Pin1 (see also the chapter by F. Edlich and G. Fischer, this volume). Pin1 acts as a phosphorylation-dependent PPIase that specifically recognizes and isomerizes the phosphoserine-proline or phosphothreonine-proline bonds present in mitotic phosphoproteins (Yaffe et al. 1997). Hence, this enzyme establishes a direct connection between protein phosphorylation and chaperone catalyzed conformation.

For the small heat shock protein Hsp27, several specific functions in apoptotic signaling are described (see also the chapter by C. Didelot et al., this volume). Besides its specific interaction with cytochrome C (Bruey et al. 2000), Hsp27 could also be involved in targeting of IkappaB to proteasomal degradation, indicating a direct function in NFkappaB signaling (Parcellier et al. 2003).

For several small heat shock proteins, αA-, αB-crystallin, and Hsp22, a direct signaling action as protein kinase is proposed. The observation that total bovine alpha-crystallin or its isolated polypeptides can autophosphorylate serine residues by a cAMP-independent mechanism in the presence of Mg^{2+} (Kantorow and Piatigorsky 1994) and that αA-crystallin phosphorylation is increased after disintegration of the large protein complexes (Kantorow et al. 1995) first indicated an auto-kinase function of these proteins. Since small heat shock proteins do not carry the conserved catalytic domain of protein kinases, the mechanism by which some small Hsps could act as kinase are not clear. For Hsp22, also designated H11, some weak similarity to the conserved eukaryotic protein kinase domain was proposed and Hsp22-autophosphorylation was shown to depend on a lysine residue aligned to protein kinase catalytic subdomain II (Smith et al. 2000). Whether these data are sufficient for making Hsp22 a protein kinase is challenged by other findings (Kim et al. 2004a).

5
New Classes of Chaperones Potentially Involved in Signaling

5.1
Histone Chaperones

Chromatin assembly and dynamics is important for replication and transcription and is regulated by chromatin remodeling complexes and histone modification as well as by specific chaperones, which interact with histones in the context of the nucleosome. The first protein described as histone chaperone is nucleoplasmin, which is a typical acceptor for histone storage before DNA deposition (Philpott et al. 1991). However, many proteins have been identified that escort and regulate histone nuclear translocation, DNA deposition, and chromatin remodeling in a chaperone-like manner (reviewed in Loyola and Almouzni 2004). It is highly probable that histone chaperones interfere with histone modifications and chromatin remodeling and thus transfer further information into transcriptional regulation.

5.2
RNA Chaperones

Even relatively small RNAs are able to fold in many different ways because of structural promiscuity and may be trapped kinetically during folding. It is assumed that protein-RNA interactions are essential for reaching the native conformation of a RNA molecule. RNA chaperones can act in at least two different ways, by destabilizing misfolded RNA structures or by stabilizing or inducing otherwise unstable active conformations of RNA (reviewed in Herschlag 1995; Lorsch 2002). Examples for destabilizing chaperones are members of the DEAD box class of RNA-dependent ATPases, and hnRNP A1 (Pontius and Berg 1992). The existence of structure-stabilizing chaperones has been demonstrated for group I introns in neurospora crassa (Cyt-18) and yeast

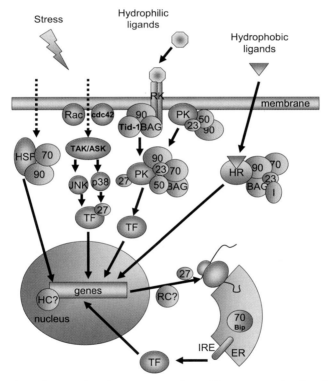

Fig. 1 Schematic and comprehensive representation of chaperone action in cellular signaling. Chaperones: *90*, Heat shock protein (Hsp)90; *70*, Hsp70; *27*, Hsp25/27; *HC*, histone chaperones; *RC*, RNA chaperones. Co-chaperones: *23*, p23; *50*, p50cdc37; *I*, immunophilin; *Tid-1*; *BAG*. Signaling molecules: *RK*, receptor kinase; *HR*, hormone receptor; *PK*, protein kinase; *TF*, transcription factor; *HSF*, heat shock factor 1; *IRE*, sensor kinase and RNase. Others: *ER*, endoplasmic reticulum

(CBP2). Taking into account the possibility that RNA chaperones could be involved in differential splicing or that RNA structure per se can act as enzymes, metabolic sensors, and translational regulators (Winkler et al. 2002), a role for RNA chaperones in cellular signaling is highly probable (Fig. 1).

6
Conclusions

It is not unexpected that the complexity of cellular signaling and the heterogeneity of established and postulated chaperone interactions at different levels makes the influence of specific chaperones on global cellular signaling difficult to mechanistically understand. Especially, the role of chaperones and heat shock on apoptosis is rather diverse and complex, leading to both pro- and anti-apoptotic effects (Chant et al. 1996; DeMeester et al. 1998; Charette et al. 2000; Xia et al. 2000; Ran et al., 2004, Mikolajczyk and Nelson 2004). This complexity becomes obvious, for example for the co-chaperone BAG-1, which not only interacts with GR, Hsp70, or raf-1, but also has a well-established role in inhibition of apoptosis by interaction with Bcl-2 (Takayama et al. 1995). On the other hand, chaperones not only influence signaling, but, vice versa, signals and signaling molecules may also affect chaperones. In addition to the paradigm of HSF1 activation in response to stress signals, recent data demonstrate that other signaling molecules, such as Fas-associated factor 1, may regulate chaperone activity of heat shock proteins, such as Hsp70 (Kim et al. 2005a). Finally, it is not surprising that mutations in Hsp-genes, such as Hsp22 and Hsp27, could have broad pathophysiological consequences probably due to impaired signaling also in humans (Evgrafov et al. 2004; Irobi et al. 2004).

References

Abravaya K, Myers MP, Murphy SP, Morimoto RI (1992) The human heat shock protein hsp70 interacts with HSF, the transcription factor that regulates heat shock gene expression. Genes Dev 6:1153–1164
Ali A, Bharadwaj S, O'Carroll R, Ovsenek N (1998) HSP90 interacts with and regulates the activity of heat shock factor 1 in Xenopus oocytes. Mol Cell Biol 18:4949–4960
Baler R, Welch WJ, Voellmy R (1992) Heat shock gene regulation by nascent polypeptides and denatured proteins: hsp70 as a potential autoregulatory factor. J Cell Biol 117:1151–1159
Bardelli A, Longati P, Albero D, Goruppi S, Schneider C, Ponzetto C, Comoglio PM (1996) HGF receptor associates with the anti-apoptotic protein BAG-1 and prevents cell death. Embo J 15:6205–6212
Bharadwaj S, Ali A, Ovsenek N (1999) Multiple components of the HSP90 chaperone complex function in regulation of heat shock factor 1 In vivo. Mol Cell Biol 19:8033–8041
Boellmann F, Guettouche T, Guo Y, Fenna M, Mnayer L, Voellmy R (2004) DAXX interacts with heat shock factor 1 during stress activation and enhances its transcriptional activity. Proc Natl Acad Sci U S A 101:4100–4105. Epub 2004 Mar 4111

Bonvini P, Dalla Rosa H, Vignes N, Rosolen A (2004) Ubiquitination and proteasomal degradation of nucleophosmin-anaplastic lymphoma kinase induced by 17-allylamino-demethoxygeldanamycin: role of the co-chaperone carboxyl heat shock protein 70-interacting protein. Cancer Res 64:3256–3264

Broemer M, Krappmann D, Scheidereit C (2004) Requirement of Hsp90 activity for IkappaB kinase (IKK) biosynthesis and for constitutive and inducible IKK and NF-kappaB activation. Oncogene 23:5378–5386

Bruey JM, Ducasse C, Bonniaud P, Ravagnan L, Susin SA, Diaz-Latoud C, Gurbuxani S, Arrigo AP, Kroemer G, Solary E, Garrido C (2000) Hsp27 negatively regulates cell death by interacting with cytochrome c. Nat Cell Biol 2:645–652

Brugge JS, Erikson E, Erikson RL (1981) The specific interaction of the Rous sarcoma virus transforming protein, pp60src, with two cellular proteins. Cell 25:363–372

Catelli MG, Binart N, Jung-Testas I, Renoir JM, Baulieu EE, Feramisco JR, Welch WJ (1985) The common 90-kd protein component of non-transformed '8S' steroid receptors is a heat-shock protein. EMBO J 4:3131–3135

Charette SJ, Lavoie JN, Lambert H, Landry J (2000) Inhibition of Daxx-mediated apoptosis by heat shock protein 27. Mol Cell Biol 20:7602–7612

Chu B, Soncin F, Price BD, Stevenson MA, Calderwood SK (1996) Sequential phosphorylation by mitogen-activated protein kinase and glycogen synthase kinase 3 represses transcriptional activation by heat shock factor-1. J Biol Chem 271:30847–30857

Citri A, Gan J, Mosesson Y, Vereb G, Szollosi J, Yarden Y (2004) Hsp90 restrains ErbB-2/HER2 signalling by limiting heterodimer formation. EMBO Rep 5:1165–1170

Cowan JL, Morley SJ (2004) The proteasome inhibitor, MG132, promotes the reprogramming of translation in C2C12 myoblasts and facilitates the association of hsp25 with the eIF4F complex. Eur J Biochem 271:3596–3611

Cutforth T, Rubin GM (1994) Mutations in Hsp83 and cdc37 impair signaling by the sevenless receptor tyrosine kinase in Drosophila. Cell 77:1027–1036

Dai Q, Zhang C, Wu Y, McDonough H, Whaley RA, Godfrey V, Li HH, Madamanchi N, Xu W, Neckers L, Cyr D, Patterson C (2003) CHIP activates HSF1 and confers protection against apoptosis and cellular stress. EMBO J 22:5446–5458

Dai R, Frejtag W, He B, Zhang Y, Mivechi NF (2000) c-Jun NH2-terminal kinase targeting and phosphorylation of heat shock factor-1 suppress its transcriptional activity. J Biol Chem 275:18210–18218

De Carcer G (2004) Heat shock protein 90 regulates the metaphase-anaphase transition in a polo-like kinase-dependent manner. Cancer Res 64:5106–5112

De Nardo D, Masendycz P, Ho S, Cross M, Fleetwood AJ, Reynolds EC, Hamilton JA, Scholz GM (2005) A central role for the Hsp90middle dotCdc37 molecular chaperone module in interleukin-1 receptor-associated-kinase-dependent signaling by toll-like receptors. J Biol Chem 280:9813–9822. Epub 2005 Jan 9812

Den Engelsman J, Bennink EJ, Doerwald L, Onnekink C, Wunderink L, Andley UP, Kato K, de Jong WW, Boelens WC (2004) Mimicking phosphorylation of the small heat-shock protein alphaB-crystallin recruits the F-box protein FBX4 to nuclear SC35 speckles. Eur J Biochem 271:4195–4203

Diehl JA, Yang W, Rimerman RA, Xiao H, Emili A (2003) Hsc70 regulates accumulation of cyclin D1 and cyclin D1-dependent protein kinase. Mol Cell Biol 23:1764–1774

Donze O, Abbas-Terki T, Picard D (2001) The Hsp90 chaperone complex is both a facilitator and a repressor of the dsRNA-dependent kinase PKR. EMBO J 20:3771–3780

Evgrafov OV, Mersiyanova I, Irobi J, Van Den Bosch L, Dierick I, Leung CL, Schagina O, Verpoorten N, Van Impe K, Fedotov V, Dadali E, Auer-Grumbach M, Windpassinger C, Wagner K, Mitrovic Z, Hilton-Jones D, Talbot K, Martin JJ, Vasserman N, Tverskaya S, Polyakov A, Liem RK, Gettemans J, Robberecht W, De Jonghe P, Timmerman V (2004) Mutant small heat-shock protein 27 causes axonal Charcot-Marie-Tooth disease and distal hereditary motor neuropathy. Nat Genet 36:602–606. Epub 2004 May 2002

Freeman BC, Felts SJ, Toft DO, Yamamoto KR (2000) The p23 molecular chaperones act at a late step in intracellular receptor action to differentially affect ligand efficacies. Genes Dev 14:422–434

Freeman BC, Yamamoto KR (2002) Disassembly of transcriptional regulatory complexes by molecular chaperones. Science 296:2232–2235

Froesch BA, Takayama S, Reed JC (1998) BAG-1L protein enhances androgen receptor function. J Biol Chem 273:11660–11666

Grammatikakis N, Lin JH, Grammatikakis A, Tsichlis PN, Cochran BH (1999) p50(cdc37) acting in concert with Hsp90 is required for Raf-1 function. Mol Cell Biol 19:1661–1672

Guay J, Lambert H, Gingras-Breton G, Lavoie JN, Huot J, Landry J (1997) Regulation of actin filament dynamics by p38 map kinase-mediated phosphorylation of heat shock protein 27. J Cell Sci 110:357–368

Guzey M, Takayama S, Reed JC (2000) BAG1L enhances trans-activation function of the vitamin D receptor. J Biol Chem 275:40749–40756

Herschlag D (1995) RNA chaperones and the RNA folding problem. J Biol Chem 270:20871–20874

Hjorth-Sorensen B, Hoffmann ER, Lissin NM, Sewell AK, Jakobsen BK (2001) Activation of heat shock transcription factor in yeast is not influenced by the levels of expression of heat shock proteins. Mol Microbiol 39:914–923

Hoffmann JH, Linke K, Graf PC, Lilie H, Jakob U (2004) Identification of a redox-regulated chaperone network. EMBO J 23:160–168. Epub 2003 Dec 2011

Hohfeld J, Jentsch S (1997) GrpE-like regulation of the hsc70 chaperone by the anti-apoptotic protein BAG-1. EMBO J 16:6209–6216

Holmberg CI, Hietakangas V, Mikhailov A, Rantanen JO, Kallio M, Meinander A, Hellman J, Morrice N, MacKintosh C, Morimoto RI, Eriksson JE, Sistonen L (2001) Phosphorylation of serine 230 promotes inducible transcriptional activity of heat shock factor 1. EMBO J 20:3800–3810

Irobi J, Van Impe K, Seeman P, Jordanova A, Dierick I, Verpoorten N, Michalik A, De Vriendt E, Jacobs A, Van Gerwen V, Vennekens K, Mazanec R, Tournev I, Hilton-Jones D, Talbot K, Kremensky I, Van Den Bosch L, Robberecht W, Van Vandekerckhove J, Broeckhoven C, Gettemans J, De Jonghe P, Timmerman V (2004) Hot-spot residue in small heat-shock protein 22 causes distal motor neuropathy. Nat Genet 36:597–601. Epub 2004 May 2002

Ito H, Kamei K, Iwamoto I, Inaguma Y, Nohara D, Kato K (2001) Phosphorylation-induced change of the oligomerization state of alpha B-crystallin. J Biol Chem 276:5346–5352. Epub 2000 Nov 5328

Jakob U, Muse W, Eser M, Bardwell JC (1999) Chaperone activity with a redox switch. Cell 96:341–352

Kantorow M, Piatigorsky J (1994) Alpha-crystallin/small heat shock protein has autokinase activity. Proc Natl Acad Sci U S A 91:3112–3116

Kantorow M, Horwitz J, van Boekel MA, de Jong WW, Piatigorsky J (1995) Conversion from oligomers to tetramers enhances autophosphorylation by lens alpha A-crystallin. Specificity between alpha A- and alpha B-crystallin subunits. Alpha-crystallin/small heat shock protein has autokinase activity. J Biol Chem 270:17215–17220

Kato K, Ito H, Kamei K, Inaguma Y, Iwamoto I, Saga S (1998) Phosphorylation of alphaB-crystallin in mitotic cells and identification of enzymatic activities responsible for phosphorylation. J Biol Chem 273:28346–28354

Kim HJ, Song EJ, Lee YS, Kim E, Lee KJ (2005a) Human Fas-associated factor 1 interacts with heat shock protein 70 and negatively regulates chaperone activity. J Biol Chem 280:8125–8133. Epub 2004 Dec 8113

Kim J, Nueda A, Meng YH, Dynan WS, Mivechi NF (1997) Analysis of the phosphorylation of human heat shock transcription factor-1 by MAP kinase family members. J Cell Biochem 67:43–54

Kim MV, Seit-Nebi AS, Gusev NB (2004a) The problem of protein kinase activity of small heat shock protein Hsp22 (H11 or HspB8). Biochem Biophys Res Commun 325:649–652

Kim SA, Yoon JH, Lee SH, Ahn SG (2005b) Polo-like kinase 1 phosphorylates HSF1 and mediates its nuclear translocation during heat stress. J Biol Chem 20:20

Kim SW, Chao TH, Xiang R, Lo JF, Campbell MJ, Fearns C, Lee JD (2004b) Tid1, the human homologue of a Drosophila tumor suppressor, reduces the malignant activity of ErbB-2 in carcinoma cells. Cancer Res 64:7732–7739

Kimura Y, Rutherford SL, Miyata Y, Yahara I, Freeman BC, Yue L, Morimoto RI, Lindquist S (1997) Cdc37 is a molecular chaperone with specific functions in signal transduction. Genes Dev 11:1775–1785

Kline MP, Morimoto RI (1997) Repression of the heat shock factor 1 transcriptional activation domain is modulated by constitutive phosphorylation. Mol Cell Biol 17:2107–2115

Knauf U, Newton EM, Kyriakis J, Kingston RE (1996) Repression of human heat shock factor 1 activity at control temperature by phosphorylation. Genes Dev 10:2782–2793

Kosano H, Stensgard B, Charlesworth MC, McMahon N, Toft D (1998) The assembly of progesterone receptor-hsp90 complexes using purified proteins. J Biol Chem 273:32973–32979

Lambert H, Charette SJ, Bernier AF, Guimond A, Landry J (1999) HSP27 multimerization mediated by phosphorylation-sensitive intermolecular interactions at the amino terminus. J Biol Chem 274:9378–9385

Le Boeuf F, Houle F, Huot J (2004) Regulation of vascular endothelial growth factor receptor 2-mediated phosphorylation of focal adhesion kinase by heat shock protein 90 and Src kinase activities. J Biol Chem 279:39175–39185. Epub 32004 Jul 39106

Lorsch JR (2002) RNA chaperones exist and DEAD box proteins get a life. Cell 109:797–800

Louvion JF, Abbas-Terki T, Picard D (1998) Hsp90 is required for pheromone signaling in yeast. Mol Biol Cell 9:3071–3083

Loyola A, Almouzni G (2004) Histone chaperones, a supporting role in the limelight. Biochim Biophys Acta 1677:3–11

McMillan DR, Xiao X, Shao L, Graves K, Benjamin IJ (1998) Targeted disruption of heat shock transcription factor 1 abolishes thermotolerance and protection against heat-inducible apoptosis. J Biol Chem 273:7523–7528

Mikolajczyk M, Nelson MA (2004) Regulation of stability of cyclin-dependent kinase CDK11p110 and a caspase-processed form, CDK11p46, by Hsp90. Biochem J 384:461–467

Miyata Y, Yahara I (1992) The 90-kDa heat shock protein, HSP90, binds and protects casein kinase II from self-aggregation and enhances its kinase activity. J Biol Chem 267:7042–7047

Miyata Y, Nishida E (2004) CK2 controls multiple protein kinases by phosphorylating a kinase-targeting molecular chaperone, Cdc37. Mol Cell Biol 24:4065–4074

Morimoto RI (1993) Cells in stress: transcriptional activation of heat shock genes. Science 259:1409–1410

Morimoto RI (2002) Dynamic remodeling of transcription complexes by molecular chaperones. Cell 110:281–284

Musch MW, Kapil A, Chang EB (2004) Heat shock protein 72 binds and protects dihydrofolate reductase against oxidative injury. Biochem Biophys Res Commun 313:185–192

Nollen EA, Morimoto RI (2002) Chaperoning signaling pathways: molecular chaperones as stress-sensing 'heat shock' proteins. J Cell Sci 115:2809–2816

Oppermann H, Levinson AD, Levintow L, Varmus HE, Bishop JM, Kawai S (1981) Two cellular proteins that immunoprecipitate with the transforming protein of Rous sarcoma virus. Virology 113:736–751

Ozaki M, Deshpande SS, Angkeow P, Suzuki S, Irani K (2000) Rac1 regulates stress-induced, redox-dependent heat shock factor activation. J Biol Chem 275:35377–35383

Parcellier A, Schmitt E, Gurbuxani S, Seigneurin-Berny D, Pance A, Chantome A, Plenchette S, Khochbin S, Solary E, Garrido C (2003) HSP27 is a ubiquitin-binding protein involved in I-kappaBalpha proteasomal degradation. Mol Cell Biol 23:5790–5802

Park KJ, Gaynor RB, Kwak YT (2003) Heat shock protein 27 association with the I kappa B kinase complex regulates tumor necrosis factor alpha-induced NF-kappa B activation. J Biol Chem 278:35272–35278. Epub 32003 Jun 35225

Philpott A, Leno GH, Laskey RA (1991) Sperm decondensation in Xenopus egg cytoplasm is mediated by nucleoplasmin. Cell 65:569–578

Pontius BW, Berg P (1992) Rapid assembly and disassembly of complementary DNA strands through an equilibrium intermediate state mediated by A1 hnRNP protein. J Biol Chem 267:13815–13818

Pratt WB, Galigniana MD, Harrell JM, DeFranco DB (2004) Role of hsp90 and the hsp90-binding immunophilins in signalling protein movement. Cell Signal 16:857–872

Rogalla T, Ehrnsperger M, Preville X, Kotlyarov A, Lutsch G, Ducasse C, Paul C, Wieske M, Arrigo AP, Buchner J, Gaestel M (1999) Regulation of Hsp27 oligomerization, chaperone function, and protective activity against oxidative stress/tumor necrosis factor alpha by phosphorylation. J Biol Chem 274:18947–18956

Sanchez ER, Toft DO, Schlesinger MJ, Pratt WB (1985) Evidence that the 90-kDa phosphoprotein associated with the untransformed L-cell glucocorticoid receptor is a murine heat shock protein. J Biol Chem 260:12398–12401

Satyal SH, Chen D, Fox SG, Kramer JM, Morimoto RI (1998) Negative regulation of the heat shock transcriptional response by HSBP1. Genes Dev 12:1962–1974

Schneikert J, Hubner S, Martin E, Cato AC (1999) A nuclear action of the eukaryotic cochaperone RAP46 in downregulation of glucocorticoid receptor activity. J Cell Biol 146:929–940

Shi Y, Mosser DD, Morimoto RI (1998) Molecular chaperones as HSF1-specific transcriptional repressors. Genes Dev 12:654–666

Silverstein AM, Galigniana MD, Chen MS, Owens-Grillo JK, Chinkers M, Pratt WB (1997) Protein phosphatase 5 is a major component of glucocorticoid receptor hsp90 complexes with properties of an FK506-binding immunophilin. J Biol Chem 272:16224–16230

Sitia R, Molteni SN (2004) Stress, protein (mis)folding, and signaling: the redox connection. Sci STKE 2004: pe27

Smith CC, Yu YX, Kulka M, Aurelian L (2000) A novel human gene similar to the protein kinase (PK) coding domain of the large subunit of herpes simplex virus type 2 ribonucleotide reductase (ICP10) codes for a serine-threonine PK and is expressed in melanoma cells. J Biol Chem 275:25690–25699

Soncin F, Zhang X, Chu B, Wang X, Asea A, Ann Stevenson M, Sacks DB, Calderwood SK (2003) Transcriptional activity and DNA binding of heat shock factor-1 involve phosphorylation on threonine 142 by CK2. Biochem Biophys Res Commun 303:700–706

Song J, Takeda M, Morimoto RI (2001) Bag1-Hsp70 mediates a physiological stress signalling pathway that regulates Raf-1/ERK and cell growth. Nat Cell Biol 3:276–282

Stokoe D, Engel K, Campbell DG, Cohen P, Gaestel M (1992) Identification of MAPKAP kinase 2 as a major enzyme responsible for the phosphorylation of the small mammalian heat shock proteins. FEBS Lett 313:307–313

Suzuki A, Sugiyama Y, Hayashi Y, Nyu-i N, Yoshida M, Nonaka I, Ishiura S, Arahata K, Ohno S (1998) MKBP, a novel member of the small heat shock protein family, binds and activates the myotonic dystrophy protein kinase. J Cell Biol 140:1113–1124

Takayama S, Bimston DN, Matsuzawa S, Freeman BC, Aime-Sempe C, Xie Z, Morimoto RI, Reed JC (1997) BAG-1 modulates the chaperone activity of Hsp70/Hsc70. EMBO J 16:4887–4896

Takayama S, Sato T, Krajewski S, Kochel K, Irie S, Millan JA, Reed JC (1995) Cloning and functional analysis of BAG-1: a novel Bcl-2-binding protein with anti-cell death activity. Cell 80:279–284

Wang HG, Takayama S, Rapp UR, Reed JC (1996) Bcl-2 interacting protein, BAG-1, binds to and activates the kinase Raf-1. Proc Natl Acad Sci U S A 93:7063–7068

Wang X, Grammatikakis N, Siganou A, Stevenson MA, Calderwood SK (2004) Interactions between extracellular signal-regulated protein kinase 1, 14-3-3epsilon, and heat shock factor 1 during stress. J Biol Chem 279:49460–49469. Epub 42004 Sep 49410

Winkler W, Nahvi A, Breaker RR (2002) Thiamine derivatives bind messenger RNAs directly to regulate bacterial gene expression. Nature 419:952–956. Epub 2002 Oct 2016

Winklhofer KF, Reintjes A, Hoener MC, Voellmy R, Tatzelt J (2001) Geldanamycin restores a defective heat shock response in vivo. J Biol Chem 276:45160–45167. Epub 42001 Sep 45126

Xia W, Voellmy R, Spector NL (2000) Sensitization of tumor cells to fas killing through overexpression of heat-shock transcription factor 1. J Cell Physiol 183:425–431

Xu W, Mimnaugh E, Rosser MF, Nicchitta C, Marcu M, Yarden Y, Neckers L (2001) Sensitivity of mature Erbb2 to geldanamycin is conferred by its kinase domain and is mediated by the chaperone protein Hsp90. J Biol Chem 276:3702–3708. Epub 2000 Nov 3708

Yaffe MB, Schutkowski M, Shen M, Zhou XZ, Stukenberg PT, Rahfeld JU, Xu J, Kuang J, Kirschner MW, Fischer G, Cantley LC, Lu KP (1997) Sequence-specific and phosphorylation-dependent proline isomerization: a potential mitotic regulatory mechanism. Science 278:1957–1960

Yan LJ, Christians ES, Liu L, Xiao X, Sohal RS, Benjamin IJ (2002) Mouse heat shock transcription factor 1 deficiency alters cardiac redox homeostasis and increases mitochondrial oxidative damage. EMBO J 21:5164–5172

Young JC, Hartl FU (2002) Chaperones and transcriptional regulation by nuclear receptors. Nat Struct Biol 9:640–642

Zeiner M, Gehring U (1995) A protein that interacts with members of the nuclear hormone receptor family: identification and cDNA cloning. Proc Natl Acad Sci U S A 92:11465–11469

Zeiner M, Niyaz Y, Gehring U (1999) The hsp70-associating protein Hap46 binds to DNA and stimulates transcription. Proc Natl Acad Sci U S A 96:10194–10199

Zeke T, Morrice N, Vazquez-Martin C, Cohen PT (2005) Human protein phosphatase 5 dissociates from heat-shock proteins and is proteolytically activated in response to arachidonic acid and the microtubule-depolymerizing drug nocodazole. Biochem J 385:45–56

Zou J, Guo Y, Guettouche T, Smith DF, Voellmy R (1998) Repression of heat shock transcription factor HSF1 activation by HSP90 (HSP90 complex) that forms a stress-sensitive complex with HSF1. Cell 94:471–480

Chaperoning of Glucocorticoid Receptors

W. B. Pratt · Y. Morishima (✉) · M. Murphy · M. Harrell

Department of Pharmacology, University of Michigan Medical School, 1301 MSRB III, Ann Arbor MI, 48109-0632, USA
ymo@umich.edu

1	Introduction	112
2	Hsp90 and GR Ligand-Binding Activity	113
2.1	Hsp90 Acts on the GR Ligand-Binding Domain	113
2.2	The Cleft Hypothesis for Hsp90 Action	114
2.3	Steroid Binding Within the Cleft Changes the Dynamics of Receptor Binding to Hsp90	115
3	GR–Hsp90 Heterocomplex Assembly	116
3.1	The Hsp90/Hsp70-Based Chaperone Machinery	116
3.2	The Five-Protein Heterocomplex Assembly System	119
3.3	Priming of the GR by Hsp70	119
3.4	Opening of the Steroid-Binding Cleft by Hsp90	121
4	Hsp90 and GR Trafficking	122
4.1	Nuclear Translocation	122
4.2	GR Trafficking Within the Nucleus	125
5	Roles of Hsp90 and Hsp70 in Glucocorticoid Receptor Turnover	128
	References	130

Abstract A multiprotein hsp90/hsp70-based chaperone machinery functions as a 'cradle-to-grave' system for regulating the steroid binding, trafficking and turnover of the glucocorticoid receptor (GR). In an ATP-dependent process where hsp70 and hsp90 act as essential chaperones and Hop, hsp40, and p23 act as nonessential co-chaperones, the machinery assembles complexes between the ligand binding domain of the GR and hsp90. During GR–hsp90 heterocomplex assembly, the hydrophobic ligand-binding cleft is opened to access by steroid, and subsequent binding of steroid within the cleft triggers a transformation of the receptor such that it engages in more dynamic cycles of assembly/disassembly with hsp90 that are required for rapid dynein-dependent translocation to the nucleus. Within the nucleus, the hsp90 chaperone machinery plays a critical role both in GR movement to transcription regulatory sites and in the disassembly of regulatory complexes as the hormone level declines. The chaperone machinery also plays a critical role in stabilization of the GR to ubiquitylation and proteasomal degradation. The initial GR interaction with hsp70 appears to be critical for the triage between hsp90 heterocomplex assembly and preservation of receptor function vs CHIP-dependent ubiquitylation and proteasomal degradation. The hsp90 chaperone machinery is ubiquitous and functionally conserved among eukaryotes, and it is possible that all physiologically significant actions of hsp90 require the hsp70-dependent assembly of client protein–hsp90 heterocomplexes.

Keywords Glucocorticoid receptor · Hsp90 · Hsp70 · Immunophilins · Dynein

1
Introduction

In 1981, v-Src was the first signaling protein reported to be in heterocomplex with the abundant (1%–2% of cytosolic protein) protein chaperone hsp90 (Brugge et al. 1981; Oppermann et al. 1981). But, it was not until 1985, when steroid receptor–hsp90 heterocomplexes were reported (Sanchez et al. 1985; Schuh et al. 1985; Catelli et al. 1985), that the chaperone was considered to regulate a signaling pathway, a notion that was confirmed in vivo by Picard et al. in 1990. The discovery of 'client' proteins that are regulated by hsp90 accelerated markedly after 1994 when the ansamycin antibiotic geldanamycin was reported to inhibit hsp90 function (Whitesell et al. 1994). Hsp90 is a member of a very limited family of proteins, the GHKL family, which possess a unique binding pocket for ATP (Dutta and Inouye 2000), and the inhibitors geldanamycin and radicicol bind to this nucleotide site (Stebbins et al. 1997; Prodromou et al. 1997) and prevent hsp90 from achieving its ATP-dependent conformation, thus blocking hsp90 action. These hsp90 inhibitors have been shown to affect a wide variety of signaling pathways involved in endocrine responses, cell cycling, tumorigenesis, and apoptosis. At the time of this writing, almost 150 transcription factors and protein kinases involved in signal transduction have been shown to be hsp90 client proteins (reviewed by Pratt and Toft 2003). These signaling proteins are assembled into heterocomplexes with hsp90 by a multiprotein chaperone machinery in which both hsp90 and hsp70 are essential for conformational change in the client protein (Pratt and Toft 1997, 2003). Because the client protein-bound hsp90 can bind other proteins, such as Cdc37 and the TPR (tetratricopeptide repeat) domain immunophilins, the formation of heterocomplexes with hsp90 can affect the state of the client protein in a variety of ways. In this chapter, we focus on the glucocorticoid receptor (GR) to discuss how the dynamic assembly of hsp90 heterocomplexes affects ligand binding activity, cytoplasmic-nuclear translocation, receptor cycling within the nucleus, and receptor turnover.

Figure 1 presents a diagram of the core GR–hsp90–immunophilin heterocomplex as it is immunoadsorbed with anti-GR antibody from lysates of hormone-free cells. In these native heterocomplexes, one molecule of GR is bound to a dimer of hsp90 and variable but substoichiometric amounts of hsp70 (reviewed in Pratt and Toft 1997). It is not known whether one or two molecules of p23 are bound to the GR-associated hsp90 dimers. The complexes contain one of several TPR domain immunophilins that bind to a TPR acceptor site at the C-terminus of hsp90. The immunophilins are characterized by their peptidylprolyl isomerase (PPIase) domains, which are the binding sites of the immunosuppressant drugs FK506 (FKBPs) or cyclosporine A (CyPs).

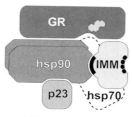

Fig. 1 The core GR–hsp90–immunophilin complex as it is immunoadsorbed from cytosols prepared from hormone-free cells. One molecule of GR is shown bound to a dimer of hsp90 and one molecule of immunophilin. The immunophilin (*IMM*) binds via its TPR domain (*black crescent*) to a TPR acceptor site on hsp90. Immunophilin PPIase domain (*dotted crescent*). The steroid structure in the GR indicates that the ligand-binding cleft is open and can be accessed by steroid

The steroid receptor complexes contain one of three immunophilins (FKBP51, FKBP52, or CyP-40) or PP5, a protein phosphatase that contains both a TPR domain and a PPIase homology domain (Silverstein et al. 1997). Because the TPR proteins can exchange for binding to the common acceptor site on hsp90, a single GR–hsp90 complex can theoretically be associated over time with more than one immunophilin. However, it has been shown that, at any point in time, the TPR domain proteins exist in separate GR–hsp90 complexes (Owens-Grillo et al. 1995; Renoir et al. 1995). Although the number of immunophilin molecules in a GR–hsp90 complex is somewhat controversial (one vs two), most observations are consistent with a GR–hsp90–immunophilin stoichiometry of 1:2:1 (see discussion in Pratt and Toft 2003).

2
Hsp90 and GR Ligand-Binding Activity

2.1
Hsp90 Acts on the GR Ligand-Binding Domain

Hsp90 binds to the GR ligand-binding domain (LBD), which is located in the C-terminal one-third of the receptor (Pratt et al. 1988). In the case of the GR and some other members of the nuclear receptor family (e.g., mineralocorticoid and aryl hydrocarbon receptors), the LBD must be in a complex with hsp90 for the receptor to have high-affinity ligand-binding activity (Bresnick et al. 1989; Hutchison et al. 1992). When the GR is stripped of its associated hsp90, it immediately loses its ability to bind steroid, and steroid-binding activity is regenerated when GR–hsp90 heterocomplexes are reassembled by the hsp90/hsp70-based chaperone machinery (Morishima et al. 2000b). Steroid ligands bind deep in a hydrophobic cleft that appears to be collapsed in the unliganded, hsp90-free receptor, such that the receptor must change its conformation to allow entry of the ligand (Gee and Katzenellenbogen 2001). The hsp90/hsp70-based chaperone machinery carries out an ATP-dependent opening of the steroid binding cleft, and once the cleft is open, it remains open as long as the metastable complex with hsp90 is maintained. In addition to opening the steroid binding cleft, the conformational changes in the receptor that occur with hsp90 heterocomplex assembly increase the sensitivity of the GR LBD to attack by thiol derivatizing agents and trypsin (Stancato et al. 1996; Modarress et al. 1997). This ability of the chaperone machinery to facilitate the opening of hydrophobic clefts may reflect a primary interaction with proteins in general. For example, assembly of complexes with hsp90 has a similar effect on the conformation of neuronal nitric-oxide synthase, where the hydrophobic heme-binding cleft in the apo-enzyme is opened to facilitate heme entry (Billecke et al. 2002, 2004).

Most interpretations of chaperone action are based on the ability of the chaperones to interact with non-native states of proteins to protect folding intermediates from inappropriate interactions and aggregation (Gething and Sambrook 1992). Thus, mechanistic interpretations of chaperone-dependent effects are often based on the assumption that a protein substrate has changed from an incompletely folded state toward a properly folded, minimal energy conformation. Hsp90 has clearly been shown to bind to non-native protein and promote refolding in vitro (Wiech et al. 1992), and it has been proposed that, in the stressed cell, hsp90 binds to proteins at very early stages of unfolding (Jakob et al. 1995). However, when hsp90 and hsp70 are acting together as essential components of the chaperone machinery, they may act on proteins that are in a native folding state. There is no evidence, for example, that the hsp90-free GR LBD is in a non-native or even a near-native conformation when it interacts with the chaperone machinery. The first component to interact with the receptor is hsp70 (Morishima et al. 2000b), and it is the ATP-dependent conformation of hsp70 that binds to the hsp90-free GR rather than the ADP-dependent conformation that has affinity for binding hydrophobic peptides (Kanelakis et al. 2002). Thus, we conceive of the chaperone machinery as acting on a GR LBD that is in its native, minimal energy conformation and that it then opens the hydrophobic ligand-binding cleft to yield a metastable complex in which hsp90 maintains an open cleft conformation of the GR that reflects a very early stage of unfolding (Pratt and Toft 2003).

2.2
The Cleft Hypothesis for Hsp90 Action

Consistent with this cleft opening model is the finding that there is a very focal site of hsp90 attack on the surface of the GR at the steroid-binding cleft. Early studies using C-terminal truncations of the steroid receptors suggested that multiple regions throughout the GR and PR (progesterone receptor) LBDs are involved in hsp90 complex formation (Cadepond et al. 1991; Schowalter et al. 1991). However, this approach appears to have generated artifactual data because the truncated LBDs were not in native conformation and there was no functional endpoint for an hsp90 effect. The Simons laboratory took a different approach in which the entire GR LBD and N-terminal LBD truncations were fused to another protein to maintain their stability in vivo, allowing them to determine the N-terminus of a functional LBD with steroid-binding activity (Xu et al. 1996). It was found that a seven-amino acid segment at the N-terminus of the GR LBD is required for hsp90 binding and for steroid binding activity (Xu et al. 1998). This region that is required for LBD–hsp90 heterocomplex assembly lies on the surface of the LBD at the opening of the hydrophobic steroid binding cleft (Giannoukos et al. 1999), and hsp90 binding requires the presence but no defined composition of this segment (Kaul et al. 2002).

The model in which the chaperone machinery binds at the opening of the steroid binding cleft on the LBD may explain how the machinery can form hsp90 heterocomplexes with such a wide variety of signaling proteins regardless of their size, shape, or sequence. Although it is not yet clear whether the chaperone machinery has the ability to recognize cleft openings on other signaling proteins, virtually all proteins in native conformation have regions where the hydrophobic surfaces of the protein interior merge with their hydrophilic exterior. This may be the general topological feature with which hsp70 interacts in priming the receptor for binding of hsp90 as the chaperone machinery assembles signaling protein–hsp90 heterocomplexes.

2.3
Steroid Binding Within the Cleft Changes the Dynamics of Receptor Binding to Hsp90

The complexes that are formed between the unliganded GR and hsp90 are constantly being assembled and disassembled in the intact cell at 37 °C, but they have been called persistent or stable complexes because they are stable enough at 0°–4 °C to permit their purification and analysis with biochemical techniques. Binding of steroid deep within the ligand-binding cleft facilitates a temperature-dependent collapse or closing of the cleft with loss of the GR's ability to form persistent complexes with hsp90. This liganded form of the GR is said to be transformed or activated because it can now move to the nucleus and be transcriptionally active (Pratt and Toft 1997). The interactions between the steroid and the hydrophobic binding pocket favor cleft closure, with the result that the interaction of the LBD with the hsp90 chaperone machinery becomes much more dynamic. The liganded, transformed GR now engages in dynamic heterocomplex assembly/disassembly with hsp90 that is important for GR movement to the nucleus and, subsequently, within the nucleus (reviewed by Pratt et al. 2004). We will describe the role of dynamic GR–hsp90 heterocomplex assembly/disassembly in GR movement later.

This steroid-dependent and temperature-dependent change in receptor interaction with hsp90 (Sanchez et al. 1987) is the key initial event in signal transduction by steroids. But, even before the temperature-dependent transformation, ligand binding to the GR has been shown to alter receptor conformation at 0°–4°C such that there is a change in the immunophilin composition of the GR–hsp90–immunophilin heterocomplexes. Davies et al. (2002) have shown that binding of steroid to the GR in intact cells at 0°–4°C causes a hormone-induced loss of the immunophilin FKBP51 from receptor complexes and a corresponding increase in FKBP52. The immunophilins in steroid receptor heterocomplexes bind via their TPR domains to hsp90 (reviewed in Pratt and Toft 2003), but FKBP52 also can bind directly to the GR in a region containing the NL1 nuclear localization signal (Silverstein et al. 1999). The NL1 is buried or masked when the unliganded GR is bound to hsp90 and removal of hsp90 opens up the NL1 (Scherrer et al. 1993). The likely explanation for this hormone-induced immunophilin exchange is that binding of steroid to the GR favors closure of the steroid binding cleft, promoting enough of a conformational change in the receptor to expose the NL1. The exposure of a second binding site for FKBP52 increases the affinity of the GR–hsp90 complex for FKBP52 and favors the immunophilin exchange.

The effect of hsp90 on steroid access to the ligand-binding cleft and the effect of steroid binding within the cleft on the dynamics of GR–hsp90 heterocomplex assembly/disassembly are not understood in greater detail than that presented here. However, it is clear that an understanding of these interactions will be critical to achieving a mechanistic understanding of how hsp90 regulates signal transduction by the receptor. The regulation conferred on the LBD by its interaction with the chaperone machinery is transferable. Thus, fusion of a steroid receptor LBD to another transcription factor, such as the viral E1A protein, may yield a chimera in which transcriptional activation is regulated by steroid binding (reviewed by Picard 1993). It is clear that the LBD in such cases confers hormone-regulated binding to hsp90 onto the chimeric protein (Scherrer et al. 1993).

3
GR–Hsp90 Heterocomplex Assembly

3.1
The Hsp90/Hsp70-Based Chaperone Machinery

The notion that hsp70 and hsp90 might function together germinated in 1989 when Kost et al. (1989) showed that both chaperones co-purified with the PR and Sanchez et al. (1990) showed that the two chaperones existed together in cytosolic complexes independent of binding to steroid receptors. Cell-free hsp90 heterocomplex assembly experiments were then performed in which immunoadsorbed PR or GR that was salt-stripped of hsp90 was incubated with rabbit reticulocyte lysate and the receptors became complexed with hsp90 (Smith et al. 1990; Scherrer et al. 1990). Because the GR was converted to the steroid-binding state, it was clear that the system in reticulocyte lysate was making the appropriate conformational change in the receptor when the complex with hsp90 was formed (Scherrer et al. 1990; Hutchison et al. 1992). The concept of a chaperone machinery evolved when a high-molecular-weight complex containing both hsp90 and hsp70 was isolated from reticulocyte lysate and shown to assemble GR–hsp90 heterocomplexes that had normal steroid-binding activity (Scherrer et al. 1992).

An important advance came when a 60-kDa protein found in PR–hsp90 complexes assembled by reticulocyte lysate under ATP-limiting conditions (Smith et al. 1992) was shown to bind both hsp90 and hsp70 to form a tripartite complex containing both chaperones (Smith et al. 1993). It was then shown that hsp70 is absolutely required for assembly of functional GR–hsp90 complexes in reticulocyte lysate (Hutchison et al. 1994a) and that all of the components that are required to form a GR–hsp90 heterocomplex are co-immunoadsorbed from lysate with hsp90 (Hutchison et al. 1994b), consistent with the existence of a preformed machinery for assembly. The 60-kDa protein that links hsp90 and hsp70 to form the machinery is now called Hop (Hsp organizing protein) and the yeast ortholog is Sti1. Hop binds independently via an N-terminal TPR domain to hsp70 and via a central TPR domain to hsp90 to form hsp90–Hop–hsp70 complexes (Chen et al. 1996). The native machinery immunoadsorbed from reticulocyte lysate with anti-Hop antibody also contains some of the hsp70 co-chaperone hsp40 (Dittmar et al. 1998).

The chaperone machinery forms spontaneously upon mixing of the purified proteins, and the complexes immunoadsorbed with anti-Hop antibody convert the immunoadsorbed, hsp90-free GR to the steroid-binding state when the two immune pellets are incubated together in the presence of ATP (Dittmar and Pratt 1997; Dittmar et al. 1998). The small hsp90 co-chaperone p23 (Johnson and Toft 1994) is not part of the native chaperone machinery, but to form stable GR–hsp90 heterocomplexes, p23 must be present to bind to the complexes and stabilize them once they are formed (Dittmar and Pratt 1997; Dittmar et al. 1997). Under conditions where client protein–hsp90 heterocomplexes are not being formed (0°–4°C, no ATP supplementation), all of the Hop and roughly 30% of the hsp90 in reticulocyte lysate is present in hsp90–Hop–hsp70 heterocomplexes (Murphy et al. 2001). Analysis of the native chaperone machinery by native gel electrophoresis and by cross-linking was consistent with an hsp90:Hop:hsp70 stoichiometry of 2:1:1 (Murphy et al. 2001). However, the stoichiometry of hsp90–Hop–hsp70 complexes made by mixing purified proteins was 2:2:1, suggesting that Hop may be present as a dimmer (Hernandez et al. 2002b). Hop does not just function as a passive linker between hsp90 and hsp70, it influences the conformational state and function of each chaperone protein (Hernandez et al. 2002b; Carrigan et al. 2004).

A major question that remains unresolved is whether the machinery in vivo attaches to client proteins as a preformed heterocomplex assembly machine, or whether the machine is assembled in a sequence of steps on the client protein.

It may be that both can occur. An argument for attachment of the preformed chaperone machinery to the client protein is that the immunoadsorbed machinery from reticulocyte lysate can convert the immunoadsorbed GR to the steroid binding state (Dittmar and Pratt 1997; Dittmar et al. 1998). An argument in support of sequential attachment of hsp70 and hsp90 to the client protein comes from studies of the time-course of PR–hsp90 heterocomplex assembly in reticulocyte lysate where hsp70 is bound earlier than hsp90 (Smith

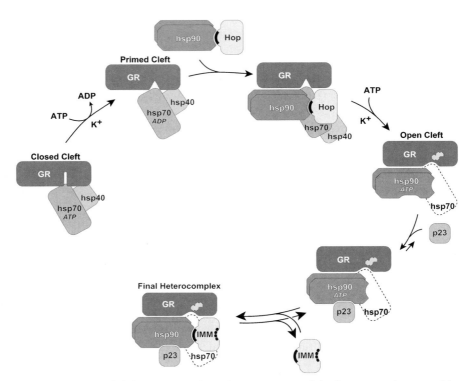

Fig. 2 Mechanism of cleft opening and GR–hsp90–immunophilin heterocomplex assembly. The ATP-dependent conformation of hsp70 binds initially to the GR and in an ATP-, K$^+$-, and hsp40-dependent step or steps a GR–hsp70 complex is formed that is primed to interact with hsp90. After hsp90 binding, there is a second ATP- and K$^+$-dependent step that is rate-limiting and leads to opening of the steroid-binding cleft, enabling access by steroid (indicated by the steroid structure). During GR–hsp90 heterocomplex assembly in cells and in reticulocyte lysate, Hop and some of the hsp70 dissociate during or at the end of the cleft-opening step. The GR-bound hsp90 is now in its ATP-dependent conformation and can be bound by p23, which stabilizes the chaperone in that conformation, preventing disassembly of the GR–hsp90 heterocomplex. When Hop dissociates, the TPR domain immunophilins or PP5 can bind reversibly to the TPR acceptor site on GR-bound hsp90. The hsp40 and Hop components of the five-protein assembly system have been omitted from later steps for simplicity. TPR domains (*black crescents*)

1993). The study of stepwise assembly of GR–hsp90 complexes with the purified proteins of the machinery (Morishima et al. 2000b) has been particularly useful in establishing a sequence of assembly events leading to the mechanistic scheme shown in Fig. 2.

3.2
The Five-Protein Heterocomplex Assembly System

The GR–hsp90 heterocomplex assembly system was originally reconstituted from reticulocyte lysate (Dittmar et al. 1996), and we now use a minimal system of five purified proteins—hsp90, hsp70, Hop, hsp40, p23—for efficient assembly of stable receptor–hsp90 heterocomplexes (Dittmar et al. 1998; Kosano et al. 1998). Of the five proteins, only hsp70 and hsp90 are essential for opening the ligand-binding cleft to permit steroid binding: this was shown by trapping the steroid-bound GR during the assembly reaction (Morishima et al. 2000a). Hop and hsp40 act as nonessential co-chaperones that increase the efficiency of GR–hsp90 complex assembly (Morishima et al. 2000a). p23 acts both in the purified five-protein system (Dittmar et al. 1997) and in vivo (Morishima et al. 2003) to stabilize GR–hsp90 heterocomplexes once they are formed. Thus, although p23 is not essential for opening the ligand-binding cleft, it is essential for assembly of GR–hsp90 heterocomplexes that are stable, such that they can be washed and then bound with steroid. It is important to note that the chaperone machinery has been isolated from yeast (Chang and Lindquist 1994), where selected gene disruptions have provided evidence for physiological roles of the yeast orthologs of the five-protein system in steroid receptor regulation (Nathan and Lindquist 1995; Chang et al. 1997; Caplan et al. 1995; Kimura et al. 1995; Bohen 1998).

3.3
Priming of the GR by Hsp70

As illustrated in Fig. 2, the first step in GR–hsp90 heterocomplex assembly is the ATP-dependent priming of the GR to form a GR–hsp70 complex that can be washed free of unbound hsp70 and incubated with purified hsp90 and Hop in a second ATP-dependent step (Morishima et al. 2000b). The product of the second step is a GR–hsp90 complex in which the receptor has an open steroid binding cleft that can be accessed by steroid. If the hsp70 and hsp90 steps are reversed in order, there is no assembly. We have shown that ATP-bound hsp70 binds directly to the hsp90-free, non-steroid-binding GR, whereas hsp70 that has been converted to its ADP-bound conformation by preincubation with hsp40 (YDJ-1) does not (Kanelakis et al. 2002). Thus, we have assumed that hsp70 is the first component of the chaperone machinery to contact the GR, and the role of hsp40 in priming is to promote the ATPase activity of GR-bound hsp70.

In studies with the PR, the Toft laboratory found that hsp40 (YDJ-1) first binds with high affinity to the PR in a 1:1 molar ratio, forming the PR–hsp40 complex that can be washed and subsequently bound with hsp70 (Hernandez et al. 2002a). Thus, it is argued that the first step in PR–hsp90 assembly is binding of hsp40, which targets the receptor for hsp70 binding. This is clearly not the case with the GR (Murphy et al. 2003), and inasmuch as hsp40 is not essential for PR–hsp90 heterocomplex assembly (Kosano et al. 1998), hsp40 is not required for targeting the PR for hsp70 binding. The PR is different from the GR in that the PR must be heated after dissociation of hsp90 to eliminate steroid-binding activity, and in the experiments of Hernandez et al. (2002a), the PR is incubated with hsp40 (YDJ-1) while the steroid binding is being inactivated. Thus, in the case of PR, hsp40 is exposed to an open steroid-binding cleft with the potential for interaction with exposed hydrophobic cleft interior, whereas it is the closed cleft form of the GR LBD that is exposed to hsp40. An analogy can be made with DnaJ, which has an affinity for unfolded proteins (Langer et al. 1992; Szabo et al. 1994). Such a chaperone function of hsp40 (YDJ-1) would favor binding to a hydrophobic region in the receptor, forming a ternary ATP–hsp70–LBD–hsp40 complex in which the J domain of hsp40 is interacting also with receptor-bound ATP–hsp70 to promote ATP hydrolysis. Han and Christen (2003) have proposed this mechanism for DnaJ and DnaK bound to substrate, and it seems likely that such ternary complexes form with both the GR and PR LBDs during the priming reaction. In such a complex, hsp40 is not targeting hsp70 to the receptor LBD, but the formation of the ternary complex would improve the efficiency of co-chaperone interaction of hsp40 with receptor-bound hsp70. The likely site for this focal attack is the region where the hydrophobic interior of the steroid-binding cleft merges with the receptor surface.

It is important to note that incubation of immunoadsorbed, chaperone-free GR with hsp90 does not yield a complex as it does with hsp70. Experiments examining the stimulation of hsp90 ATPase activity by purified, bacterially expressed GR LBD fragment imply that some kind of specific recognition must occur. Hsp90 ATPase activity is stimulated up to 200-fold by the GR LBD, and unfolded or partially folded non-client proteins do not affect ATPase activity (McLaughlin et al. 2002). The solubilized form of the GR LBD fragment used in such experiments is denatured, and neither the full-length GR nor the LBD fragment solubilized in this manner can be converted to the steroid-binding state by reticulocyte lysate or by the purified five-protein assembly system. In the absence of the physiological endpoint of generating steroid-binding activity, it is unclear where this observation of hsp90 interaction with a denatured LBD fragment fits in the assembly pathway. It is likely that the LBD stimulation of hsp90's ATPase activity models either the interaction of hsp90 with the hsp70-primed GR or with the GR at a more advanced stage of cleft opening as the hydrophobic cleft interior becomes exposed to the aqueous environment.

The initial priming step in assembly is MgATP-, K^+-, and hsp40-dependent, and the GR-bound hsp70 at the end of the priming reaction is in both ADP-bound and ATP-bound states (Morishima et al. 2000b, 2001). The ATPase activity of hsp70 is K^+-dependent (O'Brien and McKay 1995) and hsp40 stimulates hsp70 ATPase activity (Minami et al. 1996). The requirement for a continuous high level of ATP and the dependence upon hsp40 and K^+ show that both ATP binding and ATPase activity are required to produce primed GR–hsp70 complexes. This requirement has led to the proposal that, once hsp70 is bound to the GR, the chaperone may oscillate back and forth between ATP-bound and ADP-bound configurations, performing an iterative function during the priming step (Morishima et al. 2001).

The priming step is very rapid and the hsp70-bound GR produced in this step has no steroid binding activity (Morishima et al. 2000b). Examination of the primed GR–hsp70 complexes by atomic force microscopy revealed the most common stoichiometry to be 1:1, with complexes of 1:2 also being present (Murphy et al. 2003). A major common theme that emanates from both GR–hsp90 and PR–hsp90 heterocomplex assembly is the very focal nature of the process. Thus, hsp40 binds to the PR in a molar ratio of 1:1 (Hernandez et al. 2002a) and hsp70 binds as a monomer or dimer to the GR LBD. Again, this is consistent with a focal attack of the chaperone machinery on the steroid binding cleft.

3.4
Opening of the Steroid-Binding Cleft by Hsp90

As indicated in Fig. 2, the primed GR–hsp70 complex binds hsp90 and there is more binding if Hop is present (Morishima et al. 2000b). The binding of hsp90 is rapid, and it is the subsequent ATP-dependent opening of the steroid binding cleft that is rate-limiting in the overall assembly process (Kanelakis et al. 2002). In this cleft opening step, the GR-bound hsp90 is converted to its ATP-dependent conformation. To have an open steroid binding cleft, the receptor-bound hsp90 must assume its ATP-dependent conformation (Grenert et al. 1999) and it is only the ATP-dependent conformation of hsp90 that binds p23 (Sullivan et al. 1997). Like the priming step with hsp70, this second, cleft opening step is also K^+-dependent, suggesting that the GR-bound hsp70 must be converted from its ATP-dependent to its ADP-dependent conformation while it cooperates with hsp90 to activate steroid binding activity (Morishima et al. 2001).

The cleft-opening reaction is not understood in greater detail, but in reticulocyte lysate, it seems that a series of events are occurring. For example, much of the hsp70 and all of the Hop dissociate during this second step (Smith 1993). Reticulocyte lysate contains the TPR domain immunophilins and PP5; thus dissociation of Hop from the GR-bound hsp90 yields an open TPR acceptor site, permitting binding of FKBP51, FKBP52, CyP-40, or PP5.

It is important to emphasize that the purified five-protein system is a minimal system for efficient assembly of stable GR–hsp90 heterocomplexes. Reticulocyte lysate contains other components that likely play important roles in a more complex assembly system. For example, Aha proteins (activator of hsp90 ATPase) are hsp90 co-chaperones present in cell lysates (Panaretou et al. 2002), and addition of purified Aha protein to the five-protein system accelerates the rate of generation of steroid-binding activity. This would be consistent with an iterative process in which GR-bound hsp90 switches between its ATP-bound and ADP-bound conformations during the ATP-dependent cleft opening process.

Small amounts of both Hip (hsc70-interacting protein) and BAG-1 (Bcl-2-associated gene product-1) are present in the hsp90–Hop–hsp70–hsp40 complexes immunoadsorbed from reticulocyte lysate with antibody against Hop (Kanelakis et al. 1999). Hip and BAG-1 are co-chaperones that compete with each other for binding to the ATPase domain of hsp70; thus, they must exist in separate chaperone machinery complexes. Although Hip was regarded as essential for receptor–hsp90 heterocomplex assembly in reticulocyte lysate (Prapapanich et al. 1998), it is not present in the five-protein assembly system and it does not affect GR–hsp90 assembly by that system (Kanelakis et al. 2000). Nelson et al. (2004) have reported that introduction of Hip into yeast enhances hormone-dependent activation of a reporter gene by the GR, but how it works is unclear. At physiological levels, BAG-1 does not affect the rate of GR–hsp90 assembly by the five-protein system, but at high levels, it is inhibitory (Kanelakis et al. 1999). BAG-1(RAP46) is associated with the transformed GR and travels to the nucleus with it (Zeiner and Gehring 1995; Schneikert et al. 2000). BAG-1 has a nonspecific DNA-binding domain that is required for its inhibition of GR-dependent transactivation (Schmidt et al. 2003), but again, the mechanism of the inhibition is not clear.

4
Hsp90 and GR Trafficking

4.1
Nuclear Translocation

In hormone-free cells, the steroid receptors shuttle continuously between the cytoplasm and the nucleus, and the unliganded receptor at steady state may be predominantly cytoplasmic, like the GR, or nuclear, like the estrogen receptor (reviewed by DeFranco et al. 1995). The GR has been particularly useful for studying the mechanism of retrograde movement because steroid-dependent transformation of the GR triggers its rapid ($t_{1/2}$ ~4.5 min) translocation to the nucleus (reviewed by Pratt et al. 2004). A model for rapid retrograde GR movement along microtubular tracks (Fig. 3a) was first proposed in 1993

(Pratt 1993). At that time, it was known that vesicles move along microtubules and that cytoplasmic dynein is the motor protein responsible for movement in the retrograde direction (reviewed by Hirokawa 1998). The notion was that protein solutes might move in a similar fashion, with the unique addition being that dynamic assembly of heterocomplexes with hsp90 facilitates linkage of the steroid receptors to the motor system. The hsp90-binding immunophilin FKBP52, which was known to localize to microtubules and to exist in cytosolic complexes with cytoplasmic dynein (Czar et al. 1994), was originally envisaged as the linker to the motor system. Subsequently, it has been shown that other hsp90-binding TPR domain proteins, such as CyP-40 and PP5, also serve as linkers to the dynein motor system (Galigniana et al. 2002). The tools used to uncouple GR–hsp90 heterocomplexes at various sites in vitro and to impede GR retrograde movement in vivo are shown in Fig. 3b.

The entire complex shown in Fig. 3a can be immunoadsorbed from cell lysates prepared with buffer containing paclitaxel and GTP to stabilize microtubules, and these complexes can be assembled under cell-free conditions by incubating immunoadsorbed, stripped GR with similar lysates (Harrell et al. 2005). As diagramed in Fig. 3b, assembly of complexes in cell lysates containing the hsp90 inhibitor geldanamycin prevents GR association with hsp90 and all other components of the heterocomplex; competition with a TPR domain fragment of PP5 during assembly yields a GR–hsp90 complex without immunophilins or dynein; and competition with the PPIase domain fragment of FKBP52 yields a GR–hsp90–immunophilin complex without dynein (Silverstein et al. 1999; Galigniana et al. 2001; Harrell et al. 2005). FKBP52 forms complexes with dynein even when its PPIase activity is inhibited by FK506; thus, the PPIase domain is acting as a protein interaction domain to determine binding to the motor protein complex (Galigniana et al. 2001). Both the TPR domain interaction with hsp90 and the PPIase domain interaction with dynein are evolutionarily conserved in plant immunophilins, suggesting that linkage of hsp90 through immunophilins to dynein is fundamental to the biology of the eukaryotic cell (Harrell et al. 2002).

Riggs et al. (2003) have shown that FKBP52 expression in yeast potentiates GR-dependent reporter gene activation. The potentiation was due to an increase in GR hormone-binding affinity that required both the hsp90 binding activity and the PPIase activity of FKBP52. These observations suggest that FKBP52 also acts as a peptidylprolyl isomerase on the GR to create a high-affinity steroid-binding form of the receptor. This action is specific to the GR, as opposed to the estrogen receptor, and is specific to FKBP52 vs other hsp90-bound immunophilins (Riggs et al. 2003). Thus, although the PPIase domain of FKBP52 may function in a dual capacity as an enzyme for protein folding and as a site for protein interaction with dynein, it is likely that the latter function is the more general one. A general role for immunophilin PPIase domains in movement is supported by the observation that the non-hsp90-binding immunophilin CyP-A exists in a native heterocomplex with dynein

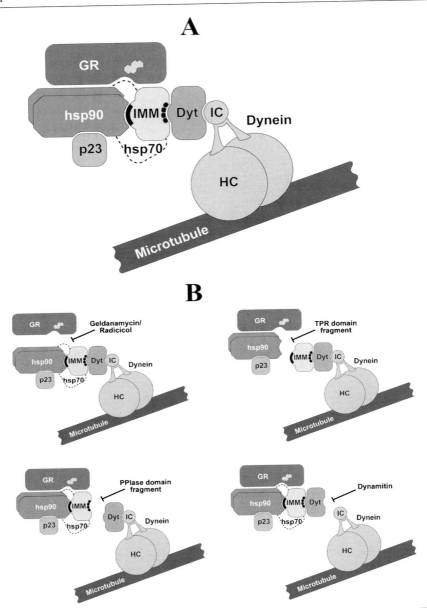

Fig. 3a,b The GR movement system. **a** Illustration of the complete movement system. The PPIase domain (*dotted crescent*) of the immunophilin binds to the dynamitin (*Dyt*) subunit of the dynein-associated dynactin complex. Dynein is a large multisubunit complex (~1.2 MDa) comprised of two heavy chains (*HC*) that have the processive motor activity, three intermediate chains (*IC*), and some light chains, which are not shown. **b** Sites of uncoupling by the hsp90 inhibitor geldanamycin, by the TPR domain fragment of PP5, by the PPIase domain fragment of FKBP52, and by dynamitin that requires the CyP-A PPIase domain, but not its PPIase activity (Galigniana et al. 2005).

Dynein is a large multisubunit complex and it links to vesicles and organelles indirectly via another multisubunit complex, dynactin (Hirokawa 1998). In the model illustrated in Fig. 3a, we speculate that dynamitin, a 50-kDa subunit of dynactin, may be the component of the dynein–dynactin motor protein complex that interacts directly with the immunophilin. Dynamitin is a component of native GR–hsp90 complexes (Harrell et al. 2005), and both purified FKBP52 and CyP-A bind directly to immunopurified *myc*-dynamitin in a PPIase domain-specific manner (Galigniana et al. 2004b, 2005).

Treatment of cells with geldanamycin slows by an order of magnitude the rate of ligand-dependent nuclear translocation of both endogenous GR (Czar et al. 1997) and a green fluorescent protein fusion with the GR (GFP-GR) (Galigniana et al. 1998). The rapid hsp90-dependent movement requires intact cytoskeleton, and the slow movement that occurs when hsp90 is inhibited appears to represent diffusion (Galigniana et al. 1998). In neurites where proteins cannot move by random diffusion alone, retrograde GFP-GR movement is blocked by geldanamycin (Galigniana et al. 2004a), suggesting that the hsp90-dependent movement machinery is required for retrograde movement in axons and dendrites. Consistent with a broad utilization of the movement machinery by hsp90 client proteins, geldanamycin has been shown to inhibit retrograde movement of the androgen receptor (Georget et al. 2002; Thomas et al. 2004), the aryl hydrocarbon receptor (Kazlauskas et al. 2000; 2001), and the tumor suppressor protein p53 (Galigniana et al. 2004b).

The PPIase domain fragment competes for immunophilin binding to the dynein–dynactin motor complex (Fig. 3b), and overexpression of the fragment slows steroid-dependent translocation of GFP-GR to the nucleus to the same extent as treatment of cells with geldanamycin (Galigniana et al. 2001). Similarly, overexpression of dynamitin slows GR translocation to the nucleus (Harrell et al. 2004). Overexpression of dynamitin impedes movement by dissociating the dynein–dynactin motor complex from cargo (Hirokawa 1998). In the case of the GR, overexpressed dynamitin binds to the immunophilin PPIase domain, and because the expressed dynamitin is in great excess of dynein–dynactin, the great majority of the GR–hsp90–immunophilin complexes bind to free dynamitin that is not associated with the dynein–dynactin motor system.

4.2
GR Trafficking Within the Nucleus

When the GR arrives at the nuclear membrane, it binds to importin-α and passes through the nuclear pores by an importin-dependent process (Savory et al. 1999; Tanaka et al. 2003; Freedman and Yamamoto 2004). The form of the GR that passes across the pore is not established, but it is clear that the GR can pass into the nucleus as a GR–hsp90 heterocomplex (Kang et al. 1994), and it is likely that the receptor is normally transferred across the pores as a GR–hsp90–immunophilin heterocomplex (see discussion in Pratt et al. 2004). Examination of GFP-GR localization in nuclei of living cells shows the receptor accumulating in punctate foci throughout the nucleus excluding nucleoli (Htun et al. 1996; Nishi et al. 2001). A similar punctate localization has been observed for all of the steroid receptors (reviewed by Baumann et al. 1999). For several of the receptors that localize to the nucleus in the absence of ligand, formation of the very discrete foci is agonist-dependent. Thus, there is the impression that the receptors move to 'staging areas' from which they move to discrete foci upon ligand-dependent transformation (Baumann et al. 1999). The results of experiments performed by DeFranco and his colleagues have shown that GR–hsp90 heterocomplex cycling occurs within the nucleus (reviewed in DeFranco et al. 1998). A model of nuclear GR–hsp90 cycling is presented in Fig. 4.

Using permeabilized cells to examine in vitro nuclear export of the GR, Yang et al. (1997) showed that GR released from chromatin could recycle to chromatin upon rebinding hormone without exiting the nucleus. Liu and DeFranco (1999) then showed that geldanamycin inhibits recycling of these hormone-withdrawn GRs to the hormone-binding state and inhibits GR release from chromatin during hormone withdrawal. Additional support for chaperone-dependent GR release from chromatin was provided by Freeman and Yamamoto (2002) who transfected liver cells with p23 that was targeted to bind near hormone response elements and showed a dramatic reduction in GR-dependent transcriptional activation. In chromatin immunoprecipitation assays, it was shown that both p23 and hsp90 localized to glucocorticoid response elements in a hormone-dependent manner, and in experiments in vitro, p23 inhibited transcriptional activation by preformed regulatory complexes. Thus, it was suggested that p23 promotes disassembly of transcriptional regulatory complexes via a direct chaperone effect on the GR (Freeman and Yamamoto 2002). However, it has been shown that p23 expression in vivo affects client proteins, including the GR, by acting as a co-chaperone to affect hsp90 (Oxelmark et al. 2003; Morishima et al. 2003; Wochnik et al. 2004). Taken together, these observations strongly suggest a role for the hsp90/hsp70-based chaperone machinery in the termination of transcriptional activation by the GR as free hormone levels decline (Fig. 4). Inasmuch as the DNA-binding activity of BAG-1 is required for its inhibition of GR-dependent transactivation (Schmidt et al. 2003), this hsp70 co-chaperone may play a critical role in GR release from chromatin.

The use of GFP fusion proteins and the technique of fluorescence recovery after photobleaching (FRAP) in live cells has demonstrated that a number of transcription factors interact in a very dynamic fashion with their regulatory sites (reviewed by Hager et al. 2002). Hager and his colleagues have examined the dynamics of GFP-GR binding to an artificial amplified array of mouse mammary tumor virus (MMTV) reporter elements on chromosome 4

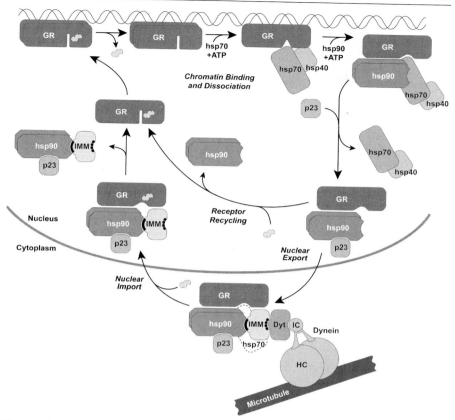

Fig. 4 Model for the recycling of GRs in the nucleus by the hsp90/hsp70-based chaperone machinery. After dissociation of hormone, the GRs are released from high-affinity chromatin-binding sites by the chaperone machinery. The chaperone machinery is depicted here as acting in two ATP-dependent steps, the first involving hsp70 and the second hsp90, as in Fig. 2. The nonessential co-chaperone Hop, which brings the chaperones together into an hsp90–Hop–hsp70–hsp40 complex, has been omitted to simplify presentation. Nuclear GR that has been recycled into GR–hsp90–p23 complexes can bind hormone without exiting the nucleus and be recycled to the chromatin-bound state. It is not known when GR dimer to monomer conversion occurs, but the stoichiometry of the final cytosolic complex is one molecule of GR bound to a dimer of hsp90

in a mouse cell line (McNally et al. 2000). The array includes 800–1,200 binding sites for the GR that forms a single patch of bright fluorescence in cells expressing GFP-GR that are treated with dexamethasone. Photobleaching experiments showed that the hormone-bound GFP-GR exchanges rapidly with the chromosomal regulatory sites, with a half maximal time for fluorescence recovery of approximately 5 s. Stavreva et al. (2004) have shown that hsp90, hsp70, and p23 co-localize with the GFP-GR at the arrays and that inhibition of hsp90 with geldanamycin or radicicol accelerated GFP-GR exchange at the MMTV promoter sites. In contrast to the conclusion drawn above from the experiments of Freeman and Yamamoto (2002), faster GFP-GR exchange with geldanamycin implies that the chaperone machinery normally stabilizes GR binding rather than promoting its removal. Although there is controversy regarding the effect of chaperones on binding of GR to the promoter, as Stavreva et al. (2004) have noted, both their results and those of Freeman and Yamamoto (2002) support a role for chaperones in GR binding at a promoter.

There is also direct evidence that the hsp90/hsp70-based chaperone system may be required for general GR mobility within the nucleus. Elbi et al. (2004) have developed a permeabilized cell system in which transcriptionally active nuclei are depleted of soluble factors required for GR nuclear mobility. GR nuclear mobility was restored on incubation with reticulocyte lysate and the activity of reticulocyte lysate was inhibited by geldanamycin. Most importantly, ATP-dependent movement of the GFP-GR back into the photobleached area was restored by incubation with the purified five-protein chaperone system. How the chaperone machinery restores mobility is not known, but these observations provide clear evidence that the heterocomplex assembly is critical for GR trafficking within the nucleus.

5
Roles of Hsp90 and Hsp70 in Glucocorticoid Receptor Turnover

As reviewed in detail elsewhere in this volume, treatment of cells with geldanamycin or radicicol increases the degradation rate of many (probably all) client proteins that form persistent heterocomplexes with hsp90 (Neckers et al. 1999). Increased turnover was first demonstrated for some receptor tyrosine kinases that were degraded via the ubiquitin-proteasome pathway (Sepp-Lorenzino et al. 1995). Inhibition of GR–hsp90 heterocomplex assembly by treatment of cells with geldanamycin or radicicol leads to a rapid decline in the amount of GR protein, a decline that is inhibited by lactacystin and other proteasome inhibitors (Whitesell and Cook 1996; Segnitz and Gehring 1997). Not only does formation of heterocomplexes with hsp90 protect client proteins against ubiquitin-dependent degradation, but the ubiquitylation of many proteins requires hsp70 (Bercovich et al. 1997). Because mechanistic studies with the purified chaperone machinery have led to considerable definition of the sequential roles of hsp70 and hsp90 in GR–hsp90 heterocomplex assembly, the GR itself may prove to be a unique tool for studying chaperone-dependent ubiquitylation.

Of particular interest with regard to GR ubiquitylation is CHIP (carboxy terminus of hsc70-interacting protein), a member of the RING-domain family of E3 ubiquitin ligases. CHIP is a 35-kDa protein that binds via an amino-terminal TPR domain to both hsc/hsp70 and hsp90 (Ballinger et al. 1999;

Connell et al. 2001). CHIP possesses a carboxy-terminal U-box that interacts with the UBCH5 family of E2 ubiquitin-conjugating enzymes (Jiang et al. 2001), and it induces ubiquitylation of the GR (Connell et al. 2001). It is currently thought that the RING-type E3 enzymes, like CHIP, act as bridging proteins to bring the ubiquitin-charged E2 enzyme into the vicinity of the substrate (Pickart 2004). However, it is not known if CHIP itself contacts the receptor, and it is thought that the chaperones target CHIP to the receptor (Hohfeld et al. 2001).

Clearly the key question is how the triage decision is made for the GR to proceed on to GR–hsp90 heterocomplex assembly or to proceed toward ubiquitylation and proteasomal degradation. Because CHIP binds to the TPR acceptor site on hsp90, it has been argued that CHIP's interaction with GR-bound hsp90 causes the GR to be ubiquitylated and degraded (Connell et al. 2001). Such a model, however, is counterintuitive, inasmuch as inhibition of client protein–hsp90 assembly with geldanamycin promotes degradation. Binding of CHIP to hsp90 eliminates p23 binding (Connell et al. 2001), indicating that CHIP binding does not favor the ATP-dependent conformation of hsp90, which is the form of the chaperone in the GR–hsp90 heterocomplex. CHIP has been co-immunoadsorbed with GR–hsp90 complexes (Connell et al. 2001), but it is likely that this CHIP is interacting with GR-bound hsp70 in the complexes. In cells, it has been shown that both hsp70 and CHIP undergo steroid-dependent, retrograde, geldanamycin-sensitive movement with the GR and the AR (Galigniana et al. 2004a; Thomas et al. 2004). Thus, it seems that CHIP can bind to GR-bound hsp70 in fully assembled GR–hsp90 heterocomplexes but the presence of hsp90 mitigates against CHIP-dependent ubiquitylation of the receptor.

In the case of the hsp90 client proteins, like the GR, it seems most likely that the decision to proceed toward hsp90 heterocomplex assembly and preservation of client protein function vs CHIP-dependent degradation is made by the hsp70 component of the chaperone machinery when it initially interacts with the protein. As reviewed above, hsp70 engages in a very focal interaction with the steroid-binding cleft (Murphy et al. 2003), and we return to the cleft hypothesis of chaperone machinery action to see how such a choice might be made. When hsp70 binds to the undamaged, chaperone-free GR, the ligand binding cleft is closed and the chaperone binds in its ATP-dependent form (Kanelakis et al. 2002). This undamaged GR can proceed to the primed cleft state (Fig. 2) where the GR-bound hsp70 may preferentially bind Hop and hsp90 to form a GR–hsp90 heterocomplex with an open ligand-binding cleft. However, at early stages of unfolding, hydrophobic moieties that are normally internalized in the properly folded cleft will become exposed on the receptor surface. It is the ADP-dependent conformation of hsp70 that has high affinity for binding hydrophobic peptides, and it is this state of hsp70 that is likely to interact with the unfolding GR. The hsp70-bound, unfolding GR may not be able to proceed to the primed cleft state; rather, it binds CHIP and proceeds with chaperone-dependent ubiquitylation. Thus, in this model, the triage decision is made by the nature of the hsp70 interaction with the receptor and not by hsp90, Hop, or CHIP. The hsp70 co-chaperone BAG-1 may also play a role in degradation in that it associates with the proteasome through an ubiquitin-like domain (Luders et al. 2000) and stimulates CHIP-induced degradation of the GR (Demand et al. 2001). The study of hsp70 interaction with the GR ligand binding cleft in native and unfolding conformations may help to focus the important concepts offered by Hohfeld, Cyr, and Patterson regarding the roles of the chaperones and their co-chaperones in making such protein quality control triage decisions (Hohfeld et al. 2001; Cyr et al. 2002).

Acknowledgements The work of the authors described herein was supported by National Institutes of Health grants CA28010 from the National Cancer Institute and DK31573 from the National Institute of Diabetes and Digestive and Kidney Diseases.

References

Ballinger CA, Connell P, Wu Y, Hu Z, Thompson LJ, Yin LY, Patterson C (1999) Identification of CHIP, a novel tetratricopeptide repeat-containing protein that interacts with heat shock proteins and negatively regulates chaperone functions. Mol Cell Biol 19:4535–4545

Baumann CT, Lim CS, Hager GL (1999) Intracellular localization and trafficking of steroid receptors. Cell Biochem Biophys 31:119–127

Bercovich B, Stancovski I, Mayer A, Blumenfeld N, Laszlo A, Schwartz AL, Ciechanover A (1997) Ubiquitin-dependent degradation of certain protein substrates in vitro requires the molecular chaperone hsc70. J Biol Chem 272:9002–9010

Billecke SS, Bender AT, Kanelakis KC, Murphy PJM, Lowe ER, Kamada Y, Pratt WB, Osawa Y (2002) Hsp90 is required for heme binding and activation of apo-neuronal nitric-oxide synthase. J Biol Chem 277:20504–20509

Billecke SS, Draganov DI, Morishima Y, Murphy PJM, Dunbar AY, Pratt WB, Osawa Y (2004) The role of hsp90 in heme-dependent activation of apo-neuronal nitric-oxide synthase. J Biol Chem 279:30252–30258

Bohen SP (1998) Genetic and biochemical analysis of p23 and ansamycin antibiotics in the function of hsp90-dependent signaling proteins. Mol Cell Biol 18:3330–3339

Bresnick EH, Dalman FC, Sanchez ER, Pratt WB (1989) Evidence that the 90-kDa heat shock protein is necessary for the steroid binding conformation of the L cell glucocorticoid receptor. J Biol Chem 264:4992–4997

Brugge JS, Erikson E, Erikson RL (1981) The specific interaction of the Rous sarcoma virus transformed protein, pp60$^{v\text{-}src}$, with two cellular proteins. Cell 25:363–372

Cadepond F, Schweizer-Groyer G, Segard-Maurel I, Jibard N, Hollenberg SM, Giguere V, Evans RM, Baulieu EE (1991) Heat shock protein 90 as a critical factor in maintaining glucocorticosteroid receptor in a nonfunctional state. J Biol Chem 266:5834–5841

Caplan AJ, Langley E, Wilson EM, Vidal J (1995) Hormone-dependent transactivation by the human androgen receptor is regulated by a dnaJ protein. J Biol Chem 270:5251–5257

Carrigan PE, Nelson GM, Roberts PJ, Stoffer J, Riggs DL, Smith DF (2004) Multiple domains of the co-chaperone Hop are important for hsp70 binding. J Biol Chem 279:16185–16193

Catelli MG, Binart N, Jung-Testas I, Renoir JM, Baulieu EE, Feramisco JR, Welch WJ (1985) The common 90-kd protein component of nontransformed "8S" steroid receptors is a heat-shock protein. EMBO J 4:3131–3135

Chang HCJ, Lindquist S (1994) Conservation of hsp90 macromolecular complexes in *Saccharomyces cerevisiae*. J Biol Chem 269:24983–24988

Chang HCJ, Nathan DF, Lindquist S (1997) In vivo analysis of the hsp90 cochaperone Sti1 (p60). Mol Cell Biol 17:318–325

Chen S, Prapapanich V, Rimerman RA, Honore B, Smith DF (1996) Interactions of p60, a mediator of progesterone receptor assembly, with heat shock proteins hsp90 and hsp70. Mol Endocrinol 10:682–693

Connell P, Ballinger CA, Jiang J, Wu Y, Thompson LJ, Hohfeld J, Patterson C (2001) The co-chaperone CHIP regulates protein triage decisions mediated by heat-shock proteins. Nature Cell Biol 3:93–96

Cyr DM, Hohfeld J, Patterson C (2002) Protein quality control: U-box-containing E3 ubiquitin ligases join the fold. Trends Biochem Sci 27:368–375

Czar MJ, Owens-Grillo JK, Yem AW, Leach KL, Deibel MR, Welsh MJ, Pratt WB (1994) The hsp56 immunophilin component of untransformed steroid receptor complexes is localized both to microtubules in the cytoplasm and to the same nonrandom regions within the nucleus as the steroid receptor. Mol Endocrinol 8:1731–1741

Czar MJ, Galigniana MD, Silverstein AM, Pratt WB (1997) Geldanamycin, a heat shock protein 90-binding benzoquinone ansamycin, inhibits steroid-dependent translocation of the glucocorticoid receptor from the cytoplasm to the nucleus. Biochemistry 36:7776–7785

Davies TH, Ning YM, Sanchez ER (2002) A new first step in activation of steroid receptors. Hormone-induced switching of FKBP51 and FKBP52 immunophilins. J Biol Chem 277:4597–4600

DeFranco DB, Madan AP, Tang Y, Chandran UR, Xiao N, Yang J (1995) Nuclear cytoplasmic shuttling of steroid receptors. Vitam Horm 51:315–338

DeFranco DB, Ramakrishnan C, Tang Y (1998) Molecular chaperones and subcellular trafficking of steroid receptors. J Steroid Biochem Mol Biol 65:51–58

Demand J, Alberti S, Patterson C, Hohfeld J (2001) Cooperation of a ubiquitin domain protein and an E3 ubiquitin ligase during chaperone/proteasome coupling. Curr Biol 11:1569–1577

Dittmar KD, Pratt WB (1997) Folding of the glucocorticoid receptor by the reconstituted hsp90-based chaperone machinery. The initial hsp90–p60–hsp70-dependent step is sufficient for creating the steroid binding conformation. J Biol Chem 272:13047–13054

Dittmar KD, Hutchison KA, Owens-Grillo JK, Pratt WB (1996) Reconstitution of the steroid receptor–hsp90 heterocomplex assembly system of rabbit reticulocyte lysate. J Biol Chem 271:12833–12839

Dittmar KD, Demady DR, Stancato LF, Krishna P, Pratt WB (1997) Folding of the glucocorticoid receptor by the heat shock protein (hsp) 90-based chaperone machinery. The role of p23 is to stabilize receptor–hsp90 heterocomplexes formed by hsp90–p60–hsp70. J Biol Chem 272:21213–21220

Dittmar KD, Banach M, Galigniana MD, Pratt WB (1998) The role of DnaJ-like proteins in glucocorticoid receptor–hsp90 heterocomplex assembly by the reconstituted hsp90–p60–hsp70 foldosome complex. J Biol Chem 273:7358–7366

Dutta R, Inouye M (2000) GHKL, an emergent ATPase/kinase superfamily. Trends Biochem Sci 25:24–28

Elbi C, Walker DA, Romero G, Sullivan WP, Toft DO, Hager GL, DeFranco DB (2004) Molecular chaperones function as steroid receptor nuclear mobility factors. Proc Natl Acad Sci U S A 101:2876–2881

Freedman ND, Yamamoto KR (2004) Importin 7 and importin α/importin β are nuclear import receptors for the glucocorticoid receptor. Mol Biol Cell 15:2276–2286

Freeman BC, Yamamoto KR (2002) Disassembly of transcriptional regulatory complexes by molecular chaperones. Science 296:2232–2235

Galigniana MD, Scruggs JL, Herrington J, Welsh MJ, Carter-Su C, Housley PR, Pratt WB (1998) Heat shock protein 90-dependent (geldanamycin-inhibited) movement of the glucocorticoid receptor through the cytoplasm to the nucleus requires intact cytoskeleton. Mol Endocrinol 12:1903–1913

Galigniana MD, Radanyi C, Renoir JM, Housley PR, Pratt WB (2001) Evidence that the peptidylprolyl isomerase domain of the hsp90-binding immunophilin FKBP52 is involved in both dynein interaction and glucocorticoid receptor movement to the nucleus. J Biol Chem 276:14884–14889

Galigniana MD, Harrell JM, Murphy PJM, Chinkers M, Radanyi C, Renoir JM, Zhang M, Pratt WB (2002) Binding of hsp90-associated immunophilins to cytoplasmic dynein: direct binding and in vivo evidence that the peptidylprolyl isomerase domain is a dynein interaction domain. Biochemistry 41:13602–13610

Galigniana MD, Harrell JM, Housley PR, Patterson C, Fisher SK, Pratt WB (2004a) Retrograde transport of the glucocorticoid receptor in neurites requires dynamic assembly of complexes with the protein chaperone hsp90 and is linked to the CHIP component of the machinery for proteasomal degradation. Mol Brain Res 123:27–36

Galigniana MD, Harrell JM, O'Hagen HM, Ljungman M, Pratt WB (2004b) Hsp90-binding immunophilins link p53 to dynein during p53 transport to the nucleus. J Biol Chem 279:22483–22489

Galigniana MD, Morishima Y, Gallay PA, Pratt WB (2004) Cyclophilin-A is bound through its peptidylprolyl isomerase domain to the cytoplasmic dynein motor protein complex. J Biol Chem 279:55754–55759; e-pub Oct 20, 2004

Gee AC, Katzenellenbogen JA (2001) Probing conformational changes in the estrogen receptor: evidence for a partially unfolded intermediate facilitating ligand binding and release. Mol Endocrinol 15:421–428

Georget V, Terouanne B, Nicolas JC, Sultan C (2002) Mechanism of antiandrogen action: key role of hsp90 in conformational change and transcriptional activity of the androgen receptor. Biochemistry 41:11824–11831

Gething MJ, Sambrook J (1992) Protein folding in the cell. Nature 355:33–45

Giannoukos G, Silverstein AM, Pratt WB, Simons SS (1999) The seven amino acids (547–553) of rat glucocorticoid receptor required for steroid and hsp90 binding contain a functionally independent LXXLL motif that is critical for steroid binding. J Biol Chem 274:36527–36536

Grenert JP, Johnson BD, Toft DO (1999) The importance of ATP binding and hydrolysis by hsp90 in formation and function of protein heterocomplexes. J Biol Chem 274:17525–17533

Hager GL, Elbi C, Becker M (2002) Protein dynamics in the nuclear compartment. Curr Opin Genet Dev 12:137–141

Han W, Christen P (2003) Mechanism of the targeting action of DnaJ in the DnaK molecular chaperone system. J Biol Chem 278:19038–19043

Harrell JM, Kurek I, Breiman A, Radanyi C, Renoir JM, Pratt WB, Galigniana MD (2002) All of the protein interactions that link steroid receptor–hsp90–immunophilin heterocomplexes to cytoplasmic dynein are common to plant and animal cells. Biochemistry 41:5581–5587

Harrell JM, Murphy PJM, Morishima Y, Chen H, Mansfield JF, Galigniana MD, Pratt WB (2004) Evidence for glucocorticoid receptor transport on microtubules by dynein. J Biol Chem 279:54647–54654; e-pub Oct 13, 2004

Hernandez MP, Chadli A, Toft DO (2002a) Hsp40 binding is the first step in the hsp90 chaperoning pathway for the progesterone receptor. J Biol Chem 277:11873–11881

Hernandez MP, Sullivan WP, Toft DO (2002b) The assembly and intermolecular properties of the hsp70-Hop-hsp90 molecular chaperone complex. J Biol Chem 277:38294–38304

Hirokawa N (1998) Kinesin and dynein superfamily proteins and the mechanism of organelle transport. Science 279:519–526

Hohfeld J, Cyr DM, Patterson C (2001) From the cradle to the grave: molecular chaperones that may choose between folding and degradation. EMBO Rep 2:885–890

Htun H, Barsony J, Renyi I, Gould DL, Hager GL (1996) Visualization of glucocorticoid receptor translocation and intranuclear organization in living cells with a green fluorescent protein chimera. Proc Natl Acad Sci U S A 93:4845–4850

Hutchison KA, Czar MJ, Scherrer LC, Pratt WB (1992) Monovalent cation selectivity for ATP-dependent association of the glucocorticoid receptor with hsp70 and hsp90. J Biol Chem 267:14047–14053

Hutchison KA, Dittmar KD, Czar MJ, Pratt WB (1994a) Proof that hsp70 is required for assembly of the glucocorticoid receptor into a heterocomplex with hsp90. J Biol Chem 269:5043–5049

Hutchison KA, Dittmar KD, Pratt WB (1994b) All of the factors required for assembly of the glucocorticoid receptor into a functional heterocomplex with heat shock protein 90 are preassociated in a self-sufficient protein folding structure, a "foldosome". J Biol Chem 269:27894–27899

Jakob U, Lilie H, Meyer I, Buchner J (1995) Transient interaction of hsp90 with early unfolding intermediates of citrate synthase. J Biol Chem 270:7288–7294

Jiang J, Ballinger CA, Wu Y, Dai Q, Cyr DM, Hohfeld J, Patterson C (2001) CHIP is a U-box-dependent E3 ubiquitin ligase: identification of hsc70 as a target for ubiquitylation. J Biol Chem 276:42938–42944

Johnson JL, Toft DO (1994) A novel chaperone complex for steroid receptors involving heat shock proteins, immunophilins, and p23. J Biol Chem 269:24989–24993

Kanelakis KC, Morishima Y, Dittmar KD, Galigniana MD, Takayama S, Reed JC, Pratt WB (1999) Differential effects of the hsp70-binding protein BAG-1 on glucocorticoid receptor folding by the hsp90-based chaperone machinery. J Biol Chem 274:34134–34140

Kanelakis KC, Murphy PJM, Galigniana MD, Morishima Y, Takayama S, Reed JC, Toft DO, Pratt WB (2000) hsp70 interacting protein Hip does not affect glucocorticoid receptor folding by the hsp90-based chaperone machinery except to oppose the effect of BAG-1. Biochemistry 39:14314–14321

Kanelakis KC, Shewach DS, Pratt WB (2002) Nucleotide binding states of hsp70 and hsp90 during sequential steps in the process of glucocorticoid receptor–hsp90 heterocomplex assembly. J Biol Chem 277:33698–33703

Kang KI, Devin J, Cadepond F, Jibard N, Guiochon-Mantel A, Baulieu EE, Catelli MG (1994) In vivo functional protein-protein interaction: nuclear targeted hsp90 shifts cytoplasmic steroid receptor mutants into the nucleus. Proc Natl Acad Sci U S A 91:340–344

Kaul S, Murphy PJM, Chen J, Brown L, Pratt WB, Simons SS (2002) Mutations at positions 547–553 of rat glucocorticoid receptors reveal that hsp90 binding requires the presence, but not defined composition, of a seven-amino acid sequence at the amino terminus of the ligand binding domain. J Biol Chem 277:36223–36232

Kazlauskas A, Poellinger L, Pongratz I (2000) The immunophilin-like protein XAP2 regulates ubiquitination and subcellular localization of the dioxin receptor. J Biol Chem 275:41317–41324

Kazlauskas A, Sundstrom S, Poellinger L, Pongratz I (2001) The hsp90 chaperone complex regulates intracellular localization of the dioxin receptor. Mol Cell Biol 21:2594–2607

Kimura Y, Yahara I, Lindquist S (1995) Role of the protein chaperone YDJ1 in establishing hsp90-mediated signal transduction pathways. Science 268:1362–1365

Kosano H, Stensgard B, Charlesworth MC, McMahon N, Toft DO (1998) The assembly of progesterone receptor-hsp90 complexes using purified proteins. J Biol Chem 273:32973–32979

Kost SL, Smith DF, Sullivan WP, Welch WJ, Toft DO (1989) Binding of heat shock proteins to the avian progesterone receptor. Mol Cell Biol 9:3829–3838

Langer T, Lu C, Echols H, Flanagan J, Hayer MK, Hartl FU (1992) Successive action of DnaK, DnaJ and GroEL along the pathway of chaperone-mediated protein folding. Nature 356:683–689

Liu J, DeFranco DB (1999) Chromatin recycling of glucocorticoid receptors: implications for multiple roles of heat shock protein 90. Mol Endocrinol 13:355–365

Luders J, Demand J, Hohfeld J (2000) The ubiquitin-related BAG-1 provides a link between the molecular chaperones Hsc70/Hsp70 and the proteasome. J Biol Chem 275:4613–4617

McLaughlin SH, Smith HW, Jackson SE (2002) Stimulation of the weak ATPase activity of human hsp90 by a client protein. J Mol Biol 315:787–798

McNally JG, Muller WG, Walker D, Wolford R, Hager GL (2000) The glucocorticoid receptor: rapid exchange with regulatory sites in living cells. Science 287:1262–1265

Minami Y, Hohfeld J, Ohtsuka K, Hartl FU (1996) Regulation of the heat-shock protein 70 reaction cycle by the mammalian DnaJ homolog, hsp40. J Biol Chem 271:19617–19624

Modarress KJ, Opoku J, Xu M, Sarlis NJ, Simons SS (1997) Steroid-induced conformational changes at ends of the hormone-binding domain in the rat glucocorticoid receptor are independent of agonist versus antagonist activity. J Biol Chem 272:23986–23994

Morishima Y, Kanelakis KC, Murphy PJM, Shewach DS, Pratt WB (2001) Evidence for iterative ratcheting of receptor-bound hsp70 between its ATP and ADP conformations during assembly of glucocorticoid receptor–hsp90 heterocomplexes. Biochemistry 40:1109–1116

Morishima Y, Kanelakis KC, Silverstein AM, Dittmar KD, Estrada L, Pratt WB (2000a) The hsp organizer protein Hop enhances the rate of but is not essential for glucocorticoid receptor folding by the multiprotein hsp90-based chaperone system. J Biol Chem 275:6894–6900

Morishima Y, Murphy PJM, Li DP, Sanchez ER, Pratt WB (2000b) Stepwise assembly of a glucocorticoid receptor–hsp90 heterocomplex resolves two sequential ATP-dependent events involving first hsp70 and then hsp90 in opening of the steroid binding pocket. J Biol Chem 275:18054–18060

Morishima Y, Kanelakis KC, Murphy PJM, Lowe ER, Jenkins GJ, Osawa Y, Sunahara RK, Pratt WB (2003) The hsp90 cochaperone p23 is the limiting component of the multiprotein hsp90/hsp70-based chaperone system in vivo where it acts to stabilize the client protein–hsp90 complex. J Biol Chem 278:48754–48763

Murphy PJM, Kanelakis KC, Galigniana MD, Morishima Y, Pratt WB (2001) Stoichiometry, abundance, and functional significance of the hsp90/hsp70-based multiprotein chaperone machinery in reticulocyte lysate. J Biol Chem 276:30092–30098

Murphy PJM, Morishima Y, Chen H, Galigniana MD, Mansfield JF, Simons SS, Pratt WB (2003) Visualization and mechanism of assembly of a glucocorticoid receptor–hsp70 complex that is primed for subsequent hsp90-dependent opening of the steroid binding cleft. J Biol Chem 278:34764–34773

Nathan DF, Lindquist S (1995) Mutational analysis of Hsp90 function: interactions with a steroid receptor and a protein kinase. Mol Cell Biol 15:3917–3925

Neckers L, Schulte TW, Mimnaugh E (1999) Geldanamycin as a potential anti-cancer agent: its molecular target and biochemical activity. Invest New Drugs 17:361–373

Nelson GM, Prapapanich V, Carrigan PE, Roberts PJ, Riggs DL, Smith DF (2004) The heat shock protein 70 cochaperone Hip enhances functional maturation of glucocorticoid receptor. Mol Endocrinol 18:1620–1630

Nishi M, Ogawa H, Ito T, Matsuda KI, Kawata M (2001) Dynamic changes in subcellular localization of mineralocorticoid receptor in living cells: in comparison with glucocorticoid receptor using dual-color labeling with green fluorescent protein spectral variants. Mol Endocrinol 15:1077–1092

O'Brien MC, McKay DB (1995) How potassium affects the activity of the molecular chaperone Hsc70. I. Potassium is required for optimal ATPase activity. J Biol Chem 270:2247–2250

Oppermann H, Levinson W, Bishop JM (1981) A cellular protein that associates with the transforming protein of Rous sarcoma virus is also a heat-shock protein. Proc Natl Acad Sci U S A 78:1067–1071

Owens-Grillo JK, Hoffmann K, Hutchison KA, Yem AW, Deibel MR, Handschumacher RE, Pratt WB (1995) The cyclosporin A-binding immunophilin CyP-40 and the FK506-binding immunophilin hsp56 bind to a common site on hsp90 and exist in independent cytosolic heterocomplexes with the untransformed glucocorticoid receptor. J Biol Chem 270:20479–20484

Oxelmark E, Knoblauch R, Arnal S, Su LF, Schapira M, Garabedian MJ (2003) Genetic dissection of p23, an hsp90 cochaperone, reveals a distinct surface involved in estrogen receptor signaling. J Biol Chem 278:36547–36555

Panaretou B, Siligardi G, Meyer P, Maloney A, Sullivan JK, Singh S, Millson SH, Clarke PA, Naaby-Hansen S, Stein R, Cramer R, Mollapour M, Workman P, Piper PW, Pearl LH, Prodromou C (2002) Activation of the ATPase activity of hsp90 by the stress-regulated cochaperone Aha1. Mol Cell 10:1307–1318

Picard D (1993) Steroid-binding domains for regulating the functions of heterologous proteins in cis. Trends Cell Biol 3:278–280

Picard D, Khursheed B, Garabedian MJ, Fortin MG, Lindquist S, Yamamoto KR (1990) Reduced levels of hsp90 compromise steroid receptor action in vivo. Nature 348:166–168

Pickart CM (2004) Back to the future with ubiquitin. Cell 116:181–190

Prapapanich V, Chen S, Smith DF (1998) Mutation of Hip's carboxy-terminal region inhibits a transitional stage of progesterone receptor assembly. Mol Cell Biol 18:944–952

Pratt WB (1993) The role of heat shock proteins in regulating the function, folding, and trafficking of the glucocorticoid receptor. J Biol Chem 268:21455–21458

Pratt WB, Toft DO (1997) Steroid receptor interactions with heat shock protein and immunophilin chaperones. Endocr Rev 18:306–360

Pratt WB, Toft DO (2003) Regulation of signaling protein function and trafficking by the hsp90/hsp70-based chaperone machinery. Exp Biol Med 228:111–133

Pratt WB, Jolly DJ, Pratt DV, Hollenberg SM, Giguere V, Cadepond FM, Schweizer-Groyer G, Catelli MG, Evans RM, Baulieu EE (1988) A region in the steroid binding domain determines formation of the non-DNA-binding, 9 S glucocorticoid receptor complex. J Biol Chem 263:267-273

Pratt WB, Galigniana MD, Harrell JM, DeFranco DB (2004) Role of hsp90 and the hsp90-binding immunophilins in signaling protein movement. Cell Signal 16:857-872

Prodromou C, Roe SM, O'Brien R, Ladbury JE, Piper PW, Pearl LH (1997) Identification and structural characterization of the ATP/ADP-binding site in the hsp90 molecular chaperone. Cell 90:65-75

Renoir JM, Mercier-Bodard C, Hoffmann K, Le Bihan S, Ning YM, Sanchez ER, Handschumacher RE, Baulieu EE (1995) Cyclosporin A potentiates the dexamethasone-induced mouse mammary tumor virus-chloramphenicol acetyltransferase activity in LMCAT cells: a possible role for different heat shock protein-binding immunophilins in glucocorticosteroid receptor-mediated gene expression. Proc Natl Acad Sci U S A 92:4977-4981

Riggs DL, Roberts PJ, Chirillo SC, Cheung-Flynn J, Prapapanich V, Ratajczak T, Gaber R, Picard D, Smith DF (2003) The hsp90-binding peptidylprolyl isomerase FKBP52 potentiates glucocorticoid signaling in vivo. EMBO J 22:1158-1167

Sanchez ER, Toft DO, Schlesinger MJ, Pratt WB (1985) Evidence that the 90-kDa phosphoprotein associated with the untransformed L-cell glucocorticoid receptor is a murine heat-shock protein. J Biol Chem 260:12398-12401

Sanchez ER, Meshinchi S, Tienrungroj W, Schlesinger MJ, Toft DO, Pratt WB (1987) Relationship of the 90-kDa murine heat shock protein to the untransformed and transformed states of the L cell glucocorticoid receptor. J Biol Chem 262:6986-6991

Sanchez ER, Faber LE, Henzel WJ, Pratt WB (1990) The 56-59-kilodalton protein identified in untransformed steroid receptor complexes is a unique protein that exists in cytosol in a complex with both the 70- and 90-kilodalton heat shock proteins. Biochemistry 29:5145-5152

Savory JGA, Hsu B, Laquian IR, Giffin W, Reich T, Hache RJG, Lefebvre YA (1999) Discrimination between NL1- and NL2-mediated nuclear localization of the glucocorticoid receptor. Mol Cell Biol 19:1025-1037

Scherrer LC, Dalman FC, Massa E, Meshinchi S, Pratt WB (1990) Structural and functional reconstitution of the glucocorticoid receptor-hsp90 complex. J Biol Chem 265:21397-21400

Scherrer LC, Hutchison KA, Sanchez ER, Randall SK, Pratt WB (1992) A heat shock protein complex isolated from rabbit reticulocyte lysate can reconstitute a functional glucocorticoid receptor-hsp90 complex. Biochemistry 31:7325-7329

Scherrer LC, Picard D, Massa E, Harmon JM, Simons SS, Yamamoto, KR, Pratt WB (1993) Evidence that the hormone binding domain of steroid receptors confers hormonal control on chimeric proteins by determining their hormone-regulated binding to heat-shock protein 90. Biochemistry 32:5381-5386

Schmidt U, Wochnik GM, Rosenhagen MC, Young JC, Hartl FU, Holsboer F, Rein T (2003) Essential role of the unusual DNA-binding motif of BAG-1 for inhibition of the glucocorticoid receptor. J Biol Chem 278:4926-4931

Schneikert J, Hubner S, Langer G, Petri T, Jaattela M, Reed J, Cato ACB (2000) Hsp70-RAP46 interaction in downregulation of DNA binding by glucocorticoid receptor. EMBO J 19:6508-6516

Schowalter DB, Sullivan WP, Maihle NJ, Dobson ADW, Conneely OM, O'Malley BW, Toft DO (1991) Characterization of progesterone receptor binding to the 90- and 70-kDa heat shock proteins. J Biol Chem 266:21165-21173

Schuh S, Yonemoto W, Brugge J, Bauer VJ, Riehl RM, Sullivan WP, Toft DO (1985) A 90,000-dalton binding protein common to both steroid receptors and the Rous sarcoma virus transforming protein, pp60$^{v\text{-}src}$. J Biol Chem 260:14292–14296

Segnitz B, Gehring U (1997) The function of steroid hormone receptors is inhibited by the hsp90-specific compound geldanamycin. J Biol Chem 272:18694–18701

Sepp-Lorenzino L, Ma Z, Lebwohl DE, Vinitsky A, Rosen N (1995) Herbimycin A induces the 20 S proteasome- and ubiquitin-dependent degradation of receptor tyrosine kinases. J Biol Chem 270:16580–16587

Silverstein AM, Galigniana MD, Chen MS, Owens-Grillo JK, Chinkers M, Pratt WB (1997) Protein phosphatase 5 is a major component of glucocorticoid receptor–hsp90 complexes with properties of an FK506-binding immunophilin. J Biol Chem 272:16224–16230

Silverstein AM, Galigniana MD, Kanelakis KC, Radanyi C, Renoir JM, Pratt WB (1999) Different regions of the immunophilin FKBP52 determine its association with the glucocorticoid receptor, hsp90, and cytoplasmic dynein. J Biol Chem 274:36980–36986

Smith DF (1993) Dynamics of heat shock protein 90-progesterone receptor binding and the disactivation loop model for steroid receptor complexes. Mol Endocrinol 7:1418–1429

Smith DF, Schowalter DB, Kost SL, Toft DO (1990) Reconstitution of progesterone receptor with heat shock proteins. Mol Endocrinol 4:1704–1711

Smith DF, Stensgard BA, Welch WJ, Toft DO (1992) Assembly of progesterone receptor with heat shock proteins and receptor activation are ATP mediated events. J Biol Chem 267:1350–1356

Smith DF, Sullivan WP, Marion TN, Zaitsu K, Madden B, McCormick DJ, Toft DO (1993) Identification of a 60-kilodalton stress-related protein, p60, which interacts with hsp90 and hsp70. Mol Cell Biol 13:869–876

Stancato LF, Silverstein AM, Gitler C, Groner B, Pratt WB (1996) Use of the thiol-specific derivatizing agent N-iodoacetyl-3-[^{125}I]iodotyrosine to demonstrate conformational differences between the unbound and hsp90-bound glucocorticoid receptor hormone binding domain. J Biol Chem 271:8831–8836

Stavreva DA, Muller WG, Hager GL, Smith CL, McNally JG (2004) Rapid glucocorticoid receptor exchange at a promoter is coupled to transcription and regulated by chaperones and proteasomes. Mol Cell Biol 24:2682–2697

Stebbins CE, Russo AA, Schneider C, Rosen N, Hartl FU, Pavletich NP (1997) Crystal structure of an Hsp90-geldanamycin complex: targeting of a protein chaperone by an antitumor agent. Cell 89:239–250

Sullivan W, Stensgard B, Caucutt G, Bartha B, McMahon N, Alnemri ES, Litwack G, Toft D (1997) Nucleotides and two functional states of hsp90. J Biol Chem 272:8007–8012

Szabo A, Langer T, Schroder H, Flanagan J, Bukau B, Hartl FU (1994) The ATP hydrolysis-dependent reaction cycle of the *Escherichia coli* hsp70 system—DnaK, DnaJ, and GrpE. Proc Natl Acad Sci U S A 91:10345–10349

Tanaka M, Nishi M, Morimoto M, Sugimoto T, Kawata M (2003) Yellow fluorescent protein-tagged and cyan fluorescent protein-tagged imaging analysis of glucocorticoid receptor and importins in single living cells. Endocrinology 144:4070–4079

Thomas M, Dadgar N, Aphale A, Harrell JM, Kunkel R, Pratt WB, Lieberman AP (2004) Androgen receptor acetylation site mutations cause trafficking defects, misfolding, and aggregation similar to expanded glutamine tracts. J Biol Chem 279:8389–8395

Whitesell L, Cook P (1996) Stable and specific binding of heat shock protein 90 by geldanamycin disrupts glucocorticoid receptor function in intact cells. Mol Endocrinol 10:705–712

Whitesell L, Mimnaugh EG, De Costa B, Myers CE, Neckers LM (1994) Inhibition of heat shock protein HSP90-pp60^{v-src} heteroprotein complex formation by benzoquinone ansamycins: essential role for stress proteins in oncogenic transformation. Proc Natl Acad Sci U S A 91:8324–8328

Wiech H, Buchner J, Zimmermann R, Jakob U (1992) Hsp90 chaperones protein folding in vitro. Nature 358:169–170

Wochnik GM, Young JC, Schmidt U, Holsboer F, Hartl FU, Rein T (2004) Inhibition of GR-mediated transcription by p23 requires interaction with hsp90. FEBS Lett 560:35–38

Xu M, Chakraborti PK, Garabedian MJ, Yamamoto KR, Simons SS (1996) Modular structure of glucocorticoid receptor domains is not equivalent to functional independence. J Biol Chem 271:21430–21438

Xu M, Dittmar KD, Giannoukos G, Pratt WB, Simons SS (1998) Binding of hsp90 to the glucocorticoid receptor requires a specific 7-amino acid sequence at the amino terminus of the hormone-binding domain. J Biol Chem 273:13918–13924

Yang J, Liu J, DeFranco DB (1997) Subnuclear trafficking of glucocorticoid receptors in vitro: chromatin recycling and nuclear export. J Cell Biol 137:523–538

Zeiner M, Gehring U (1995) A protein that interacts with members of the nuclear hormone receptor family: identification and cDNA cloning. Proc Natl Acad Sci U S A 92:11465–11469

Heat Shock Response: Lessons from Mouse Knockouts

E. S. Christians[1] (✉) · I. J. Benjamin[2]

[1] Centre de Biologie du Développement, UMR5547, 118 route de Narbonne, 31062 Toulouse, France
Elisabeth.Christians@cict.fr

[2] The University of Utah Health Sciences Center, 30 North 1900 East, Room 4A100, Salt Lake City UT, 84132-2401, USA

1	Introduction to Heat Shock Response and Mouse Models for Loss of Function	140
1.1	Heat Shock Response: Definition	140
1.2	Mouse Knockouts: So far, a Strategy to Study Loss of Function	141
2	HSF Knockouts	142
2.1	HSF and Regulation of the Heat Shock Response	143
2.2	HSF and Thermotolerance	144
2.3	HSF, Inflammation, Immunoreaction, and Heat Shock Response	144
2.4	Regulators of Heat Shock Response and Reproduction	145
2.4.1	HSF and Female Reproduction	145
2.4.2	HSF and Male Reproduction	146
3	HSP and Heat Shock-Dependent Gene Knockouts	146
3.1	Hsp Knockouts	146
3.2	Non-Hsp HSF-Dependent Genes	148
4	HSF Knockouts: Future Directions	149
	References	149

Abstract Organisms are endowed with integrated regulatory networks that transduce and amplify incoming signals into effective responses, ultimately imparting cell death and/or survival pathways. As a conserved cytoprotective mechanism from bacteria to humans, the heat shock response has been established as a paradigm for inducible gene expression, stimulating the interests of biologists and clinicians alike to tackle fundamental questions related to the molecular switches, lineage-specific requirements, unique and/or redundant roles, and even efforts to harness the response therapeutically. Gene targeting studies in mice confirm HSF1 as a master regulator required for cell growth, embryonic development, and reproduction. For example, sterility of Hsf1-null female but not null male mice established strict requirements for maternal HSF1 expression in the oocyte. Yet Hsf2 knockouts by three independent laboratories have not fully clarified the role of mammalian HSF2 for normal development, fertility, and postnatal neuronal function. In contrast, Hsf4 knockouts have provided a consistent demonstration for HSF4's critical role during lens formation. In the future, molecular analysis of HSF knockout mice will bring new insights to HSF interactions, foster better understanding of gene regulation at the genome level, lead to a better integration of the HSF pathway in life beyond heat shock, the classical laboratory challenge.

Keywords HSF · Transcription · Development · Disease

1
Introduction to Heat Shock Response and Mouse Models for Loss of Function

1.1
Heat Shock Response: Definition

First discovered in flies upon heat or chemical treatment (see historical article, Ritossa 1996), the heat shock response can be defined as a rapid and transient induction of a set of conserved classes of polypeptides historically called heat shock proteins (Hsps). This response is highly conserved from bacteria to animals and humans. As a paradigm of transcriptional regulation, the heat shock response has been extensively investigated and described in numerous review papers. Readers will find additional detailed information in Morimoto and Santoro 1998; Pirkkala et al. 2001; Voellmy 2004; Wu 1995.

In this chapter, we intend to adhere to a rather limited definition of the heat shock response in order to better define the significance of mouse models targeting specific heat shock-related genes. When living organisms are submitted to severe modifications of either their internal and/or external environment such as elevated temperature and chemical exposure, distinct macromolecular alterations in protein structure and function ensue, depending on the organism's ability to elicit efficient protective mechanisms. It should be noted here that eukaryotic cells have also evolved compartment-specific mechanisms for handling stressful conditions. The most extensively studied, the unfolded protein response (UPR) endoplasmic reticulum (ER) stress response involves an intricate pathway for induction of stress proteins in response to protein unfolding in the endoplasmic reticulum (Ma and Hendershot 2004). Both systems serve complementary roles depending on the stress conditions.

This heat shock response corresponds to the activation of the heat shock transcriptional factor, which binds specific consensus sequences (HSE, heat shock element) located in the regulatory regions of specific genes (heat shock genes coding for Hsps). Thus, the heat shock response could be staged to a three-step pathway: stress → HSF → HSP. Recently, however, Trinklein and collaborators have updated analysis of HSE motifs and developed an extensive genome-wide investigation of HSF target genes (Trinklein et al. 2004). Although previously suggested by several studies (e.g., HSF-dependent expression of MDR, Kim et al. 1997; HSF-dependent repression of TNF-α, Singh et al. 2002), those authors have clearly demonstrated that the third step of the heat shock pathway is more diversified than expected. Within the transcriptome of mammalian cells, numerous genes, classified as Hsp or not, are dependent on HSF in a direct or indirect manner. Indeed, there is no pervasive general stress response and only few genes are repeatedly recruited in response to different stresses, as described in the study published by Murray and collaborators (Murray et al. 2004).

Beyond the heat shock (HS) response itself, thermotolerance is an important phenomenon that depends on the molecular changes induced by HS response and explains how organisms can cope with successive stresses and enhance survival (Joyeux-Faure et al. 2003; Pespeni et al. 2005).

In this review, we will discuss how mouse knockout models have provided new insights into the current knowledge acquired on HSF(s) and HSF target genes (including Hsp and non-Hsp) and their involvement in the heat shock response and thermotolerance.

1.2
Mouse Knockouts: So far, a Strategy to Study Loss of Function

While heat shock response is highly conserved through evolution, it has evolved at numerous levels from the number of heat shock-related regulators to target genes and mechanisms of transcriptional activation (e.g., RNA pol II pausing in flies; Lis 1998).

From an abundant literature, we already know that the heat shock response exhibits cell–tissue–organ and stress-specific patterns (e.g., Murray et al. 2004; Xiao et al. 1999). Beyond the artificial experimental conditions used for such studies, more robust approaches were needed to fully understand the mechanisms and function of heat shock response in mammals. In this manner, the full repertoire of the heat shock response integrates the physiology of multicellular organisms that display homeostatic protection mechanisms. In particular, experimental strategies based on loss of function such as mouse knockout are invaluable for the careful and extended analysis of the resulting phenotype, especially those associated with (pathological) physiological challenges.

The technology for mouse knockout was first developed at the end of 1980s and since then, hundreds of genes have been targeted, generating numerous mutant mice. By using homologous recombination, a portion of the endogenous gene is replaced by a cassette containing genes for selection and/or reporter to follow expression. Consequently, the endogenous gene is either interrupted with or without deletion. Additional refinements of this procedure enable highly specific changes, e.g. at the level of a single amino acid. Indeed, the NIH supports a comprehensive knockout mouse consortium to make available a null mutant mouse model per gene, for corresponding information (Austin et al. 2004). Present information on published knockout models can be retrieved from the following website (http://www.informatics.jax.org/).

Nevertheless, various parameters can interfere with the loss of function of the targeted gene. First, when the genome was incompletely known, it was possible that the deleted region included not only the desired gene but also some not yet discovered genes (e.g., αB-crystallin/HspB5 and MKBP/HspB2; Brady et al. 2001). Second, long-distance effects have also been reported, indicating that genomic modifications can impact activity of genes located at remote sites (e.g., Pham et al. 1996). Third, the genetic background can influence the phenotype of mouse knockout as different forms of modifier genes can change the overall effect of the loss of function (e.g., Wolfer et al. 2002). These caveats become important when using and analyzing results obtained with mouse knockout models.

2
HSF Knockouts

In contrast to yeast and flies, vertebrates and in particular mammals count at least three heat shock factors (HSFs: HSF1, 2, 4). The question of their

Table 1 HSF knockout mice

HSF	Line ID	Major phenotypes/studies
Hsf1	Hsf1^{tm1Anak}	Increased cell death in spermatogonia
		Reduced cell death in spermatocytes
		Reduced sensory hair cell survival after acoustic overexposure
		Impaired IgG production
	Hsf1^{tm1Ijb}	Absence of stress response
		Lack of thermotolerance
		Partial embryonic lethality
		Placenta defect
		Reduced size
		Female infertility
		Reduced viability to endotoxinic shock
	Hsf1^{tm1Miv}	Absence of stress response
		Lack of thermotolerance
		Normal viability
		Female infertility
Hsf2	Hsf2^{tm1Ijb}	Viable, fertile
	Hsf2^{tm1Miv}	Reduced viability
		Partial male and female infertility
		Central nervous system defect
	Hsf2^{tm1Mmr}	Viable
		Reduced female fertility
		Brain abnormalities
Hsf4	Hsf4^{tm1Anak}	Viable
		Cataract, lens abnormalities
	Hsf4^{tm1Miv}	Viable
		Cataract, lens abnormalities

specific roles was addressed by several laboratories exploiting the homologous recombination to induce specific HSF loss of function (Pirkkala et al. 2001; Voellmy 2004). The first HSF knockout was published in 1998 and the last ones (HSF4), recently in 2004, providing the scientific community with valuable models to study HSF function under health and disease conditions. After initial characterization, analyses of HSF models are far from being complete (Table 1).

2.1
HSF and Regulation of the Heat Shock Response

McMillan et al. (1998) demonstrated that HSF1 is essential to activate the heat shock response and to induce transcription of the Hsp25/27, Hsp70, and to a lesser extent the Hsp60 gene (McMillan et al. 1998). Xiao et al. (1999) investigated the HS response in intact organisms by Western blotting of a defined series of Hsps (Hsp25/27, Hsp60, Hsp70, Hsc70, Hsp90 α and β) to similarly conclude that, in absence of HSF1, those genes did not exhibit any induction (Xiao et al. 1999). Comparable data were obtained in cells and organs from the two other Hsf1 knockouts: Hsp25/27, Hsp70, and Hsp105 require HSF1 to be induced by heat stress (Inouye et al. 2003; Zhang et al. 2002).

In contrast to well-defined experimental stressful conditions such as elevated temperature, Hsp regulation by HSF1 under physiological conditions remains unclear, with contrary conclusions among studies (Inouye et al. 2003; Yan et al. 2002; Zheng and Li 2004). Multiple factors such as culture conditions in case of cell culture studies and the health status or genetic background in the intact organisms might influence basal expression of HSPs. However, Trinklein et al. (2004) have examined gene expression with and without HS using $Hsf1^{+/+}$ and $Hsf1^{-/-}$ fibroblasts (MEFs, mouse embrionic fibroblasts) and have unambiguously demonstrated that absence of HSF affects gene expression in nonstress conditions. Future studies are needed to reconcile the apparent discrepancy.

In HSF2-deficient cells, heat shock clearly induced Hsp genes such as Hsp25/27, Hsp70 confirming that HSF2 is not a major regulator of heat shock response (McMillan et al. 2002). Nevertheless, when a more detailed analysis was performed, HSF2 loss of function was associated with altered sensitivity to heat shock, exemplified by induction at a lower temperature and increased Hsp25/27 response (Paslaru et al. 2003). Other studies have indicated that HSF2 can interact with HSF1 and bookmark Hsp genes to further stimulate heat shock response (He et al. 2003; Xing et al. 2005). Therefore, the demonstration of HSF2 positive or/and negative effect on HS response will require more detailed investigation, taking into account parameters such as stress, cell type, cell cycle, and target gene.

How different members of HSF family are engaged physiologically could be that the approach organisms fine tune response to stress, making HSF knockouts the ideal models to investigate such functional interactions. Double knockouts, HSF1-HSF4, have already suggested a potential antagonistic effect of these two factors on FGF expression (Fujimoto et al. 2004) in addition to changes on Hsp expression, in particular Hsp25/27 (Fujimoto et al. 2004). The heat shock stress response is awaiting further analysis in this double knockout as well as in *Hsf1-Hsf2* double knockout (Fujimoto et al. 2004; Wang et al. 2004).

2.2
HSF and Thermotolerance

As mentioned previously, thermotolerance arising from the heat shock response is an important protective mechanism against stress-induced apoptosis and cell death. The absence of HSF1 abrogates such protection in mouse embryonic fibroblasts and expression of Hsp25 and Hsp70 is insufficient to restore it (Luft et al. 2001). So Hsp70 is alone is an inadequate surrogate for the heat shock response and other HSF1 target genes as those revealed by the study published by Trinklein and collaborators are likely to be essential to stress protection (Trinklein et al. 2004).

Using $Hsf1^{-/-}$ mice ($Hsf1^{tm1Ijb}$), Wirth et al. (2004) contributed additional information on the stress tolerance phenomenon in lung. When the protective effect of a mild thermal stress against an acute toxic exposure undertaken by intranasal instillation of cadmium, a proven HSF1 activator, was studied, Wirth et al. (2003) demonstrated that HSF1 is the only essential factor for cross-tolerance and this protection exists only within a defined range of stress severity. Not only does thermotolerance depend on the activation of numerous genes, but also the HSF1 pathway most likely exhibits tissue-specific protection in vivo.

2.3
HSF, Inflammation, Immunoreaction, and Heat Shock Response

Although fever or elevated temperature is a common feature in inflammation and immunoreaction, the potential link with HSF1 pathway remains unapproachable until Hsf1 knockout mice become available.

Cytokines such as TNF-α are important mediators of inflammation and infection. Expression of TNF-α is controlled at many steps: transcription, mRNA stability, and excretion. Both in vivo and in vitro experiments indicated that HSF1 is a transcriptional repressor of this cytokine, which contains a HSE-like sequence in its regulatory 3' region (Singh et al. 2002). HSF1 is also involved in immunoglobulin production through cytokine regulation. In contrast to the repression of IL-1β or TNF-α, HSF1 is an activator for IL-6 and CCL5, which promotes antigen-specific IgG2a production (Inouye et al. 2004).

Important in both innate and adaptive immunity, dendritic cells (DCs) can be activated by heat shock in absence of other exogenous inflammatory stimuli, triggering the expression of CD80, CD86, CD40, MHCII, and MHCI, markers of maturation. Heat shock activates DCs what leads to the formation of peculiar aggresomes made of polyubiquitinated proteins and consequent Hsp70 induction, even in absence of HSF1. This process is not HSF1-dependent but appears to be under the control of another pathway that remains to be elucidated (Zheng et al. 2003). In contrast, HSF1 plays an important role in antigen cross-presentation (Ag XP), as recently demonstrated by Zheng and Li (2004). This role is mediated by the HSF1 regulation of Hsp70-Hsp90 expression under normal conditions.

No data are available regarding HSF2 or 4 roles in inflammation or immunity. As HSF4 is abundant in lungs and based on the fact that this factor can interact with HSF1 functions, it would be interesting to test further the pathophysiological consequences of HSF4 deficiency in either innate or adaptive immunity, or both.

2.4
Regulators of Heat Shock Response and Reproduction

While HSFs are involved in defensive and protective mechanisms such as inflammation-immunity in cells and organisms, it was intuitively expected that HSF1 and HSF2 knockouts would lead to severe phenotypes affecting female and male fertility. If there is any link between reproduction and stress, then protective mechanisms must be recruited against even very small environmental changes that could impact on the gonad's ability to produce perfect gametes as required for efficient reproduction.

2.4.1
HSF and Female Reproduction

In $Hsf1^{-/-}$ females, ovulated oocytes can be fertilized but the resulting embryos are blocked and die after one or two cleavages. This phenotype of very early embryonic lethality depends on female genotype and not offspring genotype, meaning that $Hsf1$ knockout corresponds to a maternal effect mutation (Christians et al. 2000). HSF mutation previously reported in *Drosophila melanogaster* also affected the very early stage of oogenesis, but Jedlicka and collaborators concluded that this was not caused by a reduced expression of the main Hsp, and the ultimate explanation for this phenotype remains presently unsettled (Jedlicka et al. 1997).

In contrast to the severe phenotype observed in HSF1-deficient females, $Hsf2^{-/-}$ females were only affected in some knockout lines showing a reduced number of pups or a more severe defect leading to the presence of large hemorrhagic follicles with nonovulated oocytes (Kallio et al. 2002; McMillan et al. 2002; Wang et al. 2003).

$Hsf4^{-/-}$ females are fertile, indicating that even if HSF4 is expressed during oogenesis—which remains to be determined—there is no functional interference with the abundantly expressed HSF1 (Fujimoto et al. 2004).

2.4.2
HSF and Male Reproduction

Hsf1-null males exhibit normal fertility unless they are challenged by a stress such as heat shock (Izu et al. 2004). Discordant data were obtained regarding the fertility of HSF2-deficient males: spermatogenesis in Hsf2tm1^{Miv} Hsf2tm1^{Mmr} males seemed to be severely affected, with 80% and 58% reductions in sperm count, respectively (Kallio et al. 2002; Wang et al. 2003). In contrast, studying Hsf2tm1^{IJB} male fertility, we could not detect such dramatic alterations but only a potential faster aging process with the appearance of degenerated tubules in males older than 3 months (McMillan et al. 2002). As mentioned above, discordant results among knockout mouse lines are not uncommon.

HSF4-deficient males remain fertile, indicating that HSF4 is not involved in spermatogenesis and/or redundant roles are subserved by other HSFs (Fujimoto et al. 2004).

3
HSP and Heat Shock-Dependent Gene Knockouts

Until recently, heat shock response was generally described as inducing Hsp expression and most analyses were limited to members of Hsp70 family. Accumulating data showing that numerous non-Hsp genes are differentially expressed after HS have shifted our view of the heat shock response: namely, this cannot be restricted to Hsp70 alone. Therefore, the following paragraph attempts an integration between Hsp knockout and heat shock-dependent gene knockout using the microarray database published by Trinklein et al. (2004) as the primary source.

3.1
Hsp Knockouts

While there are several multigenic Hsp families, the gene-targeting approach has been repeatedly applied to Hsp70.1 and Hsp70.3, known as the major inducible Hsp genes in mice (Table 2). These two genes are located in a tandem array on the major histocompatibility complex (MHC) region on the mouse chromosome 17.

Chronologically, the first Hsp70.1 knockout was created in Korea by Seo and collaborators (Lee et al. 2001). With this animal model, they first showed that infarction volume fololwing focal cerebral ischemia was increased by 30% in Hsp70.1-null mice relative to wild type mice, demonstrating the protective function of Hsp70.1 against cell death and the absence of a complete compensatory effect of the second inducible Hsp70.3. Absence of active redundancy between Hsp70.1 and Hsp70.3 heat shock response was also reported by Huang

Table 2 Characteristics: Hsp70.1 and Hsp70.3 knockout mice

Gene	Knockout mouse line	Phenotype/Challenge
Hsp70.1	Hspa1b^{tm1Seo}	No gross abnormality
		Increased infarction volume after focal cerebral ischemia
	Hspa1b$^{tm1.1Msk}$	No gross abnormality
	Hspa1b^{tm1Clib}	Normal, fertile
		Absence of HS preconditioning in TNF-induced lethality
Hsp70.3	Hspa1a$^{tm1.1Msk}$	No gross abnormality
Hsp70.1– Hsp70.3	Hspa1a/Hspa1b^{tm1Dix}	Reduced size
		Increased sensitivity to irradiation
		Absence of late-phase protection in ischemic preconditioning
		Genome instability

et al. (2001). In order to further characterize specific Hsp70 gene response, Lee and Seo tested Hsp70.1 expression in MEFs with various stressors such as amino acid analog, heavy metals, oxidative stress, hyperosmotic stress (azetidine, cadmium chloride, zinc chloride sodium arsenite, hydrogen peroxide, and sodium chloride) (Lee and Seo 2002). In MEFs, Hsp70.1, which is expressed at a higher level than Hsp70.3 under basal and stress conditions, is the only gene induced by hyperosmolarity. Although these authors analyzed Hsp70.1 knockout cells in this study, they did not correlate the lack of Hsp70.1 expression and the ability of the cells to survive. In an additional study, they made this link by testing the effect of osmotic stress in kidney (Shim et al. 2002). When Hsp70.1 mice were exposed to this type of challenge by adding 3% NaCl in their drinking water, they exhibited increased apoptosis in their kidneys.

Hsp70.1 response depends on the type of stress (external environment) but also on the specificity in cellular environment as, for example, oxidative stress stimulates the gene in M-1 cells but not in MEFs (Lee and Seo 2002). This suggests that final regulation of Hsp expression is made of particular combinations of stress, genes, and tissue or cell type.

The other Hsp70.1 knockouts were also exploited to demonstrate the protective role of this chaperone against TNF-α toxicity by heat preconditioning (Van Molle et al. 2002) or against heart infarction by ischemic preconditioning (Hampton et al. 2003). In the latter example, both Hsp70.1 and 3 were knocked down. Surprisingly, these animals that were deficient in the main inducible protective chaperones remained viable and fertile, exhibiting only a certain increase in abnormal meiotic chromosomal configurations. Such mild phenotype indicates that cell protection can be complemented by the other chaperones.

Beyond the Hsp70 family, very few successful attempts to disrupt Hsp genes have been reported and those include the knockouts of Hsp70.2, Hsp90β, and HspB5-2 (Brady et al. 2001; Dix et al. 1996; Voss et al. 2000). The first two exhibited developmental phenotypic traits rather than direct stress response-related conditions so they were not described in detail. Hsp70.2 is essential to spermatogenesis, specifically synaptonemal complex and CDC2 kinase chaperoning. Hsp90β is involved in the chorioallantoic fusion required for placenta development and embryo survival beyond 10 days of gestation.

In contrast, HspB5 (or CRYAB, αB-crystallin) is a chaperone induced by heat shock but also highly constitutively expressed in cardiac and skeletal muscle. In absence of HSF1, HspB5 is significantly reduced in heart, confirming that HspB5 depends on this factor not only in stress but also physiological conditions (Yan et al. 2002). A gain of function model was created for HspB5 and demonstrated protection against ischemia reperfusion (Ray et al. 2001). Brady and co-workers have reported that double knockout mice for two small Hsps, HspB5 (CRYAB) and HspB2 (MKBP), exhibit severe age-dependent degeneration. Using the model, Morrison et al. demonstrated increased necrosis and decreased contractile function in the ex vivo heart, indicating a function requirement for one or both sHSPs for acute ischemic conditions (Morrison et al. 2004).

3.2
Non-Hsp HSF-Dependent Genes

Microarray analyses compared gene expression in HSF wild type and HSF-deficient cells or animals (Inouye et al. 2004; Trinklein et al. 2004; Wang et al. 2003, 2004).

The most exhaustive set of data is provided by Trinklein et al., who combined transcriptome analysis and chromatin immunoprecipitation experiments to better identify HSF1 target genes. A brief survey of the 46 genes listed as bound by HSF1 and induced by HS reveals that 18 are identified as Hsp or ubiquitin genes or related genes. The other genes belong to very diverse series of family (e.g., relaxin H1); surprisingly among these genes, little is known regarding a potential link with the HSF1 protective pathway. Another HSF1 microarray study conducted by Inouye et al. provides very limited information regarding the microarray analysis in which HSF1 appears to regulate expression of non-Hsp genes insisting on targets related to immune response (Inouye et al. 2004).

Hsf2 knockout mice were exploited by Wang et al. to search for HSF2 direct or indirect target genes (Wang et al. 2003). Comparative microarrays between $Hsf2^{+/+}$ and $Hsf2^{-/-}$ were performed using samples collected at two different times during embryonic development (8.5 and 10.5) and adult organs (testis).

In the published lists, absence of HSF2 did not alter the expression of any Hsp genes and there are only very few common features between the different tissues tested (e.g., members of apolipoprotein family).

Further analysis of the HSF-dependent transcriptome was undertaken in testis from single and double knockout *Hsf1-Hsf2*. In this organ, expression of several dozen genes depend on HSF1 and HSF2 either directly or indirectly but very few of them appear to be Hsp. In contrast, one of the most affected genes in absence of HSFs was a member of a microtubule-associated protein family; as the named gene, MAP2 is exclusively expressed in neurons (Wang et al. 2004).

4
HSF Knockouts: Future Directions

From 1998 to 2004, the entire family of heat shock factors underwent gene disruption by investigators providing us with several loss of function animal models for HSFs. The next step to be envisioned is manipulation within the HSF gene of distinct elements of function or regulation such as alternative splicing, phosphorylation-sumoylation sites, and bisulfide bonds, leading to a better understanding of the HSF-signaling pathway.

On the HSF gene target side, nearly everything remains to be done, as only the major Hsp70 gene has so far received attention. Due to expected pleiotropic effects, the conditional gene targeting approach must be encouraged.

Together, the field of HSF regulation and Hsp function continues to mature with exciting new directions focused on small animal models of human diseases especially neurodegenerative ones. A new era is now opened to investigators who can extensively challenge these animals in order to mimic human toxicological or pathophysiological conditions as well as to test therapeutic maneuvers.

References

Austin CP, Battey JF, Bradley A, Bucan M, Capecchi M, Collins FS, Dove WF, Duyk G, Dymecki S, Eppig JT, Grieder FB, Heintz N, Hicks G, Insel TR, Joyner A, Koller BH, Lloyd KC, Magnuson T, Moore MW, Nagy A, Pollock JD, Roses AD, Sands AT, Seed B, Skarnes WC, Snoddy J, Soriano P, Stewart DJ, Stewart F, Stillman B, Varmus H, Varticovski L, Verma IM, Vogt TF, von Melchner H, Witkowski J, Woychik RP, Wurst W, Yancopoulos GD, Young SG, Zambrowicz B (2004) The knockout mouse project. Nat Genet 36:921–924

Brady JP, Garland DL, Green DE, Tamm ER, Giblin FJ, Wawrousek EF (2001) AlphaB-crystallin in lens development and muscle integrity: a gene knockout approach. Invest Ophthalmol Vis Sci 42:2924–2934

Christians E, Davis AA, Thomas SD, Benjamin IJ (2000) Maternal effect of Hsf1 on reproductive success. Nature 407:693–694

Dix DJ, Allen JW, Collins BW, Mori C, Nakamura N, Poorman-Allen P, Goulding EH, Eddy EM (1996) Targeted gene disruption of Hsp70-2 results in failed meiosis, germ cell apoptosis, and male infertility. Proc Natl Acad Sci U S A 93:3264–3268

Fujimoto M, Izu H, Seki K, Fukuda K, Nishida T, Yamada S, Kato K, Yonemura S, Inouye S, Nakai A (2004) HSF4 is required for normal cell growth and differentiation during mouse lens development. EMBO J 23:4297–4306

Hampton CR, Shimamoto A, Rothnie CL, Griscavage-Ennis J, Chong A, Dix DJ, Verrier ED, Pohlman TH (2003) Heat-shock proteins 70.1 and 70.3 are required for late-phase protection induced by ischemic preconditioning of the mouse heart. Am J Physiol Heart Circ Physiol 285:H866–H874

He H, Soncin F, Grammatikakis N, Li Y, Siganou A, Gong J, Brown SA, Kingston RE, Calderwood SK (2003) Elevated expression of heat shock factor (HSF) 2A stimulates HSF1-induced transcription during stress. J Biol Chem 278:35465–35475

Huang L, Mivechi NF, Moskophidis D (2001) Insights into regulation and function of the major stress-induced hsp70 molecular chaperone in vivo: analysis of mice with targeted gene disruption of the hsp70.1 or hsp70.3 gene. Mol Cell Biol 21:8575–8591

Inouye S, Izu H, Takaki E, Suzuki H, Shirai M, Yokota Y, Ichikawa H, Fujimoto M, Nakai A (2004) Impaired IgG production in mice deficient for heat shock transcription factor 1. J Biol Chem 279:38701–38709

Inouye S, Katsuki K, Izu H, Fujimoto M, Sugahara K, Yamada S, Shinkai Y, Oka Y, Katoh Y, Nakai A (2003) Activation of heat shock genes is not necessary for protection by heat shock transcription factor 1 against cell death due to a single exposure to high temperatures. Mol Cell Biol 23:5882–5895

Izu H, Inouye S, Fujimoto M, Shiraishi K, Naito K, Nakai A (2004) Heat shock transcription factor 1 is involved in quality-control mechanisms in male germ cells. Biol Reprod 70:18–24

Jedlicka P, Mortin MA, Wu C (1997) Multiple functions of Drosophila heat shock transcription factor in vivo. EMBO J 16:2452–2462

Joyeux-Faure M, Arnaud C, Godin-Ribuot D, Ribuot C (2003) Heat stress preconditioning and delayed myocardial protection: what is new? Cardiovasc Res 60:469–477

Kallio M, Chang Y, Manuel M, Alastalo TP, Rallu M, Gitton Y, Pirkkala L, Loones MT, Paslaru L, Larney S, Hiard S, Morange M, Sistonen L, Mezger V (2002) Brain abnormalities, defective meiotic chromosome synapsis and female subfertility in HSF2 null mice. EMBO J 21:2591–2601

Kim SH, Hur WY, Kang CD, Lim YS, Kim DW, Chung BS (1997) Involvement of heat shock factor in regulating transcriptional activation of MDR1 gene in multidrug-resistant cells. Cancer Lett 115:9–14

Lee JS, Seo JS (2002) Differential expression of two stress-inducible hsp70 genes by various stressors. Exp Mol Med 34:131–6

Lee SH, Kim M, Yoon BW, Kim YJ, Ma SJ, Roh JK, Lee JS, Seo JS (2001) Targeted hsp70.1 disruption increases infarction volume after focal cerebral ischemia in mice. Stroke 32:2905–2912

Lis J (1998) Promoter-associated pausing in promoter architecture and postinitiation transcriptional regulation. Cold Spring Harb Symp Quant Biol 63:347–356

Luft JC, Benjamin IJ, Mestril R, Dix DJ (2001) Heat shock factor 1-mediated thermotolerance prevents cell death and results in G2/M cell cycle arrest. Cell Stress Chaperones 6:326–336

Ma Y, Hendershot LM (2004) ER chaperone functions during normal and stress conditions. J Chem Neuroanat 28:51–65

McMillan DR, Christians E, Forster M, Xiao X, Connell P, Plumier JC, Zuo X, Richardson J, Morgan S, Benjamin IJ (2002) Heat shock transcription factor 2 is not essential for embryonic development, fertility, or adult cognitive and psychomotor function in mice. Mol Cell Biol 22:8005–8014

McMillan DR, Xiao X, Shao L, Graves K, Benjamin IJ (1998) Targeted disruption of heat shock transcription factor 1 abolishes thermotolerance and protection against heat-inducible apoptosis. J Biol Chem 273:7523–7528

Morimoto RI, Santoro MG (1998) Stress-inducible responses and heat shock proteins: new pharmacologic targets for cytoprotection. Nat Biotechnol 16:833–838

Morrison LE, Whittaker RJ, Klepper RE, Wawrousek EF, Glembotski CC (2004) Roles for alphaB-crystallin and HSPB2 in protecting the myocardium from ischemia-reperfusion-induced damage in a KO mouse model. Am J Physiol Heart Circ Physiol 286:H847–H855

Murray JI, Whitfield ML, Trinklein ND, Myers RM, Brown PO, Botstein D (2004) Diverse and specific gene expression responses to stresses in cultured human cells. Mol Biol Cell 15:2361–2374

Paslaru L, Morange M, Mezger V (2003) Phenotypic characterization of mouse embryonic fibroblasts lacking heat shock factor 2. J Cell Mol Med 7:425–435

Pespeni M, Hodnett M, Pittet JF (2005) In vivo stress preconditioning. Methods 35:158–164

Pham CT, MacIvor DM, Hug BA, Heusel JW, Ley TJ (1996) Long-range disruption of gene expression by a selectable marker cassette. Proc Natl Acad Sci U S A 93:13090–13095

Pirkkala L, Nykanen P, Sistonen L (2001) Roles of the heat shock transcription factors in regulation of the heat shock response and beyond. FASEB J 15:1118–1131

Ray PS, Martin JL, Swanson EA, Otani H, Dillmann WH, Das DK (2001) Transgene overexpression of alphaB crystallin confers simultaneous protection against cardiomyocyte apoptosis and necrosis during myocardial ischemia and reperfusion. FASEB J 15:393–402

Ritossa F (1996) Discovery of the heat shock response. Cell Stress Chaperones 1:97–98

Shim EH, Kim JI, Bang ES, Heo JS, Lee JS, Kim EY, Lee JE, Park WY, Kim SH, Kim HS, Smithies O, Jang JJ, Jin DI, Seo JS (2002) Targeted disruption of hsp70.1 sensitizes to osmotic stress. EMBO Rep 3:857–861

Singh IS, He JR, Calderwood S, Hasday JD (2002) A high affinity HSF-1 binding site in the 5'-untranslated region of the murine tumor necrosis factor-alpha gene is a transcriptional repressor. J Biol Chem 277:4981–4988

Trinklein ND, Murray JI, Hartman SJ, Botstein D, Myers RM (2004) The role of heat shock transcription factor 1 in the genome-wide regulation of the mammalian heat shock response. Mol Biol Cell 15:1254–1261

Van Molle W, Wielockx B, Mahieu T, Takada M, Taniguchi T, Sekikawa K, Libert C (2002) HSP70 protects against TNF-induced lethal inflammatory shock. Immunity 16:685–695

Voellmy R (2004) On mechanisms that control heat shock transcription factor activity in metazoan cells. Cell Stress Chaperones 9:122–133

Voss AK, Thomas T, Gruss P (2000) Mice lacking HSP90beta fail to develop a placental labyrinth. Development 127:1–11

Wang G, Ying Z, Jin X, Tu N, Zhang Y, Phillips M, Moskophidis D, Mivechi NF (2004) Essential requirement for both hsf1 and hsf2 transcriptional activity in spermatogenesis and male fertility. Genesis 38:66–80

Wang G, Zhang J, Moskophidis D, Mivechi NF (2003) Targeted disruption of the heat shock transcription factor (hsf)-2 gene results in increased embryonic lethality, neuronal defects, and reduced spermatogenesis. Genesis 36:48–61

Wirth D, Christians E, Li X, Benjamin IJ, Gustin P (2003) Use of Hsf1(-/-) mice reveals an essential role for HSF1 to protect lung against cadmium-induced injury. Toxicol Appl Pharmacol 192:12–20

Wolfer DP, Crusio WE, Lipp HP (2002) Knockout mice: simple solutions to the problems of genetic background and flanking genes. Trends Neurosci 25:336–340

Wu C (1995) Heat shock transcription factors: structure and regulation. Annu Rev Cell Dev Biol 11:441–469

Xiao X, Zuo X, Davis AA, McMillan DR, Curry BB, Richardson JA, Benjamin IJ (1999) HSF1 is required for extra-embryonic development, postnatal growth and protection during inflammatory responses in mice. EMBO J 18:5943–5952

Xing H, Wilkerson DC, Mayhew CN, Lubert EJ, Skaggs HS, Goodson ML, Hong Y, Park-Sarge OK, Sarge KD (2005) Mechanism of hsp70i gene bookmarking. Science 307:421–423

Yan LJ, Christians ES, Liu L, Xiao X, Sohal RS, Benjamin IJ (2002) Mouse heat shock transcription factor 1 deficiency alters cardiac redox homeostasis and increases mitochondrial oxidative damage. EMBO J 21:5164–5172

Zhang Y, Huang L, Zhang J, Moskophidis D, Mivechi N (2002) Targeted disruption of hsf1 leads to lack of thermotolerance and defines tissue-specific regulation for stress-inducible Hsp molecular chaperones. J Cell Biochem 86:376–393

Zheng H, Benjamin IJ, Basu S, Li Z (2003) Heat shock factor 1-independent activation of dendritic cells by heat shock: implication for the uncoupling of heat-mediated immunoregulation from the heat shock response. Eur J Immunol 33:1754–1762

Zheng H, Li Z (2004) Cutting edge: cross-presentation of cell-associated antigens to MHC class I molecule is regulated by a major transcription factor for heat shock proteins. J Immunol 173:5929–5933

HSFs in Development

M. Morange

Département de Biologie, Unité de Génétique Moléculaire, Ens, 46 rue d'Ulm, 75230 Paris Cedex 05, France
morange@biologie.ens.fr

1	**Introduction**	154
1.1	Heat Shock Proteins, Cell Differentiation, and Development	154
1.2	The Existence of Multiple Heat Shock Transcription Factors	154
1.3	A Hypothesis and Its Failure: Additional HSFs Control the Expression of Hsps During Differentiation and Development	155
1.4	What Does the Inactivation of the Heat Shock Transcription Factor Genes Tell Us?	155
2	**Inactivation of *Hsf1***	156
2.1	*Hsf1* Is Essential for the Stress Response	156
2.2	*Hsf1* Is Required During Embryogenesis and at the Adult Stage	156
2.3	*Hsf1* and the Quality Control of Spermatogenesis	157
2.4	*Hsf1* and the Immune Response	158
2.5	The Target Genes of HSF1	158
3	**Inactivation of *Hsf2***	159
3.1	*Hsf2* Is Involved in Fertility and Brain Development	159
3.2	Conflicting Results Obtained with *Hsf2* Inactivation and a Possible Interpretation	160
3.3	The Search for Specific Target Genes for HSF2	161
3.4	HSF2 and Chromatin Modification	162
4	**Inactivation of the *Hsf4* Gene**	162
4.1	*Hsf4*-Specific Expression in the Lens	162
4.2	HSF4 and HSF1 Cooperate in the Formation of the Lens	163
5	**Preliminary Conclusions**	163
5.1	More Results Are to Be Expected	163
5.2	HSFs Control *Hsp* and *Non-Hsp* Gene Expression	164
5.3	HSFs at the Crossroads Between Aging, Reproduction, and the Environment	164
5.4	Why HSFs and Hsps Have Been Recruited for Additional Functions	165
	References	165

Abstract Heat shock transcription factors, as well as heat shock proteins, are involved in different steps in differentiation and development, in addition to their role in adaptation to stress. This has already been demonstrated in the case of the single heat shock factor present in *Drosophila*. Over the last 6 years, similar observations have accumulated from the progressive inactivation of the different *hsf* genes in mammals, the use of double-null animals, and the slow characterization of their complex phenotypes. Although these studies are not yet complete, the data so far can be used to draw some conclusions. All *hsf* genes contribute to development in mammals and to normal functions at the adult stage, by controlling the expression of *Hsp* and *non-Hsp* genes. Reproduction, the immune response and aging are the processes that are the most deeply affected. An attractive hypothesis would be that these new functions have been recruited during evolution in order to coordinate these processes: HSFs may occupy a central place in the trade off that organisms make between reproduction and maintenance, in response to the variations in the environment.

Keywords Heat shock transcription factors · Gene inactivation · Reproduction · Gene targets · Evolution

1
Introduction

1.1
Heat Shock Proteins, Cell Differentiation, and Development

The discovery that the expression of some heat shock proteins (Hsps) is tightly correlated with specific steps in differentiation and development immediately followed the characterization of these proteins and preceded the discovery of the chaperone function shared by most of them: marked induction of the small Hsps by ecdysterone (Ireland and Berger 1982) that occurs in vivo at specific stages of *Drosophila* development, expression of the HSP70 proteins during mammalian (mouse) early zygotic genome activation (Bensaude et al. 1983), etc. Many additional data have been obtained since these initial observations. The explanations that have been proposed range between two alternative and nonexclusive hypotheses: there would be a general need for one (or many) chaperones at a specific step of development, for instance to assemble or to disassemble protein complexes during the transition from one differentiation state to another, or there exist specific developmental and differentiation targets for chaperones in addition to the numerous targets that general chaperones have.

1.2
The Existence of Multiple Heat Shock Transcription Factors

The same ambiguity emerged from the study of the heat shock transcription factors (HSFs) involved in the control of heat shock gene expression (for a review, see Pirkkala et al. 2001). These factors bind to a heat shock element (HSE) found upstream of the heat shock genes. The existence of one unique gene in yeast and *Drosophila*, the transcriptional activity of which is stimulated by heat shock and stress, suggested that these factors were involved in the increased expression of Hsps that follows a stress and is necessary for the adaptation of the organism to the stressful conditions, and in the repair of the damage that might have occurred inside the cells and organisms.

However, the picture became more complex. In vertebrates, the existence of a family of transcription factors was demonstrated: HSF1, HSF2, and HSF3 in chicken, HSF1, HSF2, and HSF4 in mammals. With the exception of HSF3, these additional factors were not clearly involved in the stress response. In addition, their expression varied strongly during differentiation and development: for instance, HSF2 is expressed at a high level during the first two-thirds of mouse embryogenesis, but at the adult stage its expression is restricted to testis, and at a very low level to brain (Sarge et al. 1994; Rallu et al. 1997).

1.3
A Hypothesis and Its Failure: Additional HSFs Control the Expression of Hsps During Differentiation and Development

A simple hypothesis was put forward in an attempt to rationalize these puzzling observations: these additional factors were in charge of the specific expression of Hsps/chaperones during differentiation and development, whereas one (HSF1) or two (HSF1 and HSF3 in birds; Tanabe et al. 1998) are involved in the stress response.

But this hypothesis was very rapidly challenged: no clear correlation exists between the expression of these additional HSFs and the variations in the expression of Hsps and chaperones during development (Rallu et al. 1997). It was also shown by Carl Wu's group that in *Drosophila*, where one unique HSF exists, this factor is required at two distinct developmental steps, oogenesis and early larval development: the involvement of *Drosophila* HSF in development is not correlated with an increase in Hsp expression at these steps, and these two additional functions can be genetically distinguished (Jedlicka et al. 1997).

1.4
What Does the Inactivation of the Heat Shock Transcription Factor Genes Tell Us?

It may be better to renounce any a priori interpretation of the high level of expression of Hsps during development, and of the existence of different HSFs, some of which are specifically expressed during embryogenesis, and to consider what experiments—in particular gene inactivation experiments—tell us about the role played by Hsps and HSFs in development.

We will focus our presentation on the inactivation (knockout) of HSF genes in mice. Experiments over the last 6 years have yielded a rich harvest of results. They force us to adopt a different vision of the function of HSFs and Hsps, and the way the organisms adapt to their environment. This will be discussed in the last section. The observations made on inactivation of some Hsps and chaperones can be easily integrated into the general landscape revealed by the inactivation of HSF genes.

2
Inactivation of *Hsf1*

2.1
Hsf1 Is Essential for the Stress Response

The initial description given by Ivor Benjamin and colleagues (see also the chapter by Christians and Benjamin, this volume) revealed a very simple phenotype (McMillan et al. 1998). In mouse cells where the *hsf1* gene has been inactivated, the basal level of Hsp gene expression is not altered, but the induced expression of Hsps after a stress is totally abolished (Xiao et al. 1999). As a consequence, there is no acquired thermotolerance and an increased heat-inducible apoptosis. More recently, it was shown that $hsf1^{-/-}$ mice are more sensitive to a cadmium injury (Wirth et al. 2003). HSF1 also protects cardiomyocytes from ischemia (Zou et al. 2003). The $hsf1^{-/-}$ mice are also much more sensitive to the toxic effects of bacterial endotoxin (Xiao et al. 1999). This is probably because the inhibitory action that HSF1 has on the expression of cytokine genes is absent (Cahill et al. 1996). HSF1 directly interacts with the nuclear factor of interleukin-6 and prevents its activating effect on IL-1β and other promoters (Xie et al. 2002). In contrast to what has been described in yeast for the unique HSF and for HSF1 in avian cells (Nakai and Ishikawa 2001), HSF1 is apparently not involved in the constitutive expression of Hsps. These initial experiments did not reveal functions (and targets) for HSF1 other than the control of the stress response and cytokine production.

2.2
Hsf1 Is Required During Embryogenesis and at the Adult Stage

Further descriptions, however, cloud the picture. The $hsf1^{-/-}$ mice are viable, but a fraction of the $hsf1^{-/-}$ embryos dies during prenatal development (Xiao et al. 1999), depending on the genetic background. Death is due to a defect in the formation of the chorioallantoic placenta. Postnatal growth is delayed. No precise molecular mechanism for the origin of the defects observed in $hsf1^{-/-}$ animals emerged from these studies.

In the same way, the development of embryos from $hsf1^{-/-}$ mothers is blocked at a very early stage of development, during or immediately after the transcriptional activation of the zygotic genome (Christians et al. 2000). This blockade does not prevent the early expression of *Hsp70* genes, which is the earliest sign of the zygotic genome activation. It demonstrates that *Hsp* genes are probably not the HSF1 targets responsible for this early developmental arrest.

More recently, the same group has shown that the inactivation of the *hsf1* gene reduces the level of Hsp25, αB-crystallin and Hsp70, but not Hsp90 in the heart (Yan et al. 2002). As a consequence, the level of glucose 6-phosphate dehydrogenase activity is diminished, leading to a decrease in the ratio between reduced and oxidized glutathione. The consequence is an increase in the amount of reactive oxygen species, and the oxidation of different essential mitochondrial proteins. Similar observations have been made in the kidney: HSF1 disruption leads to a decreased expression of Hsp25 and Hsp90 (but not αB-crystallin), a change in redox status, and an increase in superoxide generation, with deleterious effects on mitochondrial functions similar to those previously described in the heart (Yan et al. 2005).

2.3
Hsf1 and the Quality Control of Spermatogenesis

Quite independently, Akira Nakai's group demonstrated the involvement of HSF1 in the quality control of spermatogenesis. The first step resulted from overexpression of HSF1 in mice (Nakai et al. 2000). A permanently-active HSF1 protein was obtained by deletion of the regulatory region, which inhibits the transcriptional activation domain located in the C-terminal part of the factor. Although the promoter selected (the human β-actin promoter) is considered ubiquitously active, the transgenic mice that were obtained expressed HSF1 at a high level in only three organs: the testis, heart, and stomach.

Infertility resulted from the arrest of spermatogenesis at the pachytene stage of spermatocytes, concomitant with a massive death of cells at this differentiation stage. The authors related these effects to the well-known inhibition in humans of spermatogenesis by acute febrile diseases or crypto-orchidism, as well as by sauna or wearing close-fitting underwear. All these conditions result in an increase in the temperature of the testes, which in normal conditions are maintained at a temperature lower than the core body temperature. Activation of HSF1 in these conditions would be responsible for the observed effects.

The physiological significance of this HSF1-induced cell death was made more explicit by looking at the consequence of *hsf1* gene inactivation in mice (Izu et al. 2003). Whereas the most immature germ cells are more sensitive to heat treatment in HSF1-null mice—in agreement with the normal protective effect of HSF1 and the heat shock response—opposite effects are seen at the pachytene stage of spermatocytes. Absence of HSF1 leads to a decrease in cell death, the mirror effect of that observed in animals overexpressing HSF1.

The pachytene stage of spermatocytes was already known as a checkpoint in differentiation of male germ cells: many mutations affecting spermatogenesis result in cell death at this specific stage of differentiation. The interpretation by Akira Nakai is that conditions leading to the abnormal accumulation of misfolded proteins lead to the activation of HSF1. At the pachytene stage of spermatocytes, this activation will be interpreted as a sign of defect and will induce the initiation of the cell death program by a molecular process that is still unknown. The similarity with the control exerted by p53 on the integrity of the genome is worth noting. The role of HSF1 in the quality control of spermatogenesis is not mediated by a variation in the level of Hsps. The synthesis of the major Hsps during spermatogenesis is not affected in the HSF1-null mice, in normal conditions or after a hyperthermic treatment.

2.4
Hsf1 and the Immune Response

$Hsf1^{-/-}$ mice have also been used to test the recurring hypothesis that Hsps have a place in the immune response at the antigen-presentation stage by their peptide-binding capacities. It is not within the scope of this review to examine the numerous and conflicting data on this topic. What is important for us is that *hsf1* gene inactivation is shown to downregulate the cross-presentation of MHC class I-associated antigens, in parallel with a decrease in the level of Hsp90 in the dendritic cells, liver, and embryonic fibroblasts (Zheng and Li 2004). The constitutive expression of HSP70 seen in the fibroblasts of the wild-type mice was absent in the fibroblasts derived from HSF1-null mice.

Mice deficient in HSF1 exhibit a decrease in the T cell-dependent B cell response with an impaired Ig production (Inouye et al. 2004). This results from a reduced proliferation of the spleen cells, the consequence of a decreased level of interleukin-6 and CCL5. By chromatin immunoprecipitation, HSF1 was shown to bind to the interleukin-6 promoter and to activate the transcription of the *Il-6* gene.

2.5
The Target Genes of HSF1

The results obtained with HSF1 raise two questions concerning this factor's mechanism of action. The first is to know whether HSF1 contributes to the basal expression of Hsps and chaperones in the absence of stress. As we have seen, the observations are conflicting. In some cases, when the absence of HSF1 leads to a decrease in Hsp-expression, such as in the heart or the skin, it is not possible to exclude the possibility that the stress response is constitutively induced as a response to the permanent insults to which these tissues are submitted (ROS in the case of the heart, chemical agents from the environment for the skin). The activation of mouse HSF1 in the liver and intestine by the xenobiotic compounds present in food has been demonstrated (Katsuki et al. 2004). Another possibility is raised by the observations made in *Caenorhabditis elegans*. In this organism, the unique HSF is involved in the basal expression of the small Hsps. This constitutive activation requires the presence of another transcription factor (DAF-16), which also binds to the regulatory sequences found upstream of the *sHsp* genes (Hsu et al. 2003). As we shall see, this co-activation results in a decreased formation of protein aggregates, the abundance of which increases with age, and a prolonged life-span.

The observations also suggest that HSF1 controls the activity of genes distinct from the HSP and chaperone genes in the formation of the placenta, in the capacity of oocytes to allow an early development of the embryo, during spermatogenesis as well as in the immune response. Unfortunately, in all these cases but one, the nature of the targets remains unknown: the only documented case is the activation of IL-6 gene transcription.

3
Inactivation of *Hsf2*

3.1
Hsf2 Is Involved in Fertility and Brain Development

When gene inactivation was initiated, the functions of HSF2 were unknown. HSF2 is reversibly inactivated by heat shock (Mathew et al. 2001), but activated after a blockade of the proteolytic proteasome pathway (Mathew et al. 1998). In this latter case, however, it was later shown that all the HSFs were activated in the same conditions, HSF1 even more than HSF2 (Pirkkala et al. 2000). In addition, it is difficult to discard the hypothesis that proteasome inhibition is responsible for an overall stabilization of short-life proteins, among which HSFs, and that the observed effects have nothing to do with a normal physiological response. Two isoforms of HSF2 with different transcriptional capacities have been described (Fiorenza et al. 1995; Goodson et al. 1995). From what was known of the situations in which this factor was expressed at a high level, it was imagined that HSF2 might have a role in spermatogenesis (Sarge et al. 1994; Alastalo et al. 1998), in differentiation and development (Sistonen et al. 1992; Rallu et al. 1997), in particular of the nervous system but also of the heart (Eriksson et al. 2000).

The first reported *hsf2* gene inactivation resulted from the insertion in exon 5 of a β-geo gene, generating a chimeric HSF2-βgeo protein, with an intact HSE-binding domain, but lacking an integral oligomerization domain (Kallio et al. 2002). The homozygotes are viable and have an apparently normal behavior, but exhibit two defects. The first consists in structural abnormalities in the adult brain, with an enlargement of the vesicles and a reduction in the size of the hippocampus and striatum. The second concerns spermatogenesis and oogenesis. The size of the testis is reduced, and the structure of the seminiferous tubules is altered: these modifications are the consequence of the disruption of spermatogenesis, with increased apoptosis at the late pachytene stage of meiotic prophase. The synaptonemal complex, which contributes to the pairing of homologue chromosomes, is altered, with the presence of loop-like structures indicative of defective synapsis. Despite these defects, the fertility of males is not significantly reduced.

In contrast, female fertility is decreased in $hsf2^{-/-}$ animals. The size of litters is reduced, independently of the genotype of the partner. Ovarian function is affected with the presence of hemorrhagic large follicles and a 60-fold increase in luteinizing hormone receptor. Ovulation is perturbed, but can be returned to a normal level by treatments used to obtain superovulation. Nevertheless, even in these conditions, the eggs have an abnormal shape and the rate of embryonic death is abnormally high. The nature of the target genes of HSF2, the underexpression of which might be responsible for the observed alterations, remains unknown. The level of expression of the major *Hsp* genes was checked by RT-PCR in different tissues of the $hsf2^{-/-}$ embryos and adults and found not to be different from the level measured in wild-type animals. The level of HSP70-2, a member of the *Hsp70* gene family specifically expressed during the formation of the male germ cells (Sarge et al. 1994), was halved. Hsp70-2 is essential for spermatogenesis (Dix et al. 1996) and gel shift assays suggest it harbors HSE sequences in its promoter. The significance of the result obtained with HSP70-2 is dubious, since the alteration of spermatogenesis in $hsf2^{-/-}$ animals might be indirectly responsible for this decrease of limited amplitude.

3.2
Conflicting Results Obtained with *Hsf2* Inactivation and a Possible Interpretation

These results were not reproduced by Ivor Benjamin's group (McMillan et al. 2002), who was unable to detect any deficiency in fertility, although some degeneration of the seminiferous tubules was observed in the testis of more than three-month-old $hsf2^{-/-}$ mice. No brain abnormalities were observed, and the $hsf2^{-/-}$ animals successively passed a battery of behavioral tests.

This type of discrepancy between the results obtained by different laboratories inactivating the same gene is not unusual in the knockout field. The obvious and less interesting possibility is that the defects that are observed do not result from gene inactivation: a second copy of the transgene has integrated at some other place in the genome, and the effects result from this insertion process and not from gene inactivation. Another possibility is that gene inactivation has resulted in the modification of expression of genes located close to the inactivated gene. Or, on the contrary, the inactivation may be incomplete, the chimeric protein keeping part of its activity and interfering with normal cellular functions. A more interesting hypothesis is that the difference results from the genetic background, and the existence of modifier genes that may increase or mask the consequence of gene inactivation. The influence of genetic background on the consequences of gene inactivation has been amply documented in several well-described instances.

It is probably the right explanation in the case of *hsf2*, since the initial observations on brain abnormalities and the alteration of germ line differentiation were independently reproduced by a third group, that of Mivechi (Wang et al. 2003), but with a stronger phenotype than the one observed by the first group. The presence of enlarged brain vesicles was confirmed, but shown to be already present in embryos, leading to an increased prenatal rate of death for the $hsf2^{-/-}$ animals. Spermatogenesis was reduced, leading to a significant reduction of fertility in $hsf2^{-/-}$ males. The target genes of HSF2 were looked for in E10.5 embryos and adult testis by using microarrays. The results, confirmed by RT-PCR, characterize genes whose expression is increased or decreased in $hsf2^{-/-}$ animals. None of these genes belong to the HSP/chaperone family. Nothing proves that the genes whose expression varies are direct targets of the HSF2 factor.

The same group has recently published the results of the double inactivation of the *hsf2* and *hsf1* genes (Wang et al. 2004). Spermatogenesis is arrested around the spermatocyte stage, and the males are infertile. There is a deep reduction in pre- and postmeiotic gene transcription. No data were provided on what happened in females, or in other parts of the organism (in particular in the central nervous system).

3.3
The Search for Specific Target Genes for HSF2

To summarize briefly the observations made so far on HSF2, one might say that this factor is required for spermatogenesis, oogenesis, and brain development, and that no data allow us to correlate these modifications with a change in Hsp/chaperone expression.

HSF1 and HSF2 do not have the same affinity for different heat shock gene promoters (Kroeger and Morimoto 1994; Manuel et al. 2002; Trinklein et al. 2004a). But the small differences sharply contrast with the apparently complete absence of alteration of *Hsp* gene expression in the $hsf2^{-/-}$ animals— without any concomitant variation in the level and activity of the other HSFs. One possibility would be that the target genes of HSF2 are totally distinct from those of HSF1, for instance because HSF2 would always cooperate with another transcription factor to activate its target genes. The present data are clearly insufficient to guess what this factor might be. The thioredoxin gene, activated by HSF2 in the K562 erythroleukemia cell line when its differentiation is induced by hemin, does not seem to be a direct target of HSF2 (Leppä et al. 1997a). The RANK Ligand gene, coding for a critical osteoclastogenic factor, is a target of HSF2 (Roccisana et al. 2004). The gene coding for p35, a cyclin-like protein specifically expressed in postmitotic nervous cells, has been shown by chromatin immunoprecipitation to bind to the HSF2 promoter, and its level of expression is reduced in $hsf2^{-/-}$ animals (Mezger, unpublished results). This protein would be an interesting candidate to relate brain abnormalities observed in $hsf2^{-/-}$ animals to the absence of HSF2.

Whatever the discoveries might be in the future, ex vivo and in vitro observations agree on the fact that the different isoforms of HSF2 are weakly active transcriptional regulators. Other functions should perhaps be sought for this factor. The possibility that HSF2 interferes (positively or negatively) with HSF1 activation has been frequently advanced (Sistonen et al. 1994; Leppä et al. 1997b; He et al. 2003; Paslaru et al. 2003), for instance at the level of the stress granules that form after stress (Alastalo et al. 2003); it does not give clues to the way HSF2 might act during development. The conserved modification of HSF2 by sumoylation is perhaps the sign that the subnuclear localization of this factor is important for its function, but it does not tell us what this function is (Goodson et al. 2001; Hilgarth et al. 2004; Le Goff et al. 2004).

3.4
HSF2 and Chromatin Modification

The recent observation of Kevin Sarge that HSF2 is involved in gene bookmarking during mitosis (Xing et al. 2005) casts new light on HSF2 function and allows reinterpretation of previous observations. Some years ago, the same group had already shown that HSF2 interacts with the phosphoserine-phosphothreonine phosphatase 2A (PP2A) (Hong and Sarge 1999), but the significance of this result remained unknown. The compaction of the genome during mitosis is due to the action of an enzymatic complex called condensin, which is itself activated by phosphorylation by the Cdc2-cyclinB kinase. Some genes escape this condensing and inactivation process, such as the *Hsp70* gene. Kevin Sarge's group has demonstrated that HSF2 is responsible for this lack of inactivation: the sumoylated form of HSF2 interacts with the CAP-G subunit of condensin, which is dephosphorylated and inactivated by the complex between HSF2 and PP2A. Although the authors amply discuss the role that HSF2 might play through this process in cellular protection against stress, they do not consider what it might tell us about the role of HSF2 during differentiation. It is highly tempting to imagine that HSF2 might have a similar function during development, preventing the chromatin condensation and inactivation of some genes which have to be active, or at least activatable.

4
Inactivation of the *Hsf4* Gene

4.1
Hsf4-Specific Expression in the Lens

HSF4 was described in 1997 in humans as a permanently active repressor of heat shock genes, mainly expressed in the heart, muscles, brain, and pancreas (Nakai et al. 1997). Further studies showed that this factor, also present in mice, exists in two different forms, one inactive and the other active for transcription (Tanabe et al. 1999).

The first clue to its in vivo function came from the study of Chinese and Danish families suffering from congenital cataracts. In both cases, the mutations responsible for the disease were mapped to the *hsf4* locus (Bu et al. 2002). They affect the DNA-binding domain of HSF4, and probably prevent its interaction with DNA.

Somasundaram and Bhat (2004) described the high level of expression of *hsf4* at the postnatal stage in the rat lens, whereas *hsf1* and *hsf2* are mainly expressed at the fetal stage. The authors demonstrated the specific affinity of HSF4 for αB-crystallin gene promoter, confirming earlier observations from the same laboratory (Somasundaram and Bhat 2000).

4.2
HSF4 and HSF1 Cooperate in the Formation of the Lens

The demonstration of the role of *hsf4* in the development of the lens was provided by the very elegant study of Akira Nakai's group (Fujimoto et al. 2004). HSF4 is expressed in both epithelial and fiber cells. *Hsf4*-null mice are viable, fertile, have the same weight as the wild-type animals, and exhibit no major abnormalities of the brain, lung, testis, or ovary. The animals suffer from cataract. The lens fiber cells contain inclusion-like structures rich in αA- and αB-crystallins, which can be detected as early as 2 days after birth. The authors show that the main direct targets of HSF4 are the γ-crystallin genes. But HSF4 also represses the *fgf* (1 and 4) gene expression in epithelial cells, which proliferate in its absence, whereas HSF1 has an opposite effect. In the double-null animals, the level of expression of FGFs and the number of epithelial cells returns to normal, demonstrating the antagonistic effects of HSF1 and HSF4 anticipated by Zhang et al. (2001), and suggesting a contrario a role so far not seen for HSF1 in the formation of the lens (and of other tissues such as the lung).

This study demonstrates the complexity of interactions between different HSFs: antagonistic for the expression of FGFs or some Hsps, but synergistic for the expression of γ-crystallins. In addition, it shows that the definition of the different forms of HSFs as inhibitors or activators has only a limited value. The complementary study of Nahid Mivechi's group confirmed the involvement of HSF4 in lens development, but attributed it to the regulation of the 25-kDa heat shock protein (Min et al. 2004).

5
Preliminary Conclusions

5.1
More Results Are to Be Expected

Our conclusions can only be preliminary. Analysis of the phenotype of the knockout animals is long and tedious work, and initial descriptions frequently omit important aspects of the phenotype. The analysis of double-null animals has revealed some characteristics that suggest the presence in single-null animals of defects that remain unnoticed (Fujimoto et al. 2004; Wang et al. 2004). In addition, the probable existence of redundancy between the different HSFs allows us to predict that the characterization of double- and triple-null animals will reveal defects so far undetected. In contrast with what was previously thought, the list of HSFs is not closed. Distantly related HSFs, one of which, HSFY, is specifically expressed in spermatogenesis, were recently described (Shinka et al. 2004; Tessari et al. 2004). There are some experimental arguments to implicate deletion of the *hsfy* gene in cases of azoospermia and oligospermia.

5.2
HSFs Control *Hsp* and *Non-Hsp* Gene Expression

Part of the function of HSFs stems from their regulation of the expression of chaperone and *Hsp* genes. This is true for HSF1 in normal constitutive conditions, in contrast to the first descriptions made. But many of the effects are due to the regulation of genes unrelated to the *Hsp* genes, such as growth factor genes. The rules governing the binding of HSF to HSE, and the transcriptional activation of the downstream genes remain largely unknown (Trinklein et al. 2004b). More global roles, such as the regulation of the chromatin state in the case of HSF2, must also be considered.

5.3
HSFs at the Crossroads Between Aging, Reproduction, and the Environment

There is something common behind the different phenotypes revealed by the knockout experiments. The first is the involvement of HSFs at different phases of reproduction: spermatogenesis and oogenesis (HSF1 and HSF2), but also early development (HSF1) and late development (HSF1 and HSF2). A second observation, only poorly documented so far, is the link between HSFs and aging: reduced spermatogenesis and the formation of cataract are age-linked processes that occur prematurely in *hsf*-mutated animals. The alteration of HSF1 activation with age has been amply documented (Shamovsky and Gershon 2004). And we have seen that in nematodes the reduction in the synthesis of the unique HSF has clearly been linked with aging, the decrease in the synthesis of the small Hsps, and the subsequent formation of intracellular protein aggregates. Finally, HSF1 is also involved in the immune response.

The trio—aging, reproduction, response to environmental stress—is a very interesting one from the evolutionary point of view. Today, it is widely admitted by evolutionary biologists that there is a trade-off by organisms between the resources they attribute to their maintenance—the control of aging—and reproduction. This trade-off depends upon the environment which is (or is not) favorable to reproduction. The fact that HSFs might be a link between adaptation to the environment, the control of reproduction and, indirectly, aging is therefore quite significant.

5.4
Why HSFs and Hsps Have Been Recruited for Additional Functions

A reasonable evolutionary scenario would be to accord prime importance to the adaptive response to the environment. The progressive recruitment during evolution, by the insertion of HSE sequences in their promoters, of target genes unrelated to the immediate stress response but involved in reproduction and the maintenance of organisms, would have been positively selected if it was a way to adapt these processes to the environment. Duplication of *hsf* genes would have favored this process, although the observations of *Drosophila* show that it was not a sine qua non. In a similar way, HSF1 has acquired during evolution a direct protective effect against heat shock, independent of the activation of heat shock genes (Inouye et al. 2003). In a parallel and nonexclusive way, chaperones were also recruited to accomplish specific tasks during reproduction, development, and aging in addition to their normal chaperone function. The role attributed by Susan Lindquist to Hsp90—capacitor of gene mutations—points in the same direction (Rutherford and Lindquist 1998; Queitsch et al. 2002). Further characterization of knockout animals will confirm or reject this model.

References

Alastalo TP, Lönnstrom M, Leppä S, Kaarniranta K, Pelto-Huikko M, Sistonen L, Parvinen M (1998) Stage-specific expression and cellular localization of the heat shock factor 2 isoforms in the rat seminiferous epithelium. Exp Cell Res 240:16–27

Alastalo TP, Hellesuo M, Sandqvist A, Hietakangas V, Kallio M, Sistonen L (2003) Formation of nuclear stress granules involves HSF2 and coincides with the nucleolar localization of Hsp70. J Cell Sci 116:3557–3570

Bensaude O, Babinet C, Morange M, Jacob F (1983) Heat shock proteins, first major products of zygotic gene activity in mouse embryo. Nature 305:331–333

Bu L, Jin Y, Shi Y, Chu R, Ban A, Eiberg H, Andres L, Jiang H, Zheng G, Qian M, Cui B, Xia Y, Liu J, Hu L, Zhao G, Hayden MR, Kong X (2002) Mutant DNA-binding domain of HSF4 is associated with autosomal dominant lamellar and Marner cataract. Nat Genet 31:276–278

Cahill CM, Waterman WR, Xie Y, Auron PE, Calderwood SK (1996) Transcriptional repression of the *prointerleukin 1β* gene by heat shock factor 1. J Biol Chem 271:24874–24879

Christians E, Davis AA, Thomas SD, Benjamin IJ (2000) Maternal effect of *Hsf1* on reproductive success. Nature 407:693–694

Dix DJ, Allen JW, Collins BW, Mori C, Nakamura N, Poorman-Allen P, Goulding EH, Eddy EM (1996) Targeted gene disruption of *Hsp70-2* results in failed meiosis, germ cell apoptosis, and male infertility. Proc Natl Acad Sci U S A 93:3264–3268

Eriksson M, Jokinen E, Sistonen L, Leppä S (2000) Heat shock factor 2 is activated during mouse heart development. Int J Dev Biol 44:471–477

Fiorenza MT, Farkas T, Dissing M, Kolding D, Zimarino V (1995) Complex expression of murine heat shock transcription factors. Nucleic Acids Res 23:467–474

Fujimoto M, Izu H, Seki K, Fukuda K, Nishida T, Yamada SI, Kato K, Yonemura S, Inouye S, Nakai A (2004) HSF4 is required for normal cell growth and differentiation during mouse lens development. EMBO J 23:4297–4306

Goodson ML, Park-Sarge OK, Sarge KD (1995) Tissue-dependent expression of heat shock factor 2 isoforms with distinct transcriptional activities. Mol Cell Biol 15:5288–5293

Goodson ML, Hong Y, Rogers R, Matunis MJ, Park-Sarge OK, Sarge KD (2001) Sumo-1 modification regulates the DNA binding activity of heat shock transcription factor 2, a promyelocytic leukemia nuclear body associated transcription factor. J Biol Chem 276:18513–18518

He H, Soncin F, Grammatikakis N, Li Y, Siganou A, Gong J, Brown SA, Kingston RE, Calderwood SK (2003) Elevated expression of heat shock factor (HSF) 2A stimulates HSF1-induced transcription during stress. J Biol Chem 278:35465–35475

Hilgarth RS, Murphy LA, O'Connor CM, Clark JA, Park-Sarge OK, Sarge KD (2004) Identification of *Xenopus* heat shock transcription factor-2: conserved role of sumoylation in regulating deoxyribonucleic acid-binding activity of heat shock transcription factor-2 proteins. Cell Stress Chaperones 9:214–220

Hong Y, Sarge KD (1999) Regulation of protein phosphatase 2A activity by heat shock transcription factor 2. J Biol Chem 274:12967–12970

Hsu AL, Murphy CT, Kenyon C (2003) Regulation of aging and age-related disease by DAF-16 and heat-shock factor. Science 300:1142–1145

Inouye S, Katsuki K, Izu H, Fujimoto M, Sugahara K, Yamada S, Shinkai Y, Oka Y, Katoh Y, Nakai A (2003) Activation of heat shock genes is not necessary for protection by heat shock transcription factor 1 against cell death due to a single exposure to high temperatures. Mol Cell Biol 23:5882–5895

Inouye S, Izu H, Takaki E, Suzuki H, Shirai M, Yokota Y, Ichikawa H, Fujimoto M, Nakai A (2004) Impaired IgG production in mice deficient for heat shock transcription factor 1. J Biol Chem 279:38701–38709

Ireland RC, Berger EM (1982) Synthesis of low molecular weight heat shock peptides stimulated by ecdysterone in a cultured *Drosophila* cell line. Proc Natl Acad Sci U S A 79:855–859

Izu H, Inouye S, Fujimoto M, Shiraishi K, Naito K, Nakai A (2004) Heat shock transcription factor 1 is involved in quality-control mechanisms in male germ cells. Biol Reprod 70:18–24

Jedlicka P, Mortin MA, Wu C (1997) Multiple functions of *Drosophila* heat shock transcription factor *in vivo*. The EMBO J 16:2452–2462

Kallio M, Chang Y, Manuel M, Alastalo TP, Rallu M, Gitton Y, Pirkkala L, Loones MT, Paslaru L, Larney S, Hiard S, Morange M, Sistonen L, Mezger V (2002) Brain abnormalities, defective meiotic chromosome synapsis and female subfertility in HSF2 null mice. EMBO J 21:2591–2601

Katsuki K, Fujimoto M, Zhang XY, Izu H, Takaki E, Tanizawa Y, Inouye S, Nakai A (2004) Feeding induces expression of heat shock proteins that reduce oxidative stress. FEBS Lett 571:187–191

Kroeger PE, Morimoto RI (1994) Selection of new HSF1 and HSF2 DNA-binding sites reveals difference in trimer cooperativity. Mol Cell Biol 14:7592–7603

Le Goff P, Le Drean Y, Le Peron C, Le Jossic-Corcos C, Ainouche A, Michel D (2004) Intracellular trafficking of heat shock factor 2. Exp Cell Res 294:480–493

Leppä S, Pirkkala L, Saarento H, Sarge KD, Sistonen L (1997a) Overexpression of HSF2-β inhibits hemin-induced heat shock gene expression and erythroid differentiation in K562 cells. J Biol Chem 272:15293–15298

Leppä S, Pirkkala L, Chow SC, Eriksson JE, Sistonen L (1997b) Thioredoxin is transcriptionally induced upon activation of heat shock factor 2. J Biol Chem 272:30400–30404

Manuel M, Rallu M, Loones MT, Zimarino V, Mezger V, Morange M (2002) Determination of the consensus binding sequence for the purified embryonic heat shock factor 2. Eur J Biochem 269:2527–2537

Mathew A, Mathur SK, Morimoto RI (1998) Heat shock response and protein degradation: regulation of HSF2 by the ubiquitin-proteasome pathway. Mol Cell Biol 18:5091–5098

Mathew A, Mathur SK, Jolly C, Fox SG, Kim S, Morimoto RI (2001) Stress-specific activation and repression of heat shock factors 1 and 2. Mol Cell Biol 21:7163–7171

McMillan DR, Xiao X, Shao L, Graves K, Benjamin IJ (1998) Targeted disruption of heat shock transcription factor 1 abolishes thermotolerance and protection against heat-inducible apoptosis. J Biol Chem 273:7523–7528

McMillan DR, Christians E, Forster M, Xiao X, Connell P, Plumier JC, Zuo X, Richardson J, Morgan S, Benjamin IJ (2002) Heat shock transcription factor 2 is not essential for embryonic development, fertility, or adult cognitive and psychomotor function in mice. Mol Cell Biol 22:8005–8014

Min JN, Zhang Y, Moskophidis D, Mivechi NF (2004) Unique contribution of heat shock transcription factor 4 in ocular lens development and fiber cell differentiation. Genesis 40:205–217

Nakai A, Tanabe M, Kawazoe Y, Inazawa J, Morimoto RI, Nagata K (1997) HSF4, a new member of the human heat shock factor family which lacks properties of a transcriptional activator. Mol Cell Biol 17:469–481

Nakai A, Suzuki M, Tanabe M (2000) Arrest of spermatogenesis in mice expressing an active heat shock transcription factor 1. EMBO J 19:1545–1554

Nakai A, Ishikawa T (2001) Cell cycle transition under stress conditions controlled by vertebrate heat shock factors. EMBO J 20:2885–2895

Paslaru L, Morange M, Mezger V (2003) Phenotypic characterization of mouse embryonic fibroblasts lacking heat shock factor 2. J Cell Mol Med 7:425–435

Pirkkala L, Alastalo TP, Zuo X, Benjamin IJ, Sistonen L (2000) Disruption of heat shock factor 1 reveals an essential role in the ubiquitin proteolytic pathway. Mol Cell Biol 20:2670–2675

Pirkkala L, Nykänen P, Sistonen L (2001) Roles of the heat shock transcription factors in regulation of the heat shock response and beyond. FASEB J 15:1118–1131

Queitsch C, Sangster TA, Lindquist S (2002) Hsp90 as a capacitor of phenotypic variation. Nature 417:618–624

Rallu M, Loones MT, Lallemand Y, Morimoto RI, Morange M, Mezger V (1997) Function and regulation of heat shock factor 2 during mouse embryogenesis. Proc Natl Acad Sci U S A 94:2392–2397

Roccisana JL, Kawanabe N, Kajiya H, Koide M, Roodman GD, Reddy SV (2004) Functional role for heat shock factors in the transcriptional regulation of human RANK ligand gene expression in stromal/osteoblast cells. J Biol Chem 279:10500–10507

Rutherford SL, Lindquist (1998) HSP90 as a capacitor for morphological evolution. Nature 396:336–342

Sarge KD, Park-Sarge OK, Kirby JD, Mayo KE, Morimoto RI (1994) Expression of heat shock factor 2 in mouse testis: potential role as a regulator of heat shock gene expression during spermatogenesis. Biol Reprod 50:1334–1343

Shamovsky I, Gershon D (2004) Novel regulatory factors of HSF-1 activation: facts and perspectives regarding their involvement in the age-associated attenuation of the heat shock response. Mech Ageing Dev 125:767–775

Shinka T, Sato Y, Chen G, Naroda T, Kinoshita K, Unemi Y, Tsuji K, Toida K, Iwamoto T, Nakahori Y (2004) Molecular characterization of heat shock-like factor encoded on the human Y chromosome, and implications for male infertility. Biol Reprod 71:297–306

Sistonen L, Sarge KD, Phillips B, Abravaya K, Morimoto RI (1992) Activation of heat shock factor 2 during hemin-induced differentiation of human erythroleukemia cells. Mol Cell Biol 12:4104–4111

Sistonen L, Sarge KD, Morimoto RI (1994) Human heat shock factors 1 and 2 are differentially activated and can synergistically induce Hsp70 gene transcription. Mol Cell Biol 14:2087–2099

Somansundaram T, Bhat SP (2000) Canonical heat shock element in the αB-*crystallin* gene shows tissue-specific and developmentally controlled interactions with heat shock factor. J Biol Chem 275:17154–17159

Somansundaram T, Bhat SP (2004) Developmentally dictated expression of heat shock factors: exclusive expression of HSF4 in the postnatal lens and its specific interaction with αB-crystallin heat shock promoter. J Biol Chem 279:44497–44503

Tanabe M, Kawazoe Y, Takeda S, Morimoto RI, Nagata K, Nakai A (1998) Disruption of the *HSF3* gene results in the severe reduction of heat shock gene expression and loss of thermotolerance. EMBO J 17:1750–1758

Tessari A, Salata E, Ferlin A, Bartoloni L, Slongo ML, Foresta C (2004) Characterization of *HSFY*, a novel *AZFb* gene on the Y chromosome with a possible role in human spermatogenesis. Mol Hum Reprod 10:253–258

Trinklein ND, Chen WC, Kingston RE, Myers RM (2004a) Transcriptional regulation and binding of heat shock factor 1 and heat shock factor 2 to 32 human heat shock genes during thermal stress and differentiation. Cell Stress Chaperones 9:21–28

Trinklein ND, Murray JI, Hartman SJ, Botstein D, Myers RM (2004b) The role of heat shock transcription factor 1 in the genome-wide regulation of the mammalian heat shock response. Mol Biol Cell 15:1254–1261

Wang G, Zhang J, Moskophidis D, Mivechi NF (2003) Targeted disruption of the heat shock transcription factor (*hsf*)-2 gene results in increased embryonic lethality, neuronal defects, and reduced spermatogenesis. Genesis 36:48–61

Wang G, Ying Z, Jin X, Tu N, Zhang Y, Phillips M, Moskophidis D, Mivechi NF (2004) Essential requirement for both *hsf1* and *hsf2* transcriptional activity in spermatogenesis and male fertility. Genesis 38:66–80

Wirth D, Christians E, Li X, Benjamin IJ, Gustin P (2003) Use of $Hsf1^{-/-}$ mice reveals an essential role for HSF1 to protect lung against cadmium-induced injury. Toxicol Appl Pharmacol 192:12–20

Xiao X, Zuo X, Davis AA, Mcmillan DR, Curry BB, Richardson JA, Benjamin IJ (1999) HSF1 is required for extra-embryonic development, postnatal growth and protection during inflammatory responses in mice. EMBO J 18:5943–5952

Xie Y, Chen C, Stevenson MA, Auron PE, Calderwood SK (2002) Heat shock factor 1 represses transcription of the *IL-1β* gene through physical interaction with the nuclear factor of interleukin 6. J Biol Chem 277:11802–11810

Xing H, Wilkerson DC, Mayhew CN, Lubert EJ, Skaggs HS, Goodson ML, Hong Y, Park-Sarge OK, Sarge KD (2005) Mechanism of Hsp70i gene bookmarking. Science 307:421–423

Yan LJ, Christians ES, Liu L, Xiao X, Sohal RS, Benjamin IJ (2002) Mouse heat shock transcription factor 1 deficiency alters cardiac redox homeostasis and increases mitochondrial oxidative damage. EMBO J 21:5164–5172

Yan LJ, Rajasekaran NS, Sathyanarayanan S, Benjamin IJ (2005) Mouse *HSF1* disruption perturbs redox state and increases mitochondrial oxidative stress in kidney. Antioxid Redox Signal 7:465–471

Zhang Y, Frejtag W, Dai R, Mivechi NF (2001) Heat shock factor-4 (HSF-4a) is a repressor of HSF-1 mediated transcription. J Cell Biochem 82:692–703

Zheng H, Li Z (2004) Cross-presentation of cell-associated antigens to MHC class I molecule is regulated by a major transcription factor for heat shock proteins. J Immunol 173:5929–5933

Zou Y, Zhu W, Sakamoto M, Qin Y, Akazawa H, Toko H, Mizukami M, Takeda N, Minamino T, Takano H, Nagai T, Nakai A, Komuro I (2003) Heat shock transcription factor 1 protects cardiomyocytes from ischemia/reperfusion injury. Circulation 108:3024–3030

Heat Shock Proteins: Endogenous Modulators of Apoptotic Cell Death

C. Didelot · E. Schmitt · M. Brunet · L. Maingret · A. Parcellier · C. Garrido (✉)

Faculty of Medicine and Pharmacy, INSERM U-517, 7 Boulevard Jeanne d'Arc, 21079 Dijon, France
cgarrido@u-bourgogne.fr

1	Apoptosis or Programmed Cell Death	171
2	Heat Shock Protein Determinants for the Apoptotic Cellular Response to Stress	174
2.1	Hsp27: An Anti-apoptotic Protein	175
2.2	Hsp70: A Potent Anti-apoptotic Protein	177
2.3	Hsp90: Anti-apoptotic Proteins	180
2.4	Hsp60: A Pro-apoptotic Chaperon?	182
2.5	Hsp110: A Role in Apoptosis?	183
3	Are the Different Functions of Hsps in Apoptosis Specific and/or Related to Their Chaperone Activity?	183
4	Heat Shock Proteins and Cancer	185
5	Hsps as Pharmacological Targets in Apoptosis Modulation for Cancer Therapy	188
	References	189

Abstract The highly conserved heat shock proteins (Hsps) accumulate in cells exposed to heat and a variety of other stressful stimuli. Hsps, that function mainly as molecular chaperones, allow cells to adapt to gradual changes in their environment and to survive in otherwise lethal conditions. The events of cell stress and cell death are linked and Hsps induced in response to stress appear to function at key regulatory points in the control of apoptosis. Hsps include anti-apoptotic and pro-apoptotic proteins that interact with a variety of cellular proteins involved in apoptosis. Their expression level can determine the fate of the cell in response to a death stimulus, and apoptosis-inhibitory Hsps, in particular Hsp27 and Hsp70, may participate in carcinogenesis. This review summarizes the apoptosis-regulatory function of Hsps.

Keywords Heat shock proteins · Apoptosis · Tumorigenicity · Cancer therapy

1
Apoptosis or Programmed Cell Death

Apoptosis or programmed cell death is a highly regulated, energy-dependent form of cell death with a characteristic morphological appearance that involves cellular shrinkage and chromatin condensation. Apoptosis is responsible for the removal of unwanted or supernumerary cells during development, as well as in adult homeostasis (Jacobson et al. 1997). Apoptosis is also the predominant form of cell death triggered by cytotoxic drugs in tumor cells (Solary et al. 2000).

There are two pathways leading to apoptosis: the intrinsic pathway and the extrinsic (death receptor) pathway (Fig. 1). In the intrinsic pathway, the mitochondrion fulfills a dual function, (a) as an integrator of multiple pro-apoptotic signaling cascades or damage pathways, and (b) as a coordinator of the catabolic reactions culminating in apoptosis. In response to multiple apoptotic signals of different origins (Ferri and Kroemer 2001), the outer mitochondrial membrane becomes permeabilized, resulting in the release of molecules normally confined to the intermembrane space. Such proteins translocate from mitochondria to the cytosol in a reaction that is controlled by Bcl-2 and Bcl-2-related proteins (Kroemer and Reed 2000). Various molecular mechanisms have been proposed to account for the permeabilization of the outer mitochondrial membrane. These include pore formation in the external mitochondrial membrane by proteins such as Bax (alone or in combination with these proteins) and physical disruption of the outer membrane as a result of mitochondrial matrix swelling (resulting from the formation of nonspecific pores in the inner membrane and/or increased net influx of ions and water) (Marzo et al. 1998; Zamzami and Kroemer 2001). Mitochondrial intermembrane molecules include cytochrome c, apoptosis-inducing factor (AIF), endonuclease G (EndoG), Omi/HtrA2, and second mitochondria-derived activator of caspases (Smac), also called DIABLO. Cytochrome c, once in the cytosol, interacts with Apaf-1 (apoptotic protease activation factor-1), thereby triggering the ATP-dependent oligomerization of Apaf-1, while exposing its CARD domain (caspase recruitment domain) (Li et al. 1997; Hu et al. 1999). Oligomerized Apaf-1 then binds to cytosolic procaspase-9 in a homotopic interaction involving the CARD domain of caspase-9, thereby leading to the formation of the so-called apoptosome, the caspase-9 activation complex. Activated caspase-9 triggers the proteolytic maturation of pro-caspase-3, setting on the activation in the cytosol of a caspase cascade that leads to the limited proteolytic cleavage of intracellular, structural, and regulatory proteins, leading to membrane blebbing, chromatin condensation, and nuclear DNA fragmentation (Li et al. 1997). In contrast to cytochrome c, AIF and EndoG directly translocate to the nucleus and trigger caspase-independent nuclear changes (Susin et al. 1999; Joza et al. 2001). Smac/DIABLO and Htra2/Omi activate apoptosis by neutralizing the inhibitory activity of the IAPs (inhibitory apoptotic proteins) that associate with and inhibit caspases (Du et al. 2000) (Fig. 1).

The extrinsic pathway involves plasma membrane death receptors. These receptors (TNF-R1, CD95/APO-1/Fas, TRAIL-R1, TRAIL-R2, DR3, DR6, etc.) belong to the superfamily of TNF receptors (Nagata 1997). Death receptors contain an intracytoplasmic domain called the death domain. Upon ligation of TNF, Fas, or related death receptors, a complex protein known as the death-

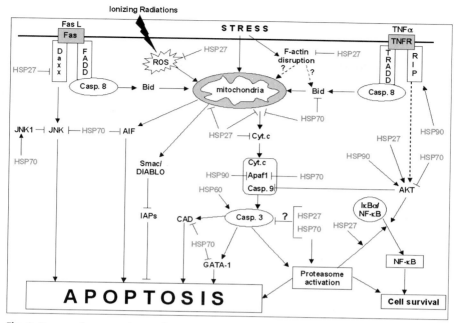

Fig. 1 Proposed apoptosis regulatory function of Hsps. The intrinsic and the extrinsic pathways of apoptosis are indicated, as well as the different Hsps targets (see text). →, induction; ⊣, inhibition; *ROS*, reactive oxygen species

inducing signaling complex (DISC) form at the cytosolic C-terminus of the receptor. This complex includes adaptor cytosolic proteins such as TRADD (TNF receptor death domain protein) or FADD (Fas-associated death domain protein) that recruit procaspase-8 (and often procaspase-10), thereby provoking their proteolytic autoactivation to generate active caspase-8 (and perhaps caspase-10). Downstream of caspase-8 and -10, two alternative pathways can trigger apoptotic cell death. One involves the direct activation of other caspases (in the so-called type 1 cells), while the other requires the intervention of mitochondria (in the so-called type 2 cells) and therefore converges to the above-mentioned mitochondrial apoptotic pathway (Scaffidi et al. 1999). In type 2 cells, caspase-8 cleaves and activates the pro-apoptotic Bcl2 family protein Bid, which then translocates to mitochondria and triggers permeabilization of the outer membrane.

One of the best-studied mediators of apoptosis are the caspases (Thornberry and Lazebnik 1998). Caspases are cysteine proteases expressed in virtually all animal cells. They are synthesized as inactive proenzymes (procaspases) and can be classified into two main groups according to the length of their N-terminal prodomain. Procaspases with a short prodomain (procaspase-3, -6 and -7) are effectors of apoptotic cell death by cleaving essential cellular substrates. Procaspases with a long prodomain (procaspase-8, -9 and -10) are usually the initiators of a caspase cascade. The serine proteases calpains and cathepsins have also been involved in apoptotic pathways, either working in synergy with caspases or by inducing a caspase-independent apoptotic cell death (Jaattela 2002). In addition or alternatively (Ferri et al. 2000; Susin et al. 2000), proteins such as AIF or EndoG may act independently of caspases and constitute a direct molecular link between outer mitochondrial membrane permeabilization and nuclear chromatin condensation. Indeed, in several paradigms of cell death, inhibition of caspases will lead to abortive, presumably AIF-dependent nuclear apoptosis (Daugas et al. 2000; Hisatomi et al. 2001; Joza et al. 2001; Loeffler et al. 2001).

The apoptotic process is tightly regulated. An abnormal increase in apoptosis leading to the unwarranted demise of cells is involved in several pathological processes such as myocardial infarction, stroke, neurodegenerative disease, and AIDS (Kroemer and Reed 2000). In contrast, a deficit in apoptosis is involved in cancer development. Proteins of the Bcl2 family, IAPs, and recently heat shock proteins have been demonstrated to control the apoptotic process at different key points. In this chapter, we will discuss the potential apoptosis modulating functions of different Hsps, putting special emphasis on the nature of their molecular partners and their consequent role in tumorigenicity.

2
Heat Shock Protein Determinants for the Apoptotic Cellular Response to Stress

Stress or heat shock proteins (Hsps) were first discovered in 1962 (Ritossa 1962) as a set of highly conserved proteins whose expression was induced by different kinds of stress. It has subsequently been shown that most Hsps have strong cytoprotective effects and behave as molecular chaperones for other cellular proteins. Inappropriate activation of signaling pathways could occur during acute or chronic stress as a result of protein misfolding, protein aggregation, or disruption of regulatory complexes. The action of chaperones, through their properties in protein homeostasis, is thought to restore balance. For instance, after severe exposure to elevated temperatures, heat-damaged proteins are sequestered through interactions with chaperones and are either refolded to the native state or degraded through the proteasome. Mammalian Hsps have been classified into five families according to their molecular size: Hsp110, Hsp90, Hsp70, Hsp60, and the small Hsps (Table 1). Each family of Hsps is composed of members expressed either constitutively or regulated inductively, and/or targeted to different subcellular compartments (Table 1). For example, Hsp60, HSC70, and Hsp90 are constitutively expressed in mammalian cells, while Hsp27 and Hsp70 are strongly induced by different stresses such as heat, oxidative stress, or anticancer drugs. Some of the

Table 1 Main Hsps involved in apoptosis

HSP	Location	Apoptotic function	Main apoptotic targets[a]
HSP27	Cytosol/nucleus	Anti-apoptotic	−: ROS, Q/cytochome c
			+: NF-κB, AKT
HSP60	Mitochondria	Pro-apoptotic	+: Caspase 3
HSP70	Cytosol/nucleus	Anti-apoptotic	−: AIF. Apaf1, CAD, DR4, DR5, cathepsins, JNK
			+:AKT
HSP90	Cytosol/nucleus	Anti-apoptotic	−: Apaf1, calpains
			+: AKT, RIP1
HSP110	Cytosol/nucleus	Anti-apoptotic	?

[a] A minus sign denotes negative regulation, a plus sign indicates positive regulation

important housekeeping functions attributed to the molecular chaperones include:

- Import of proteins into cellular compartments
- Folding of proteins in the cytosol, endoplasmic reticulum, and mitochondria
- Degradation of unstable proteins
- Dissolution of protein complexes
- Prevention of protein aggregation
- Control of regulatory proteins
- Refolding of misfolded proteins (Bukau and Horwich 1998; Young et al. 2004)

It is now well accepted that Hsps function at multiple points in the apoptotic signaling pathway (Fig. 1) and regulate apoptosis. Hsp27, Hsp70, and Hsp90 are anti-apoptotic, while Hsp60 and Hsp10 are pro-apoptotic. A titration must occur among the different apoptotic regulatory molecules and this balance will determine the fate of the stressed cells.

2.1
Hsp27: An Anti-apoptotic Protein

Hsp27 belongs to the subfamily of small Hsps, a group of proteins that vary in size from 15 to 30 kDa and share sequence homologies and biochemical properties such as phosphorylation and oligomerization. Hsp27 can form oligomers up to 1,000 kDa. The dimer of Hsp27 seems to be the building block for the multimeric complexes. Hsp27 oligomerization is a dynamic process that depends on the phosphorylation status of the protein and exposure to stress (Bruey et al. 2000a; Garrido 2002). Hsp27 can be phosphorylated at three serine residues and its dephosphorylation enhances oligomerization. This phosphorylation is a reversible process catalyzed by the MAPKAP kinases 2 and 3 in response to a variety of stresses, including differentiating agents, mitogens, inflammatory cytokines such as tumor necrosis factor alfa (TNFα) and interleukin-1 beta (IL-1β), hydrogen peroxide, and other oxidants. Hsp27 is expressed in many cell types and tissues at specific stages of development and differentiation (Garrido et al. 1998). Hsp27 is an ATP-independent powerful chaperone, its main chaperone function being protection against protein aggregation (Ehrnsperger et al. 1997).

Overexpressed Hsp27 protects against apoptotic cell death (Fig. 1) triggered by various stimuli, including hyperthermia, oxidative stress, staurosporine, ligation of the Fas/Apo-1/CD95 death receptor, and cytotoxic drugs (Garrido et al. 1996; Mehlen et al. 1996b). Several of these stimuli induce Hsp27 (and Hsp70) overexpression, providing an example of how pro-apoptotic stimuli, delivered below a threshold level, can elicit protective responses. Hsp27 has been shown to interact and inhibit components of both stress- and receptor-induced apoptotic pathways. We have demonstrated that Hsp27 could prevent the formation of the apoptosome and the subsequent activation of caspases (Garrido et al. 1999). It does so by directly sequestering cytochrome c when released from the mitochondria into the cytosol (Bruey et al. 2000a). The heme group of cytochrome c is necessary but not sufficient for this interaction, which involves amino-acids 51 and 141 of Hsp27 and requires dimerization of the stress protein. Actually, we have shown both in vitro and in vivo that the postmitochondrial anti-apoptotic effect of Hsp27 involved large, nonphosphorylated oligomers of Hsp27 (Bruey et al. 2000b).

At higher Hsp27 intracellular levels, the protein has also been shown to interfere with caspase activation upstream of the mitochondria. This effect seems to be related to the ability of Hsp27 to stabilize actin microfilaments (Lavoie et al. 1993). Hsp27 binds to F-actin to prevent disruption of the cytoskeleton resulting from heat shock, actin filament disrupting agent cytochalasin D, and other stresses (Guay et al. 1997). For example, in L929 murine fibrosarcoma cells exposed to cytochalasin D or staurosporine, overexpression of Hsp27 prevents the cytoskeletal disruption and Bid intracellular redistribution that precede cytochrome c release (Paul et al. 2002). More recently, Hsp27 has been shown to inhibit the mitochondrial release of Smac and thereby to confer resistance of multiple myeloma cells to dexamethasone (Chauhan et al. 2003a). The ability of Hsp27 to interact with caspase-3 is a more controversial issue (Pandey et al. 2000a; Concannon et al. 2001). Initially identified in cell-free extracts from 293T cells, this interaction was not confirmed when studied in breast cancer cells (Kamradt et al. 2001). In addition, interaction of Hsp27 with caspase-3 was found to disappear when cells were exposed to cytotoxic drugs, and thus may not account for the cytoprotective function of the protein (Pandey et al. 2000a).

Hsp27 has been shown to increase the anti-oxidant defense of cells by decreasing reactive oxygen species cell content (Mehlen et al. 1996a) and to neutralize the toxic effects of oxidized proteins (Rogalla et al. 1999). This latter effect may occur more specifically in neuronal cells in which the Hsp27 protective effect does not depend on its interaction with cytochrome c and involve phosphorylated Hsp27 (Wyttenbach et al. 2002).

Hsp27 also inhibits apoptosis by regulating upstream signaling pathways. Survival factors, such as nerve growth factor or platelet-derived growth factor, inhibit apoptosis by activating the phosphatidylinositol 3-kinase pathway (PI3-K). Activated PI3-K phosphorylated inositol lipids in the plasma membrane that attract the serine/threonine kinase AKT/PKB. AKTs target multiple proteins of the apoptotic machinery, including Bad and caspase-9 (Datta et al. 1997; Cardone et al. 1998; Biggs et al. 1999; Ozes et al. 1999). Hsp27 has been shown to bind the protein kinase AKT, an interaction that is necessary for AKT activation in stressed cells (Konishi et al. 1997; Rane et al. 2003). In turn, AKT could phosphorylate Hsp27, thus leading to the disruption of Hsp27–AKT complexes (Rane et al. 2003). Hsp27 also affects one of the Fas-mediated apoptotic pathways. The phosphorylated form of Hsp27 directly interacts with DAXX. This latter protein connects Fas signaling to the protein kinase Ask1, which mediates a caspase-independent cell death (Charette et al. 2000). Recently, we have demonstrated that under stress conditions Hsp27 increases IκBα ubiquitination/degradation, which results in an increase in NF-κB activity and increased survival (Parcellier et al. 2003).

In conclusion, Hsp27 can interact with different partners implicated in the apoptotic process (Table 1 and Fig. 1). Interestingly, the relations shown above suggest that the phosphorylation status of Hsp27, which at least modulates the level of oligomerization of the protein in vitro, might distinguish between the mitochondrial inhibition of apoptosis that can be performed by a non-phosphorylated form of Hsp27 and the death receptor pathway, inhibition that involves the phosphorylated protein.

2.2
Hsp70: A Potent Anti-apoptotic Protein

The Hsp70 family constitutes the most conserved and best studied class of Hsps. Human cells contain several Hsp70 family members, including stress-inducible Hsp70, constitutively expressed HSC70, and GRP78, localized in the endoplasmic reticulum (Jaattela 1999). Under normal conditions, Hsp70 proteins function as ATP-dependent molecular chaperones by assisting the folding of newly synthesized polypeptides, the assembly of multiprotein complexes, and the transport of proteins across cellular membranes (Beckmann et al. 1990; Shi and Thomas 1992). Under various stress conditions, the synthesis of stress-inducible Hsp70 enhances the ability of stressed cells to cope with increased concentrations of unfolded or denatured proteins (Nollen et al.

1999). Like Hsp27, Hsp70 has been shown to increase the tumorigenicity of cancer cells in rodent models (Jaattela 1995). In contrast, downregulation of Hsp70 induces cell death or increases their sensibility to die depending on the cellular model. As a result, in experimental models, Hsp70 downregulation strongly decreases tumorigenicity (Schmitt et al. 2003). Furthermore, gene removal studies show that Hsp70 plays an important role in apoptosis. Recently, a mouse line lacking *hsp70.1* and *hsp70.3* was generated. These cells are very sensitive to apoptosis induced by a wide range of lethal stimuli (Schmitt et al. 2003). The testis-specific isoform of Hsp70 (*hsp70.2*), when ablated, results in germ cell apoptosis. Germ cells show G_2/M arrest and death in late pachytene cells, indicating an important role for *hsp70.2* in meiosis regulation (Dix et al. 1996).

Overexpression of Hsp70, like Hsp27, can inhibit apoptosis and thereby increase the survival of cells exposed to many different lethal stimuli (Jaattela et al. 1992; Mosser et al. 1997). Indeed, overexpression of Hsp70 protects cells from stress-induced apoptosis, both upstream and downstream of the caspase cascade activation (Fig. 1). Preliminary data suggest that Hsp70 could protect the cells from energy deprivation and/or ATP depletion associated with cell death (Wong et al. 1998). Elevated levels of Hsp70, attained in transient or stable transfections, reduce or block caspase activation and suppress mitochondrial damage and nuclear fragmentation (Buzzard et al. 1998). These findings were supported by Li et al. (Li et al. 2000) who found Hsp70 inhibited apoptosis downstream of the release of cytochrome c and upstream of the activation of caspase-3. This anti-apoptotic effect was explained by the Hsp70-mediated modulation of the apoptosome. Indeed, Hsp70 has been demonstrated to directly bind to Apaf-1, thereby preventing the recruitment of procaspase-9 to the apoptosome (Beere et al. 2000; Saleh et al. 2000). The ATPase domain of Hsp70 was described to be necessary for this interaction (Mosser et al. 2000; Saleh et al. 2000). Other relations have shown that Hsp70 interacts with procaspase-3 and procaspase-7 and prevents their maturation, thereby inhibiting caspase-dependent apoptotic signaling (Komarova et al. 2004). However, these results have been contradicted by a recent study in which the authors demonstrate that the inhibition of caspase-dependent apoptosis by Hsp70 results from an inhibition of cytochrome c release from the mitochondria and not from any direct effect of Hsp70 in caspase activation. They explain this contradictory result by showing that it is a high salt concentration but not Hsp70 that inhibits caspase activation (Steel et al. 2004).

Hsp70 can also prevent caspase-independent apoptosis pathways. Hsp70 prevents cell death in conditions in which caspase activation does not occur due to the addition of exogenous caspase inhibitors (Creagh et al. 2000). We have recently observed that overexpression of Hsp70 protects Apaf-$1^{-/-}$ cells from apoptotic cell death induced by serum withdrawal (Ravagnan et al. 2001), indicating that the cytochrome c/Apaf-1/caspase was not the sole pathway of the anti-apoptotic action of Hsp70. Indeed, it appears that Hsp70 directly binds to AIF and inhibits AIF-induced chromatin condensation. Hsp70 was found to neutralize the apoptogenic effects of AIF in cell-free systems, in intact cells microinjected with recombinant Hsp70 and/or AIF protein, as well as in cells transiently transfected with AIF cDNA. Hsp70 inhibited apoptosis induced by overexpression of both full-length AIF (which has to transit mitochondria to become apoptogenic) (Susin et al. 1999; Loeffler et al. 2001) and AIF lacking the mitochondrial localization sequence (AIFΔ1–100). Of note, endogenous level of Hsp70 seem to be sufficiently high to control AIF-mediated apoptosis since downregulation of Hsp70 by an anti-sense construct sensitized the cells to serum withdrawal and AIF (Ravagnan et al. 2001).

Hsp70 can also rescue cells from a later phase of apoptosis better than any known survival-enhancing drug or protein. In TNFα-induced apoptosis, Hsp70 does not preclude the activation of caspase-3 but prevents downstream morphological changes that are characteristic of dying cells such as activation of phospholipase A2 and changes in nuclear morphology (Jaattela et al. 1998). During the final phases of apoptosis, chromosomal DNA is digested by the Dnase CAD (caspase-activated Dnase) following activation by caspase-3. The enzymatic activity and proper folding of CAD has been reported to be regulated by Hsp70, its co-chaperone Hsp40, and ICAD, the inhibitor of CAD. ICAD recognizes an intermediate folding state conferred by Hsp70-Hsp40 (Sakahira and Nagata 2002). It has also been reported in TCR-stimulated T cells that Hsp70 binds CAD and enhances its activity (Liu et al. 2003). Another nuclear target of caspase-3 is the transcription factor GATA-1. We have demonstrated in hematopoietic cells that Hsp70 can protect GATA-1 from caspase-3 cleavage. As a consequence, cells do not die by apoptosis but instead differentiate (Zermati et al. 2001).

Hsp70 has also been proposed to act on the apoptotic pathway at early steps, for instance by preventing JNK activation (Meriin et al. 1999). Indeed Park et al. (2001) have shown that Hsp70 binds to and functions as a natural inhibitory protein of c-Jun N-terminal Kinase (JNK1). The ATPase domain of Hsp70 was dispensable for this binding (Mosser et al. 2000). However, JNK inhibition achieved by this binding was not sufficient for the prevention of apoptosis (Mosser et al. 2000), putting a question mark over the role of JNK itself in apoptosis. Hsp70 also appears to affect the Bid-dependent apoptotic pathway. Hsp70 is able to inhibit TNF-induced cell death. However, this protective effect is lost in Bid homozygous-deleted MEF cells (Mosser et al. 2000). It has been suggested that Hsp70, by inhibiting JNK activation, could affect Bid-dependent cell death by a mechanism that is not fully clear (Gabai et al. 2000). In apoptosis induced by hyperosmolarity, Hsp70 has been found to modulate JNK and ERK phosphorylation (Lee et al. 2004).

Hsp70 has recently been shown to act at the death receptors level to mediate Bcr-Abl-mediated resistance to apoptosis in human acute leukemia cells. Hsp70 binds to the death receptors DR4 and DR5, thereby inhibiting TRAIL-induced assembly and activity of death-inducing signaling complex (DISC)

(Guo et al. 2004). Exposure of hematopoietic cells to TNF induces the activity of the pro-apoptotic double-stranded RNA-dependent protein kinase (PKR). An inhibitor of PKR is the Fanconi anemia complementation group C gene product (FANCC). Hsp70 interacts with the FANCC protein via its ATPase domain and, together with Hsp40, inhibits TNF-induced apoptosis through the ternary complex Hsp70, FANCC, and PKR (Pang et al. 2001; Pang et al. 2002). Hsp70 has been shown to bind to nonphosphorylated protein kinase C (PKC) via the kinase's nonphosphorylated carboxyl-terminus, priming the kinase for new phosphorylation and stabilizing the protein. In a similar manner, Hsp70 binds AKT, resulting in its stabilization (Gao and Newton 2002).

Hsp70 has also been shown to associate to the pro-apoptotic proteins p53 and c-myc (Jolly and Morimoto 2000). However, the functional impact of these interactions in Hsp70 survival effects remains elusive (Jolly and Morimoto 2000). Yet another apoptosis regulatory protein interacting with Hsp70 is Bag-1. Bag-1 has been reported to function as co-chaperone of Hsp70, and simultaneously regulates the activities of proteins such as Bcl-2 and Raf-1. It has been shown that Hsp70/Bag-1 regulates Raf-1/ERK kinase and cell growth in response to stress (Song et al. 2001; Gotz et al. 2004). Whether the Hsp70-Bag-1 interaction is important for Hsp70-mediated apoptosis regulation is unknown. Finally, Hsp70 is also involved in the inhibition of cathepsins, and lysosome proteases are also involved in apoptosis (Nylandsted et al. 2004).

The data discussed above show that Hsp70 inhibits apoptosis by interacting with several pro-apoptotic effectors including AIF, Apaf1, CAD, and perhaps signaling molecules such as JNK-1, p53, and c-myc. At present, it is not clear whether these neutralizing interactions synergize to determine the broad cytoprotective effect of Hsp70.

2.3
Hsp90: Anti-apoptotic Proteins

Prominent members of the Hsp90 family of proteins are Hsp90α, Hsp90β, and Grp94 (Sreedhar et al. 2004). The two Hsp90 isoforms are essential for the viability of eukaryotic cells. They are rather abundant constitutively, make up 1%–2% of cytosolic proteins, and can be further stimulated in their expression level by stress. Hsp90 associates with a number of signaling proteins, including ligand-dependent transcription factors such as steroid receptor (Nathan and Lindquist 1995), ligand-independent transcription factors such as MyoD (Shaknovich et al. 1992), tyrosine kinases such as v-Src (Hartson and Matts 1994), and serine/threonine kinases such as Raf-1 (Wartmann and Davis 1994). The main chaperon role of Hsp90 is to promote the conformational maturation of these receptors and signal transducing kinases. Hsp90, like Hsp70 and Hsp60, binds ATP and undergoes a conformational change upon ATP binding needed to facilitate the refolding of denatured proteins. Although most studies indicated an anti-apoptotic function for Hsp90, two reports suggest a pro-apoptotic function for this chaperone. Indeed, overexpression of Hsp90 has been demonstrated to increase the rate of apoptosis in the monoblastoid cell line U937 following induction with TNFα and cycloheximide (Galea-Lauri et al. 1996). More recently, it has been demonstrated that treatment of human embryonic fibroblasts with the Hsp90α inhibitor geldanamycin increases the resistance of these cells to nicotine (Wu et al. 2002). These results, which contradict most reports (see below and Fig. 1), may reflect a difference in signaling among the different apoptotic stimuli or between the two Hsp90 isoforms.

Hsp90 overexpression in U937 cells can inhibit apoptosis induced by staurosporin and can prevent the activation of caspases in cytosolic extracts treated with cytochrome c (Pandey et al. 2000b). Pandey et al. reported that Hsp90 inhibited apoptosis as a result of a negative effect on Apaf-1 function. Hsp90 directly binds Apaf-1 and inhibits its oligomerization and further recruitment of procaspase-9 (Pandey et al. 2000b). The anti-apoptotic action of Hsp90 is also reflected by its capacity to interact with phosphorylated serine/threonine kinase AKT/PKB, a protein that generates a survival signal in response to growth factor stimulation. Phosphorylated AKT can phosphorylate the Bcl-2 family protein Bad and caspase-9 (Cardone et al. 1998), leading to their inactivation and to cell survival. However, AKT has been also shown to phosphorylate IκB kinase, which results in promotion of NF-κB-mediated inhibition of apoptosis (Ozes et al. 1999). A role for Hsp90 in AKT pathway was first suggested by studies using an Hsp90 inhibitor that promoted apoptosis in HEK293T and resulted in suppressed AKT activity (Basso et al. 2002). A direct interaction between AKT and Hsp90 was latter on reported (Sato et al. 2000; Basso et al. 2002). Binding of Hsp90 protects AKT from protein phosphatase 2A (PP2A)-mediated dephosphorylation (Sato et al. 2000). When this interaction was prevented by Hsp90 inhibitors, AKT was dephosphorylated and destabilized and the likelihood of apoptosis increased (Sato et al. 2000). Additional studies showed that another chaperone participates in the AKT-Hsp90 complex, namely Cdc37 (Basso et al. 2002). Together this complex protects AKT from proteasome degradation. In human endothelial cells during early phases of high glucose exposure, apoptosis can be prevented by Hsp90 through augmentation of the interaction between eNOS and Hsp90 and recruitment of the activated AKT (Lin et al. 2005).

Hsp90 has also been shown to interact with and stabilize the receptor-interacting protein (RIP). Upon ligation of TNFR-1, RIP-1 is recruited to the receptor and promotes the activation of NF-κB and JNK. Degradation of RIP-1 in the absence of Hsp90 precludes activation of NF-κB mediated by TNFα and sensitizes cells to apoptosis (Lewis et al. 2000). Another route by which Hsp90 can affect NF-κB survival activity is via the IKK complex. This complex is composed of two catalytic and one regulatory subunit, and recently it was determined that Hsp90 and Cdc37 were also present, with the association mediated through the kinase domain of the catalytic subunits (Chen et al. 2002). The Hsp90 inhibitor geldanamycin abolishes this complex and prevents

TNF-induced activation of IKK and NF-κB, highlighting the role of Hsp90 in NF-κB activation following TNFα exposure.

Another pathway of cell survival in which Hsp90 can be involved is p53. It has been demonstrated that P53 repress Hsp90β gene expression in UV irradiated cells (Zhang et al. 2004). Other Hsp90 client proteins through which this chaperone could participate in cell survival are the transcription factors Her2 and HIF1α (Hur et al. 2002; Munster et al. 2002).

Finally, Hsp90 proteins may also protect apoptosis by inhibiting the action of the calcium-dependent protease calpains. The Hsp90 family protein Grp94 was shown to cleave calpain and to protect human neuroblastoma cells from hypoxia/reoxygenation-induced apoptosis involving calpains (Bando et al. 2003).

In conclusion, Hsp90 seems to have different molecular partners depending on the apoptotic stimuli, the effect of the protein being predominantly anti-apoptotic (Fig. 1). It should be noted that most studies do not differentiate between the α and β isoforms of Hsp90.

2.4
Hsp60: A Pro-apoptotic Chaperon?

Mammalian Hsp60, also called chaperonin, is mostly contained within the mitochondrial matrix, although it has also been detected in extra-mitochondrial sites (Soltys and Gupta 1996). Hsp60 participates in the folding of mitochondrial proteins and facilitates the proteolytic degradation of misfolded or denatured proteins in an ATP-dependent manner. The chaperone function of Hsp60 is regulated by Hsp10, which binds to Hsp60 and regulates substrate binding and ATPase activity. In the presence of ADP, two Hsp10 molecules bind to one Hsp60 molecule. Hsp60 and Hsp10 do not always act as a single functional unit: only newly imported proteins are severely affected by inactivation of Hsp10 (Bukau and Horwich 1998).

A pro-apoptotic role for Hsp60 and Hsp10 during apoptosis has recently been demonstrated by two independent groups. It was shown in Hela and Jurkat cells that activation of caspase-3 by camptothecin or staurosporine occurs simultaneously with Hsp60 and Hsp10 release from the mitochondria (Fig. 1). Further, the authors demonstrate in vivo and in vitro that Hsp60 and Hsp10 associate to procaspase-3 and favor its activation by cytochrome c in an ATP-dependent manner, suggesting that the chaperone function of Hsp60 is involved in this process (Samali et al. 1999; Xanthoudakis et al. 1999). In contrast with these results, it has recently been shown that overexpression of Hsp60 and/or Hsp10 by stable transfection in cardiac myocytes using an adenoviral vector increases the survival rate of cardiac myocytes undergoing ischemia/reperfusion injury (Lin et al. 2001). Reduction in Hsp60 is sufficient to precipitate apoptosis in myocytes (Kirchhoff et al. 2002). Cytosolic Hsp60 has also been shown to be complexed with the pro-apoptotic protein Bax (Gupta and Knowlton 2002). Under hypoxic conditions, Hsp60 and Bax dissociate, whereupon Bax translocates to the mitochondria and induces apoptosis. The interaction of Hsp60 and Bax may therefore prevent apoptosis. Accordingly, recent studies show that reducing Hsp60 expression with antisense oligonucleotides in cardiomyocytes correlated with an increase in Bax and reduction in Bcl2 and resulted in induction of apoptosis (Kirchhoff et al. 2002; Shan et al. 2003). However, in human esophageal squamous cell carcinoma, expression of Hsp60 has been defined as a good prognostic indicator, again reflecting the ambiguous role of this chaperone in the tumor cell death process (Faried et al. 2004).

These contradictory results could be reconciled on the basis of experimental variables. Increased expression of Hsp60 obtained by stable transfection, shown to have a protective effect against apoptosis, is unlikely to occur under physiological conditions, Hsp60 being mainly a constitutive mitochondrial protein. However, these results could also reflect a difference in signaling between different cell types and in response to different stress.

2.5
Hsp110: A Role in Apoptosis?

The Hsp110 family is represented mainly by the constitutively expressed Hsp110α (also called Hsp105) and the inductive Hsp110β, which is an alternative splicing form of Hsp110 α. They can be phosphorylated by casein kinase II (Ishihara et al. 2000). These ATP-binding chaperones have not been identified in prokaryotes and, interestingly, knock down of Hsp110 in *Drosophila* is lethal. Hsp110 proteins seem related to the Hsp70 family proteins: they are located in the nucleus and cytoplasm often associated with Hsp70, which suggests that Hsp110 functions cooperatively with HSC/Hsp70 or that it regulates the function of these Hsp70 chaperones. Hsp110 is abundant in the brain, human testis, and sperm. Although the role of these proteins under normal conditions or physiological stresses is not known, they protect heat-denatured proteins, protein aggregation, and confer cellular thermoresistance (Oh et al. 1997; Yamagishi et al. 2000). Hsp110 as well as Apg-2, another Hsp110 family member, has been shown to inhibit apoptosis induced by mutant HSCO proteins (Gotoh et al. 2004). More studies are needed to determine whether the role on apoptosis of these chaperones has physiological relevance.

3
Are the Different Functions of Hsps in Apoptosis Specific and/or Related to Their Chaperone Activity?

The heat shock proteins as chaperones serve as cellular safeguards to protect the network of protein–protein interactions that sense stress signals and relay them to the apoptotic machinery. Conformational alteration among the protein components of the cell-death machinery is central to their activation. As a result, it is not surprising that Hsps, which can influence the aggregation, assembly, transport and folding of other proteins, may directly affect the execution of apoptotic signaling pathways. Since Hsps can act at multiple points in the apoptotic pathways to ensure that stress-induced damage does not inappropriately trigger cell death, it has been difficult to understand whether stress proteins have a specificity in this process.

The specificity needed for ATP-dependent chaperones to play a wide-ranging role in apoptosis may be determined by the presence of a given co-chaperone that may modulate the affinity of the Hsp for one or another substrate. This seems to be the case for Hsp60 and its co-chaperone Hsp10, proteins that associate with procaspase-3 and stimulate pro-enzyme maturation. Another example is Hsp70 and its co-chaperone Hsp40, needed to modulate the last step of apoptosis, DNA fragmentation. In addition, some co-chaperones may have a role in apoptosis on their own. This is the case of CHIP (C-terminus of HSC70 interacting protein), which can regulate cell death in some types of Parkinson's disease (Imai et al. 2002) and BAG1 (bcl2-associated atherogenic 1), a regulator of apoptosis by virtue of its ability to bind Bcl2 (Takayama et al. 1998).

In the case of small Hsps such as Hsp27 or α-crystallins, ATP-independent chaperones for which no co-chaperones have been reported, the phosphorylation and/or oligomerization status of the proteins seem to modulate the affinity of the proteins for a given substrate. Hsp27 oligomerization is a highly dynamic process modulated by the phosphorylation of the protein. We have demonstrated in vitro and in vivo, that for the general anti-apoptotic effect of Hsp27, caspase-dependent, large nonphosphorylated oligomers of Hsp27 were the active form of the protein (Bruey et al. 2000b). In contrast, for Hsp27 thermotolerant function small oligomers of Hsp27 were necessary. These results suggest that the oligomerization/phosphorylation of the protein alters Hsp27 conformation and hence determines its capacity to interact with different apoptotic proteins and its survival effects.

However, the specificity of Hsps could also be explained if Hsp's apoptotic function is independent of their chaperone properties. Like other chaperones, Hsp70 possesses a docking site for interaction with a substrate (peptide-binding domain, PBD), as well as an ATP-binding domain (ABD), both of which are required for the protein refolding (foldase) activity of Hsp70. It appears that the structural features of Hsp70 required for inhibition of Apaf-1 and AIF inhibition are distinct (Ravagnan et al. 2001). The in vitro interaction of Hsp70 and Apaf-1 relies on ATP hydrolysis (Saleh et al. 2000) and transfection with Hsp70ΔABD (an Hsp70 mutant in which the ATP-binding domain, necessary for the foldase activity of the protein, has been removed) fails to prevent caspase activation (Mosser et al. 2000). Accordingly, Hsp70ΔABD fails to co-immunoprecipitate with Apaf-1 (Ravagnan et al. 2001) and fails to prevent presumably Apaf-1- and caspase-dependent (Yoshida et al. 1998) cell death induced by etoposide, cisplatin, or doxorubicin (Ravagnan et al. 2001). These observations are in line with the assumption that the chaperone activity of Hsp70 is involved in its effect on Apaf-1 (Beere et al. 2000; Saleh et al. 2000). In contrast, Hsp70ΔABD interacts with AIF (Ravagnan et al. 2001) and can protect cells against AIF overexpression, serum withdrawal, menadione, staurosporine (Ravagnan et al. 2001), or thermal stress (Li et al. 1992). Moreover, these observations indicate that the foldase activity of Hsp70 is not required for the neutralization of AIF. It is possible that the binding of denatured proteins to the remaining substrate binding domain of Hsp70 is sufficient for the inhibition of protein aggregation. Indeed, the deletion of the substrate domain abolishes the protective effect of Hsp70 against heat, its association with AIF and, as a result, its inhibitory effect of the AIF-apoptotic pathway (Ravagnan et al. 2001). As for Hsp27, it appears that the chaperon activity and anti-apoptotic function could also be dissociated. Hsp27 is a chaperon that prohibits the aggregation and promotes the refolding of denatured proteins *in* vitro in an ATP-independent manner. It has been shown that the 33 amino acids of N-terminal region adjacent to the highly conserved α-crystallin domain of the protein was dispensable for its chaperone activity in vitro (Guo and Cooper 2000). However, we have shown that this region was essential for cytochrome c binding and for caspase-inhibitory properties of the protein (Bruey et al. 2000a). In conclusion, Hsps might exert the apoptosis-regulatory function, at least in part, through protein–protein interactions not directly related to their chaperon function. Future studies will unravel the fine mechanisms of such interactions.

4
Heat Shock Proteins and Cancer

The ability of heat shock proteins to influence the cell's fate through modulation of numerous control points, together with the fact that cells or tissues from a wide range of tumors have been shown to express unusually high levels of one or more Hsps, might endow these proteins with the unusual capacity to contribute in a decisive way and at multiple points in the process of tumorigenesis. Experimental models support the role of Hsps in tumorigenesis since Hsp27 and Hsp70 have been shown to increase the tumorigenic potential of rodent cells in syngeneic host (Jaattela 1995; Garrido et al. 1998). For Hsp27 and Hsp70, their tumorigenic potential seems to correlate with their anti-apoptotic abilities. Rat colon cancer cells engineered to express human Hsp27 were observed to form more aggressive tumors in syngenic animals than control cells and the increase in tumorigenicity correlated with a reduced rate of tumor cell apoptosis (Garrido et al. 1998). Overexpressed Hsp27 did not increase the tumorigenicity of these rat colon cancer cells, nor did it affect their survival rate in vivo when inoculated into immunodeficient animals, suggesting that Hsp27 somehow subverts the tumor-specific immune response. Hsp27 overexpression was also reported to increase the metastatic potential of human breast cancer cells inoculated into athymic (nude) mice (Lemieux et al. 1997). Conversely, Hsp27 antisense oligonucleotide-induced silencing in prostate tumors enhances apoptosis and delays tumor progression (Rocchi et al. 2004). Concerning Hsp70, antisense constructs of Hsp70 have been shown to increase the cell's sensitivity to apoptosis and to eradicate tumors (glioma, breast, and colon carcinomas) in several models (Buzzard et al. 1998; Li et al. 2000; Saleh et al. 2000). Thus, Hsp27 and Hsp70 contribute to tumorigenesis, at least in part, through their cytoprotective activity.

Clinically, in a number of cancers such as breast cancer, ovarian cancer, osteosarcoma, endometrial cancer, and leukemias, an increased level of Hsp27, relative to its level in nontransformed cells, has been detected (Garrido et al. 1998). In ovarian tumors, Hsp27 expression increases with the stage of the tumor (Geisler et al. 2004). In addition, the pattern of Hsp27 phosphorylation in tumor cells is different from that observed in primary nontransformed cells (Ciocca et al. 1993). Consequently, the diversity of Hsp27 isoforms may also represent a useful tumor marker, as recently demonstrated in human renal cell carcinomas (Sarto et al. 2004). Increased expression of Hsp70 has been reported in high-grade malignant tumors such as endometrial cancer, osteosarcoma, and renal cell tumors (Santarosa et al. 1997; Nanbu et al. 1998). Hsp90 and Hsp60 are also overexpressed in breast tumors, lung cancer, leukemias, and Hodgkin's disease (Yufu et al. 1992; Hsu and Hsu 1998). Hsp90 has been shown to be overexpressed in B-cell non-Hodgkin lymphomas compared to normal B cells (Ghobrial et al. 2005). Hsp110 protein members are highly expressed in hepatocarcinomas and probably contribute to hepatocarcinogenesis by acting with other Hsps to inhibit apoptosis induced by mutant proteins (Gotoh et al. 2004). Overexpression of the small heat shock protein α-B-crystallin has been observed in glial tumors such as astrocytoma, glioblastoma, and oligodendroglioma (Aoyama et al. 1993), and in renal carcinoma tumors (Pinder et al. 1994). The molecular basis for overexpression of Hsps in tumors is not completely understood but may have different etiologies and be tumor-specific. For example, in some tumors it may be due to a suboptimal cellular environment in the poorly vascularized hypoxic tumor or due to the growth conditions within the solid tumor (Garrido et al. 1997). In other tumors, oncogenic mutations could create an increased requirement for chaperone activity toward abnormally folded protein variants. Another possibility is the occurrence of gain-of-function mutations in transcription factors that increase heat shock promoter activity. In adenocarcinoma cell lines, an increased level of heat shock transcription factor 1 (HSF1) was associated with an increased Hsp70 and Hsp27 protein level (Hoang et al. 2000). In breast carcinomas, the signal transducer and activator of transcription 3 (Stat3) is constitutively activated. It has been demonstrated that Stat3 upregulates Hsp27 in those cancer cells

(Song et al. 2004). Hsp70 is very abundant in Bcr-Abl human leukemia cells and the GATA-RE element found in hsp70 promotor is necessary for this accumulation (Ray et al. 2004). The c-myc proto-oncogene directly activates Hsp90 transcription in different tumor cell models (Teng et al. 2004).

Could Hsps be used for diagnostic or as prognostic markers ? The ability of the Hsps to prevent apoptosis induced by several anticancer drugs as well as other stimuli also explain how these proteins could limit the efficacy of cancer therapy. High expression of Hsp27 and Hsp70 in breast, endometrial, or gastric cancer has been associated with metastasis, poor prognosis, and resistance to chemotherapy or radiation therapy (Ciocca et al. 1993; Vargas-Roig et al. 1998; Brondani Da Rocha et al. 2004). An overexpression of Hsp90α is associated with poor prognosis in breast cancer, pancreatic carcinoma, and leukemias (Yufu et al. 1992; Gress et al. 1994). In prostate cancer, Hsp27 is an independent predictor of clinical outcome. A low expression of Hsp27 is associated with a delay in tumor progression (Rocchi et al. 2004). Similar conclusions were reached in patients with ovarian carcinoma. Decreased Hsp27 staining was related to decreased survival (Geisler et al. 2004). Recently, it has been demonstrated that evaluation of the soluble level of Hsp27 in human serum may be useful to distinguish Hsp27 levels in breast cancers (De and Roach 2004). Hsp70, along with PSA, are good tumor markers to identify patients with early-stage prostate cancer (Abe et al. 2004). However, Hsp27 or Hsp70 are not universal markers of poor prognosis. Even though Hsp70 levels correlate with malignancy in osteosarcoma and renal cell tumors, its expression is paradoxically associated with improved prognosis (Santarosa et al. 1997; Trieb et al. 1998). In oral squamous cell carcinoma, cases with reduced expression of Hsp27 that were more aggressive and poorly differentiated were found (Lo Muzio et al. 2004). Similar contradictory results have been obtained for Hsp27 in breast tumors treated with estrogens. In breast cancers, Hsp27 expression has been associated with increased invasiveness but decreased cell motility (Lemieux et al. 1997). The hsp27 gene contains an imperfect estrogen-responsive element (Oesterreich et al. 1997) and can be induced by estrogen treatment in breast cancer cells (Ciocca et al. 1993). Overexpressed Hsp27 in a serie of breast cancers has been correlated with the expression of estrogen receptors, small tumor size, and a low proliferation index (Hurlimann et al. 1993). Despite this, clinicopathological studies attempting to correlate Hsp27 protein level in breast cancers after hormone therapy with tumor progression and clinical outcome have provided contradictory results (Ciocca et al. 1993; Nakopoulou et al. 1995). It is therefore possible that the chemoprotective effects of Hsps, under certain circumstances, may be bypassed by a variety of other modulators of drug resistance in human tumors.

5
Hsps as Pharmacological Targets in Apoptosis Modulation for Cancer Therapy

Constitutively high Hsp expression is a property of and essential for the survival of at least some cancers. Neutralizing Hsps is therefore an attractive strategy for anticancer therapy. Accordingly, Hsp90 can be inhibited by the benzoquinone ansamycin antibiotic geldanamycin and its analog 17-AAG (17-allylamino-17-deemethoxygeldanamycin), two drugs that are currently undergoing clinical trials for anti-cancer activity (Neckers 2002; Neckers and Ivy 2003). The fact that geldanamycin and 17-AAG selectively kills cancer cells has been rationalized by assuming that tumor cells, as compared to their normal counterparts, would exhibit a stressed phenotype, with an enhanced dependency on the cytoprotective action of Hsp90. In tumors, Hsp90 is present entirely in multichaperone complexes with high ATPase activity, whereas Hsp90 formed in normal tissues is in an uncomplexed state (Kamal et al. 2003). Similarly, we and others have extensively reported that Hsp70 antisense constructs have chemosensitizing properties and may even kill cancer cell lines (in the context of adenoviral infection) in the absence of additional stimuli (Nylandsted et al. 2000; Gurbuxani et al. 2003; Zhao and Shen 2005). The cytotoxic effect of Hsp70 downmodulation is particularly strong in transformed cells, yet undetectable in normal, nontransformed cell lines or primary cells (Schmitt et al. 2003). Studies in Bcr-Abl human leukemia cells show that Hsp70 is a promising therapeutic target for reversing drug resistance, probably through its ability to inhibit apoptosis upstream and downstream of the mitochondria (Guo et al. 2004; Ray et al. 2004). Unfortunately, thus far no small molecules that would selectively inhibit Hsp70 are available. Since Hsp70 blocks apoptosis at the postmitochondrial level by inactivating the apoptosome as well as AIF, strategies targeting Hsp70 may be especially effective in overcoming tumor cell resistance. We have recently demonstrated that rationally engineered decoy targets of Hsp70 derived from AIF can sensitize cancer cells to apoptosis induction by neutralizing Hsp70 function. These AIF-derived peptides all carry the AIF region from amino acids 150–228, previously defined as required for Hsp70 binding (Gurbuxani et al. 2003). These constructs bind to Hsp70 but lack an apoptotic function. Experiments using different cancer cell lines (leukemia, colon cancer, breast cancer, and cervix cancer) demonstrate that certain of these AIF derivative inhibitors of Hsp70 sharply increase the sensitivity of cancer cells to chemotherapy in vitro. This effect was merely related to their ability to neutralize endogenous Hsp70 since this pro-apoptotic activity was lost in Hsp70-negative cells (Schmitt et al. 2003). In vivo, in a syngeneic rat colon cancer cell model, these inhibitors, called ADD70 (for AIF-derived decoy for Hsp70), decreased the size of the tumors and provoked their total regression after treatment with the anticancer agent cisplatin (Schmitt et al., unpublished results). Therefore a positive strategy aimed at interfering with Hsp70, as opposed to negative strategies based on antisense constructs or RNA interference, is feasible for chemosensitization, at least in vitro and in vivo in experimental models. The future will tell whether a similar strategy may allow for the chemosensitization of Hsp70-expressing human tumors.

Concerning Hsp27, phosphorothioate Hsp27 antisense oligonucleotides have been demonstrated in prostate cancer to enhance apoptosis and delay tumor progression (Rocchi et al. 2004). Paclitaxel, by inhibiting Hsp27 expression, seems to overcome drug resistance to etoposide, colcemid, and vincristine in ovarian and uterine cancer cells in vitro (Tanaka et al. 2004). Antisense strategies have also demonstrated that lymphomas and multiple myelomas can be sensitized to chemotherapeutic drugs such as dexamethasone and the inhibitor of proteasome Velcade (PS-341). In dexamethasone-resistant cell lines, Hsp27 is overexpressed. Its downregulation by siRNA restores the apoptotic response to dexamethasone by triggering caspase activation (Chauhan et al. 2003a). We have demonstrated that Hsp27 participates in protein ubiquitination/proteasomal degradation and that this effect contributes to its protective functions by enhancing the activity of proteins such as NF-κB (Parcellier et al. 2003). Velcade, currently tested in clinical trials involving multiple myelomas, has been shown to induce apoptosis in several cancer cell lines in vitro. Hsp27 confers Velcade resistance and an Hsp27 antisense approach sensitizes cells to Velcade-induced apoptosis (Chauhan et al. 2003a, 2003b). It is therefore tempting to conclude that a combinational therapy using Velcade together with an inhibitor of Hsp27 will increase the chemosensitization effect of both products. These findings demonstrate the advantage of developing novel therapeutic drugs targeting Hsps to improve patient outcome in different cancers.

References

Abe M, Manola JB, Oh WK, Parslow DL, George DJ, Austin CL, Kantoff PW (2004) Plasma levels of heat shock protein 70 in patients with prostate cancer: a potential biomarker for prostate cancer. Clin Prostate Cancer 3:49–53

Aoyama A, Steiger RH, Frohli E, Schafer R, von Deimling A, Wiestler OD, Klemenz R (1993) Expression of alpha B-crystallin in human brain tumors. Int J Cancer 55:760–764

Bando Y, Katayama T, Kasai K, Taniguchi M, Tamatani M, Tohyama M (2003) GRP94 (94 kDa glucose-regulated protein) suppresses ischemic neuronal cell death against ischemia/reperfusion injury. Eur J Neurosci 18:829–840

Basso AD, Solit DB, Chiosis G, Giri B, Tsichlis P, Rosen N (2002) Akt forms an intracellular complex with heat shock protein 90 (Hsp90) and Cdc37 and is destabilized by inhibitors of Hsp90 function. J Biol Chem 277:39858–39866

Beckmann RP, Mizzen LE, Welch WJ (1990) Interaction of Hsp 70 with newly synthesized proteins: implications for protein folding and assembly. Science 248:850–854

Beere HM, Wolf BB, Cain K, Mosser DD, Mahboubi A, Kuwana T, Tailor P, Morimoto RI, Cohen GM, Green DR (2000) Heat-shock protein 70 inhibits apoptosis by preventing recruitment of procaspase-9 to the Apaf-1 apoptosome. Nat Cell Biol 2:469–475

Biggs WH 3rd, Meisenhelder J, Hunter T, Cavenee WK, Arden KC (1999) Protein kinase B/Akt-mediated phosphorylation promotes nuclear exclusion of the winged helix transcription factor FKHR1. Proc Natl Acad Sci U S A 96:7421–7426

Brondani Da Rocha A, Regner A, Grivicich I, Pretto Schunemann D, Diel C, Kovaleski G, Brunetto De Farias C, Mondadori E, Almeida L, Braga Filho A, Schwartsmann G (2004) Radioresistance is associated to increased Hsp70 content in human glioblastoma cell lines. Int J Oncol 25:777–785

Bruey JM, Ducasse C, Bonniaud P, Ravagnan L, Susin SA, Diaz-Latoud C, Gurbuxani S, Arrigo AP, Kroemer G, Solary E, Garrido C (2000a) Hsp27 negatively regulates cell death by interacting with cytochrome c. Nat Cell Biol 2:645–652

Bruey JM, Paul C, Fromentin A, Hilpert S, Arrigo AP, Solary E, Garrido C (2000b) Differential regulation of Hsp27 oligomerization in tumor cells grown in vitro and in vivo. Oncogene 19:4855–4863

Bukau B, Horwich AL (1998) The Hsp70 and Hsp60 chaperone machines. Cell 92:351–366

Buzzard KA, Giaccia AJ, Killender M, Anderson RL (1998) Heat shock protein 72 modulates pathways of stress-induced apoptosis. J Biol Chem 273:17147–17153

Cardone MH, Roy N, Stennicke HR, Salvesen GS, Franke TF, Stanbridge E, Frisch S, Reed JC (1998) Regulation of cell death protease caspase-9 by phosphorylation. Science 282:1318–1321

Charette SJ, Lavoie JN, Lambert H, Landry J (2000) Inhibition of Daxx-mediated apoptosis by heat shock protein 27. Mol Cell Biol 20:7602–7612

Chauhan D, Li G, Hideshima T, Podar K, Mitsiades C, Mitsiades N, Catley L, Tai YT, Hayashi T, Shringarpure R, Burger R, Munshi N, Ohtake Y, Saxena S, Anderson KC (2003a) Hsp27 inhibits release of mitochondrial protein Smac in multiple myeloma cells and confers dexamethasone resistance. Blood 102:3379–3386

Chauhan D, Li G, Shringarpure R, Podar K, Ohtake Y, Hideshima T, Anderson KC (2003b) Blockade of Hsp27 overcomes Bortezomib/proteasome inhibitor PS-341 resistance in lymphoma cells. Cancer Res 63:6174–6177

Chen G, Cao P, Goeddel DV (2002) TNF-induced recruitment and activation of the IKK complex require Cdc37 and Hsp90. Mol Cell 9:401–410

Ciocca DR, Oesterreich S, Chamness GC, McGuire WL, Fuqua SA (1993) Biological and clinical implications of heat shock protein 27,000 (Hsp27): a review. J Natl Cancer Inst 85:1558–1570

Concannon CG, Orrenius S, Samali A (2001) Hsp27 inhibits cytochrome c-mediated caspase activation by sequestering both pro-caspase-3 and cytochrome c. Gene Expr 9:195–201

Creagh EM, Carmody RJ, Cotter TG (2000) Heat shock protein 70 inhibits caspase-dependent and -independent apoptosis in Jurkat T cells. Exp Cell Res 257:58–66

Datta SR, Dudek H, Tao X, Masters S, Fu H, Gotoh Y, Greenberg ME (1997) Akt phosphorylation of BAD couples survival signals to the cell-intrinsic death machinery. Cell 91:231–241

Daugas E, Susin SA, Zamzami N, Ferri KF, Irinopoulou T, Larochette N, Prevost MC, Leber B, Andrews D, Penninger J, Kroemer G (2000) Mitochondrio-nuclear translocation of AIF in apoptosis and necrosis. FASEB J 14:729–739

De AK, Roach SE (2004) Detection of the soluble heat shock protein 27 (hsp27) in human serum by an ELISA. J Immunoassay Immunochem 25:159–170

Dix DJ, Allen JW, Collins BW, Mori C, Nakamura N, Poorman-Allen P, Goulding EH, Eddy EM (1996) Targeted gene disruption of Hsp70-2 results in failed meiosis, germ cell apoptosis, and male infertility. Proc Natl Acad Sci U S A 93:3264–3268

Du C, Fang M, Li Y, Li L, Wang X (2000) Smac, a mitochondrial protein that promotes cytochrome c-dependent caspase activation by eliminating IAP inhibition. Cell 102:33–42

Ehrnsperger M, Graber S, Gaestel M, Buchner J (1997) Binding of non-native protein to Hsp25 during heat shock creates a reservoir of folding intermediates for reactivation. EMBO J 16:221–229

Faried A, Sohda M, Nakajima M, Miyazaki T, Kato H, Kuwano H (2004) Expression of heat-shock protein Hsp60 correlated with the apoptotic index and patient prognosis in human oesophageal squamous cell carcinoma. Eur J Cancer 40:2804–2811

Ferri KF, Jacotot E, Blanco J, Este JA, Zamzami N, Susin SA, Xie Z, Brothers G, Reed JC, Penninger JM, Kroemer G (2000) Apoptosis control in syncytia induced by the HIV type 1-envelope glycoprotein complex: role of mitochondria and caspases. J Exp Med 192:1081–1092

Ferri KF, Kroemer G (2001) Organelle-specific initiation of cell death pathways. Nat Cell Biol 3: E255–E263

Gabai VL, Yaglom JA, Volloch V, Meriin AB, Force T, Koutroumanis M, Massie B, Mosser DD, Sherman MY (2000) Hsp72-mediated suppression of c-Jun N-terminal kinase is implicated in development of tolerance to caspase-independent cell death. Mol Cell Biol 20:6826–6836

Galea-Lauri J, Richardson AJ, Latchman DS, Katz DR (1996) Increased heat shock protein 90 (hsp90) expression leads to increased apoptosis in the monoblastoid cell line U937 following induction with TNF-alpha and cycloheximide: a possible role in immunopathology. J Immunol 157:4109–4118

Gao T, Newton AC (2002) The turn motif is a phosphorylation switch that regulates the binding of Hsp70 to protein kinase C. J Biol Chem 277:31585–31592

Garrido C (2002) Size matters: of the small Hsp27 and its large oligomers. Cell Death Differ 9:483–485

Garrido C, Bruey JM, Fromentin A, Hammann A, Arrigo AP, Solary E (1999) Hsp27 inhibits cytochrome c-dependent activation of procaspase-9. FASEB J 13:2061–2070

Garrido C, Fromentin A, Bonnotte B, Favre N, Moutet M, Arrigo AP, Mehlen P, Solary E (1998) Heat shock protein 27 enhances the tumorigenicity of immunogenic rat colon carcinoma cell clones. Cancer Res 58:5495–5499

Garrido C, Mehlen P, Fromentin A, Hammann A, Assem M, Arrigo AP, Chauffert B (1996) Inconstant association between 27-kDa heat-shock protein (Hsp27) content and doxorubicin resistance in human colon cancer cells. The doxorubicin-protecting effect of Hsp27. Eur J Biochem 237:653–659

Garrido C, Ottavi P, Fromentin A, Hammann A, Arrigo AP, Chauffert B, Mehlen P (1997) Hsp27 as a mediator of confluence-dependent resistance to cell death induced by anticancer drugs. Cancer Res 57:2661–2667

Geisler JP, Tammela JE, Manahan KJ, Geisler HE, Miller GA, Zhou Z, Wiemann MC (2004) Hsp27 in patients with ovarian carcinoma: still an independent prognostic indicator at 60 months follow-up. Eur J Gynaecol Oncol 25:165–168

Ghobrial IM, McCormick DJ, Kaufmann SH, Leontovich AA, Loegering DA, Dai NT, Krajnik KL, Stenson MJ, Melhem MF, Novak AJ, Ansell SM, Witzig TE (2005) Proteomic analysis of mantle cell lymphoma by protein microarray. Blood 105:3722–3730

Gotoh K, Nonoguchi K, Higashitsuji H, Kaneko Y, Sakurai T, Sumitomo Y, Itoh K, Subjeck JR, Fujita J (2004) Apg-2 has a chaperone-like activity similar to Hsp110 and is overexpressed in hepatocellular carcinomas. FEBS Lett 560:19–24

Gotz R, Kramer BW, Camarero G, Rapp UR (2004) BAG-1 haplo-insufficiency impairs lung tumorigenesis. BMC Cancer 4:85

Gress TM, Muller-Pillasch F, Weber C, Lerch MM, Friess H, Buchler M, Beger HG, Adler G (1994) Differential expression of heat shock proteins in pancreatic carcinoma. Cancer Res 54:547–551

Guay J, Lambert H, Gingras-Breton G, Lavoie JN, Huot J, Landry J (1997) Regulation of actin filament dynamics by p38 map kinase-mediated phosphorylation of heat shock protein 27. J Cell Sci 110:357–368

Guo F, Sigua C, Bali P, George P, Fiskus W, Scuto A, Annavarapu S, Mouttaki A, Sondarva G, Wei S, Wu J, Djeu J, Bhalla K (2004) Mechanistic role of heat shock protein 70 in Bcr-Abl mediated resistance to apoptosis in human acute leukemia cells. Blood 105:1246–1255

Guo Z, Cooper LF (2000) An N-terminal 33-amino-acid-deletion variant of hsp25 retains oligomerization and functional properties. Biochem Biophys Res Commun 270:183–189

Gupta S, Knowlton AA (2002) Cytosolic heat shock protein 60, hypoxia, and apoptosis. Circulation 106:2727–2733

Gurbuxani S, Schmitt E, Cande C, Parcellier A, Hammann A, Daugas E, Kouranti I, Spahr C, Pance A, Kroemer G, Garrido C (2003) Heat shock protein 70 binding inhibits the nuclear import of apoptosis-inducing factor. Oncogene 22:6669–6678

Hartson SD, Matts RL (1994) Association of Hsp90 with cellular Src-family kinases in a cell-free system correlates with altered kinase structure and function. Biochemistry 33:8912–8920

Hisatomi T, Sakamoto T, Murata T, Yamanaka I, Oshima Y, Hata Y, Ishibashi T, Inomata H, Susin SA, Kroemer G (2001) Relocalization of apoptosis-inducing factor in photoreceptor apoptosis induced by retinal detachment in vivo. Am J Pathol 158:1271–1278

Hoang AT, Huang J, Rudra-Ganguly N, Zheng J, Powell WC, Rabindran SK, Wu C, Roy-Burman P (2000) A novel association between the human heat shock transcription factor 1 (HSF1) and prostate adenocarcinoma. Am J Pathol 156:857–864

Hsu PL, Hsu SM (1998) Abundance of heat shock proteins (hsp89, hsp60, and hsp27) in malignant cells of Hodgkin's disease. Cancer Res 58:5507–5513

Hu Y, Benedict MA, Ding L, Nunez G (1999) Role of cytochrome c and dATP/ATP hydrolysis in Apaf-1-mediated caspase-9 activation and apoptosis. EMBO J 18:3586–3595

Hur E, Kim HH, Choi SM, Kim JH, Yim S, Kwon HJ, Choi Y, Kim DK, Lee MO, Park H (2002) Reduction of hypoxia-induced transcription through the repression of hypoxia-inducible factor-1alpha/aryl hydrocarbon receptor nuclear translocator DNA binding by the 90-kDa heat-shock protein inhibitor radicicol. Mol Pharmacol 62:975–982

Hurlimann J, Gebhard S, Gomez F (1993) Oestrogen receptor, progesterone receptor, pS2, ERD5, Hsp27 and cathepsin D in invasive ductal breast carcinomas. Histopathology 23:239–248

Imai Y, Soda M, Hatakeyama S, Akagi T, Hashikawa T, Nakayama KI, Takahashi R (2002) CHIP is associated with Parkin, a gene responsible for familial Parkinson's disease, and enhances its ubiquitin ligase activity. Mol Cell 10:55–67

Ishihara K, Yasuda K, Hatayama T (2000) Phosphorylation of the 105-kDa heat shock proteins, Hsp105alpha and Hsp105beta, by casein kinase II. Biochem Biophys Res Commun 270:927–931

Jaattela M (1995) Over-expression of hsp70 confers tumorigenicity to mouse fibrosarcoma cells. Int J Cancer 60:689–693

Jaattela M (1999) Heat shock proteins as cellular lifeguards. Ann Med 31:261–271

Jaattela M (2002) Programmed cell death: many ways for cells to die decently. Ann Med 34:480–488

Jaattela M, Wissing D, Bauer PA, Li GC (1992) Major heat shock protein hsp70 protects tumor cells from tumor necrosis factor cytotoxicity. EMBO J 11:3507–3512

Jaattela M, Wissing D, Kokholm K, Kallunki T, Egeblad M (1998) Hsp70 exerts its anti-apoptotic function downstream of caspase-3-like proteases. EMBO J 17:6124–6134

Jacobson MD, Weil M, Raff MC (1997) Programmed cell death in animal development. Cell 88:347–354

Jolly C, Morimoto RI (2000) Role of the heat shock response and molecular chaperones in oncogenesis and cell death. J Natl Cancer Inst 92:1564–1572

Joza N, Susin SA, Daugas E, Stanford WL, Cho SK, Li CY, Sasaki T, Elia AJ, Cheng HY, Ravagnan L, Ferri KF, Zamzami N, Wakeham A, Hakem R, Yoshida H, Kong YY, Mak TW, Zuniga-Pflucker JC, Kroemer G, Penninger JM (2001) Essential role of the mitochondrial apoptosis-inducing factor in programmed cell death. Nature 410:549–554

Kamal A, Thao L, Sensintaffar J, Zhang L, Boehm MF, Fritz LC, Burrows FJ (2003) A high-affinity conformation of Hsp90 confers tumour selectivity on Hsp90 inhibitors. Nature 425:407–410

Kamradt MC, Chen F, Cryns VL (2001) The small heat shock protein alpha B-crystallin negatively regulates cytochrome c- and caspase-8-dependent activation of caspase-3 by inhibiting its autoproteolytic maturation. J Biol Chem 276:16059–16063

Kirchhoff SR, Gupta S, Knowlton AA (2002) Cytosolic heat shock protein 60, apoptosis, and myocardial injury. Circulation 105:2899–2904

Komarova EY, Afanasyeva EA, Bulatova MM, Cheetham ME, Margulis BA, Guzhova IV (2004) Downstream caspases are novel targets for the antiapoptotic activity of the molecular chaperone hsp70. Cell Stress Chaperones 9:265–275

Konishi H, Matsuzaki H, Tanaka M, Takemura Y, Kuroda S, Ono Y, Kikkawa U (1997) Activation of protein kinase B (Akt/RAC-protein kinase) by cellular stress and its association with heat shock protein Hsp27. FEBS Lett 410:493–498

Kroemer G, Reed JC (2000) Mitochondrial control of cell death. Nat Med 6:513–519

Lavoie JN, Gingras-Breton G, Tanguay RM, Landry J (1993) Induction of Chinese hamster Hsp27 gene expression in mouse cells confers resistance to heat shock. Hsp27 stabilization of the microfilament organization. J Biol Chem 268:3420–3429

Lee JS, Lee JJ, Seo JS (2004) Hsp70 deficiency results in activation of c-jun N-terminal kinase, extracellular signal-regulated kinase, and caspase-3 in hyperosmolarity-induced apoptosis. J Biol Chem 280:6634–6641

Lemieux P, Oesterreich S, Lawrence JA, Steeg PS, Hilsenbeck SG, Harvey JM, Fuqua SA (1997) The small heat shock protein hsp27 increases invasiveness but decreases motility of breast cancer cells. Invasion Metastasis 17:113–123

Lewis J, Devin A, Miller A, Lin Y, Rodriguez Y, Neckers L, Liu ZG (2000) Disruption of hsp90 function results in degradation of the death domain kinase, receptor-interacting protein (RIP), and blockage of tumor necrosis factor-induced nuclear factor-kappaB activation. J Biol Chem 275:10519–10526

Li CY, Lee JS, Ko YG, Kim JI, Seo JS (2000) Heat shock protein 70 inhibits apoptosis downstream of cytochrome c release and upstream of caspase-3 activation. J Biol Chem 275:25665–25671

Li GC, Li L, Liu RY, Rehman M, Lee WM (1992) Heat shock protein hsp70 protects cells from thermal stress even after deletion of its ATP-binding domain. Proc Natl Acad Sci U S A 89:2036–2040

Li P, Nijhawan D, Budihardjo I, Srinivasula SM, Ahmad M, Alnemri ES, Wang X (1997) Cytochrome c and dATP-dependent formation of Apaf-1/caspase-9 complex initiates an apoptotic protease cascade. Cell 91:479–489

Lin KM, Lin B, Lian IY, Mestril R, Scheffler IE, Dillmann WH (2001) Combined and individual mitochondrial Hsp60 and Hsp10 expression in cardiac myocytes protects mitochondrial function and prevents apoptotic cell deaths induced by simulated ischemia-reoxygenation. Circulation 103:1787–1792

Lin LY, Lin CY, Ho FM, Liau CS (2005) Up-regulation of the association between heat shock protein 90 and endothelial nitric oxide synthase prevents high glucose-induced apoptosis in human endothelial cells. J Cell Biochem 94:194–201

Liu QL, Kishi H, Ohtsuka K, Muraguchi A (2003) Heat shock protein 70 binds caspase-activated DNase and enhances its activity in TCR-stimulated T cells. Blood 102:1788–1796

Lo Muzio L, Leonardi R, Mariggio MA, Mignogna MD, Rubini C, Vinella A, Pannone G, Giannetti L, Serpico R, Testa NF, De Rosa G, Staibano S (2004) Hsp 27 as possible prognostic factor in patients with oral squamous cell carcinoma. Histol Histopathol 19:119–128

Loeffler M, Daugas E, Susin SA, Zamzami N, Metivier D, Nieminen AL, Brothers G, Penninger JM, Kroemer G (2001) Dominant cell death induction by extramitochondrially targeted apoptosis-inducing factor. FASEB J 15:758–767

Marzo I, Brenner C, Zamzami N, Jurgensmeier JM, Susin SA, Vieira HL, Prevost MC, Xie Z, Matsuyama S, Reed JC, Kroemer G (1998) Bax and adenine nucleotide translocator cooperate in the mitochondrial control of apoptosis. Science 281:2027–2031

Mehlen P, Kretz-Remy C, Preville X, Arrigo AP (1996a) Human hsp27, Drosophila hsp27 and human alphaB-crystallin expression-mediated increase in glutathione is essential for the protective activity of these proteins against TNFalpha-induced cell death. EMBO J 15:2695–2706

Mehlen P, Schulze-Osthoff K, Arrigo AP (1996b) Small stress proteins as novel regulators of apoptosis. Heat shock protein 27 blocks Fas/APO-1- and staurosporine-induced cell death. J Biol Chem 271:16510–16514

Meriin AB, Yaglom JA, Gabai VL, Zon L, Ganiatsas S, Mosser DD, Sherman MY (1999) Protein-damaging stresses activate c-Jun N-terminal kinase via inhibition of its dephosphorylation: a novel pathway controlled by Hsp72. Mol Cell Biol 19:2547–2555

Mosser DD, Caron AW, Bourget L, Denis-Larose C, Massie B (1997) Role of the human heat shock protein hsp70 in protection against stress-induced apoptosis. Mol Cell Biol 17:5317–5327

Mosser DD, Caron AW, Bourget L, Meriin AB, Sherman MY, Morimoto RI, Massie B (2000) The chaperone function of hsp70 is required for protection against stress-induced apoptosis. Mol Cell Biol 20:7146–7159

Munster PN, Marchion DC, Basso AD, Rosen N (2002) Degradation of HER2 by ansamycins induces growth arrest and apoptosis in cells with HER2 overexpression via a HER3, phosphatidylinositol 3′-kinase-AKT-dependent pathway. Cancer Res 62:3132–3137

Nagata S (1997) Apoptosis by death factor. Cell 88:355–365

Nakopoulou L, Lazaris AC, Baltas D, Giannopoulou I, Kavantzas N, Tzonou A (1995) Prognostic evaluation of oestrogen-regulated protein immunoreactivity in ductal invasive (NOS) breast cancer. Virchows Arch 427:33–40

Nanbu K, Konishi I, Mandai M, Kuroda H, Hamid AA, Komatsu T, Mori T (1998) Prognostic significance of heat shock proteins Hsp70 and Hsp90 in endometrial carcinomas. Cancer Detect Prev 22:549–555

Nathan DF, Lindquist S (1995) Mutational analysis of Hsp90 function: interactions with a steroid receptor and a protein kinase. Mol Cell Biol 15:3917–3925

Neckers L (2002) Hsp90 inhibitors as novel cancer chemotherapeutic agents. Trends Mol Med 8: S55–S61

Neckers L, Ivy SP (2003) Heat shock protein 90. Curr Opin Oncol 15:419–424

Nollen EA, Brunsting JF, Roelofsen H, Weber LA, Kampinga HH (1999) In vivo chaperone activity of heat shock protein 70 and thermotolerance. Mol Cell Biol 19:2069–2079

Nylandsted J, Gyrd-Hansen M, Danielewicz A, Fehrenbacher N, Lademann U, Hoyer-Hansen M, Weber E, Multhoff G, Rohde M, Jaattela M (2004) Heat shock protein 70 promotes cell survival by inhibiting lysosomal membrane permeabilization. J Exp Med 200:425–435

Nylandsted J, Rohde M, Brand K, Bastholm L, Elling F, Jaattela M (2000) Selective depletion of heat shock protein 70 (Hsp70) activates a tumor-specific death program that is independent of caspases and bypasses Bcl-2. Proc Natl Acad Sci U S A 97:7871–7876

Oesterreich S, Lee AV, Sullivan TM, Samuel SK, Davie JR, Fuqua SA (1997) Novel nuclear matrix protein HET binds to and influences activity of the Hsp27 promoter in human breast cancer cells. J Cell Biochem 67:275–286

Oh HJ, Chen X, Subjeck JR (1997) Hsp110 protects heat-denatured proteins and confers cellular thermoresistance. J Biol Chem 272:31636–31640

Ozes ON, Mayo LD, Gustin JA, Pfeffer SR, Pfeffer LM, Donner DB (1999) NF-kappaB activation by tumour necrosis factor requires the Akt serine-threonine kinase. Nature 401:82–85

Pandey P, Farber R, Nakazawa A, Kumar S, Bharti A, Nalin C, Weichselbaum R, Kufe D, Kharbanda S (2000a) Hsp27 functions as a negative regulator of cytochrome c-dependent activation of procaspase-3. Oncogene 19:1975–1981

Pandey P, Saleh A, Nakazawa A, Kumar S, Srinivasula SM, Kumar V, Weichselbaum R, Nalin C, Alnemri ES, Kufe D, Kharbanda S (2000b) Negative regulation of cytochrome c-mediated oligomerization of Apaf-1 and activation of procaspase-9 by heat shock protein 90. EMBO J 19:4310–4322

Pang Q, Christianson TA, Keeble W, Koretsky T, Bagby GC (2002) The anti-apoptotic function of Hsp70 in the interferon-inducible double-stranded RNA-dependent protein kinase-mediated death signaling pathway requires the Fanconi anemia protein, FANCC. J Biol Chem 277:49638–49643

Pang Q, Keeble W, Christianson TA, Faulkner GR, Bagby GC (2001) FANCC interacts with Hsp70 to protect hematopoietic cells from IFN-gamma/TNF-alpha-mediated cytotoxicity. EMBO J 20:4478–4489

Parcellier A, Schmitt E, Gurbuxani S, Seigneurin-Berny D, Pance A, Chantome A, Plenchette S, Khochbin S, Solary E, Garrido C (2003) Hsp27 is a ubiquitin-binding protein involved in I-kappaBalpha proteasomal degradation. Mol Cell Biol 23:5790–5802

Park HS, Lee JS, Huh SH, Seo JS, Choi EJ (2001) Hsp72 functions as a natural inhibitory protein of c-Jun N-terminal kinase. EMBO J 20:446–456

Paul C, Manero F, Gonin S, Kretz-Remy C, Virot S, Arrigo AP (2002) Hsp27 as a negative regulator of cytochrome C release. Mol Cell Biol 22:816–834

Pinder SE, Balsitis M, Ellis IO, Landon M, Mayer RJ, Lowe J (1994) The expression of alpha B-crystallin in epithelial tumours: a useful tumour marker? J Pathol 174:209–215

Rane MJ, Pan Y, Singh S, Powell DW, Wu R, Cummins T, Chen Q, McLeish KR, Klein JB (2003) Heat shock protein 27 controls apoptosis by regulating Akt activation. J Biol Chem 278:27828–27835

Ravagnan L, Gurbuxani S, Susin SA, Maisse C, Daugas E, Zamzami N, Mak T, Jaattela M, Penninger JM, Garrido C, Kroemer G (2001) Heat-shock protein 70 antagonizes apoptosis-inducing factor. Nat Cell Biol 3:839–843

Ray S, Lu Y, Kaufmann SH, Gustafson WC, Karp JE, Boldogh I, Fields AP, Brasier AR (2004) Genomic mechanisms of p210BCR-ABL signaling: induction of heat shock protein 70 through the GATA response element confers resistance to paclitaxel-induced apoptosis. J Biol Chem 279:35604–35615

Ritossa F (1962) A new puffing pattern induced by heat shock and DNP in *Drosophila*. Experimentia 18:571–573

Rocchi P, So A, Kojima S, Signaevsky M, Beraldi E, Fazli L, Hurtado-Coll A, Yamanaka K, Gleave M (2004) Heat shock protein 27 increases after androgen ablation and plays a cytoprotective role in hormone-refractory prostate cancer. Cancer Res 64:6595–6602

Rogalla T, Ehrnsperger M, Preville X, Kotlyarov A, Lutsch G, Ducasse C, Paul C, Wieske M, Arrigo AP, Buchner J, Gaestel M (1999) Regulation of Hsp27 oligomerization, chaperone function, and protective activity against oxidative stress/tumor necrosis factor alpha by phosphorylation. J Biol Chem 274:18947–18956

Sakahira H, Nagata S (2002) Co-translational folding of caspase-activated DNase with Hsp70, Hsp40, and inhibitor of caspase-activated DNase. J Biol Chem 277:3364–3370

Saleh A, Srinivasula SM, Balkir L, Robbins PD, Alnemri ES (2000) Negative regulation of the Apaf-1 apoptosome by Hsp70. Nat Cell Biol 2:476–483

Samali A, Cai J, Zhivotovsky B, Jones DP, Orrenius S (1999) Presence of a pre-apoptotic complex of pro-caspase-3, Hsp60 and Hsp10 in the mitochondrial fraction of jurkat cells. EMBO J 18:2040–2048

Santarosa M, Favaro D, Quaia M, Galligioni E (1997) Expression of heat shock protein 72 in renal cell carcinoma: possible role and prognostic implications in cancer patients. Eur J Cancer 33:873–877

Sarto C, Valsecchi C, Magni F, Tremolada L, Arizzi C, Cordani N, Casellato S, Doro G, Favini P, Perego RA, Raimondo F, Ferrero S, Mocarelli P, Galli-Kienle M (2004) Expression of heat shock protein 27 in human renal cell carcinoma. Proteomics 4:2252–2260

Sato S, Fujita N, Tsuruo T (2000) Modulation of Akt kinase activity by binding to Hsp90. Proc Natl Acad Sci U S A 97:10832–10837

Scaffidi C, Schmitz I, Zha J, Korsmeyer SJ, Krammer PH, Peter ME (1999) Differential modulation of apoptosis sensitivity in CD95 type I and type II cells. J Biol Chem 274:22532–22538

Schmitt E, Parcellier A, Gurbuxani S, Cande C, Hammann A, Morales MC, Hunt CR, Dix DJ, Kroemer RT, Giordanetto F, Jaattela M, Penninger JM, Pance A, Kroemer G, Garrido C (2003) Chemosensitization by a non-apoptogenic heat shock protein 70-binding apoptosis-inducing factor mutant. Cancer Res 63:8233–8240

Shaknovich R, Shue G, Kohtz DS (1992) Conformational activation of a basic helix-loop-helix protein (MyoD1) by the C-terminal region of murine Hsp90 (Hsp84). Mol Cell Biol 12:5059–5068

Shan YX, Liu TJ, Su HF, Samsamshariat A, Mestril R, Wang PH (2003) Hsp10 and Hsp60 modulate Bcl-2 family and mitochondria apoptosis signaling induced by doxorubicin in cardiac muscle cells. J Mol Cell Cardiol 35:1135–1143

Shi Y, Thomas JO (1992) The transport of proteins into the nucleus requires the 70-kilodalton heat shock protein or its cytosolic cognate. Mol Cell Biol 12:2186–2192

Solary E, Droin N, Bettaieb A, Corcos L, Dimanche-Boitrel MT, Garrido C (2000) Positive and negative regulation of apoptotic pathways by cytotoxic agents in hematological malignancies. Leukemia 14:1833–1849

Soltys BJ, Gupta RS (1996) Immunoelectron microscopic localization of the 60-kDa heat shock chaperonin protein (Hsp60) in mammalian cells. Exp Cell Res 222:16–27

Song H, Ethier SP, Dziubinski ML, Lin J (2004) Stat3 modulates heat shock 27 kDa protein expression in breast epithelial cells. Biochem Biophys Res Commun 314:143–150

Song J, Takeda M, Morimoto RI (2001) Bag1-Hsp70 mediates a physiological stress signalling pathway that regulates Raf-1/ERK and cell growth. Nat Cell Biol 3:276–282

Sreedhar AS, Kalmar E, Csermely P, Shen YF (2004) Hsp90 isoforms: functions, expression and clinical importance. FEBS Lett 562:11–15

Steel R, Doherty JP, Buzzard K, Clemons N, Hawkins CJ, Anderson RL (2004) Hsp72 inhibits apoptosis upstream of the mitochondria and not through interactions with Apaf-1. J Biol Chem 279:51490–51499

Susin SA, Daugas E, Ravagnan L, Samejima K, Zamzami N, Loeffler M, Costantini P, Ferri KF, Irinopoulou T, Prevost MC, Brothers G, Mak TW, Penninger J, Earnshaw WC, Kroemer G (2000) Two distinct pathways leading to nuclear apoptosis. J Exp Med 192:571–580

Susin SA, Lorenzo HK, Zamzami N, Marzo I, Snow BE, Brothers GM, Mangion J, Jacotot E, Costantini P, Loeffler M, Larochette N, Goodlett DR, Aebersold R, Siderovski DP, Penninger JM, Kroemer G (1999) Molecular characterization of mitochondrial apoptosis-inducing factor. Nature 397:441–446

Takayama S, Krajewski S, Krajewska M, Kitada S, Zapata JM, Kochel K, Knee D, Scudiero D, Tudor G, Miller GJ, Miyashita T, Yamada M, Reed JC (1998) Expression and location of Hsp70/Hsc-binding anti-apoptotic protein BAG-1 and its variants in normal tissues and tumor cell lines. Cancer Res 58:3116–3131

Tanaka Y, Fujiwara K, Tanaka H, Maehata K, Kohno I (2004) Paclitaxel inhibits expression of heat shock protein 27 in ovarian and uterine cancer cells. Int J Gynecol Cancer 14:616–620

Teng SC, Chen YY, Su YN, Chou PC, Chiang YC, Tseng SF, Wu KJ (2004) Direct activation of Hsp90A transcription by c-Myc contributes to c-Myc-induced transformation. J Biol Chem 279:14649–14655

Thornberry NA, Lazebnik Y (1998) Caspases: enemies within. Science 281:1312–1316

Trieb K, Lechleitner T, Lang S, Windhager R, Kotz R, Dirnhofer S (1998) Heat shock protein 72 expression in osteosarcomas correlates with good response to neoadjuvant chemotherapy. Hum Pathol 29:1050–1055

Vargas-Roig LM, Gago FE, Tello O, Aznar JC, Ciocca DR (1998) Heat shock protein expression and drug resistance in breast cancer patients treated with induction chemotherapy. Int J Cancer 79:468–475

Wartmann M, Davis RJ (1994) The native structure of the activated Raf protein kinase is a membrane-bound multi-subunit complex. J Biol Chem 269:6695–6701

Wong HR, Menendez IY, Ryan MA, Denenberg AG, Wispe JR (1998) Increased expression of heat shock protein-70 protects A549 cells against hyperoxia. Am J Physiol 275:L836–L841

Wu YP, Kita K, Suzuki N (2002) Involvement of human heat shock protein 90 alpha in nicotine-induced apoptosis. Int J Cancer 100:37–42

Wyttenbach A, Sauvageot O, Carmichael J, Diaz-Latoud C, Arrigo AP, Rubinsztein DC (2002) Heat shock protein 27 prevents cellular polyglutamine toxicity and suppresses the increase of reactive oxygen species caused by huntingtin. Hum Mol Genet 11:1137–1151

Xanthoudakis S, Roy S, Rasper D, Hennessey T, Aubin Y, Cassady R, Tawa P, Ruel R, Rosen A, Nicholson DW (1999) Hsp60 accelerates the maturation of pro-caspase-3 by upstream activator proteases during apoptosis. EMBO J 18:2049–2056

Yamagishi N, Nishihori H, Ishihara K, Ohtsuka K, Hatayama T (2000) Modulation of the chaperone activities of Hsc70/Hsp40 by Hsp105alpha and Hsp105beta. Biochem Biophys Res Commun 272:850–855

Yoshida H, Kong YY, Yoshida R, Elia AJ, Hakem A, Hakem R, Penninger JM, Mak TW (1998) Apaf1 is required for mitochondrial pathways of apoptosis and brain development. Cell 94:739–750

Young JC, Agashe VR, Siegers K, Hartl FU (2004) Pathways of chaperone-mediated protein folding in the cytosol. Nat Rev Mol Cell Biol 5:781–791

Yufu Y, Nishimura J, Nawata H (1992) High constitutive expression of heat shock protein 90 alpha in human acute leukemia cells. Leuk Res 16:597–605

Zamzami N, Kroemer G (2001) The mitochondrion in apoptosis: how Pandora's box opens. Nat Rev Mol Cell Biol 2:67–71

Zermati Y, Garrido C, Amsellem S, Fishelson S, Bouscary D, Valensi F, Varet B, Solary E, Hermine O (2001) Caspase activation is required for terminal erythroid differentiation. J Exp Med 193:247–254

Zhang Y, Wang JS, Chen LL, Cheng XK, Heng FY, Wu NH, Shen YF (2004) Repression of hsp90beta gene by p53 in UV irradiation-induced apoptosis of Jurkat cells. J Biol Chem 279:42545–42551

Zhao ZG, Shen WL (2005) Heat shock protein 70 antisense oligonucleotide inhibits cell growth and induces apoptosis in human gastric cancer cell line SGC-7901. World J Gastroenterol 11:73–78

Protein Aggregation as a Cause for Disease

T. Scheibel · J. Buchner (✉)

Department Chemie, Lehrstuhl für Biotechnologie, Technische Universität München, Lichtenbergstr. 4, 85747 Garching, Germany
johannes.buchner@ch.tum.de

1	Introduction	199
2	Determinants of Protein Aggregation	200
3	Quality Control of Protein Folding and the Role of Molecular Chaperones in Disease	204
4	Mechanisms of Protein Aggregate Mitigation and Clearance	207
4.1	Formation of Inclusion Bodies, Aggresomes, Lewis Bodies, and Russell Bodies	207
4.2	Protein Disaggregation by Hsp100 Proteins	208
5	Protein Aggregation as a Cause for Disease	209
5.1	Conformational Diseases	209
5.2	Protein Deposits and Amyloids	210
5.3	Toxic Species	212
6	Conclusions	214
	References	214

Abstract The ability of proteins to fold into a defined and functional conformation is one of the most fundamental processes in biology. Certain conditions, however, initiate misfolding or unfolding of proteins. This leads to the loss of functional protein or it can result in a wide range of diseases. One group of diseases, which includes Alzheimer's, Parkinson's, Huntington's disease, and the transmissible spongiform encephalopathies (prion diseases), involves deposition of aggregated proteins. Normally, such protein aggregates are not found in properly functioning biological systems, because a variety of mechanisms inhibit their formation. Understanding the nature of these protective mechanisms together with the understanding of factors reducing or deactivating the natural protection machinery will be crucial for developing strategies to prevent and treat these disastrous diseases.

Keywords Alzheimer · Amyloid · Huntington · Parkinson · Prion · Protein deposits · Quality control · Toxic species

1
Introduction

Protein aggregation is an unwanted side reaction in vitro that often causes technical problems in pharmaceutical and biotechnological processes. In vivo, protein aggregation can have detrimental effects, since it is critically involved in a variety of potentially lethal diseases (Table 1). Unless specifically noted, we will apply the term "aggregation" to processes involving the formation of insoluble protein precipitates that are pathological in nature. This is in contrast to the insolubility of the native state due to protein concentrations exceeding the solubility limit or the intermolecular association involved in the formation of native oligomers. It should be mentioned that in many cases of pathological aggregation, first soluble aggregates are formed, which become insoluble when they exceed a certain size. Many of the protein aggregation diseases discussed in this article give rise to deposits in the form of amyloid fibrils and plaques.

2
Determinants of Protein Aggregation

Typically, specific intermolecular interactions between hydrophobic surfaces of structural subunits in partially folded or unfolded intermediates are responsible for the formation of aggregates. An important consequence is that aggregation will be favored by factors and conditions that favor the population of these intermediates, and that aggregation will be influenced by the properties of the intermediates (Uversky 2003; Jaenicke 1995; Goldberg et al. 1991). The initial stages of aggregation are quite specific, in the sense that they involve the interaction of surface elements of one molecule with matching

Table 1 Aggregating proteins that cause disease

Protein	Disease
α-Synuclein	Parkinson's disease (PD); multisystem atrophy
α_1-Antitrypsin	α_1-Antitrypsin deficiency; emphysema cirrhosis
β2-Microglobulin	Hemodialysis amyloidosis; prostatic amyloid
ABri peptide	ABri cerebral amyloidosis
Amyloid precursor protein (APP) Aβ-peptide 1–42, 1–43	Alzheimer's disease (AD); Down's syndrome
Androgen receptor (polyQ disorder)	Spinal and bulbar muscular atrophy
Antithrombin	Antithrombin deficiency; thromboembolic disease
Apolipoprotein A1	Familial amyloid polyneuropathy II; familial visceral amyloid
Ataxin (polyQ disorder)	Spinocerebellar ataxia
Athropin-1 (polyQ disorder)	Dentatorubral and pallidoluysian atrophy
Atrial natriuretic factor (ANF)	Atrial amyloidosis
C1-Inhibitor	C_1-inhibitor deficiency angioedema

Table 1 (continue)

Protein	Disease
Calcitonin	Medullary carcinoma of the thyroid
Crystallins	Cataracts
Cystatin C	Hereditary cerebral angiopathy
Cytokeratin	Lichen amyloidosis; seborrheic keratosis; macular amyloidosis
Fibrillin	Marfan disease
Fibrinogen	Hereditary renal amyloidosis
Gelsolin	Finnish hereditary systemic amyloidosis
Glial fibrillary acidic protein (GFAP)	Alexander disease (AXD)
Hemoglobin	Sickle cell anemia
Huntingtin	Huntington's disease (HD)
Insulin	Injection-localized amyloidosis
Immunoglobulin heavy chain	AH amyloidosis
Immunoglobulin light chain	Primary systemic amyloidosis; Nodular amyloidosis
Islet amyloid polypeptide (IAPP)	Type II diabetes
Keratoepithelin	Corneal lattice dystrophy type 1
Lactadherin	AMed amyloidosis
Lactoferrin	Gelatinous drop-like corneal dystrophy (GDLD)
Lithostathine	AD
Lung surfactant-associated protein C (SP-C)	Pulmonary alveolar proteinosis
Lysozyme	Familial visceral amyloidosis; hereditary non-neuropathic systemic amyloidosis
Medin	Aortic medial amyloid
Neuronal intermediate filaments (IF)	Charcot-Marie-Tooth (CMT) disease; amyotrophic lateral sclerosis (ALS)
p53	Cancer
Phenylalanine hydroxylase	Phenylketonuria
Prion protein (PrP)	Creutzfeldt-Jakob disease (CJD); fatal familial insomnia; Gerstmann-Sträussler-Scheinker's disease; kuru; new variant CJD; Sporadic insomnia; transmissible spongiform encephalopathies (TSEs)
Prolactin	APro amyloidosis
Rhodopsin	Retinitis pigmentosa
Serum amyloid A protein	Reactive secondary systemic amyloidosis; chronic inflammatory disease
Tau	AD; frontotemporal dementia; Pick's disease
Transthyretin	Familial amyloid neuropathy I; familial cardiac amyloid; senile systemic amyloidosis

hydrophobic surface areas of a neighboring molecule (Goldberg et al. 1991). The three-dimensional propagation of this process leads to large aggregates. Initially, the aggregates will be soluble, but eventually their size will exceed the solubility limit. Solvent-exposed hydrophobic surfaces will further minimize their solubility. Folding intermediates are more prone to aggregate than the unfolded state, because in the unfolded state the hydrophobic side chains are scattered relatively randomly in many small hydrophobic regions. In the partially folded intermediates, there will be large patches of contiguous surface hydrophobicity, leading to a much stronger propensity for aggregation (Fig. 1). In the productive folding process, these surfaces of folding intermediates will interact in an intramolecular manner to form the native conformation.

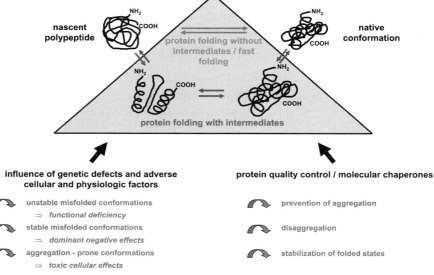

Fig. 1 Relationship between protein quality control and disease-causing protein conformations. Folding intermediates are more prone to aggregation than the nascent polypeptide or the native conformation, because in the nascent chain the hydrophobic side-chains are scattered relatively randomly in many small hydrophobic regions, and in the native conformation they are usually securely packed inside the structure. In partially folded intermediates, large patches of contiguous surface hydrophobicity are present, leading to a much stronger propensity for aggregation. Adverse cellular and physiologic factors influence protein folding pathways and probably lead to accumulation of folding intermediates. In addition, genetic defects such as single point mutations can propagate unstably misfolded conformations followed by functional deficiency, stably misfolded conformations showing dominant negative effects, or aggregation-prone conformations that might cause toxicity. Usually, the protein quality control system involving molecular chaperone machineries and proteolytic processes takes care of such nonfunctional conformations to clear the cell before damage occurs. However, in case the quality control system gets overwhelmed, the accumulation of folding intermediates might cause severe diseases

Aggregation often appears to be irreversible, but this is a reflection of the very slow rates of disaggregation and the fact that the equilibrium lies far in favor of the aggregate rather than its soluble monomeric form (Fink 1998). Under certain conditions aggregates, including amyloid deposits in vivo, can be reversed (Deyoung et al. 1993; Tennent et al. 1995). In practice, however, once insoluble aggregates form, the process is irreversible under native-like conditions in vitro and in vivo.

Circumstances that lead to the population of partially folded intermediates, especially if their concentration is high, are thus likely to lead to aggregation. In most instances of aggregation, there is a kinetic competition between aggregation and folding. While folding is favored at low protein concentrations, high concentrations of folding polypeptide chains will favor aggregation as a higher-order process (Kiefhaber et al. 1991) The aggregation potential of a protein is influenced by genetic defects and mutations that lead to destabilization of the native state relative to the partially folded intermediate. In the context of physiological aggregation, the role of post-translational processing might also be critical. Further, adverse cellular and physiological factors or environmental conditions influence the aggregation behavior of proteins (Fig. 1).

Among the cellular conditions that have been identified to act as determinants for protein aggregation diseases, elevated temperature is of key importance (Uversky et al. 2001; Lomas et al. 1992). In addition to increased temperature, altered pH (Dobson 2001) and decreased ATP levels may inhibit the acquisition of the folded state of these proteins and promote the transition to partially folded intermediates, thus increasing the aggregation propensity and pathogenecity (Gregersen et al. 2003). Further, oxidative stress has been shown to contribute to the pathogenesis of many protein aggregation diseases (Butterfield and Kanski 2001). Oxidative stress develops when oxidative phosphorylation and the cell's antioxidative capacity become overloaded. In these situations, reactive oxygen species (ROS) are generated and damage the cells and its DNA, RNA, lipids, and proteins. Misfolded and partly folded protein structures may be particularly susceptible to oxidative modifications, which may promote unfolding and thus increase the susceptibility to further modifications that elevate the level of stress response (Dukan et al. 2000). Further determining factors of the extent and rate of protein aggregation are ionic strength, protein concentration, co-solutes (e.g., ligands that interact selectively with either the native or the non-native conformation of the protein or the aggregated form), and the presence or absence of various molecular chaperones.

The propensity of a given protein to aggregate correlates in part with the lifetime of partially folded intermediates and with the exposure of hydrophobic amino acids. Longer-lived intermediates are more likely to cause aggregation, since there is a greater chance of interaction with another partially folded intermediate. In general, molecular chaperones prevent protein aggregation by binding partially folded states. In case of an unphysiologically high concentration of folding intermediates, the chaperones may become saturated. Thus, not enough free chaperones would be available and aggregation would be increased.

Even though the exact mechanisms and ordered events may be quite different in the various protein aggregation diseases, the endpoint is chronic stress and eventually cell death. Therefore the cellular ability to cope with misfolded and damaged proteins and further inherited defects in components of the stress-response systems and cell aging are highly relevant for the development of protein aggregation diseases.

3
Quality Control of Protein Folding and the Role of Molecular Chaperones in Disease

Protein aggregation in the cell is intimately tied to protein folding and stability. These intrinsic properties of proteins are modified by molecular chaperones. Accumulation of abnormally folded proteins as a result of a variety of stress situations, including hyperthermia, viral infection, ischemia, anoxia, oxidative stress, and exposure to heavy metals, triggers the heat shock response, which results in the expression of heat shock proteins (Hsps) in many cellular systems. Constitutively expressed Hsps function as molecular chaperones and participate in protein synthesis, protein folding, protein transport, and protein translocation processes, and, upon stress, prevent irreversible aggregation of proteins.

Molecular chaperones represent groups of phylogenetically conserved unrelated protein families that transiently bind to proteins when they are in unfolded or partially folded conformations. Chaperones have been defined as proteins that assist the correct folding of other polypeptides in vivo, but are not components of the final structures when they are performing their biological functions. Further, molecular chaperones do not provide specific steric information for the folding of a target protein but rather inhibit unproductive interactions that lead to protein misfolding and aggregation, allowing proteins to fold more efficiently into their native conformation.

The role of chaperones in protein folding has been summarized in a number of articles (Walter and Buchner 2002; Fink 1998; Beissinger and Buchner 1998; Mathew and Morimoto 1998; Bukau and Horwich 1998; Hartl 1996) and will not be extensively reviewed in this chapter. In general, ATP binding and hydrolysis is used by most molecular chaperones to switch between high- and low-affinity states for non-native proteins: one of the best-studied chaperone families are the Hsp70 chaperones. In human cells, Hsp70 paralogs are found in the cytosol, mitochondria, and the endoplasmic reticulum (ER). Hsp70 is able to bind and release exposed hydrophobic peptide segments of folding intermediates in an ATP-dependent manner, which gives the polypeptide a time window to resume folding after it is released. Binding of Hsp70 counteracts aggregation as the sticky hydrophobic stretches are not available to interactions with other folding intermediates while the polypeptide is bound to the chaperone. The heat shock response and stress proteins are therefore involved in defense mechanisms against cellular stress. The importance of molecular chaperones for cell viability is illustrated by the fact that deletions of the underlying genes are often lethal or they cause severe cellular defects.

Upregulation of stress proteins is an important step in the prevention of protein aggregation and misfolding after stress. However, the chaperone system functions closely together with the proteolytic machinery to determine the fate of proteins within cells. Malfunctions in this quality control system can have pathogenic consequences and may lead to cell death. The balance between the cellular capacity to eliminate misfolded and damaged proteins and the tendency of the particular protein to evade the system is a determining factor in the development and severity of protein aggregation diseases. In healthy and young cells, misfolded and damaged proteins are eliminated by the protein quality control systems involving Hsps and proteases. Hsps can prevent protein aggregation, facilitate refolding and support proteolytic degradation by targeting nonreparable proteins, e.g., for the ubiquitin-proteasome pathway (UPP).

UPP involvement in protein aggregation should be viewed not simply as an isolated degradation machinery but rather as a complex cascade linked both to ubiquitin-dependent processes and to molecular chaperone systems. The UPP is the major nonlysosomal degradation route for proteins, including short-lived, misfolded, and damaged polypeptides (Glickman and Ciechanover 2002). Most proteasome-mediated degradation is ubiquitin-dependent such that proteins that are destined for degradation must become conjugated to a polyubiquitin chain to be recognized by the proteasome. Ubiquitin conjugation/ubiquitination is a highly regulated process in which a ubiquitin-activating enzyme (E1) first activates and transfers ubiquitin to a ubiquitin-conjugating enzyme (E2), which then acts in concert with one of many ubiquitin protein ligases (E3) to transfer ubiquitin to a lysine residue on the target substrate. A chain of at least four ubiquitin moieties is required for substrate recognition by the 26S proteasome complex (Weissman 2001). The proteasome complex consists of a 20S proteolytic core and typically a 19S cap. The 19S cap cleaves ubiquitin moieties from the substrate, unfolds the polypeptide, and feeds it through a narrow channel to the proteolytic chamber of the 20S core (Berke and Paulson 2003; Yao and Cohen 2002; Zwickl et al. 2001; Braun et al. 1999). The 19S cap further serves chaperoning function and refolds and prevents degradation of substrates in some cases. In addition, various molecular chaperones are linked to the UPP by binding non-native proteins and mediating their refolding or degradation. Failure of the UPP to satisfactorily clear unwanted proteins results in the accumulation of abnormal proteins, the aggregation of such proteins, and disruption of cellular homeostasis and integrity (Sherman and Goldberg 2001). In particular, mutations either in E3 ubiquitin-ligase enzymes, or in protein substrates of E3, can be sufficient to cause disease (Table 2). However, reasons for UPP failures are diverse, including the production of abnormal proteins that resist and inhibit proteolysis, defects in protein ubiquitination, reduced deubiquitination, and proteasomal dysfunction. Taken together, it is clear that altered function of the UPP is sufficient to cause familial cases of neurodegenerative disorders (Layfield et al. 2003).

Importantly, in addition to changes in the UPP, other, protein-specific proteases can play crucial roles in causing protein aggregation diseases. Presenilin-1 and presenilin-2 are proteases associated with γ-secretase complexes, which process the final steps in amyloid precursor protein (APP) maturation. Mutations in the enzymes lead to the increased formation of the Aβ-peptide, which confers susceptibility to amyloid formation in Alzheimer's disease (AD) patients (Forloni et al. 2002; Weihofen et al. 2002; Aguzzi and Haass 2003).

If the quality control systems are overwhelmed, as may be the case in cells of patients with inherited defects of the defense systems and in aged cells, aberrant proteins may accumulate and cause protein aggregation diseases. In aged cells, the resistance to oxidative stress as well as the capability to induce the activity of the protein quality control systems are decreased, which leads to difficulties for the cells in maintaining native protein conformations and elimination of misfolded and damaged proteins (Soti and Csermely 2000). Although the molecular mechanisms for these disabilities are still poorly defined, they may contribute significantly to the pathogenesis of many of the age-related protein aggregation diseases.

Table 2 The ubiquitin-proteasome pathway (UPP) and disease

Protein	Relationship to UPP	Disease
α-Synuclein	Substrate	PD
Cystic fibrosis transmembrane regulator (CFTR)	Substrate	Cystic fibrosis
Huntingtin	Substrate	HD
p53	Substrate	cancer
Parkin	Ubiquitin ligase (E3)	PD
Presenilin-1	Substrate	AD
Prion protein (PrP)	Substrate	TSEs
Tau	Substrate	Frontotemporal dementia
Ubiquitin	Co-factor	AD

4
Mechanisms of Protein Aggregate Mitigation and Clearance

4.1
Formation of Inclusion Bodies, Aggresomes, Lewis Bodies, and Russell Bodies

Cells avoid accumulating potentially toxic aggregates by mechanisms such as those discussed above, including the suppression of aggregate formation by molecular chaperones and the degradation of misfolded proteins by proteases. Once formed, aggregates tend to be refractory to proteolysis and accumulate in inclusion bodies. The term "inclusion bodies" has been applied to the intracellular foci into which aggregated proteins are sequestered. They are usually present in low numbers, most often only one or two per cell. Inclusion bodies could form as a consequence of the self-assembly of non-native monomers into growing polymers (Fink 1998; Speed et al. 1996). According to this model, aggregation is seeded at a single or limited number of nucleation sites. Since the addition of monomer to existing aggregates is strongly favored by the thermodynamic stability of the aggregate, a starter aggregate will likely dominate and limit the number of inclusion bodies in the cell. Therefore, inclusion bodies reflect giant aggregates in this model. Such models assume that monomers diffuse to the site of inclusion body formation, a mechanism that might serve in the relatively small, restricted environment of a bacterium or a subcellular organelle, but is probably inadequate over the much larger distances encountered in animal cells such as neurons. Alternatively, inclusion bodies are assumed to be aggregates of aggregates, in which individual aggregates coalesce into a single or limited number of foci to form the inclusion (Kopito 2000).

Most studies of inclusion body formation have focused on prokaryotic systems, in which inclusion bodies are an important source of recombinant protein for technical production (Lange and Rudolph 2005; Mayer and Buchner 2004). However, in human and animal cells, cytoplasmic inclusions have been observed in a number of disease states. Evidence suggests that the formation of cytoplasmic inclusion bodies in mammalian cells requires active, retrograde transport of misfolded proteins on microtubules (Garcia-Mata et al. 1999; Johnston et al. 1998). Therefore, mammalian cells seem to have a microtubule-based apparatus for the sequestration of protein aggregates within the cytoplasm. To distinguish microtubule-dependent cytoplasmic inclusion bodies from those formed in cells or organelles that lack a cytoskeleton, they have been named aggresomes (Kopito 2000).

Bacterial inclusion bodies are observed upon overexpression of a protein and are therefore highly enriched for a single protein species. In addition to a major aggregated protein species, aggresomes are further enriched in molecular chaperones, including Hsp70, Hsp40, the chaperonin CCT, and additionally 19S and 26S proteasome subunits (Garcia-Mata et al. 1999; Wigley et al. 1999). Aggresomes are part of a general cellular response to the formation of aggregated proteins. The aggresome pathway likely cleans the cytoplasm of aggregated proteins and delivers them to the microtubule organizing center. The formation of aggresomes in the cytosol may reflect a cellular defense mechanism that acts in addition to molecular chaperones and proteases. It is speculated that induction of an autophagic response may serve to eliminate these aggresomes (Kopito 2000). An aggresome-like structure has been found in patients suffering from Parkinson's disease (PD), and in this specific case the inclusions have been named Lewis bodies (LBs). LBs form in the cytoplasm of neurons and show a dense core surrounded by a clear halo. They are usually present in the substantia nigra and locus ceruleus of PD patients (Pollanen et al. 1993).

Certain secretory proteins such as heavy chains might be intrinsically difficult to dislocate to the cytosol for degradation either because they are engaged in interactions with ER chaperones, or because they form aggregates (Mancini et al. 2000). In case of proteins that are aggregating along the secretory pathway, it has been found that these transport-incompetent proteins are condensed in dilated cisternae of the ER and are called Russell bodies (RBs) (Kopito and Sitia 2000). Structurally, RBs are generally separated from the normal ER network, suggesting a sorting mechanism facilitating the segregation of insoluble aggregates that might otherwise disrupt the secretory pathway. Such intracisternal granules are eventually fused with lysosomes (Tooze et al. 1990). Autophagic vacuoles may be formed when cell division is not sufficient to avoid levels of accumulation incompatible with the maintenance of the proper cellular architecture.

4.2
Protein Disaggregation by Hsp100 Proteins

At least eubacteria, yeast, and plants have found an additional way to deal with protein aggregates. As detailed in recent studies, the Hsp100 family (with Hsp104 in *Saccharomyces cerevisiae* and ClpB in *Escherichia coli*) seem to act as a disaggregation machinery that is able to solubilize and, in combination with Hsp70, reactivate aggregated proteins (Horwich 2004; Lee et al. 2004; Maurizi and Xia 2004; Queitsch et al. 2000; Mogk et al. 1999; Glover and Lindquist 1998). The disassembly of aggregates or complexes, whether for the purpose of rescuing function or degrading proteins, is a fundamentally different challenge in comparison to preventing aggregation.

Hsp100 proteins are poorly represented in animal genomes and therefore may not be immediately considered as useful tools in solving problems imposed by disease-associated protein aggregation. However, there may be other proteins for disassembling aggregates. Together with the prevention of protein aggregation, this function might be of importance in the future for fighting pathogenic aggregates.

5
Protein Aggregation as a Cause for Disease

5.1
Conformational Diseases

Despite all cellular protection mechanisms, protein aggregation plays an increasing role in health with age, especially in the light of the increasing life span in Western civilizations. Carrell and Lomas (1997) proposed a new group of disorders, the conformational diseases, which have an important impact on public health, such as AD, PD, Huntington's disease (HD), transmissible spongiform encephalopathies (TSEs) or prion diseases, cystic fibrosis, sickle cell anemia, and other less common conditions such as those summarized in Table 1 (Cairns et al. 2004; Scheibel 2004; Johansson et al. 2004; Aguzzi and Haass 2003; Stojanovic et al. 2003; Ishimaru et al. 2003; Bates 2003; Dawson and Dawson 2003; Stirling et al. 2003; Sandilands et al. 2002; Dobson 2001; Merlini et al. 2001; Avilla 2000; Clark and Muchowski 2000; Damas and Saraiva 2000; Kelly 1998; Carrell and Lomas 1997; Kielty and Shuttleworth 1994; Nishio et al. 1983; see also the chapter by Winklhofer and Tatzelt, this volume). Conformational diseases are described as conditions in which a constituent protein undergoes a structural change that results in self-association, aggregation, and tissue deposition (Carrell and Lomas 1997). Conformational diseases can be caused by gene sequence alterations. It is interesting to note that approximately half of the mutations in genetically based conformational disorders change a single amino acid in the polypeptide chain (Krawczak et al. 2000).

In the case of protein aggregation disorders, the formation of oligomers and aggregates exerts a toxic gain-of-function effect on the cell, and cell damage or death is decisive for the clinical phenotype (Dobson 2001). Although the endpoint reflecting the accumulation of aggregated protein is similar for the paradigmatic examples of α-1-antitrypsin deficiency, HD, PD, AD, and prion disease, the pathogenesis in these five diseases is quite different.

In α-1-antitrypsin deficiency, a mutation hinders the proper folding of the protein in the ER of liver cells. The misfolded protein tends to form oligomers that are targeted for degradation (Carrell and Lomas 2002). In heterozygous carriers and in homozygous patients with the lung form of the disease, the capacity of degradation components of the protein quality control system is sufficient to cope with the accumulated protein. However, owing to a yet unexplained decrease in the degradation capacity found in 10%–15% of homozygous patients, the protein aggregates cannot be eliminated in the liver cells of such individuals and they develop cirrhosis-like liver damage and hepatocellular carcinoma (Wu et al. 1994).

The type of pathogenesis as found in HD is shared by at least nine other inherited neurological diseases where the pathogen is a string of glutamine residues, which is part of the respective protein (Sakahira et al. 2002; Taylor et al. 2002; Wanker 2000; Perutz 1999). The mutation that causes HD is an unstable expansion of CAG (encoding glutamine) in the Huntingtin gene. The mutation is inherited in an autosomal dominant manner. In patients with HD, the glutamine repeat length may be more than 55, and the longer the repeat the more prone to aggregation is the fragment. This finding is reflected in earlier disease onset for patients with long repeats compared to patients with shorter strings of glutamine (Zoghbi and Botas 2002).

The glutamine repeat-containing proteins share the tendency to self-aggregation with other cellular proteins, among them α-synuclein and Aβ-peptide, which are the pathogens considered in a specific subset of PD and AD (Dobson 2001). Early forms of the diseases are inherited due to mutations in the respective genes, which further promote the self-aggregation of the proteins. Such gain-of-function effects contrast with those found in late-onset forms of PD and AD, where self-aggregating proteins accumulate and participate in the development of degenerative disorders due to an intrinsic conformational instability of the wild-type protein.

Prion diseases or TSEs are rare fatal neurodegenerative diseases of humans and animals (Hetz and Soto 2003; Collinge 2001; Prusiner 1998). Primary symptoms include progressive dementia and ataxia (Ironside and Bell 1997). The hallmark pathologies of TSEs are spongiform degeneration of the brain accompanied by extensive astrogliosis and accumulation of the abnormal, protease-resistant prion protein (PrP) isoform in the central nervous system, which sometimes forms plaques (Budka et al. 1995). TSEs in humans can be divided into three groups: familial, sporadic, and infectious. Human familial TSEs are all associated with different mutations in the PrP gene and include some forms of Creutzfeldt-Jakob disease (CJD), Gerstmann-Straussler-Scheinker's (GSS) syndrome, and fatal familial insomnia (FFI) (Prusiner and Scott 1997). Sporadic CJD has not been associated with any known mutation and occurs worldwide with an incidence of 0.5–1.5 new cases per 1 million people each year (Johnson and Gibbs 1998). Infectious TSE diseases include Kuru, which was propagated by ritualistic cannibalism, and iatrogenic CJD, which is spread by tissue transplantation, contamination of surgical tools, or inoculation with materials derived from CJD-infected tissues (Prusiner 1998). New variant CJD (vCJD) is a novel infectious disease, which is strongly linked to exposure to the bovine spongiform encephalopathy (BSE) agent, the most common TSE disease in cattle (Collinge 2001; 1999; Bruce 2000; Wille et al. 1996).

5.2
Protein Deposits and Amyloids

Several forms of deposits can be found for aggregating proteins. We have already discussed the formation of inclusion bodies, aggresomes, and Russell bodies. Here we would like to describe extracellular protein deposits as found in several diseases. The protein aggregate deposits involved in conformational diseases may have various ultrastructural organizations. In most instances, the misfolded protein is rich in β-sheet conformation. β-sheets are part of the secondary structures in folded proteins. They consist of peptide strands linked by hydrogen bonds between amine and carbonyl groups of adjacent strands. The highly organized oligomerization of misfolded proteins into β-sheet-rich conformations can lead to the production of so-called amyloids, which describe extracellular, fibrillar protein deposits associated with disease in humans (Scheibel and Serpell 2005; Horwich 2002). The deposits accumulate in various tissue types such as brain, vital organs (liver and spleen), or in skeletal tissue, and are defined by their tinctorial and morphological properties (Sunde and Blake 1997). The quantity of such aggregates can be almost undetectable in some cases or reach kilograms of deposited protein in systemic diseases (Dobson 2004). Each amyloid diseases, also called amyloidosis (see Table 1), involves the aggregation of one specific protein, although a range of other components, including other proteins and carbohydrates, is also incorporated into the deposits.

Amyloidosis is slow onset and degenerative and can be found not only in humans but in most mammals. It is important to mention that a single protein in addition to amyloid-like deposits can also form various other aggregate morphologies that can all cause disease. The best-described example are immunoglobulins, where light and heavy chains can aggregate and form deposits with various morphologies mostly in kidneys (Table 3) (Merlini et al. 2001; Preudhomme et al. 1994). The morphologies can be fibrillar, granular, crystalline, or tubular leading to different diseases depending on the morphologies (Table 3). In addition to diseases with extracellular deposits, there are others, notably Parkinson's and Huntington's disease, that appear to involve very similar aggregates, which are intracellular, e.g., in the nucleus or the cytoplasm, and

Table 3 Immunoglobulin-related protein aggregation diseases

Disease	Aggregate morphology	Organ
AL amyloidosis (light chain), immunotactoid glomerulopathy	Fibrillar	Systemic or localized kidney
Light and/or heavy chain deposition disease	Granular	Kidney (systemic)
Acquired Fanconi's syndrome	Crystals	Bone marrow plasma cells
Light chain cast nephropathy	Cast	Kidney
Cryoglobulinemia	Microtubular	Small vessels, kidney

these diseases are therefore not included in the strict definition of amyloidosis (Dobson 2001; Rochet and Lansbury 2000).

Although the proteins involved in amyloidosis are distinct with respect to amino acid sequence and function, the fibrils formed by different amyloidogenic proteins are structured similarly. This observation is remarkable since the soluble native forms of these proteins vary considerably. Some of the soluble starter proteins are large, some small, some are largely α-helical, some largely β-sheet. Some are intact in the fibrous form, others are at least partially degraded. Some are cross-linked with disulfide bonds and some are not. For many years, it was generally assumed that the ability to form amyloid fibrils with the characteristics described above was limited to a relatively small number of proteins, largely those seen in disease states, and that these proteins possess specific sequence motifs encoding the amyloid core structure. Recent studies have suggested, however, that the ability of polypeptide chains to form such structures is common, and indeed can be considered a generic feature of the backbones of polypeptide chains (Dobson 2004). The most direct evidence for the latter statement is that fibrils can be formed by many different proteins that are not associated with disease, including such well-known proteins as myoglobin, and also by homopolymers such as polythreonine or polylysine (Dobson 2004).

Amyloid fibrils are unusually stable, unbranched, range from 6 to 12 nm in diameter, and specifically bind the dye Congo red (CR) showing an apple-green birefringence in polarized light (Teplow 1998; Eanes and Glenner 1968). Amyloid fibrils produce a characteristic cross-β X-ray diffraction pattern, consistent with a model in which stacked β-sheets form parallel to the fibril axis with individual β-strands perpendicular to the fibril axis (Sunde and Blake 1997; Geddes et al. 1968). Remarkably, fibrils of similar appearance to those containing large proteins can be formed by peptides with just a handful of residues. One can consider that amyloid fibrils are highly organized structures (effectively one-dimensional crystals) adopted by an unfolded polypeptide chain. Therefore, polypeptide backbone chains are assumed to be a typical polymer, since similar types of structure can be formed by many types of synthetic polymers. The essential features of such structures are determined by the physicochemical properties of the polymer chain.

5.3
Toxic Species

Investigating the involvement of protein aggregation and aggregate deposits in causing a conformational disease is one of the main focuses of pathology. For Alzheimer's disease, it is still unclear how and where Aβ-peptide exerts its putative toxicity in vivo. However, the extracellular location of plaques in post-mortem brain has driven studies of extracellular toxicity (Small et al. 2001). Originally, the toxicity was thought to be a property of the fibrillar form of

Aβ-peptide, consistent with the widespread notion at the time that the amyloid fibril itself is pathogenic (Lorenzo and Yankner 1994; Yankner et al. 1990). Although neuritic plaques are a hallmark of AD, there is a poor correlation between plaque density in human postmortem material and antemortem cognitive deficits (Dickson et al. 1995; Terry et al. 1991). Strikingly, soluble material such as dimers and trimers of Aβ-peptide was observed in postmortem human material, indicating that oligomeric intermediates rather than protein fibrils or larger aggregates could be the main toxic species in AD (Zhang et al. 2002; McLean et al. 1999; Hartley et al. 1999). It appears that the toxicity of such oligomeric intermediates is related to their structure but not their sequence (Caughey and Lansbury 2003; Lansbury 1999). There is some evidence that the main toxicity of Aβ-peptide oligomers could be an intracellular event, with the intracellular toxicity exceeding the extracellular toxicity by approximately 100,000-fold (Zhang et al. 2002).

Attempts to identify the abnormal forms of PrP responsible for various TSE diseases has led to the notion that abnormal forms of PrP important for transmission and neuropathogenesis might be different. Indeed, it is possible to induce PrP-dependent neurodegeneration without the accumulation of infectivity (Caughey and Lansbury 2003). It is most likely that a highly neurotoxic entity may not be protease-resistant PrP (PrPres) as found in plaques but some other form of PrP. Several neurotoxic PrP species have been described. However, the toxic species in TSE diseases have not been identified unambiguously. Therefore, it is unclear at the moment which of the perturbations of PrP conformation and aggregation is of fundamental neuropathological significance.

For toxicity in HD, a variety of possible mechanisms is discussed. As for the previous examples, globular and protofibrillar intermediates, which form before maturation of Huntingtin fibrils, might be crucial for toxicity (Sanchez et al. 2003). An alternative possibility might involve toxicity associated with the linear addition of monomers to a nascent fibril (Thakur and Wetzel 2002). Toxicity caused by polyglutamine-rich proteins may involve recruitment of other proteins also containing short polyglutamine stretches into the aggregates. Many proteins in the cell have such polyglutamine stretches, including transcription factors and other transcriptional regulators (Ross and Poirier 2004). A completely different mechanism of polyglutamine toxicity involves the interference with the proteasome. Mutant Huntingtin is able to inhibit the proteasome, presumably because it is recognized and binds but cannot be cleaved (Bence et al. 2001). Recent studies have established a link between mutant Huntingtin, excitotoxicity and neurotrophic factors. Neurotrophic factors prevent cell death in degenerative processes but they can also enhance growth and function of neurons that are affected in Huntington's disease. The endogenous regulation of the expression of neurotrophic factors and their receptors in the striatum and its connections can be important to protect striatal cells and maintains basal ganglia connectivity (Alberch et al. 2004).

Several different aggregation intermediates with size and morphology, similar to those for the Aβ-peptide, have been described for α-synuclein. Importantly, misfolded or only partially folded α-synuclein could be more toxic to cells than its aggregates and formation of deposits could have protective properties. The pathway of assembly for intermediate forms of α-synuclein has been reported to be complex with globular and ring-like forms in addition to protofibrils (Volles et al. 2001). Fatty acids seem to promote oligomerization, suggesting that α-synuclein may aggregate via an interaction with cell membranes (Caughey and Lansbury 2003; Sharon et al. 2003).

6
Conclusions

Many important facets of protein folding diseases have been analyzed in recent years. While a number of key aspects still remain to be addressed on the molecular level, chances are high that it will be possible to successfully establish therapeutic concepts for these increasingly important diseases. Recently, an antibody has been generated that interacts with oligomeric, but not with monomeric or fibrillar forms of polyglutamine repeat proteins, Aβ-peptide, α-synuclein, and prion protein (Kayed et al. 2003). This antibody recognized material in postmortem AD brain tissue that was distinct from plaques. Interestingly, the antibody blocked cell toxicity by polyglutamine repeat proteins, Aβ-peptide, and α-synuclein in in vitro experiments (Glabe 2004; Kayed et al. 2003). The actual mechanism for this block in toxicity is uncertain, because polyglutamine and α-synuclein interact intracellularly, whereas Aβ-peptide interacts extracellularly. Nevertheless, it is tempting to speculate that a common structure of soluble nonfibrillar intermediates exists for all of these molecules and that there may be a common mechanism of pathogenesis, which will be a possible target for therapeutic approaches.

Acknowledgements The authors' work is supported by the Deutsche Forschungsgemeinschaft (DFG) and the Fonds der chemischen Industrie.

References

Aguzzi A, Haass C (2003) Games played by rogue proteins in prion disorders and Alzheimer's disease. Science 302:814–818
Alberch J, Perez-Navarro E, Canals JM (2004) Neurotrophic factors in Huntington's disease. Prog Brain Res 146:195–229
Avila J (2000) Tau aggregation into fibrillar polymers: taupathies. FEBS Lett 476:89–92
Bates G (2003) Huntingtin aggregation and toxicity in Huntington's disease. Lancet 361:1642–1644
Beissinger M, Buchner J (1998) How chaperones fold proteins. Biol Chem. 379:245–259

Bence NF, Sampat RM, Kopito RR (2001) Impairment of the ubiquitin-proteasome system by protein aggregation. Science 292:1552–1555

Berke SJ, Paulson HL (2003) Protein aggregation and the ubiquitin proteasome pathway: gaining the UPPer hand on neurodegeneration. Curr Opin Genet Dev 13:253–261

Braun BC, Glickman M, Kraft R et al (1999) The base of the proteasome regulatory particle exhibits chaperone-like activity. Nat Cell Biol 1:221–226

Bruce ME (2000) 'New variant' Creutzfeldt-Jakob disease and bovine spongiform encephalopathy. Nat Med 6:258–259

Budka H, Aguzzi A, Brown P, et al (1995) Neuropathological diagnostic criteria for Creutzfeldt-Jakob disease (CJD) and other human spongiform encephalopathies (prion diseases). Brain Pathol 5:459–466

Bukau B, Horwich AL (1998) The Hsp70 and Hsp60 chaperone machines. Cell 92:351–366

Butterfield DA, Kanski J (2001) Brain protein oxidation in age-related neurodegenerative disorders that are associated with aggregated proteins. Mech Aging Dev 122:945–962

Carrell RW, Lomas DA (1997) Conformational disease. Lancet 350:134–138

Carrell RW, Lomas DA (2002) Alpha1-antitrypsin deficiency: a model for conformational diseases. N Engl J Med 346:45–53

Caughey B, Lansbury PT (2003) Protofibrils, pores, fibrils, and neurodegeneration: separating the responsible protein aggregates from the innocent bystanders. Annu Rev Neurosci 26:267–298

Clark JI, Muchowski PJ (2000) Small heat-shock proteins and their potential role in human disease. Curr Opin Struct Biol 10:52–59

Collinge J (1999) Variant Creutzfeldt-Jakob disease. Lancet 354:317–323

Collinge J (2001) Prion diseases of humans and animals: their causes and molecular basis. Annu Rev Neurosci 24:519–550

Damas AM, Saraiva MJ (2000) Review: TTR amyloidosis-structural features leading to protein aggregation and their implications on therapeutic strategies. J Struct Biol 130:290–299

Dawson TM, Dawson VL (2003) Rare genetic mutations shed light on the pathogenesis of Parkinson disease. J Clin Invest 111:145–151

Deyoung LR, Fink AL, Dill KA (1993) Aggregation of globular proteins. Acc Chem Res 26:614–620

Dickson DW, Crystal HA, Bevona C et al (1995) Correlations of synaptic and pathological markers with cognition of the elderly. Neurobiol Aging 16:285–298

Dobson CM (2001) The structural basis of protein folding and its links with human disease. Philos Trans R Soc Lond B Biol Sci 356:133–145

Dobson CM (2004) Principles of protein folding, misfolding and aggregation. Semin Cell Dev Biol 15:3–16

Dukan S, Farewell A, Ballestros M et al (2000) Protein oxidation in response to increased transcriptional or translational errors. Proc Natl Acad Sci U S A 97:5746–5749

Eanes ED, Glenner GG (1968) X-ray diffraction studies on amyloid filaments. J Histochem Cytochem 16:673–677

Fink AL (1998) Protein aggregation: folding aggregates, inclusion bodies and amyloid. Fold Des 3:R9–R23

Forloni G, Terreni L, Bertani I et al (2002) Protein misfolding in Alzheimer's and Parkinson's disease: genetics and molecular mechanisms. Neurobiol Aging 23:957–976

Garcia-Mata R, Bebok Z, Sorscher EJ, Sztul ES (1999) Characterization and dynamics of aggresome formation by a cytosolic GFP-chimera. J Cell Biol 146:1239–1254

Geddes AJ, Parker KD, Atkins ED, Beighton E (1968) "Cross-beta" conformation in proteins. J Mol Biol 32:343–358

Glabe CG (2004) Conformation-dependent antibodies target diseases of protein misfolding. Trends Biochem Sci 29:542–547

Glickman MH, Ciechanover A (2002) The ubiquitin-proteasome proteolytic pathway: destruction for the sake of construction. Physiol Rev 82:373–428

Glover JR, Lindquist S (1998) Hsp104, Hsp70, and Hsp40: a novel chaperone system that rescues previously aggregated proteins. Cell 94:73–82

Goldberg ME, Rudolph R, Jaenicke R (1991) A kinetic study of the competition between renaturation and aggregation during the refolding of denatured-reduced egg white lysozyme. Biochemistry 30:2790–2797

Gregersen N, Bolund L, Bross P (2003) Protein misfolding, aggregation, and degradation in disease. Methods Mol Biol 232:3–16

Hartl FU (1996) Molecular chaperones in cellular protein folding. Nature 381:571–579

Hartley DM, Walsh DM, Ye CP et al (1999) Protofibrillar intermediates of amyloid beta-protein induce acute electrophysiological changes and progressive neurotoxicity in cortical neurons. J Neurosci 19:8876–8884

Hetz C, Soto C (2003) Protein misfolding and disease: the case of prion disorders. Cell Mol Life Sci 60:133–143

Horwich AL (2004) Chaperoned protein disaggregation-the ClpB ring uses its central channel. Cell 119:579–581

Horwich A (2002) Protein aggregation in disease: a role for folding intermediates forming specific multimeric interactions. J Clin Invest 110:1221–1232

Ironside JW, Bell JE (1997) Pathology of prion diseases In: Collinge J, Palmer MS (eds) Prion diseases. Oxford University Press, Oxford, pp 57–88

Ishimaru D, Andrade LR, Teixeira LS et al (2003) Fibrillar aggregates of the tumour suppressor p53 core domain. Biochemistry 42:9022–9027

Jaenicke R (1995) Folding and association versus misfolding and aggregation of proteins. Philos Trans R Soc Lond B Biol Sci 348:97–105

Johnson RT, Gibbs CJ (1998) Creutzfeldt-Jakob disease and related transmissible spongiform encephalopathies. N Engl J Med 339:1994–2004

Johansson J, Weaver TE, Tjernberg LO (2004) Proteolytic generation and aggregation of peptides from transmembrane regions: lung surfactant protein C and amyloid beta-peptide. Cell Mol Life Sci 61:326–335

Johnston JA, Ward CL, Kopito RR (1998) Aggresomes: a cellular response to misfolded proteins. J Cell Biol 143:1883–1898

Kayed R, Head E, Thompson JL et al (2003) Common structure of soluble amyloid oligomers implies common mechanism of pathogenesis. Science 300:486–489

Kelly J (1998) Alternative conformation of amyloidogenic proteins and their multi-step assembly pathways. Curr Opin Struct Biol 8:101–106

Kiefhaber T, Rudolph R, Kohler H-H, Buchner J (1991) Protein aggregation in vitro and in vivo: a quantitative model of the kinetic competition between folding and aggregation. Nat Biotechnol 9:825–829

Kielty CM, Shuttleworth CA (1994) Abnormal fibrillin assembly by dermal fibroblasts from two patients with Marfan syndrome. J Cell Biol 124:997–1004

Kopito RR (2000) Aggresomes, inclusion bodies and protein aggregation. Trends Cell Biol 10:524–530

Kopito RR, Sitia R (2000) Aggresomes and Russell bodies. Symptoms of cellular indigestion? EMBO Rep 1:225–231

Krawczak M, Chuzhanova NA, Stenson PD et al (2000) Changes in primary DNA sequence complexity influence the phenotypic consequences of mutations in human gene regulatory regions. Hum Genet 107:362–365

Kuemmerle S, Gutekunst CA, Klein AM et al (1999) Huntington aggregates may not predict neuronal death in Huntington's disease. Ann Neurol 46:842–849

Lange C, Rudolph R (2005) Production of recombinant proteins for therapy, diagnostics, and industrial research by in vitro folding. In: Buchner J, Kiefhaber T (eds) Protein folding handbook. Vol. 3. Wiley, Weinheim, pp 1245–1280

Lansbury PT (1999) Evolution of amyloid: what normal protein folding may tell us about fibrillogenesis and disease. Proc Natl Acad Sci U S A 96:3342–3344

Layfield R, Cavey JR, Lowe J (2003) Role of ubiquitin-mediated proteolysis in the pathogenesis of neurodegenerative disorders. Ageing Res Rep 2:343–356

Lee S, Sowa ME, Choi JM, Tsai FT (2004) The ClpB/Hsp104 molecular chaperone-a protein disaggregating machine. J Struct Biol 146:99–105

Lomas DA, Evans DL, Finch JT, Carrell RW (1992) The mechanism of Z alpha 1-antitrypsin accumulation in the liver. Nature 357:605–607

Lorenzo A, Yankner BA (1994) Beta-amyloid neurotoxicity requires fibril formation and is inhibited by Congo red. Proc Natl Acad Sci U S A 91:12243–12247

Mancini R, Fagioli C, Fra AM, Maggioni C, Sitia R (2000) Degradation of unassembled soluble Ig subunits by cytosolic proteasomes: evidence that retro translocation and degradation are coupled events. FASEB J 14:769–778

Mathew A, Morimoto RI (1998) Role of the heat-shock response in the life and death of proteins. Ann N Y Acad Sci 851:99–111

Maurizi MR, Xia D (2004) Protein binding and disruption by Clp/Hsp100 chaperones. Structure 12:175–183

Mayer M, Buchner J (2004) Refolding of inclusion body proteins. Methods Mol Med 94:239–254

McLean CA, Cherny RA, Fraser FW et al (1999) Soluble pool of Abeta amyloid as a determinant of severity of neurodegeneration in Alzheimer's disease. Ann Neurol 46:860–866

Merlini G, Bellotti V, Andreola A et al (2001) Protein aggregation. Clin Chem Lab Med 39:1065–1075

Mogk A, Tomoyasu T, Goloubinoff P et al (1999) Identification of thermolabile Escherichia coli proteins: prevention and reversion of aggregation by DnaK and ClpB. EMBO J 18:6934–6949

Nishio I, Tanaka T, Sun ST et al (1983) Hemoglobin aggregation in single red blood cells of sickle cell anemia. Science 220:1173–1175

Perutz MF (1999) Glutamine repeats and neurodegenerative diseases: molecular aspects. Trends Biochem Sci 24:58–63

Pollanen MS, Dickson DW, Bergeron C (1993) Pathology and biology of the Lewy body. J Neuropathol Exp Neurol 52:183–191

Preudhomme C, Vachee A, Morschauser F et al (1994) Immunoglobulin and T-cell receptor delta gene rearrangements are rarely found in myelodysplastic syndromes in chronic phase. Leuk Res 18:365–371

Prusiner SB (1998) Prions. Proc Natl Acad Sci U S A 95:13363–13383

Prusiner SB, Scott MR (1997) Genetics of prions. Annu Rev Genet 31:139–175

Queitsch C, Hong SW, Vierling E, Lindquist S (2000) Heat shock protein 101 plays a crucial role in thermotolerance in Arabidopsis. Plant Cell 12:479–492

Rochet JC, Lansbury PT (2000) Amyloid fibrillogenesis: themes and variations. Curr Opin Struct Biol 10:60–68

Ross CA, Poirier MA (2004) Protein aggregation and neurodegenerative disease. Nat Med 10 Suppl:S10–S17

Sakahira H, Breuer P, Hayer-Hartl MK, Hartl FU (2002) Molecular chaperones as modulators of polyglutamine protein aggregation and toxicity. Proc Natl Acad Sci U S A 99 [Suppl 4]:16412–16418

Sanchez I, Mahlke C, Yuan J (2003) Pivotal role of oligomerization in expanded polyglutamine neurodegenerative disorders. Nature 421:373–379

Sandilands A, Hutcheson AM, Long HA et al (2002) Altered aggregation properties of mutant gamma-crystallins cause inherited cataract. EMBO J 21:6005–6014

Scheibel T (2004) Amyloid formation of a yeast prion determinant. J Mol Neurosci 23:13–22

Scheibel T, Serpell L (2005) Physical methods for studies of fibre formation and structure. In: Buchner J, Kiefhaber T (eds) Protein folding handbook. Vol. 3. Wiley, Weinheim, pp 197–253

Sharon R, Bar-Joseph I, Frosch MP et al (2003) The formation of highly soluble oligomers of alpha-synuclein is regulated by fatty acids and enhanced in Parkinson's disease. Neuron 37:583–595

Sherman MY, Goldberg AL (2001) Cellular defences against unfolded proteins: a cell biologist thinks about neurodegenerative diseases. Neuron 29:15–32

Small DH, Mok SS, Bornstein JC (2001) Alzheimer's disease and Abeta toxicity: from top to bottom. Nat Rev Neurosci 2:595–598

Soti C, Csermely P (2000) Molecular chaperones and the aging process. Biogerontology 1:225–233

Speed MA, Wang DIC, King J (1996) Specific aggregation of partially folded polypeptide chains—the molecular basis of inclusion body composition. Nat Biotechnol 14:1283–1287

Stirling PC, Lundin VF, Leroux MR (2003) Getting a grip on non-native proteins. EMBO Rep 4:565–570

Stojanovic A, Hwang I, Khorana HG, Hwa J (2003) Retinitis pigmentosa rhodopsin mutations L125R and A164 V perturb critical interhelical interactions: new insights through compensatory mutations and crystal structure analysis. J Biol Chem 278:39020–39028

Sunde M, Blake C (1997) The structure of amyloid fibrils by electron microscopy and X-ray diffraction. Adv Protein Chem 50:123–159

Taylor JP, Hardy J, Fischbeck KH (2002) Toxic proteins in neurodegenerative disease. Science 296:1991–1995

Tennent GA, Lovat LB, Pepys MB (1995). Serum amyloid P component prevents proteolysis of the amyloid fibrils of Alzheimer disease and systemic amyloidosis. Proc Natl Acad Sci U S A 92:4299–4303

Teplow DB (1998) Structural and kinetic features of amyloid beta-protein fibrillogenesis. Amyloid 5:121–142

Terry RD, Masliah E, Salmon DP et al (1991) Physical basis of cognitive alterations in Alzheimer's disease: synapse loss is the major correlate of cognitive impairment. Ann Neurol 30:572–580

Thakur AK, Wetzel R (2002) Mutational analysis of the structural organization of polyglutamine aggregates. Proc Natl Acad Sci U S A 99:17014–17019

Tooze J, Hollinshead M, Ludwig T et al (1990) In exocrine pancreas, the basolateral endocytic pathway converges with the autophagic pathway immediately after the early endosome. J Cell Biol 111:329–345

Uversky VN (2003) Protein folding revisited. A polypeptide chain at the folding-misfolding-non folding cross-roads: which way to go? Cell Mol Life Sci. 60:1852–1871

Uversky VN, Lee HJ, Li J et al (2001) Stabilisation of partially folded conformation during alpha synuclein oligomerization in both purified and cytosolic preparations. J Biol Chem 276:43495–43498

Volles MJ, Lee SJ, Rochet JC et al (2001) Vesicle permeabilization by protofibrillar alpha-synuclein: implications for the pathogenesis and treatment of Parkinson's disease. Biochemistry 40:7812–7819

Walter S, Buchner J (2002) Molecular chaperones-cellular machines for protein folding. Angew Chem Int Ed Engl 41:1098–1113

Wanker EE (2000) Protein aggregation and pathogenesis of Huntington's disease: mechanisms and correlations. Biol Chem 381:937–942

Weihofen A, Binns K, Lemberg MK et al (2002) Identification of signal peptide peptidase, a presenilin-type aspartic protease. Science 296:2215–2218

Weissman AM (2001) Themes and variations on ubiquitylation. Nat Rev Mol Cell Biol 2:169–178

Wigley WC, Fabunmi RP, Lee MG et al (1999) Dynamic association of proteasomal machinery with the centrosome. J Cell Biol 145:481–490

Wille H, Zhang GF, Baldwin MA et al (1996) Separation of scrapie prion infectivity from PrP amyloid polymers. J Mol Biol 259:608–621

Wu Y, Whitman I, Molmenti E et al (1994) A lag in intracellular degradation of mutant alpha 1-antitrypsin correlates with the liver disease phenotype in homozygous PiZZ alpha 1-antitrypsin deficiency. Proc Natl Acad Sci U S A 91:9014–9018

Yankner BA, Caceres A, Duffy LK (1990) Nerve growth factor potentiates the neurotoxicity of beta amyloid. Proc Natl Acad Sci U S A 87:9020–9023

Yao T, Cohen RE (2002) A cryptic protease couples deubiquitination and degradation by the proteasome. Nature 419:403–407

Zhang Y, McLaughlin R, Goodyer C, LeBlanc A (2002) Selective cytotoxicity of intracellular amyloid beta peptide1–42 through p53 and Bax in cultured primary human neurons. J Cell Biol 156:519–529

Zoghbi HY, Botas J (2002) Mouse and fly models of neurodegeneration. Trends Genet 18:463–471

Zwickl P, Seemuller E, Kapelari B, Baumeister W (2001) The proteasome: a supramolecular assembly designed for controlled proteolysis. Adv Protein Chem 59:187–222

The Role of Chaperones in Parkinson's Disease and Prion Diseases

K. F. Winklhofer (✉) · J. Tatzelt (✉)

Department of Cellular Biochemistry, Max-Planck-Institute for Biochemistry,
Am Klopferspitz 18, 82152 Martinsried, Germany
tatzelt@biochem.mpg.de
winklhofer@biochem.mpg.de

1	**Parkinson's Disease**	222
1.1	α-Synuclein: Which Conformation Is the Toxic One?	224
1.2	α-Synuclein Toxicity and Molecular Chaperones	225
1.3	α-Synuclein and Protein Degradation	226
1.4	Parkin: A Link to the Ubiquitin-Proteasome System	227
1.5	Molecular Chaperones and Parkin Function	230
1.6	Inactivation of Parkin by Misfolding and Aggregation	230
1.7	DJ-1 and Chaperone Activity	231
2	**Prion Diseases or Transmissible Spongiform Encephalopathies**	232
2.1	Misfolding of the Prion Protein Is Linked to Both Neurodegeneration and Propagation of Infectious Prions	233
2.2	Biogenesis of the Prion Protein	235
2.3	Cellular Compartments of PrP Misfolding	236
2.4	Chaperones Can Induce the Formation of a PrP^{Sc}-Like Isoform In Vitro	237
2.5	Chemical Chaperones Interfere with the Formation of PrP^{Sc}	238
2.6	Chaperones in the Endoplasmic Reticulum and PrP Misfolding	239
2.7	Prion Propagation and the Cellular Stress Response	240
3	**Chaperones and Other Neurodegenerative Diseases**	241
3.1	Polyglutamine Expansion Diseases	241
3.2	Alzheimer's Disease and Tauopathies	242
4	**Mechanistic Effects of Molecular Chaperones**	243
	References	244

Abstract The etiologies of neurodegenerative diseases, such as Alzheimer's disease, Parkinson's disease, polyglutamine diseases, or prion diseases may be diverse; however, aberrations in protein folding, processing, and/or degradation are common features of these entities, implying a role of quality control systems, such as molecular chaperones and the ubiquitin-proteasome pathway. There is substantial evidence for a causal role of protein misfolding in the pathogenic process coming from neuropathology, genetics, animal modeling, and biophysics. The presence of protein aggregates in all neurodegenerative diseases gave rise to the hypothesis that protein aggregates, be it intracellular or extracellular deposits, may perturb the cellular homeostasis and disintegrate neuronal function (Table 1). More recently, however, an increasing number of studies have indicated that protein aggregates are not toxic

Table 1 Protein aggregates in neurodegenerative diseases

Disease	Aggregates	Protein
Alzheimer's disease	Amyloid plaques	Aβ peptides
	Neurofibrillary tangles	Tau
Parkinson's disease	Lewy bodies	α-Synuclein
Prion diseases	PrP^{Sc} plaques	Prion protein (PrP)
Huntington's disease	Polyglutamine aggregates	Huntingtin
Spinocerebellar ataxias	Polyglutamine aggregates	Ataxins

per se and might even serve a protective role by sequestering misfolded proteins. Specifically, experimental models of polyglutamine diseases, Alzheimer's disease, and Parkinson's disease revealed that the appearance of aggregates can be dissociated from neuronal toxicity, while misfolded monomers or oligomeric intermediates seem to be the toxic species. The unique features of molecular chaperones to assist in the folding of nascent proteins and to prevent stress-induced misfolding was the rationale to exploit their effects in different models of neurodegenerative diseases. This chapter concentrates on two neurodegenerative diseases, Parkinson's disease and prion diseases, with a special focus on protein misfolding and a possible role of molecular chaperones.

Keywords Parkinson's disease · α-Synuclein · Parkin · DJ-1 · Prion diseases · PrP · Misfolding · Aggregation · Neurodegeneration · Chaperones · Ubiquitin-proteasome system · Quality control

1
Parkinson's Disease

Parkinson's disease is the second most common neurodegenerative disease after Alzheimer's disease. The estimated prevalence of PD is 1%–2% of persons older than 60 years of age (Tanner and Ben-Shlomo 1999). The most prominent clinical signs include resting tremor, bradykinesia, muscular rigidity, and postural instability. In addition to the typical motor symptoms, various nonmotor features may develop, including autonomic dysfunction, sleep disturbances, depression, and cognitive impairment. Diagnosis of PD is based on clinical features; asymmetric manifestation of the first symptoms (resting tremor in 70% of patients) and improvement with dopaminergic therapy is highly indicative of PD; however, definite diagnosis requires autopsy.

The pathological hallmark of PD is a highly specific pattern of neuronal loss, mostly affecting the dopaminergic neurons of the substantia nigra pars compacta. Intraneuronal cytosolic inclusions, termed Lewy bodies, and dystrophic neurites (Lewy neurites) are found in the affected brain regions. Lewy bodies are composed of a dense hyaline core surrounded by a halo of radiating fibrils and contain various proteins, including α-synuclein, ubiquitin, neurofilaments, and heat shock proteins. The occurrence of Lewy bodies is not restricted to PD; some other neurodegenerative entities, such as multiple system atrophy and dementia with Lewy bodies, are also characterized by the presence of α-synuclein aggregates and therefore are classified as α-synucleinopathies.

The demise of dopaminergic neurons in the substantia nigra leads to a reduced innervation of the striatum, primarily the putamen, which is a major center for motor control (review in Dauer and Przedborski 2003). Symptoms occur when more than 60% of substantia nigra dopaminergic neurons and about 80% of putamenal dopamine have been lost. The discovery of dopamine deficiency was the basis for the development of symptomatic therapeutic strategies. Replacement of dopamine by its precursor L-dopa and administration of dopaminergic agonists can relieve the symptoms of PD, however, after several years of pharmacological treatment, most patients develop adverse effects, notably motor fluctuations and dyskinesias. Surgical interventions such as deep brain stimulation offers hope for patients whose motor complications can not adequately be controlled pharmacologically. Despite these advances, no strategies have yet been established to halt or delay the progression of the degenerating process.

The most common cause of a parkinsonian syndrome is sporadic or idiopathic PD, the etiology of which is still enigmatic. Genetic as well as environmental factors have been implicated in the etiopathogenesis of PD, and oxidative stress, mitochondrial dysfunction, and protein aggregation are consistent findings in the course of this disease. The observation that different toxins, such as MPTP (1-methyl-4-phenyl-1,2,3,6-tetrahydropyridine), 6-hydroxydopamine, and rotenone, can damage dopaminergic neurons led to the development of toxin-based animal models of PD. Of note, the MPTP monkey model proved particularly useful for the preclinical evaluation of symptomatic therapeutic strategies.

Table 2 Familial variants of PD with a mendelian pattern of inheritance

Locus	Chromosomal location	Protein	Mode of inheritance
PARK1	4q21-q23	α-Synuclein	AD
PARK2	6q25.2–27	Parkin	AR
PARK3	2p13	Unknown	AD
PARK4	4p15 (triplication)	α-Synuclein	AD
PARK5	4p14	UCH-L1	AD?
PARK6	1p35-p36	PINK1	AR
PARK7	1p36	DJ-1	AR
PARK8	12p11.2-q13.1	LRRK2/dardarin	AD
PARK9	1p36	Unknown	AR

AD, autosomal dominant; AR, autosomal recessive; UCH-L1, ubiquitin C-terminal hydrolase-L1; PINK1, PTEN-induced kinase 1; LRRK2, leucine-rich repeat kinase 2

A milestone in PD research was the identification of genes that are responsible for familial variants of this disease. The first familial form of PD identified was associated with α-synuclein, which was later shown to be a major component of Lewy bodies and Lewy neurites, the pathognomonic features of sporadic PD (Spillantini et al. 1997). Subsequently, several other monogenic forms of PD were identified (Table 2), which might facilitate establishing animal and cell culture models (reviewed in Vila and Przedborski 2004). It will now be a challenging endeavor to determine the specific role of these genes and gene products in the pathogenesis of PD.

1.1
α-Synuclein: Which Conformation Is the Toxic One?

In the late 1990s, two missense mutations in the α-synuclein gene were identified as a cause of autosomal dominant PD; the A53T substitution was found in a large Italian-American/Greek family, and the A30P substitution was found in a small German family (Kruger et al. 1998; Polymeropoulos et al. 1997). Recently, a third α-synuclein missense mutation (E46K) has been described in a Spanish family (Zarranz et al. 2004), and dominantly inherited PD in two families could be attributed to a genomic triplication of wild type α-synuclein, leading to overexpression of α-synuclein (Farrer et al. 2004; Singleton et al. 2003). Overall, mutations in the α-synuclein gene seem to be a rare cause of familial PD; however, due to the presence of α-synuclein in Lewy bodies it has advanced to a prime candidate for mechanistic studies.

α-Synuclein belongs to the synuclein family, which also includes β- and γ-synuclein and has only been described in vertebrates. α-Synuclein is a soluble natively unfolded 140-kDa protein consisting of three regions. The N-terminal domain adopts an α-helical conformation upon membrane binding (Davidson et al. 1998; Eliezer et al. 2001). The central hydrophobic domain is responsible for the fibrillogenic potential of α-synuclein (Giasson et al. 2001). The acidic C-terminal domain contains several phosphorylation sites and may confer a chaperone-like activity (Okochi et al. 2000; Park et al. 2002). It has been shown that α-synuclein can prevent the aggregation of different non-native proteins in vitro under stress conditions; however, refolding activity was not observed (Kim et al. 2000; Souza et al. 2000).

In vitro, three conformational variants of α-synuclein are populated: the unfolded monomers, protofibrils (transient β-sheet-rich oligomers) and amyloid fibrils with a characteristic cross-β structure. The A53T mutation accelerates, while the A30P mutation reduces fibril formation, suggesting that α-synuclein fibrils might not be the toxic species. Both mutations, however, promote the formation of protofibrils, either thermodynamically or kinetically (Conway et al. 2000; Volles and Lansbury 2003).

α-Synuclein is expressed at high levels in the nervous system, where it is enriched in presynaptic terminals (Maroteaux et al. 1988). The physiological function of α-synuclein is not fully understood; its close association with synaptic vesicles suggests a role in neurotransmitter release or synaptic plasticity. In α-synuclein knockout mice, which do not display an overt phenotype, an activity-dependent enhanced DA release was observed at nigrostriatal terminals, indicating that α-synuclein may act as a negative regulator of dopaminergic neurotransmission (Abeliovich et al. 2000; Cabin et al. 2002). In double knockout mice for α- and β-synuclein, moderate changes in two synaptic signaling proteins (complexins and 14-3-3 proteins) and a small decrease in brain dopamine levels were observed, but no impairment of synaptic parameters (Chandra et al. 2004). In a recent study of synaptic transmission in cultured hippocampal neurons, it was shown that α-synuclein augments neurotransmitter release from the presynaptic terminal (Liu et al. 2004). These observations are difficult to reconcile, but long-term compensatory changes in knockout mice may explain these obvious discrepancies.

Based on in vitro and cell culture experiments, different mechanisms of α-synuclein-induced toxicity have been proposed. α-Synuclein may interfere with the activity of the proteasome, thereby promoting the accumulation of misfolded and potentially toxic proteins (Petrucelli et al. 2002; Snyder et al. 2003). In dopaminergic neurons, α-synuclein complexes may sequester the 14-3-3 protein, a chaperone with an anti-apoptotic potential (Xu et al. 2002). Furthermore, α-synuclein protofibrils with pore-forming activity can permeabilize vesicular membranes in vitro (Volles and Lansbury 2002; Volles et al. 2001). Remarkably, oxidized dopamine inhibits α-synuclein fibril formation, thereby enhancing the accumulation of potentially toxic α-synuclein protofibrils (Conway et al. 2001). This finding may help to explain the selective death of dopaminergic neurons in PD.

1.2
α-Synuclein Toxicity and Molecular Chaperones

Several α-synuclein-based transgenic mouse models have been established, partly with equivocal findings (reviewed in Fernagut and Chesselet 2004; Maries et al. 2003). Unfortunately, none of these models faithfully recapitulates the pathophysiological features of PD. Notably, mice overexpressing wild type or mutant α-synuclein do not lose dopaminergic neurons in the substantia nigra. However, they made it possible to draw two essential conclusions: first, α-synuclein pathology results from a toxic gain of function, and second, neurodegeneration can occur in the presence of nonfibrillar α-synuclein. Surprisingly, *Drosophila melanogaster* overexpressing either wild type or mutant α-synuclein develop an adult-onset loss of dopaminergic neurons in the dorsomedial cluster, progressive motor deficits, and fibrillar α-synuclein inclusions (Feany and Bender 2000). Moreover, the *Drosophila* model proved particularly useful to analyze the effects of molecular chaperones on α-synuclein aggregation and toxicity. Transgenic co-expression of human Hsp70 prevents the degeneration of dopaminergic neurons induced by α-synuclein, and conversely, interference with endogenous Hsp70 activity by expressing a dominant negative Hsp70 mutant defective in ATP binding enhances α-synuclein toxicity (Auluck et al. 2002). Remarkably, Hsp70 does not affect the number, morphology, or distribution of α-synuclein aggregates, supporting the view that not the fibrillar inclusions but rather oligomeric intermediates are the toxic species. In line with these observations, α-synuclein toxicity can be efficiently reduced by feeding the flies with geldanamycin, a benzochinone ansamycin that increases chaperone levels by activation of HSF1 (Auluck and Bonini 2002; Auluck et al. 2005). The effect of Hsp70 was also analyzed in mouse models of PD. By crossing α-synuclein transgenic mice with Hsp70-overexpressing mice, a reduction of Triton X-100 insoluble α-synuclein was reported, but unfortunately, there is no information on neuropathological and behavioral features of these mice (Klucken et al. 2004). In a different approach, the Hsp70 gene was delivered to dopaminergic neurons by recombinant adeno-associated virus and was shown to protect the mouse dopaminergic system against MPTP-induced neuronal loss (Dong et al. 2005). Similarly to MPTP, rotenone is a potent inhibitor of mitochondrial complex I, but in addition it can induce the formation of Lewy body-like inclusions. Overexpression of Hsp70 in cultured cells not only protected the cells from rotenone-induced toxicity but also decreased the amount of soluble α-synuclein oligomers (Zhou et al. 2004). Based on observations in a model of peripheral neuronal cells, a protective role of the small heat shock protein Hsp27 has been proposed in cells overexpressing wild type or mutant α-synuclein (Zourlidou et al. 2004). Surprisingly, in the same study Hsp70 was only moderately effective against wild type α-synuclein and showed no protection against mutant α-synuclein.

Interestingly, the significance of chaperone pathways in PD is further strengthened by the observation that polymorphisms in the Hsp70-1 gene seem to be involved in the susceptibility to PD (Wu et al. 2004).

1.3
α-Synuclein and Protein Degradation

Several studies indicate that the proteasome might play a role in α-synuclein metabolism and toxicity; however, this aspect remains controversial (Ancolio et al. 2000; Biasini et al. 2004; Martin-Clemente et al. 2004; Rideout et al. 2001). In cell culture studies, it has been observed that α-synuclein can be degraded by the proteasome without the requirement of ubiquitination, whereas in Lewy bodies α-synuclein is oligo-ubiquitinated (Bennett et al. 1999; Tofaris et al. 2001, 2003). In other studies, the lysosomal degradation pathway was shown to be involved in the clearance of α-synuclein (Lee et al. 2004; Webb et al. 2003). Recently, α-synuclein was found to be a substrate for chaperone-mediated autophagy, which involves binding of Hsp70 and specific targeting to the lysosome (Cuervo et al. 2004). So far, however, the degradation pathway of α-synuclein is not fully understood. In particular, it is conceivable that conformational variants of α-synuclein undergo distinct proteolytic pathways.

Remarkably, aging is associated with an increase in α-synuclein protein levels in the substantia nigra and thus, it is tempting to speculate that this phenomenon is causally related to a decline in proteasomal and lysosomal activity (Li et al. 2004). In cell culture and animal models, inhibition of the proteasome enhances α-synuclein accumulation and aggregation (Fornai et al. 2003; McNaught et al. 2002; Rideout et al. 2001; Stefanis et al. 2001; Tanaka et al. 2001; Tofaris et al. 2001). On the other hand, several studies indicate that α-synuclein can decrease the activity of the proteolytic machinery and hence favor its own accumulation, resulting in a vicious circle (Lindersson et al. 2004; Petrucelli et al. 2002; Snyder et al. 2003; Stefanis et al. 2001; Tanaka et al. 2001). However, this issue is difficult to address experimentally, which is reflected by a variety of inconsistent findings. Therefore, a careful evaluation of potentially confounding parameters is mandatory for the interpretation of such studies, including differences in neuronal and non-neuronal cells, possible effects of epitope tagging of α-synuclein and transcriptional effects of proteasomal inhibitors on transgene expression after long-term treatment (Biasini et al. 2004).

1.4
Parkin: A Link to the Ubiquitin-Proteasome System

In 1998, mutations in the parkin gene were identified as a cause of autosomal recessive PD with juvenile onset in Japanese families (Kitada et al. 1998). In the meantime, a large number and a wide spectrum of parkin mutations have been described all over the world, accounting for up to 50% of all familial cases. Clinically, patients with mutations in the parkin gene show the typical signs of parkinsonism. Onset of symptoms is typically before the age of 40, but also late-onset cases have been described. Pathologically, patients with parkin mutations display a loss of dopaminergic neurons in the substantia nigra. Only a few brains of parkin-proven cases of PD have been examined pathologically, and initially, no Lewy body pathology was observed, leading to the hypothesis that functional parkin is essential for the formation of Lewy bodies. However, Lewy bodies were recently detected in patients with parkin mutations; thus, a larger number of autopsies is required for definite conclusions (Farrer et al. 2001; Sasaki et al. 2004).

The parkin gene encodes a 465-amino acid protein with an ubiquitin-like domain (UBL) at the N-terminus and a RING box close to the C-terminus, consisting of two RING finger motifs separated by a cysteine-rich in-between RING domain (IBR) (Fig. 1a). The presence of the RING box suggested that parkin acts as an E3 ubiquitin ligase, which confers substrate specificity for ubiquitination and targets substrate proteins to the 26S proteasome. Degradation of a protein via the ubiquitin-proteasome pathway is a highly or-

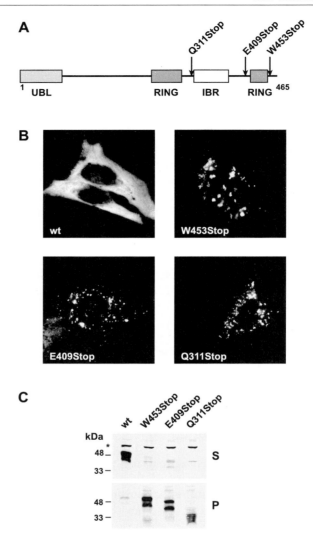

Fig. 1a–c Misfolding and aggregation of parkin. **a** Schematic representation of parkin. The location of the C-terminal mutations are indicated by arrows. *UBL*, ubiquitin-like domain; *RING*, RING finger motif; *IBR*, in-between RING fingers domain. **b** C-terminal deletion mutants of parkin are found in scattered aggregates throughout the cytosol. Human neuroblastoma SH-SY5Y cells grown on glass coverslips were transiently transfected with plasmids encoding wild type (*wt*) parkin or the indicated C-terminal deletion mutants. The expression pattern of parkin was analyzed by indirect immunofluorescence of permeabilized cells using an anti-parkin antiserum. **c** In contrast to wt parkin, the C-terminal deletion mutants are present in a detergent-insoluble conformation. Transiently transfected mouse neuroblastoma N2a cells were lysed in detergent buffer (0.1% Triton X-100 in PBS), fractionated by centrifugation and parkin present in the detergent-soluble (*S*) and -insoluble (*P*) fraction was analyzed by Western blotting using an anti-parkin antiserum

dered process and involves multiple steps, catalyzed by specific enzymes (review in Ciechanover and Brundin 2003; Varshavsky 1997). An E1 ubiquitin-activating enzyme mediates the formation of a high-energy thiol ester between the C-terminus of ubiquitin and a reactive cysteine of the E1. Ubiquitin is then transferred to a cysteine residue of an E2 ubiquitin-conjugating enzyme, which acts in concert with an E3 ubiquitin ligase to catalyze the formation of an isopeptide bond between the C-terminal glycine of ubiquitin and the ε-amino group of a lysine residue within the substrate. Additional ubiquitin moieties are added to the first ubiquitin to generate a polyubiquitin chain, which functions as a degradation signal for the 26S proteasome.

A variety of putative parkin substrates have been identified by yeast two-hybrid and co-immunoprecipitation studies, including the parkin-associated endothelin receptor-like receptor (Pael-R), CDCrel-1, synphilin-1 (review in Imai and Takahashi 2004; Kahle and Haass 2004; von Coelln et al. 2004). An attractive model proposes that loss of parkin function leads to the accumulation of substrates, which eventually may damage dopaminergic neurons. However, none of the substrates identified so far accumulates in parkin knockout models, be it mice or flies. Therefore, the physiological relevance of these substrates is still unclear (review in Kahle and Haass 2004). Nonetheless, a neuroprotective potential of parkin has been shown in various cell culture and animal models. Parkin seems to be protective against:

- Endoplasmic reticulum (ER) stress induced by the overexpression of Pael-R (Imai et al. 2001; Yang et al. 2003)

- Excitotoxicity induced by kainate (Staropoli et al. 2003)

- Proteasome inhibition and α-synuclein toxicity (Lo Bianco et al. 2004; Petrucelli et al. 2002; Yang et al. 2003)

- Ceramide-mediated cell death (Darios et al. 2003)

- Manganese-induced cell death (Higashi et al. 2004)

- Dopamine-induced apoptosis (Jiang et al. 2004)

- Cytotoxicity of an expanded polyglutamine ataxin-3 fragment (Tsai et al. 2003)

These observations may indicate a central role of parkin in neuronal viability, although a general pathway that integrates all protective effects described is difficult to propose. Notably, parkin knockout mice do not display signs of impaired neuronal integrity; therefore, it will be important to analyze whether neuronal vulnerability to stressful stimuli is increased in these animals.

1.5
Molecular Chaperones and Parkin Function

A role of chaperones and co-chaperones in the regulation of parkin activity has emerged from the observation that parkin forms a complex with Hsp70, Hdj-2, and CHIP (Imai et al. 2002). CHIP (carboxyl terminus of the Hsp70-interacting protein) was shown to promote the dissociation of Hsp70 from the parkin–Pael-R complex, thereby enhancing the E3 ligase activity of parkin. Conversely, Hsp70 inhibits parkin activity; however, when co-expressed with parkin, both CHIP and Hsp70 suppress cell death induced by the overexpression of Pael-R. These observations led the authors of this study to propose a model, suggesting that Hsp70 assists the translocation of newly synthesized Pael-R and protects Pael-R from ubiquitination by parkin. When Pael-R is overexpressed and causes ER stress, Hsp70 binds to retrotranslocated Pael-R to prevent its aggregation. Then CHIP induces the release of Hsp70 and promotes ubiquitination of Pael-R by parkin.

In a recent study, BAG5 (bcl-2-associated athanogene 5) was identified as an inhibitor of both parkin and Hsp70 activity (Kalia et al. 2004). BAG5, a member of the BAG domain-containing family whose expression is induced following injury to dopaminergic neurons, interacts with the ATPase domain of Hsp70 and reduces the chaperone activity of Hsp70. BAG5 also interacts directly with parkin. In cultured cells it decreases the E3 ligase activity and increases the sequestration of parkin in aggregates after proteasomal inhibition. As a consequence, BAG5 overexpression attenuates the protective effects of parkin on cell viability. Furthermore, targeted overexpression of BAG5 in the substantia nigra of rats enhances neuronal loss after medial forebrain bundle axotomy or MPTP treatment. However, a causal link of BAG5 effects on parkin or Hsp70 function has not been demonstrated in the animal model. Notably, BAG1, the first member of the BAG family identified, also modulates the activity of Hsp70 by binding to the ATPase domain and is upregulated following brain injury (Nollen et al. 2000; Seidberg et al. 2003). However, BAG1 exhibits anti-apoptotic activity and has been shown to promote cell survival in an in vivo stroke model (Kermer et al. 2003; Takayama et al. 1995).

1.6
Inactivation of Parkin by Misfolding and Aggregation

A wide variety of parkin mutations have been identified, including exon deletions and multiplications, missense, nonsense, and frameshift mutations (review in Mata et al. 2004). Some pathogenic mutations compromise the E3 ligase activity of parkin by impairing interactions with either substrate proteins or components of the ubiquitin proteasome system. Only recently did it become evident that parkin can also be inactivated by misfolding and aggregation. The first parkin mutant shown to be constitutively misfolded was the W453Stop mutant, which lacks 13 C-terminal amino acids (Winklhofer et al. 2003b) (Fig. 1). Interestingly, the deletion of 3 C-terminal amino acids is sufficient to interfere with the native folding of parkin and to induce the formation of cytosolic aggregates. In the meantime, alterations in the detergent solubility or cellular localization was found for different point mutants, indicating that misfolding and aggregation is not restricted to C-terminal deletion mutants (Ardley et al. 2003; Cookson et al. 2003; Gu et al. 2003; Henn et al. 2005; Muqit et al. 2004). Remarkably, even wild type parkin can be inactivated by misfolding in response to different stressful stimuli. Oxidative stress induces the aggregation of parkin in cultured cells (Winklhofer et al. 2003b), and reperfusion after transient focal cerebral ischemia decreases the amount of detergent-soluble parkin in a rat model (Mengesdorf et al. 2002). It has also been shown that nitrosative stress impairs the E3 ligase activity of parkin, however, in these studies it has not been analyzed if nitrosylation of parkin has an impact on its conformation (Chung et al. 2004; Yao et al. 2004). Formation of large perinuclear parkin aggregates has been observed in cells treated with proteasomal inhibitor (Ardley et al. 2003; Junn et al. 2002; Muqit et al. 2004). While it seems logical that after oxidative stress parkin loses its protective function (Chung et al. 2004), it is difficult to explain why parkin can protect cells against the deleterious effects of proteasomal inhibition, although it is recruited into aggregates. Presumably, methodological differences in the proteasomal inhibitor treatment may account for these discrepancies.

Based on these observations, we reasoned that molecular chaperones might interfere with the misfolding of parkin. Indeed, increased expression of Hsp70 and Hsp40 prevents stress-induced aggregation of wild type parkin and promotes folding of the W453Stop mutant in a cell culture model (Winklhofer et al. 2003b).

1.7
DJ-1 and Chaperone Activity

Mutations in the DJ-1 gene have been associated with autosomal recessive PD in 2003 (Bonifati et al. 2003). A large deletion and a missense mutation (L166P) leading to the rapid degradation of the mutant protein were the first mutations identified in patients, indicating that a loss of DJ-1 function is associated with autosomal recessive PD (review in Bonifati et al. 2004). The clinical phenotype is similar to that of parkin-related PD; however, DJ-1 mutations account for only 1%–2% of early-onset cases (Hedrich et al. 2004).

The DJ-1 gene was first cloned in 1997 in the course of a yeast two-hybrid screen for proteins interacting with c-myc (Nagakubo et al. 1997). It encodes a 189 amino acid protein, which is highly expressed in the testes, but can also be found in various tissues, including the brain (Bandopadhyay et al. 2004). The crystal structure of DJ-1 revealed that DJ-1 exists as a dimer (Honbou et al. 2003; Tao and Tong 2003; Wilson et al. 2003). Several functions have been ascribed to DJ-1. It restores the transcriptional activity of the androgen receptor (AR) following repression by PIASxα (protein inhibitor of activated STAT), which acts as a SUMO-1 (small ubiquitin-like modifier-1) ligase (Takahashi et al. 2001). DJ-1 can also positively regulate the AR-mediated transcriptional activity by binding to DJBP (DJ-1 binding protein), which seems to inhibit the AR by recruiting a histone deacetylase complex (Niki et al. 2003). Furthermore, DJ-1 was identified as the regulatory subunit of a RNA-binding protein complex (Hod et al. 1999). DJ-1 is converted into a pI variant in response to oxidative stress, suggesting a function as stress sensor (Mitsumoto et al. 2001). Based on this observation, it has been speculated that DJ-1 is linked to neurodegeneration via a role in the stress response. Structural similarities to the stress-inducible *Escherichia coli* chaperone Hsp31 led to the finding that DJ-1 has a chaperone-like activity; in vitro it prevents the heat-induced aggregation of luciferase or citrate synthase (Lee et al. 2003; Shendelman et al. 2004). Notably, this activity is maintained after oxidative stress and is not affected by ATP or ADP. In a cell culture model, it has been observed that downregulation of DJ-1 by siRNA enhances cell death induced by oxidative stress (Martinat et al. 2004; Taira et al. 2004; Yokota et al. 2003).

2
Prion Diseases or Transmissible Spongiform Encephalopathies

Prion diseases are a group of transmissible neurodegenerative disorders including Creutzfeldt-Jakob disease (CJD) and Gerstmann-Sträussler-Scheinker syndrome (GSS) in humans, scrapie in sheep and goat, bovine spongiform encephalopathy (BSE) in cattle, and chronic wasting disease (CWD) in deer (review in Aguzzi et al. 2001; Collinge 2001; Prusiner et al. 1998; Weissmann et al. 1996). A hallmark of prion diseases is the accumulation of the misfolded and proteinase K-resistant scrapie prion protein (PrP^{Sc}), which co-purifies with the infectious scrapie agent (Bolton et al. 1982). Based on these and other studies, it was proposed that PrP^{Sc} is the essential component of the transmissible agent, termed prion (acronym for proteinaceous infectious particle) (Prusiner 1982). Experimental support for this model was provided recently: recombinant PrP expressed in bacteria and subsequently misfolded in vitro transmitted the disease to transgenic mice overexpressing the C-terminal domain (aa 89–231) of PrP^C (Legname et al. 2004).

Considering a possible role of chaperones in the pathogenesis of prion diseases, different scenarios are conceivable. Given that chaperones recognize folding intermediates or misfolded conformers of PrP, such interactions could be beneficial or harmful to the cell: beneficial when chaperones promote the correct folding of PrP or target misfolded isoforms for degradation. Alternatively, an interaction of PrP with chaperones could enhance or facilitate a conformational transition into an aberrant conformation. In addition, one should consider the possibility that chaperones fail to prevent the formation and/or accumulation of aberrant PrP, either simply because they are not present in the critical cellular compartment or because they do not recognize PrP as a substrate. Interestingly, published data support any of these scenarios. In particular, different studies on pathogenic PrP mutants suggest that misfolded PrP conformers generated in the secretory pathway are not efficiently eliminated by cellular quality control pathways.

2.1
Misfolding of the Prion Protein Is Linked to Both Neurodegeneration and Propagation of Infectious Prions

A key event in prion diseases is the conformational transition of the cellular prion protein (PrP^C) into the pathogenic isoform PrP^{Sc} (Table 3). In contrast to PrP^C, PrP^{Sc} is insoluble in nonionic detergents, partially resistant to proteolytic digestion, and characterized by a high content in β-sheet secondary structure (Caughey et al. 1991; Meyer et al. 1986; Pan et al. 1993). Structural studies with recombinantly expressed PrP revealed a large flexibly disordered N-terminal region and a structured C-terminal domain (aa 126–226). This autonomously folding domain contains three α-helical regions and a short two-stranded β-sheet (Donne et al. 1997; Riek et al. 1996, 1997) (Fig. 2). The pivotal role of the prion protein for the disease was demonstrated in animal models. Mice devoid of PrP^C are resistant to scrapie infection and do not propagate proteinase K-resistant or infectious PrP^{Sc} (Büeler et al. 1993).

The low incidence of sporadic prion diseases in humans (1:1 million) indicates that spontaneous conversion of the prion protein is a very rare process. However, invading PrP^{Sc} can efficiently induce the conversion of endogenous PrP^C into the pathogenic conformation. Infectious forms in humans were linked to contaminated growth hormone derived from cadaveric pituitary gland, dura mater grafts, and corneal transplants (iatrogenic CJD), or to ritualistic cannibalism (kuru). New variant CJD (vCJD) has also been classified as an acquired prion disease, as strong evidence is accumulating for a causal association between BSE and vCJD. Familial prion diseases are invariably associated with mutations in the gene encoding the prion protein, indicating that

Table 3 Prion terminology

Prion	Proteinaceous infectious particle
PrP	Prion protein
PrP^C	Cellular prion protein, physiological isoform
PrP^{Sc}	Scrapie prion protein, pathological isoform, major component of infectious prions

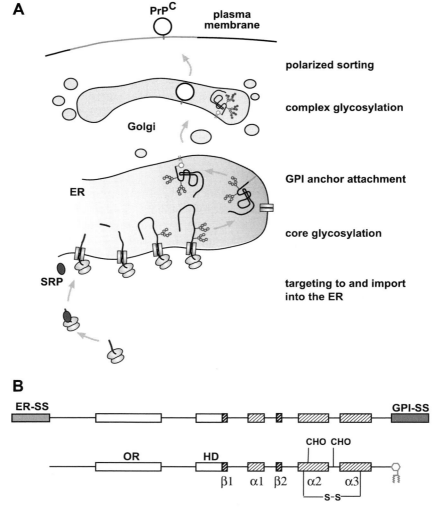

Fig. 2a,b Biogenesis of PrP^C. **a** Translation of PrP^C is initiated on free ribosomes in the cytosol. The signal recognition particle (*SRP*) binds to the N-terminal signal peptide (*ER-SS*) as well as to the ribosome and targets the nascent chain-ribosome complex to the ER membrane. During import into the ER the N-terminal signal peptide is cleaved off and PrP is modified with two N-linked glycans. Shortly after the release of PrP^C into the ER lumen, a GPI anchor is attached to the C-terminus, concomitantly with the cleavage of the C-terminal signal sequence (*GPI-SS*). On its way through the secretory pathway, the core glycans are processed into complex structures and finally PrP^C is targeted on the outer leaflet of the plasma membrane. **b** Schematic representation of the precursor (*upper panel*) and the mature form (*lower panel*) of PrP^C. The N- and C-terminal signal sequences are cleaved during import into the ER and attachment of the GPI anchor, respectively. *OC*, octarepeat region; *HD*, hydrophobic domain; $\alpha 1$–3, helical regions; $\beta 1$–2, beta-strands; *S-S*, disulfide bond; *CHO*, N-linked glycan attachment sites these mutations can favor the formation of PrP^{Sc} or at least of a neurotoxic PrP species. The pathogenic mutations identified so far are predominantly single amino acid substitutions and are mainly located in the structured C-terminal part of the protein (review in Gambetti et al. 2003; Wadsworth et al. 2003).

In the majority of prion diseases, neurodegeneration is tightly linked to the propagation of infectious PrP^{Sc}. However, transgenic mouse models revealed that these features can occur independently. Misfolded or mutated isoforms of the prion protein induced neuronal cell death in the absence of infectious PrP^{Sc} (Chiesa et al. 1998; Flechsig et al. 2003; Hegde et al. 1998; Ma et al. 2002; Muramoto et al. 1997; Shmerling et al. 1998), and propagation of infectious prions was observed without neuronal cell death (Brandner et al. 1996; Mallucci et al. 2003). Importantly, misfolding or mistargeting of PrP^C is essential for both propagation of the infectious particle and formation of a neurotoxic conformer.

2.2
Biogenesis of the Prion Protein

The biogenesis of PrP^C plays a central role in the pathogenesis of prion diseases (review in Tatzelt and Winklhofer 2004) (Fig. 2a). Interestingly, the first insights came from biochemical studies on purified PrP^{Sc} (Endo et al. 1989; Stahl et al. 1987). These studies established that the prion protein is a secretory protein characterized by extensive post-translational modifications. The C-terminus harbors a glycosylphosphatidylinositol (GPI) anchor, which targets PrP^C to the outer leaflet of the plasma membrane (Stahl et al. 1987), and two consensus sites for N-linked glycosylation, which are both occupied. PrP^C and PrP^{Sc} contain the same set of at least 52 complex bi-, tri-, and tetra-antennary glycans, although with different relative proportions of individual saccharides (Endo et al. 1989; Rudd et al. 1999; Stimson et al. 1999). Regarding a possible physiological or pathological relevance, PrP glycosylation has been shown to have an impact on the conformational transition of PrP^C into PrP^{Sc}, to influence the selective neuronal targeting of PrP^{Sc}, and to contribute to the phenomenon of strain diversity (Collinge et al. 1996; DeArmond et al. 1997; Korth et al. 2000; Lehmann and Harris 1997; Parchi et al. 1996; Priola and Lawson 2001; Taraboulos et al. 1990; Winklhofer et al. 2003a).

PrP^C is widely expressed in the brain and is concentrated in synapse-rich regions (Fournier et al. 1995; Herms et al. 1999). Studies in polarized Madin Darby canine kidney cells indicated that PrP^C is preferentially sorted to the basolateral membrane (Sarnataro et al. 2002; Uelhoff et al. 2005), while GPI-anchored proteins are usually sorted apically in these cells. This untypical sorting behavior of PrP^C is mediated by an internal hydrophobic domain (HD, ~aa 112–135) (Uelhoff et al. 2005). Interestingly, the HD was identified previously to induce misfolding or mistargeting of PrP during import into the endoplasmic reticulum (Hegde et al. 1998; Heller et al. 2003; Yost et al. 1990).

From the cell surface, PrP^C can be internalized to be degraded in a lysosomal compartment or to recycle to the plasma membrane (Shyng et al. 1993). Internalization of PrP^C is mediated mainly by the unstructured N-terminus (Kiachopoulos et al. 2004; Nunziante et al. 2003; Shyng et al. 1995; Sunyach et al. 2003); however, the precise mechanism is controversial. Two internalization pathways, via coated pits or caveolae-dependent structures, have been described (Marella et al. 2002; Peters et al. 2003; Shyng et al. 1994; Sunyach et al. 2003).

2.3
Cellular Compartments of PrP Misfolding

Regarding a possible role of chaperones in the pathogenesis of prion diseases, it is important to identify the cellular compartments involved in the formation or accumulation of pathological PrP conformers. PrP^{Sc} purified from scrapie-infected brains contains a GPI anchor and sialytated complex N-linked glycans (Endo et al. 1989; Stahl et al. 1987), indicating that fully matured PrP^C is the precursor for the formation of PrP^{Sc}. Subsequent experiments in cultured cells supported a model in which formation of PrP^{Sc} occurs after PrP^C had reached the plasma membrane or is re-internalized for degradation (Caughey and Raymond 1991). The mechanism of PrP^{Sc} formation is still enigmatic, but studies in transgenic animals favor a conversion model with a direct interaction between PrP^C and PrP^{Sc} (Prusiner et al. 1990). Thus, PrP^{Sc} propagation and accumulation seems to occur predominantly extracellularly and/or in compartments of the endocytic/lysosomal pathway. Of note, none of these locales implicated in the propagation of infectious PrP^{Sc} contains significant amounts of chaperones. However, pathogenic PrP conformers due to mutations in the PrP gene are formed during biogenesis, and in this case a role of molecular chaperones is conceivable. The first clue that alterations in the biogenesis of PrP^C could generate neurotoxic PrP conformers in the absence of infectious PrP^{Sc} derived from in vitro models analyzing the import of PrP into the endoplasmic reticulum (ER). In addition to the fully translocated secretory form, PrP can attain two different transmembrane topologies, designated ^{Ntm}PrP (N-terminus facing the ER lumen) and ^{Ctm}PrP (C-terminus facing the ER lumen). The internal hydrophobic domain (HD) of PrP was identified as a putative transmembrane domain, and mutations within this domain were found to alter the relative amount of ^{Ctm}PrP and ^{Ntm}PrP (Hegde et al. 1998; Lopez et al. 1990; Yost et al. 1990). Remarkably, the increased synthesis of ^{Ctm}PrP has been shown to coincide with progressive neurodegeneration both in GSS patients with an A117V mutation and in transgenic mice carrying a triple mutation within the HD (AV3) (Hegde et al. 1998). Studies in yeast indicated that the HD has an impact on the import of PrP^C into the ER. During post-translational ER import, the HD induced formation of misfolded and cytosolically localized PrP, which interfered with yeast viability (Heller et al. 2003).

The neurotoxic potential of cytosolic PrP was emphasized in a transgenic mouse model. Mice expressing a PrP mutant with a deleted N-terminal ER targeting signal acquire severe ataxia due to cerebellar degeneration and gliosis (Ma et al. 2002). So far, mutations within the N-terminal signal sequence of PrP, which could affect the efficiency of ER import, have not been found in patients. However, impairment of ER import and the generation of cytosolically localized PrP was observed after the expression of two pathogenic mutants, Q160Stop and W145Stop, which are linked to inherited prion diseases in humans (Heske et al. 2004). Notably, cytosolic PrP seems to be present in a small subset of neurons even under physiological conditions (Mironov et al. 2003).

A variety of pathogenic mutations in humans seem to affect the maturation of PrP in the secretory pathway. The alterations observed include incomplete glycosylation, lack of GPI anchor attachment, misfolding and abnormal membrane association (Capellari et al. 2000a, 2000b; Gu et al. 2002; Ivanova et al. 2001; Kiachopoulos et al. 2005; Lehmann and Harris 1995, 1997; Lorenz et al. 2002; Mishra et al. 2002; Negro et al. 2001; Petersen et al. 1996; Rogers et al. 1990; Singh et al. 1997; Tatzelt and Winklhofer 2004). Interestingly, immature and/or misfolded PrP conformers generated in the secretory pathway are long-lived and are obviously not efficiently eliminated by cellular quality control mechanisms. In a cell culture model, the pathogenic missense mutants T183A and F198S lack a GPI anchor, are not complex glycosylated and adopt a misfolded conformation. However, they are not efficiently subjected to ER-associated degradation (ERAD) but are rather secreted and can be re-internalized by heterologous cells (Kiachopoulos et al. 2005). Similarly, deletion of the GPI anchor signal sequence of PrP (PrPΔGPI) results in the formation of detergent-insoluble, unglycosylated PrP molecules, which are efficiently secreted into the cell culture medium (Blochberger et al. 1997; Rogers et al. 1993; Winklhofer et al. 2003c). Notably, misfolded PrPΔGPI is a preferred substrate for the formation of a proteinase K-resistant PrP isoform in the in vitro-based conversion model described below (Kocisko et al. 1994).

2.4
Chaperones Can Induce the Formation of a PrP^{Sc}-Like Isoform In Vitro

An established model system to study the folding pathway of proteins in vitro employs the *E. coli* chaperonin GroEL and its co-factor GroES. A detailed description on the mechanisms of GroEL/ES-mediated folding is found elsewhere (Bukau and Horwich 1998; Hartl and Hayer-Hartl 2002; Mayhew et al. 1996; Weissman et al. 1996). Folding is initiated by diluting chemically denatured polypeptides into physiological buffer supplemented with chaperones and cofactors. Under those conditions, recombinantly expressed (r)PrP rapidly forms aggregates if chaperones are omitted. GroEL alone prevents the aggregation of rPrP; in the presence of ATP and the co-chaperone GroES, rPrP is able to adopt a soluble conformation (Leffers et al. 2004; Stockel and Hartl 2001). This type of behavior is seen for many proteins that can not refold spontaneously and emphasizes the ability of PrP to bind to chaperones like GroEL or the yeast Hsp104 (Edenhofer et al. 1996; Schirmer and Lindquist 1997). However, an unexpected observation was made when refolded rPrP was assayed instead of denatured rPrP. In the presence of ATP, GroEL induces the conversion of refolded rPrP into a PrP^{Sc}-like conformation (Leffers et al. 2004; Stockel and Hartl 2001). A mechanistic analysis suggested that through the interaction of rPrP with GroEL, a partially unfolded conformation with unmasked hydrophobic sites is generated. In the presence of ATP and in the absence of GroES, this PrP conformer is released into solution and is highly aggregation-prone. Interestingly, low concentrations of SDS induce the aggregation of folded rPrP in a similar manner (Leffers et al. 2004). It should also be noted that GroEL/ATP-induced aggregation of folded PrP is not restricted to rPrP. The same effect can be observed when native PrP^C, i.e., PrP modified with a GPI anchor and complex N-linked glycans, is used (Leffers et al. 2004).

In a different in vitro approach, the effect of chaperones on the PrP^{Sc}-induced conversion of PrP^C to its protease-resistant form was analyzed (DebBurman et al. 1997). In a cell-free conversion assay, radiolabeled PrP^C purified from cultured cells is mixed with PrP^{Sc} purified from scrapie-infected brain, and the formation of proteinase K-resistant radiolabeled PrP is measured (Kocisko et al. 1994). GroEL or yeast Hsp104 stimulates the conversion of PrP^C to a proteinase K-resistant conformation in this system (DebBurman et al. 1997). Similarly to the effects described above (Leffers et al. 2004; Stockel and Hartl 2001), GroES shows an inhibitory effect on the conversion process promoted by GroEL (DebBurman et al. 1997).

These in vitro experiments indicate that in principle, chaperones can mediate PrP misfolding. A plausible mechanism would be that through the interaction with chaperones a transition state of PrP^C is generated, which either spontaneously misfolds, or, in case PrP^{Sc} is present, serves as a precursor for the de novo generation of PrP^{Sc}. The most intriguing questions emerging from the in vitro studies are whether such chaperone interactions occur in live cells and whether the PrP conformers generated through the interaction with chaperones are neurotoxic or infectious.

2.5
Chemical Chaperones Interfere with the Formation of PrP^{Sc}

A number of low-molecular-weight compounds, including the cellular osmolytes glycerol and trimethylamine N-oxide (TMAO) and the organic solvent dimethylsulfoxide (DMSO), have been shown to protect proteins from thermal denaturation and aggregation in vitro and to protect cells from heat-induced cytotoxicity (Back et al. 1979; Edington et al. 1989; Gekko and Koga 1983; Gekko and Timasheff 1981a, 1981b; Gerlsma and Stuur 1972; Lin et al. 1981). Based on these activities, we reasoned that chemical chaperones could stabilize the native conformation of PrP^C and thereby reduce its conversion into the pathogenic conformation. Indeed, in a cell culture model of prion diseases glycerol, TMAO, and DMSO decrease the accumulation of PrP^{Sc} in scrapie-infected (Sc)N2a cells (Tatzelt et al. 1996). Consistent with their expected mode of action, these chemical chaperones specifically interfere with the de novo synthesis of PrP^{Sc} but do not affect the pool of pre-existing PrP^{Sc} (Tatzelt et al. 1996). Molecular dynamics simulations provided an attractive model for this effect at the atomic level; TMAO stabilizes the PrP^C conformation by minimizing the level of solvent exposure of the N-terminal portion of the protein (Bennion et al. 2004).

Remarkably, the anti-prion effect of chemical chaperones is not restricted to the cell culture model. DMSO was shown to delay PrP^{Sc} accumulation and disease symptoms in prion-infected hamsters (Shaked et al. 2003) and to interfere with the formation of proteinase K-resistant PrP in a cell-free conversion assay (DebBurman et al. 1997). In addition, cell culture studies revealed that chemical chaperones can also prevent misfolding of H187R, a pathogenic PrP mutant linked to familial CJD (Gu and Singh 2004).

2.6
Chaperones in the Endoplasmic Reticulum and PrP Misfolding

Different pathogenic PrP mutations linked to inherited prion diseases in humans interfere with the maturation of PrP in the secretory pathway (review in Tatzelt and Winklhofer 2004). The ER lumen is known to contain quality control systems to prevent further transit of aberrant polypeptides (review in Ellgaard and Helenius 2003; McCracken and Brodsky 2003). Misfolded proteins in the ER can be degraded through a mechanism known as ER-associated protein degradation (ERAD). This pathway involves recognition of misfolded polypeptides by chaperones, retrograde transport into the cytosol, and subsequent proteasomal degradation (Finley et al. 1984; Hurtley and Helenius 1989; Jentsch et al. 1987; Klausner and Sitia 1990).

Several groups observed degradation of wild type and mutant PrP via the proteasomal pathway, which was interpreted as evidence for retrograde transport of PrP from the ER lumen into the cytosol (Jin et al. 2000; Ma and Lindquist 2001; Yedidia et al. 2001; Zanusso et al. 1999). This conclusion, however, was discussed controversially after cytosolic accumulation of PrP was proposed to be an experimental artifact, due to the prolonged treatment of cells with proteasomal inhibitors and overexpression of PrP driven by a viral promotor (Drisaldi et al. 2003). Similarly, we did not observe that significant fractions of the two pathogenic PrP mutants T183A and F198S were subjected to proteasomal degradation (Kiachopoulos et al. 2005). Moreover, misfolded PrP in the ER seems not to activate the unfolded protein response (UPR) pathway. Induction of the UPR leads to the upregulation of a variety of proteins, such as the ER chaperone BiP (Ma and Hendershot 2001; Patil and Walter 2001). However, none of the misfolded PrP mutants analyzed induces increased transcription of BiP in a detectable manner (Winklhofer et al. 2003c).

A possible role of Grp94 on the maturation of PrP^C emerged from studies on a defective stress response in scrapie-infected N2a cells (Tatzelt et al. 1995; Winklhofer et al. 2001). Grp94 is among the most abundant ER proteins, yet its physiological function is largely unknown (review in Argon and Simen 1999). When we treated cultured cells with geldanamycin, we observed the stabilization of a high mannose glycoform of PrP^C (Winklhofer et al. 2003a), an observation corroborated later (Ochel et al. 2003). Similarly to inhibitors of α-mannosidases, geldanamycin prevents the processing of high mannose glycoforms into complex structures. Trafficking of immature PrP^C to the outer leaflet of the plasma membrane is not impaired by geldanamycin. Moreover, in scrapie-infected cells high mannose glycoforms of PrP^C are preferred substrates for the formation of PrP^{Sc} (Winklhofer et al. 2003a). Geldanamycin, a benzochinone ansamycin derivative, has been shown to interact with Hsp90 in vitro and in vivo and to interfere with Hsp90 functions (review in Neckers 2002). Later on, geldanamycin was also reported to bind Grp94, the ER homolog of Hsp90 (Chavany et al. 1996). Since PrP^C has no cytoplasmic domain, we assumed that Grp94 might be involved in the geldanamycin-induced effects on glycosylation. Geldanamycin might either prolong a pre-existing interaction of immature PrP^C with Grp94 or impair a Grp94 function, which is required for mannose trimming (Winklhofer et al. 2003a). Interestingly, a transient interaction of high mannose glycoform of PrP^C and Grp94 was described in the absence of geldanamycin (Capellari et al. 1999).

2.7
Prion Propagation and the Cellular Stress Response

Two chaperones in the cytosol of mammalian cells, Hsp72 and Hsp28, are synthesized at high levels only after heat shock or other forms of metabolic stress (Lindquist and Craig 1988). Consequently, expression of both Hsp72 and Hsp28 is often diagnostic of the cell having initiated a stress response. When we examined the stress response in mouse neuroblastoma (N2a) cells and in their scrapie-infected counterparts (ScN2a), we observed that stress-induced expression of Hsp72 and Hsp28 is selectively impaired in ScN2a cells (Tatzelt et al. 1995). The defective stress response in ScN2a cells is linked to an accelerated deactivation of the heat shock transcription factor (HSF) 1 after stress, and can be restored pharmacologically by geldanamycin (Winklhofer et al. 2001). On the other hand, increased expression of Hsp70 was found in the brains of scrapie-infected mice (Kenward et al. 1994). However, a subsequent cell type-specific analysis revealed that Hsp72 expression in the brains of CJD patients is most prominent in Purkinje cells, a cell type that is virtually resistant to PrP^{Sc}-induced cell death (Kovacs et al. 2001). Unfortunately, it is not known whether the phenomenon of sustained viability of Purkinje cells in

CJD brains is due to their pronounced stress response or due to a lack of PrP^{Sc} propagation.

Interestingly, expression of PrP^C itself is upregulated by stress due to two heat shock elements located in the PrP gene promotor (Shyu et al. 2000, 2002). Stress-induced expression of PrP^C is not restricted to cultured cells but was also observed in the brain after focal cerebral ischemia (Weise et al. 2004). It will now be interesting to see whether upregulation of PrP^C during stress is indicative for a protective role of PrP^C function to cellular or environmental stress.

3
Chaperones and Other Neurodegenerative Diseases

3.1
Polyglutamine Expansion Diseases

The effect of molecular chaperones has been extensively studied in different models of polyglutamine expansion diseases. These hereditary neurodegenerative disorders comprise different entities, such as Huntington's disease and spinocerebellar ataxias, and are characterized by an expansion of a polyglutamine tract in the disease-associated proteins, which increases their propensity to aggregate and to form cytosolic and nuclear inclusion bodies. Different models have been proposed to explain the cytotoxicity of polyglutamine proteins. They may induce the co-aggregation of other proteins essential for cell viability, in particular transcription factors with nonpathogenic polyglutamine repeats (review in Sugars and Rubinsztein 2003). According to another model, polyglutamine expansion proteins impair the functional capacity of the ubiquitin-proteasome system (UPS) (Bence et al. 2001; Jana et al. 2001; Verhoef et al. 2002), either directly by clogging or sequestering the 26S proteasome or indirectly by affecting the activity or distribution of UPS modulators (Donaldson et al. 2003; Holmberg et al. 2004; Venkatraman et al. 2004). A recent study in a cell culture model using a GFP-reporter with a C-terminal CL1 degron indicated that aggregated proteins targeted either to the cytosol or nucleus globally impair the UPS function in both compartments. Notably, impairment of the UPS occurs prior to the formation of inclusion bodies and therefore seems to be a response to the presence of intermediate forms of protein aggregates (Bennett et al. 2005). However, a knock-in mouse model of spinocerebellar ataxia type 7 (SCA7) did not provide evidence for UPS impairment. By using transgenic mice expressing a pathogenic polyglutamine protein (ataxin-7) and a GFP-based substrate of the UPS (Ub^{G76V}-GFP), Bowman et al. analyzed the UPS activity in the most vulnerable neuronal population and showed that an increase in Ub^{G76V}-GFP protein levels is not due to an impairment of the UPS but to an increase in Ub^{G76V}-GFP mRNA levels (Bowman et al. 2005).

In cell culture models, Hsp70, Hsp40, or Hsp27 consistently suppressed polyglutamine-induced toxicity, whereas the effect on aggregate formation was not uniform and seems to depend on various factors, such as expression levels and cell type (review in Muchowski and Wacker 2005). Indeed, animal models revealed that polyglutamine-induced toxicity can be dissociated from aggregate formation. Expression of a truncated form of human ataxin-3 with an expanded polyglutamine tract in the *Drosophila* eye induces severe degeneration, which can be rescued by human Hsp70, despite the continued presence of aggregates (Warrick et al. 1999). Co-expression of Hsp70 with Hsp40/Hdj-1 synergistically suppresses ataxin-3-mediated toxicity in this model. Beneficial effects of chaperones have also been reported in a mouse model of spinocerebellar ataxia type 1 (SCA1). Crossing mice overexpressing pathogenic ataxin-1 with mice overexpressing Hsp70 mitigates the neuropathological and behavioral phenotype of SCA1, again without affecting aggregate morphology (Cummings et al. 2001). However, in transgenic mouse models of Huntington's disease, overexpression of Hsp70 alone showed only moderate effects (Hansson et al. 2003; Hay et al. 2004).

The controversial question of whether inclusion bodies are pathogenic, incidental, or even beneficial was recently addressed using an elegant automated microscope system that returns to the same neuron of interest after arbitrary intervals (Arrasate et al. 2004). Survival analysis after transfection of primary neurons with pathogenic mutant huntingtin revealed that the level of diffuse mutant huntingtin and the length of the polyglutamine expansion correlate with the risk of cell death. Inclusion body formation reduces intracellular levels of diffuse huntingtin and prolongs survival, indicating that inclusion bodies might protect neurons by decreasing the levels of toxic diffuse forms of huntingtin.

3.2
Alzheimer's Disease and Tauopathies

In Alzheimer's disease, Aβ peptides (Aβ42 and Aβ40) are the principal components of extracellular amyloid plaques. These aggregation-prone peptides are generated in the secretory pathway by the sequential action of β- and γ-secretase on the transmembrane Aβ precursor protein (APP) (review in Walter et al. 2001). In a cell culture model it has been shown that the ER-resident Hsp70 homolog BiP/Grp78 interacts with immature APP and decreases the secretion of Aβ42 and Aβ40 (Yang et al. 1998). Whether this decrease in Aβ secretion is due to an impairment of APP processing or an intracellular retention of Aβ was not addressed in this study. Using an inducible adenoviral-based system, Magrane et al. could show that targeted expression of Aβ42 in the secretory pathway of primary neurons induces apoptotic cell death, which is suppressed by overexpression of Hsp70 (Magrane et al. 2004). Although no evidence for a physical interaction of Aβ42 and Hsp70 is provided in this study, it is conceivable that Aβ42 binds to Hsp70 after its discrimination by ER quality control systems and retrotranslocation from the ER into the cytosol. Indeed, an interaction of Aβ42 with cytosolic Hsp70 has been demonstrated in a transgenic *Caenorhabditis elegans* model (Fonte et al. 2002). Increasing the chaperone activity of Hsp70 by siRNA-mediated repression of a negative regulator also protected against Aβ42-mediated toxicity in this model.

A possible role of chaperones in tauopathies emerged from recent studies. The microtubule-binding protein tau forms intraneuronal aggregates, termed neurofibrillary tangles (NFT), which occur in Alzheimer's disease and some other neurodegenerative diseases, collectively called tauopathies. Physiologically, tau promotes assembly and stability of the microtubuli network. Hyperphosphorylated tau, which is found in tauopathies, is released from microtubules and accumulates in NFT. Interestingly, different mutations in the human tau gene are associated with FTDP-17 (frontotemporal dementia with parkinsonism linked to chromosome 17), and these tau mutants are characterized by a higher propensity to form aggregates (von Bergen et al. 2001). Hsp70 and Hsp90 seem to be essential determinants in tau physiology through increasing tau solubility and its association with microtubuli (Dou et al. 2003). A CHIP-Hsp70 complex selectively ubiquitinates phosphorylated tau and protects cultured cells from phosphorylated tau-induced toxicity (Petrucelli et al. 2004; Shimura et al. 2004b). Interestingly, this protective effect is only partly due to proteasomal degradation, but seems to be mediated rather by the sequestration of insoluble ubiquitinated tau in aggregates, suggesting that soluble phosphorylated tau might be the toxic species. A protective effect of Hsp27 on tau pathology has been established in a cell culture model. Hsp27 preferentially binds to hyperphosphorylated tau, facilitating its degradation and/or dephosphorylation and suppressing tau-mediated cell death (Shimura et al. 2004a).

4
Mechanistic Effects of Molecular Chaperones

How can the protective effects of chaperones be explained? So far, the precise mechanism of their neuroprotective potential is poorly understood. Conceptually, chaperones may maintain or convert proteins in a nontoxic conformation and/or enhance the sequestration or degradation of toxic species. Surprisingly, overexpression of chaperones suppressed α-synuclein- and polyglutamine-induced toxicity in animal models without affecting the number or morphology of aggregates. However, the inclusions formed in the presence of chaperones showed different biochemical properties (Chan et al. 2000), suggesting that chaperones do not simply work by preventing protein aggregation, but exert more complex, sophisticated functions. Recent in vitro studies with polyglutamine expansion proteins indicated that Hsp70 and Hsp40 act on early intermediates in the aggregation process (Muchowski et al. 2000). By using fluorescence resonance energy transfer (FRET), Schaffar et al. could show that Hsp70 and Hsp40 interfere with an intramolecular conformational change of a soluble pathogenic polyglutamine fragment, thereby preventing the binding and inactivation of transcription factors (Schaffar et al. 2004). Notably, polyglutamine-induced toxicity seems to depend on the ongoing formation of soluble species, whereas the formation of inclusions would rather be protective by the sequestration of toxic conformers. Atomic force microscopy indicated that Hsp70 and Hsp40 cooperatively modulate protein aggregation by partitioning monomeric conformations, attenuating the formation of spherical and annular oligomers and facilitating formation of fibrillar and amorphous aggregates (Wacker et al. 2004). It will now be interesting to analyze how these insights translate into the in vivo situation and how the potential of chaperones can be used therapeutically.

Acknowledgements We thank all present and former group members, and in particular Iris H. Henn for providing Fig. 1. We are grateful to F. Ulrich Hartl for his continuous support. The authors are supported by grants from the Deutsche Forschungsgemeinschaft, the Bundesministerium für Bildung und Forschung, and from the Bayerische Staatsminister für Wissenschaft, Forschung und Kunst.

References

Abeliovich A, Schmitz Y, Farinas I, Choi-Lundberg D, Ho WH, Castillo PE, Shinsky N, Verdugo JM, Armanini M, Ryan A, Hynes M, Phillips H, Sulzer D, Rosenthal A (2000) Mice lacking alpha-synuclein display functional deficits in the nigrostriatal dopamine system. Neuron 25:239–252

Aguzzi A, Montrasio F, Kaeser PS (2001) Prions: health scare and biological challenge. Nat Rev Mol Cell Biol 2:118–126

Ancolio K, Alves da Costa C, Ueda K, Checler F (2000) Alpha-synuclein and the Parkinson's disease-related mutant Ala53Thr-alpha-synuclein do not undergo proteasomal degradation in HEK293 and neuronal cells. Neurosci Lett 285:79–82

Ardley HC, Scott GB, Rose SA, Tan NG, Markham AF, Robinson PA (2003) Inhibition of proteasomal activity causes inclusion formation in neuronal and non-neuronal cells overexpressing Parkin. Mol Biol Cell 14:4541–4556. Epub 2003 Aug 22

Argon Y, Simen BB (1999) GRP94, an ER chaperone with protein and peptide binding properties. Semin Cell Dev Biol 10:495–505

Arrasate M, Mitra S, Schweitzer ES, Segal MR, Finkbeiner S (2004) Inclusion body formation reduces levels of mutant huntingtin and the risk of neuronal death. Nature 431:805–810

Auluck PK, Bonini NM (2002) Pharmacological prevention of Parkinson disease in Drosophila. Nat Med 8:1185–1186

Auluck PK, Chan HY, Trojanowski JQ, Lee VM, Bonini NM (2002) Chaperone suppression of alpha-synuclein toxicity in a Drosophila model for Parkinson's disease. Science 295:865–868. Epub 2001 Dec 20

Auluck PK, Meulener MC, Bonini NM (2005) Mechanisms of suppression of alpha-synuclein neurotoxicity by geldanamycin in Drosophila. J Biol Chem 280:2873–2878. Epub 2004 Nov 18

Back JF, Oakenfull D, Smith MB (1979) Increased thermal stability of proteins in the presence of sugars and polyols. Biochemistry 18:5191–5199

Bandopadhyay R, Kingsbury AE, Cookson MR, Reid AR, Evans IM, Hope AD, Pittman AM, Lashley T, Canet-Aviles R, Miller DW, McLendon C, Strand C, Leonard AJ, Abou-Sleiman PM, Healy DG, Ariga H, Wood NW, de Silva R, Revesz T, Hardy JA, Lees AJ (2004) The expression of DJ-1 (PARK7) in normal human CNS and idiopathic Parkinson's disease. Brain 127:420–430. Epub 2003 Dec 08

Bence NF, Sampat RM, Kopito RR (2001) Impairment of the ubiquitin-proteasome system by protein aggregation. Science 292:1552–1555

Bennett EJ, Bence NF, Jayakumar R, Kopito RR (2005) Global impairment of the ubiquitin-proteasome system by nuclear or cytoplasmic protein aggregates precedes inclusion body formation. Mol Cell 17:351–365

Bennett MC, Bishop JF, Leng Y, Chock PB, Chase TN, Mouradian MM (1999) Degradation of alpha-synuclein by proteasome. J Biol Chem 274:33855–33858

Bennion BJ, De Marco ML, Daggett V (2004) Preventing misfolding of the prion protein by trimethylamine N-oxide. Biochem 43:12955–12963

Biasini E, Fioriti L, Ceglia I, Invernizzi R, Bertoli A, Chiesa R, Forloni G (2004) Proteasome inhibition and aggregation in Parkinson's disease: a comparative study in untransfected and transfected cells. J Neurochem 88:545–553

Blochberger TC, Cooper C, Peretz D, Tatzelt J, Griffith OH, Baldwin MA, Prusiner SB (1997) Prion protein expression in Chinese hamster ovary cells using a glutamine synthetase selection and amplification system. Protein Eng 10:1465–1473

Bolton DC, McKinley MP, Prusiner SB (1982) Identification of a protein that purifies with the scrapie prion. Science 218:1309–1311

Bonifati V, Oostra BA, Heutink P (2004) Linking DJ-1 to neurodegeneration offers novel insights for understanding the pathogenesis of Parkinson's disease. J Mol Med 82:163–174. Epub 2004 Jan 08

Bonifati V, Rizzu P, van Baren MJ, Schaap O, Breedveld GJ, Krieger E, Dekker MC, Squitieri F, Ibanez P, Joosse M, van Dongen JW, Vanacore N, van Swieten JC, Brice A, Meco G, van Duijn CM, Oostra BA, Heutink P (2003) Mutations in the DJ-1 gene associated with autosomal recessive early-onset parkinsonism. Science 299:256–259. Epub 2002 Nov 21

Bowman AB, Yoo SY, Dantuma NP, Zoghbi HY (2005) Neuronal dysfunction in a polyglutamine disease model occurs in the absence of ubiquitin-proteasome system impairment and inversely correlates with the degree of nuclear inclusion formation. Hum Mol Genet 14:679–691. Epub 2005 Jan 20

Brandner S, Isenmann S, Raeber A, Fischer M, Sailer A, Kobayashi Y, Marino S, Weissmann C, Aguzzi A (1996) Normal host prion protein necessary for scrapie-induced neurotoxicity. Nature 379:339–343

Büeler H, Aguzzi A, Sailer A, Greiner R-A, Autenried P, Aguet M, Weissmann C (1993) Mice devoid of PrP are resistant to scrapie. Cell 73:1339–1347

Bukau B, Horwich AL (1998) The Hsp70 and Hsp60 chaperone machines. Cell 92:351–366

Cabin DE, Shimazu K, Murphy D, Cole NB, Gottschalk W, McIlwain KL, Orrison B, Chen A, Ellis CE, Paylor R, Lu B, Nussbaum RL (2002) Synaptic vesicle depletion correlates with attenuated synaptic responses to prolonged repetitive stimulation in mice lacking alpha-synuclein. J Neurosci 22:8797–8807

Capellari S, Parchi P, Russo CM, Sanford J, Sy MS, Gambetti P, Petersen RB (2000a) Effect of the E200K mutation on prion protein metabolism. Comparative study of a cell model and human brain. Am J Pathol 157:613–622

Capellari S, Zaidi SI, Long AC, Kwon EE, Petersen RB (2000b) The Thr183Ala mutation, not the loss of the first glycosylation site, alters the physical properties of the prion protein. J Alzheimers Dis 2:27–35

Capellari S, Zaidi SI, Urig CB, Perry G, Smith MA, Petersen RB (1999) Prion protein glycosylation is sensitive to redox change [published erratum appears in J Biol Chem 2000 275:11538]. J Biol Chem 274:34846–34850

Caughey B, Raymond GJ (1991) The scrapie-associated form of PrP is made from a cell surface precursor that is both protease- and phospholipase-sensitive. J Biol Chem 266:18217–18223

Caughey BW, Dong A, Bhat KS, Ernst D, Hayes SF, Caughey WS (1991) Secondary structure analysis of the scrapie-associated protein PrP 27–30 in water by infrared spectroscopy. Biochemistry 30:7672–7680

Chan HY, Warrick JM, Gray-Board GL, Paulson HL, Bonini NM (2000) Mechanisms of chaperone suppression of polyglutamine disease: selectivity, synergy and modulation of protein solubility in Drosophila. Hum Mol Genet 9:2811–2820

Chandra S, Fornai F, Kwon HB, Yazdani U, Atasoy D, Liu X, Hammer RE, Battaglia G, German DC, Castillo PE, Sudhof TC (2004) Double-knockout mice for alpha- and beta-synucleins: effect on synaptic functions. Proc Natl Acad Sci U S A 101:14966–14971. Epub 2004 Oct 01

Chavany C, Mimnaugh E, Miller P, Bitton R, Nguyen P, Trepel J, Whitesell L, Schnur R, Moyer J, Neckers L (1996) p185erbB2 binds to GRP94 in vivo. Dissociation of the p185erbB2/GRP94 heterocomplex by benzoquinone ansamycins precedes depletion of p185erbB2. J Biol Chem 271:4974–4977

Chiesa R, Piccardo P, Ghetti B, Harris DA (1998) Neurological illness in transgenic mice expressing a prion protein with an insertional mutation. Neuron 21:1339–1351

Chung KK, Thomas B, Li X, Pletnikova O, Troncoso JC, Marsh L, Dawson VL, Dawson TM (2004) S-nitrosylation of parkin regulates ubiquitination and compromises parkin's protective function. Science 304:1328–1331. Epub 2004 Apr 22

Ciechanover A, Brundin P (2003) The ubiquitin proteasome system in neurodegenerative diseases: sometimes the chicken, sometimes the egg. Neuron 40:427–446

Collinge J (2001) Prion diseases of humans and animals: their causes and molecular basis. Annu Rev Neurosci 24:519–550

Collinge J, Sidle KC, Meads J, Ironside J, Hill AF (1996) Molecular analysis of prion strain variation and the aetiology of 'new variant' CJD [see comments]. Nature 383:685–690

Conway KA, Lee SJ, Rochet JC, Ding TT, Williamson RE, Lansbury PT Jr (2000) Acceleration of oligomerization, not fibrillization, is a shared property of both alpha-synuclein mutations linked to early-onset Parkinson's disease: implications for pathogenesis and therapy. Proc Natl Acad Sci U S A 97:571–576

Conway KA, Rochet JC, Bieganski RM, Lansbury PT Jr (2001) Kinetic stabilization of the alpha-synuclein protofibril by a dopamine-alpha-synuclein adduct. Science 294:1346–1349

Cookson MR, Lockhart PJ, McLendon C, O'Farrell C, Schlossmacher M, Farrer MJ (2003) RING finger 1 mutations in Parkin produce altered localization of the protein. Hum Mol Genet 12:2957–2965. Epub 2003 Sep 30

Cuervo AM, Stefanis L, Fredenburg R, Lansbury PT, Sulzer D (2004) Impaired degradation of mutant alpha-synuclein by chaperone-mediated autophagy. Science 305:1292–1295

Cummings CJ, Sun Y, Opal P, Antalffy B, Mestril R, Orr HT, Dillmann WH, Zoghbi HY (2001) Over-expression of inducible HSP70 chaperone suppresses neuropathology and improves motor function in SCA1 mice. Hum Mol Genet 10:1511–1518

Darios F, Corti O, Lucking CB, Hampe C, Muriel MP, Abbas N, Gu WJ, Hirsch EC, Rooney T, Ruberg M, Brice A (2003) Parkin prevents mitochondrial swelling and cytochrome c release in mitochondria-dependent cell death. Hum Mol Genet 12:517–526

Dauer W, Przedborski S (2003) Parkinson's disease: mechanisms and models. Neuron 39:889–909

Davidson WS, Jonas A, Clayton DF, George JM (1998) Stabilization of alpha-synuclein secondary structure upon binding to synthetic membranes. J Biol Chem 273:9443–9449

DeArmond SJ, Sanchez H, Yehiely F, Qiu Y, Ninchak-Casey A, Daggett V, Camerino AP, Cayetano J, Rogers M, Groth D, Torchia M, Tremblay P, Scott MR, Cohen FE, Prusiner SB (1997) Selective neuronal targeting in prion disease. Neuron 19:1337–1348

DebBurman SK, Raymond GJ, Caughey B, Lindquist S (1997) Chaperone-supervised conversion of prion protein to its protease-resistant form. Proc Natl Acad Sci U S A 94:13938–13943

Donaldson KM, Li W, Ching KA, Batalov S, Tsai CC, Joazeiro CA (2003) Ubiquitin-mediated sequestration of normal cellular proteins into polyglutamine aggregates. Proc Natl Acad Sci U S A 100:8892–8897. Epub 2003 Jul 11

Dong Z, Wolfer DP, Lipp HP, Bueler H (2005) Hsp70 gene transfer by adeno-associated virus inhibits MPTP-induced nigrostriatal degeneration in the mouse model of Parkinson disease. Mol Ther 11:80–88

Donne DG, Viles JH, Groth D, Mehlhorn I, James TL, Cohen FE, Prusiner SB, Wright PE, Dyson HJ (1997) Structure of the recombinant full-length hamster prion protein PrP(29–231): the N terminus is highly flexible. Proc Natl Acad Sci U S A 94:13452–13457

Dou F, Netzer WJ, Tanemura K, Li F, Hartl FU, Takashima A, Gouras GK, Greengard P, Xu H (2003) Chaperones increase association of tau protein with microtubules. Proc Natl Acad Sci U S A 100:721–726. Epub 2003 Jan 09

Drisaldi B, Stewart RS, Adles C, Stewart LR, Quaglio E, Biasini E, Fioriti L, Chiesa R, Harris DA (2003) Mutant PrP is delayed in its exit from the neither wild-type nor mutant PrP undergoes proteasomal degradation. J Biol Chem 278:21732–21743

Edenhofer F, Rieger R, Famulok M, Wendler W, Weiss S, Winnacker E-L (1996) Prion protein PrP^C interacts with molecular chaperones of the Hsp60 family. J Virol 70:4724–4728

Edington BV, Whelan SA, Hightower LE (1989) Inhibition of heat shock (stress) protein induction by deuterium oxide and glycerol: additional support for the abnormal protein hypothesis of induction. J Cell Physiol 139:219–228

Eliezer D, Kutluay E, Bussell R Jr, Browne G (2001) Conformational properties of alpha-synuclein in its free and lipid-associated states. J Mol Biol 307:1061–1073

Ellgaard L, Helenius A (2003) Quality control in the endoplasmic reticulum. Nat Rev Mol Cell Biol 4:181–191

Endo T, Groth D, Prusiner SB, Kobata A (1989) Diversity of oligosaccharide structures linked to asparagines of the scrapie prion protein. Biochemistry 28:8380–8388

Farrer M, Chan P, Chen R, Tan L, Lincoln S, Hernandez D, Forno L, Gwinn-Hardy K, Petrucelli L, Hussey J, Singleton A, Tanner C, Hardy J, Langston JW (2001) Lewy bodies and parkinsonism in families with parkin mutations. Ann Neurol 50:293–300

Farrer M, Kachergus J, Forno L, Lincoln S, Wang DS, Hulihan M, Maraganore D, Gwinn-Hardy K, Wszolek Z, Dickson D, Langston JW (2004) Comparison of kindreds with parkinsonism and alpha-synuclein genomic multiplications. Ann Neurol 55:174–179

Feany MB, Bender WW (2000) A Drosophila model of Parkinson's disease. Nature 404:394–398

Fernagut PO, Chesselet MF (2004) Alpha-synuclein and transgenic mouse models. Neurobiol Dis 17:123–130

Finley D, Ciechanover A, Varshavsky A (1984) Thermolability of ubiquitin-activating enzyme from the mammalian cell cycle mutant ts85. Cell 37:43–55

Flechsig E, Hegyi I, Leimeroth R, Zuniga A, Rossi D, Cozzio A, Schwarz P, Rulicke T, Gotz J, Aguzzi A, Weissmann C (2003) Expression of truncated PrP targeted to Purkinje cells of PrP knockout mice causes Purkinje cell death and ataxia. EMBO J 22:3095–3101

Fonte V, Kapulkin V, Taft A, Fluet A, Friedman D, Link CD (2002) Interaction of intracellular beta amyloid peptide with chaperone proteins. Proc Natl Acad Sci U S A 99:9439–9444. Epub 2002 Jun 27

Fornai F, Lenzi P, Gesi M, Ferrucci M, Lazzeri G, Busceti CL, Ruffoli R, Soldani P, Ruggieri S, Alessandri MG, Paparelli A (2003) Fine structure and biochemical mechanisms underlying nigrostriatal inclusions and cell death after proteasome inhibition. J Neurosci 23:8955–8966

Fournier JG, Escaig-Haye F, Billette de Villemeur T, Robain O (1995) Ultrastructural localization of cellular prion protein (PrPc) in synaptic boutons of normal hamster hippocampus. C R Acad Sci III 318:339–344

Gambetti P, Kong Q, Zou W, Parchi P, Chen SG (2003) Sporadic and familial CJD: classification and characterisation. Br Med Bull 66:213–239

Gekko K, Koga S (1983) Increased thermal stability of collagen in the presence of sugars and polyols. J Biochem 94:199–208

Gekko K, Timasheff SN (1981a) Mechanism of protein stabilization by glycerol: preferential hydration in glycerol-water mixtures. Biochemistry 20:4667–4676

Gekko K, Timasheff SN (1981b) Thermodynamic and kinetic examination of protein stabilization by glycerol. Biochemistry 20:4677–4686

Gerlsma SY, Stuur ER (1972) The effects of combining urea and an alcohol on the heat-induced reversible denaturation of ribonuclease. Int J Pept Proteins Res 4:372–378

Giasson BI, Murray IV, Trojanowski JQ, Lee VM (2001) A hydrophobic stretch of 12 amino acid residues in the middle of alpha-synuclein is essential for filament assembly. J Biol Chem 276:2380–2386. Epub 2000 Nov 01

Gu WJ, Corti O, Araujo F, Hampe C, Jacquier S, Lucking CB, Abbas N, Duyckaerts C, Rooney T, Pradier L, Ruberg M, Brice A (2003) The C289G and C418R missense mutations cause rapid sequestration of human Parkin into insoluble aggregates. Neurobiol Dis 14:357–364

Gu Y, Singh N (2004) Doxycycline and protein folding agents rescue the abnormal phenotype of familial CJD H187R in a cell model. Brain Res Mol Brain Res 123:37–44

Gu Y, Verghese S, Mishra RS, Xu X, Shi Y, Singh N (2002) Mutant prion protein-mediated aggregation of normal prion protein in the endoplasmic reticulum: implications for prion propagation and neurotoxicity. J Neurochem 84:10–22

Hansson O, Nylandsted J, Castilho RF, Leist M, Jaattela M, Brundin P (2003) Overexpression of heat shock protein 70 in R6/2 Huntington's disease mice has only modest effects on disease progression. Brain Res 970:47–57

Hartl FU, Hayer-Hartl M (2002) Molecular chaperones in the cytosol: from nascent chain to folded protein. Science 295:1852–1858

Hay DG, Sathasivam K, Tobaben S, Stahl B, Marber M, Mestril R, Mahal A, Smith DL, Woodman B, Bates GP (2004) Progressive decrease in chaperone protein levels in a mouse model of Huntington's disease and induction of stress proteins as a therapeutic approach. Hum Mol Genet 13:1389–1405. Epub 2004 Apr 28

Hedrich K, Djarmati A, Schafer N, Hering R, Wellenbrock C, Weiss PH, Hilker R, Vieregge P, Ozelius LJ, Heutink P, Bonifati V, Schwinger E, Lang AE, Noth J, Bressman SB, Pramstaller PP, Riess O, Klein C (2004) DJ-1 (PARK7) mutations are less frequent than Parkin (PARK2) mutations in early-onset Parkinson disease. Neurology 62:389–394

Hegde RS, Mastrianni JA, Scott MR, DeFea KA, Tremblay P, Torchia M, DeArmond SJ, Prusiner SB, Lingappa VR (1998) A transmembrane form of the prion protein in neurodegenerative disease. Science 279:827–834

Heller U, Winklhofer KF, Heske J, Reintjes A, Tatzelt J (2003) Post-translational import of the prion protein into the endoplasmic reticulum interferes with cell viability: a critical role for the putative transmembrane domain. J Biol Chem 278:36139–36147

Henn IH, Gostner JM, Tatzelt J, Winklhofer KF (2005) Pathogenic mutations inactivate parkin by distinct mechanisms. J Neurochem 92:114–122

Herms J, Tings T, Gall S, Madlung A, Giese A, Siebert H, Schurmann P, Windl O, Brose N, Kretzschmar H (1999) Evidence of presynaptic location and function of the prion protein. J Neurosci 19:8866–8875

Heske J, Heller U, Winklhofer KF, Tatzelt J (2004) The C-terminal domain of the prion protein is necessary and sufficient for import into the endoplasmic reticulum. J Biol Chem 279:5435–5443

Higashi Y, Asanuma M, Miyazaki I, Hattori N, Mizuno Y, Ogawa N (2004) Parkin attenuates manganese-induced dopaminergic cell death. J Neurochem 89:1490–1497

Hod Y, Pentyala SN, Whyard TC, El-Maghrabi MR (1999) Identification and characterization of a novel protein that regulates RNA-protein interaction. J Cell Biochem 72:435–444

Holmberg CI, Staniszewski KE, Mensah KN, Matouschek A, Morimoto RI (2004) Inefficient degradation of truncated polyglutamine proteins by the proteasome. EMBO J 23:4307–4318. Epub 2004 Oct 07

Honbou K, Suzuki NN, Horiuchi M, Niki T, Taira T, Ariga H, Inagaki F (2003) The crystal structure of DJ-1, a protein related to male fertility and Parkinson's disease. J Biol Chem 278:31380–1384. Epub 2003 Jun 08

Hurtley SM, Helenius A (1989) Protein oligomerization in the endoplasmic reticulum. Annu Rev Cell Biol 5:277–307

Imai Y, Soda M, Hatakeyama S, Akagi T, Hashikawa T, Nakayama KI, Takahashi R (2002) CHIP is associated with Parkin, a gene responsible for familial Parkinson's disease, and enhances its ubiquitin ligase activity. Mol Cell 10:55–67

Imai Y, Soda M, Inoue H, Hattori N, Mizuno Y, Takahashi R (2001) An unfolded putative transmembrane polypeptide, which can lead to endoplasmic reticulum stress, is a substrate of parkin. Cell 105:891–902

Imai Y, Takahashi R (2004) How do Parkin mutations result in neurodegeneration? Curr Opin Neurobiol 14:384–389

Ivanova L, Barmada S, Kummer T, Harris DA (2001) Mutant prion proteins are partially retained in the endoplasmic reticulum. J Biol Chem 276:42409–42421

Jana NR, Zemskov EA, Wang G, Nukina N (2001) Altered proteasomal function due to the expression of polyglutamine-expanded truncated N-terminal huntingtin induces apoptosis by caspase activation through mitochondrial cytochrome c release. Hum Mol Genet 10:1049–1059

Jentsch S, McGrath JP, Varshavsky A (1987) The yeast DNA repair gene RAD6 encodes a ubiquitin-conjugating enzyme. Nature 329:131–134

Jiang H, Ren Y, Zhao J, Feng J (2004) Parkin protects human dopaminergic neuroblastoma cells against dopamine-induced apoptosis. Hum Mol Genet 13:1745–1754. Epub 2004 Jun 15

Jin T, Gu Y, Zanusso G, Sy M, Kumar A, Cohen M, Gambetti P, Singh N (2000) The chaperone protein BiP binds to a mutant prion protein and mediates its degradation by the proteasome. J Biol Chem 275:38699–38704

Junn E, Lee SS, Suhr UT, Mouradian MM (2002) Parkin accumulation in aggresomes due to proteasome impairment. J Biol Chem 277:47870–47877. Epub 2002 Oct 2

Kahle PJ, Haass C (2004) How does parkin ligate ubiquitin to Parkinson's disease? EMBO Rep 5:681–685

Kalia SK, Lee S, Smith PD, Liu L, Crocker SJ, Thorarinsdottir TE, Glover JR, Fon EA, Park DS, Lozano AM (2004) BAG5 inhibits parkin and enhances dopaminergic neuron degeneration. Neuron 44:931–945

Kenward N, Hope J, Landon M, Mayer RJ (1994) Expression of polyubiquitin and heat-shock protein 70 genes increases in the later stages of disease progression in scrapie-infected mouse brain. J Neurochem 62:1870–1877

Kermer P, Digicaylioglu MH, Kaul M, Zapata JM, Krajewska M, Stenner-Liewen F, Takayama S, Krajewski S, Lipton SA, Reed JC (2003) BAG1 over-expression in brain protects against stroke. Brain Pathol 13:495–506

Kiachopoulos S, Heske J, Tatzelt J, Winklhofer KF (2004) Misfolding of the prion protein at the plasma membrane induces endocytosis, intracellular retention and degradation. Traffic 5:426–436

Kiachopoulos S, Bracher A, Winklhofer KF, Tatzelt J (2005) Pathogenic mutations destabilize the globular domain of the prion protein and interfere with the attachment of the glycosylphosphatidylinositol anchor. J Biol Chem 280:9320–9329

Kim TD, Paik SR, Yang CH, Kim J (2000) Structural changes in alpha-synuclein affect its chaperone-like activity in vitro. Protein Sci 9:2489–2496

Kitada T, Asakawa S, Hattori N, Matsumine H, Yamamura Y, Minoshima S, Yokochi M, Mizuno Y, Shimizu N (1998) Mutations in the parkin gene cause autosomal parkinsonism. Nature 392:605–608

Klausner RD, Sitia R (1990) Protein degradation in the endoplasmic reticulum. Cell 62:611–614

Klucken J, Shin Y, Masliah E, Hyman BT, McLean PJ (2004) Hsp70 reduces alpha-synuclein aggregation and toxicity. J Biol Chem 279:25497–25502. Epub 2004 Mar 25

Kocisko DA, Come JH, Priola SA, Chesebro B, Raymond GJ, Lansbury PT Jr, Caughey B (1994) Cell-free formation of protease-resistant prion protein. Nature 370:471–474

Korth C, Kaneko K, Prusiner SB (2000) Expression of unglycosylated mutated prion protein facilitates PrPSc formation in neuroblastoma cells infected with different prion strains. J Gen Virol 81:2555–2563

Kovacs GG, Kurucz I, Budka H, Adori C, Muller F, Acs P, Kloppel S, Schatzl HM, Mayer RJ, Laszlo L (2001) Prominent stress response of Purkinje cells in Creutzfeldt-Jakob disease. Neurobiol Dis 8:881–889

Kruger R, Kuhn W, Muller T, Woitalla D, Graeber M, Kosel S, Przuntek H, Epplen JT, Schols L, Riess O (1998) Ala30Pro mutation in the gene encoding alpha-synuclein in Parkinson's disease. Nat Genet 18:106–108

Lee HJ, Khoshaghideh F, Patel S, Lee SJ (2004) Clearance of alpha-synuclein oligomeric intermediates via the lysosomal degradation pathway. J Neurosci 24:1888–1896

Lee SJ, Kim SJ, Kim IK, Ko J, Jeong CS, Kim GH, Park C, Kang SO, Suh PG, Lee HS, Cha SS (2003) Crystal structures of human DJ-1 and Escherichia coli Hsp31, which share an evolutionarily conserved domain. J Biol Chem 278:44552–44559. Epub 2003 Aug 25

Leffers KW, Schell J, Jansen K, Lucassen R, Kaimann T, Nagel-Steger L, Tatzelt J, Riesner D (2004) The structural transition of the prion protein into its pathogenic conformation is induced by unmasking hydrophobic sites. J Mol Biol 344:839–853

Legname G, Baskakov IV, Nguyen HO, Riesner D, Cohen FE, DeArmond SJ, Prusiner SB (2004) Synthetic mammalian prions. Science 305:673–676

Lehmann S, Harris DA (1995) A mutant prion protein displays an aberrant membrane association when expressed in cultured cells. J Biol Chem 270:24589–24597

Lehmann S, Harris DA (1997) Blockade of glycosylation promotes acquisition of scrapie-like properties by the prion protein in cultured cells [published erratum appears in J Biol Chem 1998 273:5988]. J Biol Chem 272:21479–1487

Li W, Lesuisse C, Xu Y, Troncoso JC, Price DL, Lee MK (2004) Stabilization of alpha-synuclein protein with aging and familial Parkinson's disease-linked A53T mutation. J Neurosci 24:7400–7409

Lin PS, Kwock L, Hefter K (1981) Protection of heat induced cytotoxicity by glycerol. J Cell Physiol 108:439–448

Lindersson E, Beedholm R, Hojrup P, Moos T, Gai W, Hendil KB, Jensen PH (2004) Proteasomal inhibition by alpha-synuclein filaments and oligomers. J Biol Chem 279:12924–12934. Epub 2004 Jan 07

Lindquist S, Craig EA (1988) The heat-shock proteins. Annu Rev Genet 22:631–677

Liu S, Ninan I, Antonova I, Battaglia F, Trinchese F, Narasanna A, Kolodilov N, Dauer W, Hawkins RD, Arancio O (2004) Alpha-synuclein produces a long-lasting increase in neurotransmitter release. EMBO J 23:4506–4516. Epub 2004 Oct 28

Lo Bianco C, Schneider BL, Bauer M, Sajadi A, Brice A, Iwatsubo T, Aebischer P (2004) Lentiviral vector delivery of parkin prevents dopaminergic degeneration in an alpha-synuclein rat model of Parkinson's disease. Proc Natl Acad Sci U S A 101:17510–17515. Epub 2004 Dec 02

Lopez CD, Yost CS, Prusiner SB, Myers RM, Lingappa VR (1990) Unusual topogenic sequence directs prion protein biogenesis. Science 248:226–229

Lorenz H, Windl O, Kretzschmar H (2002) Cellular phenotyping of secretory and nuclear prion proteins associated with inherited prion diseases. J Biol Chem 277:8508–8516

Ma J, Lindquist S (2001) Wild-type PrP and a mutant associated with prion disease are subject to retrograde transport and proteasome degradation. Proc Natl Acad Sci U S A 98:14955–14960

Ma J, Wollmann R, Lindquist S (2002) Neurotoxicity and neurodegeneration when PrP accumulates in the cytosol. Science 298:1781–1785

Ma Y, Hendershot LM (2001) The unfolding tale of the unfolded protein response. Cell 107:827–830

Magrane J, Smith RC, Walsh K, Querfurth HW (2004) Heat shock protein 70 participates in the neuroprotective response to intracellularly expressed beta-amyloid in neurons. J Neurosci 24:1700–1706

Mallucci G, Dickinson A, Linehan J, Klohn PC, Brandner S, Collinge J (2003) Depleting neuronal PrP in prion infection prevents disease and reverses spongiosis. Science 302:871–874

Marella M, Lehmann S, Grassi J, Chabry J (2002) Filipin prevents pathological prion protein accumulation by reducing endocytosis and inducing cellular PrP release. J Biol Chem 277:25457–25464

Maries E, Dass B, Collier TJ, Kordower JH, Steece-Collier K (2003) The role of alpha-synuclein in Parkinson's disease: insights from animal models. Nat Rev Neurosci 4:727–738

Maroteaux L, Campanelli JT, Scheller RH (1988) Synuclein: a neuron-specific protein localized to the nucleus and presynaptic nerve terminal. J Neurosci 8:2804–2815

Martin-Clemente B, Alvarez-Castelao B, Mayo I, Sierra AB, Diaz V, Milan M, Farinas I, Gomez-Isla T, Ferrer I, Castano JG (2004) alpha-Synuclein expression levels do not significantly affect proteasome function and expression in mice and stably transfected PC12 cell lines. J Biol Chem 279:52984–52990. Epub 2004 Oct 04

Martinat C, Shendelman S, Jonason A, Leete T, Beal MF, Yang L, Floss T, Abeliovich A (2004) Sensitivity to oxidative stress in DJ-1-deficient dopamine neurons: an ES-derived cell model of primary parkinsonism. PLoS Biol 2:e327. Epub 2004 Oct 05

Mata IF, Lockhart PJ, Farrer MJ (2004) Parkin genetics: one model for Parkinson's disease. Hum Mol Genet 13: R127–133. Epub 2004 Feb 19

Mayhew M, da Silva AC, Martin J, Erdjument-Bromage H, Tempst P, Hartl FU (1996) Protein folding in the central cavity of the GroEL-GroES chaperonin complex. Nature 379:420–426

McCracken AA, Brodsky JL (2003) Evolving questions and paradigm shifts in endoplasmic-reticulum-associated degradation (ERAD). Bioessays 25:868–877

McNaught KS, Mytilineou C, Jnobaptiste R, Yabut J, Shashidharan P, Jennert P, Olanow CW (2002) Impairment of the ubiquitin-proteasome system causes dopaminergic cell death and inclusion body formation in ventral mesencephalic cultures. J Neurochem 81:301–306

Mengesdorf T, Jensen PH, Mies G, Aufenberg C, Paschen W (2002) Down-regulation of parkin protein in transient focal cerebral ischemia: a link between stroke and degenerative disease? Proc Natl Acad Sci U S A 99:15042–15047

Meyer RK, McKinley MP, Bowman KA, Braunfeld MB, Barry RA, Prusiner SB (1986) Separation and properties of cellular and scrapie prion proteins. Proc Natl Acad Sci U S A 83:2310–2314

Mironov AJ, Latawiec D, Wille H, Bouzamondo-Bernstein E, Legname G, Williamson RA, Burton D, DeArmond SJ, Prusiner SB, Peters PJ (2003) Cytosolic prion protein in neurons. J Neurosci 23:7183–7193

Mishra RS, Gu Y, Bose S, Verghese S, Kalepu S, Singh N (2002) Cell surface accumulation of a truncated transmembrane prion protein in Gerstmann-Straussler-Scheinker disease P102L. J Biol Chem 277:24554–24561

Mitsumoto A, Nakagawa Y, Takeuchi A, Okawa K, Iwamatsu A, Takanezawa Y (2001) Oxidized forms of peroxiredoxins and DJ-1 on two-dimensional gels increased in response to sublethal levels of paraquat. Free Radic Res 35:301–310

Muchowski PJ, Schaffar G, Sittler A, Wanker EE, Hayer-Hartl MK, Hartl FU (2000) Hsp70 and hsp40 chaperones can inhibit self-assembly of polyglutamine proteins into amyloid-like fibrils. Proc Natl Acad Sci U S A 97:7841–7846

Muchowski PJ, Wacker JL (2005) Modulation of neurodegeneration by molecular chaperones. Nat Rev Neurosci 6:11–22

Muqit MM, Davidson SM, Payne Smith MD, MacCormac LP, Kahns S, Jensen PH, Wood NW, Latchman DS (2004) Parkin is recruited into aggresomes in a stress-specific manner: over-expression of parkin reduces aggresome formation but can be dissociated from parkin's effect on neuronal survival. Hum Mol Genet 13:117–135. Epub 2003 Nov 25

Muramoto T, DeArmond SJ, Scott M, Telling GC, Cohen FE, Prusiner SB (1997) Heritable disorder resembling neuronal storage disease in mice expressing prion protein with deletion of an α-helix. Nat Med 3:750–755

Nagakubo D, Taira T, Kitaura H, Ikeda M, Tamai K, Iguchi-Ariga SM, Ariga H (1997) DJ-1, a novel oncogene which transforms mouse NIH3T3 cells in cooperation with ras. Biochem Biophys Res Commun 231:509–513

Neckers L (2002) Hsp90 inhibitors as novel cancer chemotherapeutic agents. Trends Mol Med 6: S55–S61

Negro A, Ballarin C, Bertoli A, Massimino ML, Sorgato MC (2001) The metabolism and imaging in live cells of the bovine prion protein in its native form or carrying single amino acid substitutions. Mol Cell Neurosci 17:521–538

Niki T, Takahashi-Niki K, Taira T, Iguchi-Ariga SM, Ariga H (2003) DJBP: a novel DJ-1-binding protein, negatively regulates the androgen receptor by recruiting histone deacetylase complex, and DJ-1 antagonizes this inhibition by abrogation of this complex. Mol Cancer Res 1:247–261

Nollen EA, Brunsting JF, Song J, Kampinga HH, Morimoto RI (2000) Bag1 functions in vivo as a negative regulator of Hsp70 chaperone activity. Mol Cell Biol 20:1083–1088

Nunziante M, Gilch S, Schätzl H (2003) Essential role of the prion protein N terminus in subcellular trafficking and half-life of cellular prion protein. J Biol Chem 278:3726–3734

Ochel HJ, Gademann G, Trepel J, Neckers L (2003) Modulation of prion protein structural integrity by geldanamycin. Glycobiology 13:655–660

Okochi M, Walter J, Koyama A, Nakajo S, Baba M, Iwatsubo T, Meijer L, Kahle PJ, Haass C (2000) Constitutive phosphorylation of the Parkinson's disease associated alpha-synuclein. J Biol Chem 275:390–397

Pan K-M, Baldwin M, Nguyen J, Gasset M, Serban A, Groth D, Mehlhorn I, Huang Z, Fletterick RJ, Cohen FE, Prusiner SB (1993) Conversion of α-helices into β-sheets features in the formation of the scrapie prion proteins. Proc Natl Acad Sci U S A 90:10962–10966

Parchi P, Castellani R, Capellari S, Ghetti B, Young K, Chen SG, Farlow M, Dickson DW, Sima AAF, Trojanowski JQ, Petersen RB, Gambetti P (1996) Molecular basis of phenotypic variability in sporadic Creutzfeldt-Jakob disease. Ann Neurol 39:767–778

Park SM, Jung HY, Kim TD, Park JH, Yang CH, Kim J (2002) Distinct roles of the N-terminal-binding domain and the C-terminal-solubilizing domain of alpha-synuclein, a molecular chaperone. J Biol Chem 277:28512–28520. Epub 2002 May 24

Patil C, Walter P (2001) Intracellular signaling from the endoplasmic reticulum to the nucleus: the unfolded protein response in yeast and mammals. Curr Opin Cell Biol 13:349–355

Peters PJ, Mironov AJ, Peretz D, Van Donselaar E, Leclerc E, Erpel S, DeArmond SJ, Burton DR, Williamson RA, Vey M, Prusiner SB (2003) Trafficking of prion proteins through a caveolae-mediated endosomal pathway. J Cell Biol 162:703–717

Petersen RB, Parchi P, Richardson SL, Urig CB, Gambetti P (1996) Effect of the D178N mutation and the codon 129 polymorphism on the metabolism of the prion protein. J Biol Chem 271:12661–12668

Petrucelli L, Dickson D, Kehoe K, Taylor J, Snyder H, Grover A, De Lucia M, McGowan E, Lewis J, Prihar G, Kim J, Dillmann WH, Browne SE, Hall A, Voellmy R, Tsuboi Y, Dawson TM, Wolozin B, Hardy J, Hutton M (2004) CHIP and Hsp70 regulate tau ubiquitination, degradation and aggregation. Hum Mol Genet 13:703–714. Epub 2004 Feb 12

Petrucelli L, O'Farrell C, Lockhart PJ, Baptista M, Kehoe K, Vink L, Choi P, Wolozin B, Farrer M, Hardy J, Cookson MR (2002) Parkin protects against the toxicity associated with mutant alpha-synuclein: proteasome dysfunction selectively affects catecholaminergic neurons. Neuron 36:1007–1019

Polymeropoulos MH, Lavedan C, Leroy E, Ide SE, Dehejia A, Dutra A, Pike B, Root H, Rubenstein J, Boyer R, Stenroos ES, Chandrasekharappa S, Athanassiadou A, Papapetropoulos T, Johnson WG, Lazzarini AM, Duvoisin RC, Di Iorio G, Golbe LI, Nussbaum RL (1997) Mutation in the alpha-synuclein gene identified in families with Parkinson's disease. Science 276:2045–2047

Priola SA, Lawson VA (2001) Glycosylation influences cross-species formation of protease-resistant prion protein. EMBO J 20:6692–6699

Prusiner SB (1982) Novel proteinaceous infectious particles cause scrapie. Science 216:136–144

Prusiner SB, Scott M, Foster D, Pan K-M, Groth D, Mirenda C, Torchia M, Yang S-L, Serban D, Carlson GA, Hoppe PC, Westaway D, DeArmond SJ (1990) Transgenetic studies implicate interactions between homologous PrP isoforms in scrapie prion replication. Cell 63:673–686

Prusiner SB, Scott MR, DeArmond SJ, Cohen FE (1998) Prion protein biology. Cell 93:337–348

Rideout HJ, Larsen KE, Sulzer D, Stefanis L (2001) Proteasomal inhibition leads to formation of ubiquitin/alpha-synuclein-immunoreactive inclusions in PC12 cells. J Neurochem 78:899–908

Riek R, Hornemann S, Wider G, Billeter M, Glockshuber R, Wuthrich K (1996) NMR structure of the mouse prion protein domain PrP(121-321). Nature 382:180–182

Riek R, Hornemann S, Wider G, Glockshuber R, Wuthrich K (1997) NMR characterization of the full-length recombinant murine prion protein, mPrP(23-231). FEBS Lett 413:282–288

Rogers M, Taraboulos A, Scott M, Groth D, Prusiner SB (1990) Intracellular accumulation of the cellular prion protein after mutagenesis of its Asn-linked glycosylation sites. Glycobiology 1:101–109

Rogers M, Yehiely F, Scott M, Prusiner SB (1993) Conversion of truncated and elongated prion proteins into the scrapie isoform in cultured cells. Proc Natl Acad Sci U S A 90:3182–3186

Rudd PM, Endo T, Colominas C, Groth D, Wheeler SF, Harvey DJ, Wormald MR, Serban H, Prusiner SB, Kobata A, Dwek RA (1999) Glycosylation differences between the normal and pathogenic prion protein isoforms. Proc Natl Acad Sci U S A 96:13044–13049

Sarnataro D, Paladino S, Campana V, Grassi J, Nitsch L, Zurzolo C (2002) PrPC is sorted to the basolateral membrane of epithelial cells independently of its association with rafts. Traffic 3:810–21

Sasaki S, Shirata A, Yamane K, Iwata M (2004) Parkin-positive autosomal recessive juvenile parkinsonism with alpha-synuclein-positive inclusions. Neurology 63:678–682

Schaffar G, Breuer P, Boteva R, Behrends C, Tzvetkov N, Strippel N, Sakahira H, Siegers K, Hayer-Hartl M, Hartl FU (2004) Cellular toxicity of polyglutamine expansion proteins: mechanism of transcription factor deactivation. Mol Cell 15:95–105

Schirmer EC, Lindquist S (1997) Interactions of the chaperone Hsp104 with yeast Sup35 and mammalian PrP. Proc Natl Acad Sci U S A 94:13932–13937

Seidberg NA, Clark RS, Zhang X, Lai Y, Chen M, Graham SH, Kochanek PM, Watkins SC, Marion DW (2003) Alterations in inducible 72-kDa heat shock protein and the chaperone cofactor BAG-1 in human brain after head injury. J Neurochem 84:514–521

Shaked GM, Engelstein R, Avraham I, Kahana E, Gabizon R (2003) Dimethyl sulfoxide delays PrP sc accumulation and disease symptoms in prion-infected hamsters. Brain Res 983:137–143

Shendelman S, Jonason A, Martinat C, Leete T, Abeliovich A (2004) DJ-1 is a redox-dependent molecular chaperone that inhibits alpha-synuclein aggregate formation. PLoS Biol 2:e362. Epub 2004 Oct 05

Shimura H, Miura-Shimura Y, Kosik KS (2004a) Binding of tau to heat shock protein 27 leads to decreased concentration of hyperphosphorylated tau and enhanced cell survival. J Biol Chem 279:17957–17962. Epub 2004 Feb 12

Shimura H, Schwartz D, Gygi SP, Kosik KS (2004b) CHIP-Hsc70 complex ubiquitinates phosphorylated tau and enhances cell survival. J Biol Chem 279:4869–4876. Epub 2003 Nov 10

Shmerling D, Hegyi I, Fischer M, Blättler T, Brandner S, Götz J, Rülicke T, Flechsig E, Cozzio A, von Mehring C, Hangartner C, Aguzzi A, Weissmann C (1998) Expression of animo-terminally truncated PrP in the mouse leading to ataxia and specific cerebellar lesions. Cell 93:203–214

Shyng SL, Heuser JE, Harris DA (1994) A glycolipid-anchored prion protein is endocytosed via clathrin-coated pits. J Cell Biol 125:1239–1250

Shyng SL, Huber MT, Harris DA (1993) A prion protein cycles between the cell surface and an endocytic compartment in cultured neuroblastoma cells. J Biol Chem 268:15922–15928

Shyng SL, Moulder KL, Lesko A, Harris DA (1995) The N-terminal domain of a glycolipid-anchored prion protein is essential for its endocytosis via clathrin-coated pits. J Biol Chem 270:14793–14800

Shyu WC, Harn HJ, Saeki K, Kubosaki A, Matsumoto Y, Onodera T, Chen CJ, Hsu YD, Chiang YH (2002) Molecular modulation of expression of prion protein by heat shock. Mol Neurobiol 26:1–12

Shyu WC, Kao MC, Chou WY, Hsu YD, Soong BW (2000) Heat shock modulates prion protein expression in human NT-2 cells. Neuroreport 11:771–774

Singh N, Zanusso G, Chen SG, Fujioka H, Richardson S, Gambetti P, Petersen RB (1997) Prion protein aggregation reverted by low temperature in transfected cells carrying a prion protein gene mutation. J Biol Chem 272:28461–28470

Singleton AB, Farrer M, Johnson J, Singleton A, Hague S, Kachergus J, Hulihan M, Peuralinna T, Dutra A, Nussbaum R, Lincoln S, Crawley A, Hanson M, Maraganore D, Adler C, Cookson MR, Muenter M, Baptista M, Miller D, Blancato J, Hardy J, Gwinn-Hardy K (2003) alpha-synuclein locus triplication causes Parkinson's disease. Science 302:841

Snyder H, Mensah K, Theisler C, Lee J, Matouschek A, Wolozin B (2003) Aggregated and monomeric alpha-synuclein bind to the S6' proteasomal protein and inhibit proteasomal function. J Biol Chem 278:11753–11759. Epub 2003 Jan 24

Souza JM, Giasson BI, Lee VM, Ischiropoulos H (2000) Chaperone-like activity of synucleins. FEBS Lett 474:116–119

Spillantini MG, Schmidt ML, Lee VM, Trojanowski JQ, Jakes R, Goedert M (1997) Alpha-synuclein in Lewy bodies. Nature 388:839–840

Stahl N, Borchelt DR, Hsiao K, Prusiner SB (1987) Scrapie prion protein contains a phosphatidylinositol glycolipid. Cell 51:229–240

Staropoli JF, McDermott C, Martinat C, Schulman B, Demireva E, Abeliovich A (2003) Parkin is a component of an SCF-like ubiquitin ligase complex and protects postmitotic neurons from kainate excitotoxicity. Neuron 37:735–749

Stefanis L, Larsen KE, Rideout HJ, Sulzer D, Greene LA (2001) Expression of A53T mutant but not wild-type alpha-synuclein in PC12 cells induces alterations of the ubiquitin-dependent degradation system, loss of dopamine release, and autophagic cell death. J Neurosci 21:9549–9560

Stimson E, Hope J, Chong A, Burlingame AL (1999) Site-specific characterization of the N-linked glycans of murine prion protein by high-performance liquid chromatography/electrospray mass spectrometry and exoglycosidase digestions. Biochemistry 38:4885–4895

Stockel J, Hartl FU (2001) Chaperonin-mediated de novo generation of prion protein aggregates. J Mol Biol 313:861–872

Sugars KL, Rubinsztein DC (2003) Transcriptional abnormalities in Huntington disease. Trends Genet 19:233–238

Sunyach C, Jen A, Deng J, Fitzgerald KT, Frobert Y, Grassi J, McCaffrey MW, Morris R (2003) The mechanism of internalization of glycosylphosphatidylinositol-anchored prion protein. EMBO J 22:3591–3601

Taira T, Saito Y, Niki T, Iguchi-Ariga SM, Takahashi K, Ariga H (2004) DJ-1 has a role in antioxidative stress to prevent cell death. EMBO Rep 5:213–218. Epub 2004 Jan 23

Takahashi K, Taira T, Niki T, Seino C, Iguchi-Ariga SM, Ariga H (2001) DJ-1 positively regulates the androgen receptor by impairing the binding of PIASx alpha to the receptor. J Biol Chem 276:37556–37563. Epub 2001 Jul 26

Takayama S, Sato T, Krajewski S, Kochel K, Irie S, Millan JA, Reed JC (1995) Cloning and functional analysis of BAG-1: a novel Bcl-2-binding protein with anti-cell death activity. Cell 80:279–284

Tanaka Y, Engelender S, Igarashi S, Rao RK, Wanner T, Tanzi RE, Sawa A, Dawson VL, Dawson TM, Ross CA (2001) Inducible expression of mutant alpha-synuclein decreases proteasome activity and increases sensitivity to mitochondria-dependent apoptosis. Hum Mol Genet 10:919–926

Tanner CM, Ben-Shlomo Y (1999) Epidemiology of Parkinson's disease. Adv Neurol 80:153–159

Tao X, Tong L (2003) Crystal structure of human DJ-1, a protein associated with early onset Parkinson's disease. J Biol Chem 278:31372–31379. Epub 2003 May 21

Taraboulos A, Rogers M, Borchelt DR, McKinley MP, Scott M, Serban D, Prusiner SB (1990) Acquisition of protease resistance by prion proteins in scrapie-infected cells does not require asparagine-linked glycosylation. Proc Natl Acad Sci U S A 87:8262–8266

Tatzelt J, Prusiner SB, Welch WJ (1996) Chemical chaperones interfere with the formation of scrapie prion protein. EMBO J 15:6363–6373

Tatzelt J, Winklhofer KF (2004) Folding and misfolding of the prion protein in the secretory pathway. Amyloid 11:162–172

Tatzelt J, Zuo JR, Voellmy R, Scott M, Hartl U, Prusiner SB, Welch WJ (1995) Scrapie prions selectively modify the stress response in neuroblastoma cells. Proc Natl Acad Sci U S A 92:2944–2948

Tofaris GK, Layfield R, Spillantini MG (2001) Alpha-synuclein metabolism and aggregation is linked to ubiquitin-independent degradation by the proteasome. FEBS Lett 509:22–26

Tofaris GK, Razzaq A, Ghetti B, Lilley KS, Spillantini MG (2003) Ubiquitination of alpha-synuclein in Lewy bodies is a pathological event not associated with impairment of proteasome function. J Biol Chem 278:44405–44411. Epub 2003 Aug 15

Tsai YC, Fishman PS, Thakor NV, Oyler GA (2003) Parkin facilitates the elimination of expanded polyglutamine proteins and leads to preservation of proteasome function. J Biol Chem 278:22044–22055. Epub 2003 Apr 03

Uelhoff A, Tatzelt J, Aguzzi A, Winklhofer KF, Haass C (2005) A pathogenic PrP mutation and doppel interfere with polarized sorting of the prion protein. J Biol Chem 280:5137–5140

Varshavsky A (1997) The ubiquitin system. Trends Biochem Sci 22:383–387

Venkatraman P, Wetzel R, Tanaka M, Nukina N, Goldberg AL (2004) Eukaryotic proteasomes cannot digest polyglutamine sequences and release them during degradation of polyglutamine-containing proteins. Mol Cell 14:95–104

Verhoef LG, Lindsten K, Masucci MG, Dantuma NP (2002) Aggregate formation inhibits proteasomal degradation of polyglutamine proteins. Hum Mol Genet 11:2689–2700

Vila M, Przedborski S (2004) Genetic clues to the pathogenesis of Parkinson's disease. Nat Med 10:S58–S62

Volles MJ, Lansbury PT Jr (2002) Vesicle permeabilization by protofibrillar alpha-synuclein is sensitive to Parkinson's disease-linked mutations and occurs by a pore-like mechanism. Biochemistry 41:4595–4602

Volles MJ, Lansbury PT Jr (2003) Zeroing in on the pathogenic form of alpha-synuclein and its mechanism of neurotoxicity in Parkinson's disease. Biochemistry 42:7871–7878

Volles MJ, Lee SJ, Rochet JC, Shtilerman MD, Ding TT, Kessler JC, Lansbury PT Jr (2001) Vesicle permeabilization by protofibrillar alpha-synuclein: implications for the pathogenesis and treatment of Parkinson's disease. Biochemistry 40:7812–7819

Von Bergen M, Barghorn S, Li L, Marx A, Biernat J, Mandelkow EM, Mandelkow E (2001) Mutations of tau protein in frontotemporal dementia promote aggregation of paired helical filaments by enhancing local beta-structure. J Biol Chem 276:48165–48174. Epub 2001 Oct 17

Von Coelln R, Dawson VL, Dawson TM (2004) Parkin-associated Parkinson's disease. Cell Tissue Res 318:175–184. Epub 2004 Jul 30

Wacker JL, Zareie MH, Fong H, Sarikaya M, Muchowski PJ (2004) Hsp70 and Hsp40 attenuate formation of spherical and annular polyglutamine oligomers by partitioning monomer. Nat Struct Mol Biol 11:1215–1222. Epub 2004 Nov 14

Wadsworth JD, Hill AF, Beck JA, Collinge J (2003) Molecular and clinical classification of human prion disease. Br Med Bull 66:241–254

Walter J, Kaether C, Steiner H, Haass C (2001) The cell biology of Alzheimer's disease: uncovering the secrets of secretases. Curr Opin Neurobiol 11:585–590

Warrick JM, Chan HY, Gray-Board GL, Chai Y, Paulson HL, Bonini NM (1999) Suppression of polyglutamine-mediated neurodegeneration in Drosophila by the molecular chaperone HSP70. Nat Genet 23:425–428

Webb JL, Ravikumar B, Atkins J, Skepper JN, Rubinsztein DC (2003) Alpha-synuclein is degraded by both autophagy and the proteasome. J Biol Chem 278:25009–25013. Epub 2003 Apr 28

Weise J, Crome O, Sandau R, Schulz-Schaeffer W, Bahr M, Zerr I (2004) Upregulation of cellular prion protein (PrPc) after focal cerebral ischemia and influence of lesion severity. Neurosci Lett 372:146–150

Weissman JS, Rye HS, Fenton WA, Beechem JM, Horwich AL (1996) Characterization of the active intermediate of a GroEL-GroES-mediated protein folding reaction. Cell 84:481–490

Weissmann C, Fischer M, Raeber A, Büeler H, Sailer A, Shmerling D, Rülicke T, Brandner S, Aguzzi A (1996) The role of PrP in pathogenesis of experimental scrapie. Cold Spring Harb Symp Quant Biol 61:511–522

Wilson MA, Collins JL, Hod Y, Ringe D, Petsko GA (2003) The 1.1-A resolution crystal structure of DJ-1, the protein mutated in autosomal recessive early onset Parkinson's disease. Proc Natl Acad Sci U S A 100:9256–9261. Epub 2003 Jul 10

Winklhofer KF, Heller U, Reintjes A, Tatzelt J (2003a) Inhibition of complex glycosylation increases formation of PrPSc. Traffic 4:313–322

Winklhofer KF, Henn IH, Kay-Jackson P, Heller U, Tatzelt J (2003b) Inactivation of parkin by oxidative stress and C-terminal truncations; a protective role of molecular chaperones. J Biol Chem 278:47199–47208

Winklhofer KF, Heske J, Heller U, Reintjes A, Muranji W, Moarefi I, Tatzelt J (2003c) Determinants of the in vivo-folding of the prion protein: a bipartite function of helix 1 in folding and aggregation. J Biol Chem 278:14961–14970

Winklhofer KF, Reintjes A, Hoener MC, Voellmy R, Tatzelt J (2001) Geldanamycin restores a defective heat shock response in vivo. J Biol Chem 276:45160–45167

Wu YR, Wang CK, Chen CM, Hsu Y, Lin SJ, Lin YY, Fung HC, Chang KH, Lee-Chen GJ (2004) Analysis of heat-shock protein 70 gene polymorphisms and the risk of Parkinson's disease. Hum Genet 114:236–241. Epub 2003 Nov 06

Xu J, Kao SY, Lee FJ, Song W, Jin LW, Yankner BA (2002) Dopamine-dependent neurotoxicity of alpha-synuclein: a mechanism for selective neurodegeneration in Parkinson disease. Nat Med 8:600–806

Yang Y, Nishimura I, Imai Y, Takahashi R, Lu B (2003) Parkin suppresses dopaminergic neuron-selective neurotoxicity induced by Pael-R in Drosophila. Neuron 37:911–924

Yang Y, Turner RS, Gaut JR (1998) The chaperone BiP/GRP78 binds to amyloid precursor protein and decreases Abeta40 and Abeta42 secretion. J Biol Chem 273:25552–25555

Yao D, Gu Z, Nakamura T, Shi ZQ, Ma Y, Gaston B, Palmer LA, Rockenstein EM, Zhang Z, Masliah E, Uehara T, Lipton SA (2004) Nitrosative stress linked to sporadic Parkinson's disease: S-nitrosylation of parkin regulates its E3 ubiquitin ligase activity. Proc Natl Acad Sci U S A 101:10810–10814. Epub 2004 Jul 13

Yedidia Y, Horonchik L, Tzaban S, Yanai A, Taraboulos A (2001) Proteasomes and ubiquitin are involved in the turnover of the wild-type prion protein. EMBO J 20:5383–91

Yokota T, Sugawara K, Ito K, Takahashi R, Ariga H, Mizusawa H (2003) Down regulation of DJ-1 enhances cell death by oxidative stress, ER stress, and proteasome inhibition. Biochem Biophys Res Commun 312:1342–1348

Yost CS, Lopez CD, Prusiner SB, Myers RM, Lingappa VR (1990) Non-hydrophobic extracytoplasmic determinant of stop transfer in the prion protein. Nature 343:669–672

Zanusso G, Petersen RB, Jin T, Jing Y, Kanoush R, Ferrari S, Gambetti P, Singh N (1999) Proteasomal degradation and N-terminal protease resistance of the codon 145 mutant prion protein. J Biol Chem 274:23396–23404

Zarranz JJ, Alegre J, Gomez-Esteban JC, Lezcano E, Ros R, Ampuero I, Vidal L, Hoenicka J, Rodriguez O, Atares B, Llorens V, Gomez Tortosa E, del Ser T, Munoz DG, de Yebenes JG (2004) The new mutation, E46K, of alpha-synuclein causes Parkinson and Lewy body dementia. Ann Neurol 55:164–173

Zhou Y, Gu G, Goodlett DR, Zhang T, Pan C, Montine TJ, Montine KS, Aebersold RH, Zhang J (2004) Analysis of alpha-synuclein-associated proteins by quantitative proteomics. J Biol Chem 279:39155–39164. Epub 2004 Jul 01

Zourlidou A, Payne Smith MD, Latchman DS (2004) HSP27 but not HSP70 has a potent protective effect against alpha-synuclein-induced cell death in mammalian neuronal cells. J Neurochem 88:1439–1448

Chaperoning Oncogenes: Hsp90 as a Target of Geldanamycin

L. Neckers

Urologic Oncology Branch, National Cancer Institute, 9610 Medical Ctr. Dr., Suite 300, Rockville MD, 20850, USA
len@helix.nih.gov

1	Introduction	260
2	Hsp90: A Chaperone of Oncogenes	262
2.1	Can Hsp90 Inhibitors Distinguish Between Cancer Cell and Normal Cell Hsp90?	262
2.2	Hsp90 Inhibitors Target Mutated and Chimeric Proteins Uniquely Expressed in Certain Cancers	263
2.3	Hsp90 Inhibitors Target the Androgen Receptor in Prostate Cancer	265
2.4	Hsp90 Inhibitors Exert Anti-angiogenic Activity by Promoting Oxygen- and VHL-Independent Inactivation and Degradation of HIF-1α Leading to Inhibition of VEGF Expression	265
2.5	Hsp90 Inhibitors Target MET Receptor Tyrosine Kinase	266
2.6	Combined Inhibition of Hsp90 and the Proteasome Disrupt the Endoplasmic Reticulum and Demonstrate Enhanced Toxicity Toward Cancer Cells	267
2.7	Hsp90 Inhibitors Sensitize Cancer Cells to Radiation	268
2.8	Targeting Hsp90 on the Cancer Cell Surface	269
3	Hsp90 Inhibitors that Do not Target the Amino Terminal Nucleotide Binding Site	269
4	Issues Relating to Further Development of Hsp90 Inhibitors	270
5	Conclusion	272
	References	272

Abstract Heat shock protein 90 (Hsp90) is a molecular chaperone required for the stability and function of a number of conditionally activated and/or expressed signaling proteins, as well as multiple mutated, chimeric, and/or over-expressed signaling proteins, that promote cancer cell growth and/or survival. Hsp90 inhibitors, by interacting specifically with a single molecular target, cause the inactivation, destabilization, and eventual degradation of Hsp90 client proteins, and they have shown promising anti-tumor activity in preclinical model systems. One Hsp90 inhibitor, 17-AAG, has completed Phase I clinical trial and several Phase II trials of this agent are in progress. Hsp90 inhibitors are unique in that, although they are directed toward a specific molecular target, they simultaneously inhibit multiple signaling pathways that frequently interact to promote cancer cell survival. Further, by inhibiting nodal points in multiple overlapping survival pathways utilized by cancer cells, a combination of an Hsp90 inhibitor with standard chemotherapeutic agents may dramatically increase the in vivo efficacy of the standard agent. Hsp90 inhibitors may circumvent the characteristic genetic plasticity that has allowed cancer cells to eventually evade the toxic effects of most molecularly targeted agents. The mechanism-based use of Hsp90 inhibitors, both alone and in combination with other drugs, should be effective toward multiple forms of cancer.

Keywords Heat shock protein 90 · Cancer · Molecular chaperone · Molecularly targeted therapeutics · Genetic plasticity · Oncogene · Geldanamycin

1
Introduction

Cancer is a disease of genetic instability. Although only a few specific alterations seem to be required for generation of the malignant phenotype, at least in colon carcinoma there are approximately 10,000 estimated mutations at time of diagnosis (Stoler et al. 1999; Hahn and Weinberg 2002). This genetic plasticity of cancer cells allows them to frequently escape the precise molecular targeting of a single signaling node or pathway, making them ultimately nonresponsive to molecularly targeted therapeutics. Even Gleevec (Novartis Pharmaceuticals Corp.), a well-recognized clinically active Bcr-Abl tyrosine kinase inhibitor, can eventually lose its effectiveness under intense, drug-dependent selective pressure, due to either mutation of the drug interaction site or expansion of a previously existing resistant clone (La Rosee et al. 2002). Most solid tumors at the time of detection are already sufficiently genetically diverse to resist single-agent molecularly targeted therapy (Kitano 2003). Thus, a simultaneous attack on multiple nodes of a cancer cell's web of overlapping signaling pathways should be more likely to affect survival than would inhibition of one or even a few individual signaling nodes. Given the number of key nodal proteins that are Hsp90 clients (see the website maintained by D. Picard, http://www.picard.ch/DP/downloads/Hsp90interactors.pdf), inhibition of Hsp90 may serve the purpose of collapsing, or significantly weakening, a cancer cell's safety net. Indeed, following a hypothesis first proposed by Hanahan and Weinberg several years ago (Hanahan and Weinberg 2000), genetic instability allows a cell to eventually acquire six capabilities that are characteristic of most if not all cancers. These are:

- Self-sufficiency in growth signaling
- Insensitivity to anti-growth signaling
- Ability to evade apoptosis
- Sustained angiogenesis
- Tissue invasion and metastasis
- Limitless replicative potential

As is highlighted in Fig. 1, Hsp90 plays a pivotal role in acquisition and maintenance of each of these capabilities. Several excellent reviews provide an in-depth description of the many signaling nodes regulated by Hsp90 (Goetz et al. 2003; Isaacs et al. 2003; Bagatell and Whitesell 2004; Chiosis et al. 2004; Workman 2004; Zhang and Burrows 2004).

Cancer cells survive in the face of frequently extreme environmental stress, such as hypoxia and acidosis, as well as in the face of the exogenously applied environmental stresses of chemotherapy or radiation. These stresses tend to generate free radicals that can cause significant physical damage to cellular proteins. Given the combined protective role of molecular chaperones toward damaged proteins and the dependence of multiple signal transduction pathways on Hsp90, it is therefore not surprising that molecular chaperones in general, and Hsp90 in particular, are highly expressed in most tumor cells. However, Hsp90 may be elevated in tumor cells and may provide a unique molecular target therein for an additional reason. Using *Drosophila* and *Arabidopsis* as model systems, Lindquist and colleagues have shown that an ancient function of Hsp90 may be to permit accumulation at the protein level of inherent genetic mutations, and thus the chaperone may play a pivotal role in the evolutionary process itself (Rutherford and Lindquist 1998; Queitsch et al. 2002). Extrapolating this hypothesis to genetically unstable cancer cells, it is not a great leap to think that Hsp90 may be critical to their ability to survive in the presence of an aberrantly high mutation rate.

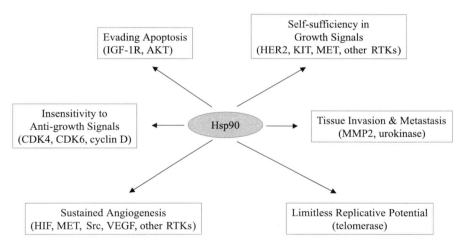

Fig. 1 Hsp90 function is implicated in establishment of each of the hallmarks of cancer as first proposed by Hanahan and Weinberg (2000). Importantly, Hsp90 function may also permit the genetic instability on which acquisition of the six hallmarks depends

2
Hsp90: A Chaperone of Oncogenes

Several recent, excellently detailed reviews of the mechanics of Hsp90 function are in the scientific literature (Prodromou and Pearl 2003; Bagatell and Whitesell 2004; Chiosis et al. 2004; Siligardi et al. 2004; Wegele et al. 2004; Zhang and Burrows 2004; Sangster and Queitsch 2005). The reader is also directed to other chapters in the current volume. For the purposes of the current update on Hsp90-directed therapeutics, suffice it to say that Hsp90 is a conformationally flexible protein that associates with a distinct set of co-chaperones in dependence on nucleotide (ATP or ADP) occupancy of an amino-terminal binding pocket in Hsp90. Nucleotide exchange and ATP hydrolysis (by Hsp90 itself, with the assistance of co-chaperones) drive the so-called Hsp90 chaperone machine to bind, chaperone, and release client proteins. The Hsp90 inhibitors currently in clinical trial (17-AAG and 17-DMAG), as well as those under development, all share the property of displacing nucleotide from the amino terminal pocket in Hsp90, and therefore short-circuiting the Hsp90 chaperone machine, much as one would stop the rotation of a bicycle wheel by inserting a stick between the spokes. Cycling of the chaperone machine is critical to its function. The Hsp90 inhibitors, by preventing nucleotide-dependent cycling, interfere with the chaperone activity of Hsp90, resulting in targeting of client proteins to the proteasome, the cell's garbage disposal, where they are degraded (Neckers 2002). Even if the proteasome is inhibited, client proteins are not rescued from Hsp90 inhibition, but instead accumulate in a misfolded, inactive form in detergent-insoluble subcellular complexes (An et al. 2000).

2.1
Can Hsp90 Inhibitors Distinguish Between Cancer Cell and Normal Cell Hsp90?

An initial concern prior to the initiation of phase I testing of 17-AAG was that inhibition of Hsp90 would be as harmful to normal cells as it seemed to be to cancer cells. This concern has turned out to be unfounded, as no Hsp90-dependent toxicities have been reported from any of the clinical trials. There are two possible explanations for this fact: first, perhaps Hsp90 in normal cells and tissues is not vital to their survival and function. Experiments with yeast suggest that eukaryotic cells can survive with markedly depleted Hsp90 levels, although they do not respond well to nutrient deprivation (Xu et al. 1999). Experiments in nontransformed mammalian cells to test the importance of Hsp90 (e.g., using siRNA techniques) have not been reported.

A second explanation for the lack of in vivo Hsp90-dependent toxicity during administration of Hsp90 inhibitors relates to drug clearance characteristics and differential binding affinity to normal and cancer cell Hsp90. Thus, 17-AAG is cleared rapidly from blood and normal tissues in a number of animal studies (and in humans), while its retention in tumors is prolonged (Eiseman et al.

2004) (to date, information is available for animal studies only). Although the mechanism for drug retention in tumors is not known, it is not unreasonable to speculate that enhanced binding to Hsp90 in tumors as compared to normal tissues may contribute to this phenomenon (Chiosis et al. 2003). This is especially the case, given the recent report that the affinity of tumor cell Hsp90 for 17-AAG and other benzoquinone ansamycins is 20- to 50-fold stronger than is the affinity of Hsp90 isolated from normal tissues (Kamal et al. 2003). An explanation for this differential affinity remains to be uncovered, but it is intriguing to speculate that either post-translational modification (e.g., phosphorylation, acetylation, or other) of Hsp90 is different in transformed as compared to nontransformed cells, resulting in altered affinity for these drugs, or perhaps that co-chaperone complexes associating with Hsp90 are distinct in tumor cells and the nature of these complexes enhance drug binding affinity to Hsp90.

While these possibilities are highly speculative, the fact that 17-AAG and other benzoquinone ansamycins bind to tumor cell Hsp90 with markedly enhanced affinity as compared to normal cell Hsp90 has been confirmed by at least one other laboratory. Whether this will prove to be the case with non-benzoquinone ansamycin Hsp90 inhibitors, or whether this is a unique property of this structural class remains to be determined.

2.2
Hsp90 Inhibitors Target Mutated and Chimeric Proteins Uniquely Expressed in Certain Cancers

Hsp90 characteristically chaperones a number of mutated or chimeric kinases that are key mediators of disease. Thus, anaplastic large cell lymphomas are characterized by expression of the chimeric protein NPM-ALK, which originates from a fusion of the nucleophosmin (*NPM*) and the membrane receptor anaplastic lymphoma kinase (*ALK*) genes. The chimeric kinase is constitutively active and capable of causing malignant transformation (Fujimoto et al. 1996). Bonvini and colleagues have shown that NPM-ALK kinase is an Hsp90 client protein, and that GA and 17-AAG destabilize the kinase and promote its proteasome-mediated degradation in several anaplastic large cell lymphoma cell lines (Bonvini et al. 2002).

FLT3 is a receptor tyrosine kinase that regulates proliferation, differentiation, and survival of hematopoietic cells. FLT3 is frequently expressed in acute myeloid leukemia, and in 20% of patients with this cancer the tumor cells express a FLT3 protein harboring an internal tandem duplication in the juxtamembrane domain. This mutation is correlated with leukocytosis and a poor prognosis (Naoe et al. 2001). Minami and colleagues have reported that Hsp90 inhibitors cause selective apoptosis of leukemia cells expressing tandemly duplicated FLT3. Further, these investigators reported that mutated FLT3 was an Hsp90 client protein and that brief treatment with multiple Hsp90 inhibitors resulted in the rapid dissociation of Hsp90 from the kinase, accompanied by the rapid loss of kinase activity together with loss of activity of several downstream FLT3 targets including MAP kinase, Akt, and Stat5a (Minami et al. 2002). Minami et al. propose that Hsp90 inhibitors should be considered as promising compounds for the treatment of acute myeloid leukemia characterized by tandemly duplicated FLT3 expression.

BCR-ABL ($p210^{Bcr-Abl}$) is an Hsp90 client protein that is also effectively inhibited by the novel tyrosine kinase inhibitor imatinib (Druker et al. 1996; An et al. 2000; Shiotsu et al. 2000). While imatinib has proven very effective in initial treatment of patients with chronic myelogenous leukemia, a majority of patients who are treated when their disease is in blast crisis stage (e.g., advanced) eventually relapse despite continued therapy (Sawyers et al. 2002). Relapse is correlated with loss of BCR-ABL inhibition by imatinib, due either to gene amplification or to specific point mutations in the kinase domain that preclude association of imatinib with the kinase (Shah et al. 2002). Gorre and colleagues have reported the very exciting finding that BCR-ABL protein, which was resistant to imatinib, remained dependent on Hsp90 chaperoning activity and thus retained sensitivity to Hsp90 inhibitors, including GA and 17-AAG. Both compounds induced the degradation of wild type and mutant BCR-ABL, with a trend indicating more potent activity toward mutated imatinib-resistant forms of the kinase (Gorre et al. 2002). These findings were recently confirmed by other investigators (Nimmanapalli et al. 2002), thus providing a rationale for the use of 17-AAG in treatment of imatinib-resistant chronic myelogenous leukemia.

Mutations in the proto-oncogene c-*kit* cause constitutive kinase activity of its product, KIT protein, and are associated with human mastocytosis and gastrointestinal stromal tumors (GIST). Although currently available tyrosine kinase inhibitors are effective in the treatment of GIST, there has been limited success in the treatment of mastocytosis. Treatment with 17-AAG of the mast cell line HMC-1.2, harboring the Asp816Val and Val560Gly KIT mutations, and the cell line HMC-1.1, harboring a single Val560Gly mutation, causes both the level and activity of KIT and downstream signaling molecules AKT and STAT3 to be downregulated following drug exposure (Fumo et al. 2004). These data were validated using Cos-7 cells transfected with wild type and mutated KIT. 17-AAG promotes cell death of both HMC mast cell lines. In addition, neoplastic mast cells isolated from patients with mastocytosis and incubated with 17-AAG ex vivo are selectively sensitive to Hsp90 inhibition as compared to the mononuclear fraction as a whole. These data provide compelling evidence that 17-AAG may be effective in the treatment of c-*kit*-related diseases, including mastocytosis, GIST, mast cell leukemia, subtypes of acute myelogenous leukemia, and testicular cancer.

2.3
Hsp90 Inhibitors Target the Androgen Receptor in Prostate Cancer

Androgen receptor continues to be expressed in the majority of hormone-independent prostate cancers, suggesting that it remains important for tumor growth and survival. Receptor overexpression, mutation, and/or post-translational modification may all be mechanisms by which androgen receptor can remain responsive either to low levels of circulating androgen or to anti-androgens. Vanaja et al. have shown that Hsp90 association is essential for the function and stability of the androgen receptor in prostate cancer cells (Vanaja et al. 2002). These investigators reported that androgen receptor levels in LNCaP cells were markedly reduced by the Hsp90 inhibitor geldanamycin (GA), as was the ability of the receptor to become transcriptionally active in the presence of synthetic androgen. In addition, Georget et al. have shown that GA preferentially destabilized androgen receptor bound to anti-androgen, thus suggesting that the clinical efficacy of anti-androgens may be enhanced by combination with an Hsp90 inhibitor (Georget et al. 2002). These investigators also reported that GA prevented the nuclear translocation of ligand-bound androgen receptor, and inhibited the transcriptional activity of nuclear-targeted receptors, implicating Hsp90 in multiple facets of androgen receptor activity. Finally, Solit and colleagues have reported that 17-AAG caused degradation of both wild type and mutant androgen receptors and inhibited both androgen-dependent and androgen-independent prostate tumor growth in nude mice (Solit et al. 2002). Importantly, these investigators also demonstrated the loss of Her2 and Akt proteins, two Hsp90 clients that are upstream post-translational activators of the androgen receptor, in the tumor xenografts taken from 17-AAG-treated animals.

2.4
Hsp90 Inhibitors Exert Anti-angiogenic Activity by Promoting Oxygen- and VHL-Independent Inactivation and Degradation of HIF-1α Leading to Inhibition of VEGF Expression

Hypoxia inducible factor-1α (HIF-1α) is a nuclear transcription factor involved in the transactivation of numerous target genes, many of which are implicated in the promotion of angiogenesis and adaptation to hypoxia (for a review, see Harris 2002). Although these proteins are normally labile and expressed at low levels in normoxic cells, their stability and activation increase several-fold in hypoxia. The molecular basis for the instability of these proteins in normoxia depends upon VHL, the substrate recognition component of an E3 ubiquitin ligase complex that targets HIF-1α for proteasome-dependent degradation (Maxwell et al. 1999). Hypoxia normally impairs VHL function, thus allowing HIF to accumulate. HIF-1α expression has been documented in diverse epithelial cancers and most certainly supports survival in the oxygen-depleted environment inhabited by most solid tumors.

VHL can also be directly inactivated by mutation or hyper-methylation, resulting in constitutive overexpression of HIF in normoxic cells. In hereditary von Hippel-Lindau disease, there is a genetic loss of VHL, and affected individuals are predisposed to an increased risk of developing highly vascular tumors in a number of organs. This is due, in large part, to deregulated HIF expression and the corresponding upregulation of the HIF target gene vascular endothelial growth factor (VEGF). A common manifestation of VHL disease is the development of clear cell renal cell carcinoma (CC-RCC) (Seizinger et al. 1988). VHL inactivation also occurs in nonhereditary, sporadic CC-RCC.

HIF-1α interacts with Hsp90 (Gradin et al. 1996), and both GA and another Hsp90 inhibitor, radicicol, reduce HIF-dependent transcriptional activity (Hur et al. 2002; Isaacs et al. 2002). Hur et al. demonstrated that HIF protein from radicicol-treated cells was unable to bind DNA, suggesting that Hsp90 is necessary for mediating the proper conformation of HIF and/or recruiting additional co-factors. Likewise, Isaacs et al. reported GA-dependent, transcriptional inhibition of VEGF. Additionally, GA downregulated HIF-1α protein expression by stimulating VHL-independent HIF-1α proteasomal degradation (Isaacs et al. 2002; Mabjeesh et al. 2002).

HIF-1α induction and VEGF expression has been associated with migration of glioblastoma cells in vitro and metastasis of glioblastoma in vivo. Zagzag et al., in agreement with the findings described above, have reported that GA blocks HIF-1α induction and VEGF expression in glioblastoma cell lines (Zagzag et al. 2003). Further, these investigators have shown that GA blocks glioblastoma cell migration, using an in vitro, assay at nontoxic concentrations. This effect on tumor cell motility was independent of p53 and PTEN status, which makes Hsp90 inhibition an attractive modality in glioblastoma, where mutations in p53 and PTEN genes are common and where tumor invasiveness is a major therapeutic challenge.

Dias et al. have recently reported that VEGF promotes elevated Bcl2 protein levels and inhibits activity of the pro-apoptotic caspase-activating protein Apaf in normal endothelial cells and in leukemia cells bearing receptors for VEGF (Dias et al. 2002). Intriguingly, these investigators show that both phenomena require VEGF-stimulated Hsp90 association (e.g., with Bcl2 and Apaf), and that GA reverses both processes. Thus, GA blocked the pro-survival effects of VEGF by both preventing accumulation of anti-apoptotic Bcl2 and blocking the inhibition of pro-apoptotic Apaf.

2.5
Hsp90 Inhibitors Target MET Receptor Tyrosine Kinase

The Met receptor tyrosine kinase is frequently overexpressed in cancer, and is involved in angiogenesis, as well as in the survival and invasive ability of cancer cells. A recent report by Maulik et al. has demonstrated a role for Met in migration and survival of small cell lung cancer (Maulik et al. 2002). Met is an Hsp90 client protein, and these investigators went on to show that GA antagonized Met activity, reduced the Met protein level, and promoted apoptosis in several small cell lung cancer cell lines, even in the presence of excess Met ligand.

Hypoxia potentiates the invasive and metastatic potential of tumor cells. In an important recent study, Pennacchietti and colleagues reported that hypoxia (via two HIF-1α response elements) transcriptionally activated the Met gene, and synergized with Met ligand in promoting tumor invasion. Further, they showed that the pro-invasive effects of hypoxia were mimicked by Met overexpression, and that inhibition of Met expression prevented hypoxia-induced tumor invasion (Pennacchietti et al. 2003). Coupled with an earlier report describing induction of HIF-1 transcriptional activity by Met ligand (Tacchini et al. 2001), these data identify the HIF-VEGF-Met axis as a critical target for intervention using Hsp90 inhibitors, either alone or in conjunction with other inhibitors of angiogenesis. As Bottaro and Liotta recently pointed out (Bottaro and Liotta 2003), the sole use of angiogenesis inhibitors to deprive tumors of oxygen might produce an unexpectedly aggressive phenotype in those cells that survived the treatment. These authors speculated that combination of Met inhibitors with anti-angiogenesis agents should therefore be beneficial. We would suggest that combination of an anti-angiogenesis drug with an Hsp90 inhibitor would not only potentiate the anti-tumor effects obtained by inhibiting angiogenesis, but would also break the HIF-Met axis by simultaneously targeting both Hsp90-dependent signaling proteins.

2.6
Combined Inhibition of Hsp90 and the Proteasome Disrupt the Endoplasmic Reticulum and Demonstrate Enhanced Toxicity Toward Cancer Cells

Proteasome-mediated degradation is the common fate of Hsp90 client proteins in cells treated with Hsp90 inhibitors (Mimnaugh et al. 1996; Schneider et al. 1996). Proteasome inhibition does not protect Hsp90 clients in the face of chaperone inhibition—instead client proteins become insoluble (An et al. 2000; Basso et al. 2002). Since the deposition of insoluble proteins can be toxic to cells (French et al. 2001; Waelter et al. 2001), interest has arisen in combining proteasome inhibition with inhibition of Hsp90, the idea being that dual treatment will lead to enhanced accumulation of insoluble proteins and trigger apoptosis. This hypothesis is particularly appealing since a small molecule proteasome inhibitor has demonstrated efficacy in early clinical trials (Aghajanian et al. 2002; L'Allemain 2002). Initial experimental support for such a hypothesis was provided by Mitsiades et al. (2002), who reported that Hsp90 inhibitors enhanced multiple myeloma cell sensitivity to proteasome inhibition. Importantly, transformed cells are more sensitive to the cytotoxic effects of this drug combination than are nontransformed cells. Thus, 3T3 fibroblasts are fully resistant to combined administration of 17-AAG and Velcade at concentrations that prove cytotoxic to 3T3 cells transformed by *HPV16* virus encoding viral proteins E6 and E7 (Mimnaugh et al. 2004). In the same study, Mimnaugh et al. demonstrated that the endoplasmic reticulum is one of the main targets of this drug combination. In the presence of combined doses of both agents that show synergistic cytotoxicity, these investigators noted a nearly complete disruption of the architecture of the endoplasmic reticulum. Since all secreted and transmembrane proteins must pass through this organelle on their route to the extracellular space, it is not surprising that a highly secretory cancer such as multiple myeloma would be particularly sensitive to combined inhibition of Hsp90 and the proteasome. One might speculate that other highly secretory cancers, including hepatocellular carcinoma and pancreatic carcinoma, would also respond favorably to this drug combination.

2.7
Hsp90 Inhibitors Sensitize Cancer Cells to Radiation

Gius and colleagues have reported that 17-AAG potentiates both the in vitro and in vivo radiation response of cervical carcinoma cells (Bisht et al. 2003). An enhanced radiation response was noted when cells were exposed to radiation within 6–48 h after drug treatment. Importantly, at 17-AAG concentrations that were themselves nontoxic, Hsp90 inhibition enhanced cell kill in response to an otherwise ineffective radiation exposure (2 Gy) by more than one log. Even at moderately effective levels of radiation exposure (4–6 Gy), addition of nontoxic amounts of 17-AAG enhanced cell kill by more than one log. Importantly, the sensitizing effects of 17-AAG observed in the cervical carcinoma cells were not seen in 3T3 cells, but were observed in *HPV16-E6* and *-E7* transformed 3T3 cells. The authors demonstrated convincingly that the effect of 17-AAG was multifactorial, since several pro-survival Hsp90 client proteins were rapidly downregulated upon drug treatment. In vitro findings were confirmed by a murine xenograft study in which the anti-tumor activity of both single and fractionated radiation exposure was dramatically enhanced by treatment with 17-AAG, either 16 h prior to single radiation exposure or on days 1 and 4 of a 6-day period during which the animals received fractionated radiation exposure. Machida and colleagues reported similar findings for lung carcinoma and colon adenocarcinoma cells in vitro (Machida et al. 2003). Thus, 17-AAG has been validated as a potential therapeutic agent that can be used at clinically relevant doses to enhance cancer cell sensitivity to radiation. It is reasonable to expect that other Hsp90 inhibitors will have a similar utility.

2.8
Targeting Hsp90 on the Cancer Cell Surface

Recently, Becker and colleagues reported that Hsp90 expression is dramatically upregulated in malignant melanoma cells as compared to benign melanocytic lesions, and that Hsp90 is expressed on the surface of seven out of eight melanoma metastases (Becker et al. 2004). Eustace et al. have identified cell surface Hsp90 to be crucial for the invasiveness of HT-1080 fibrosarcoma cells in vitro (Eustace and Jay 2004; Eustace et al. 2004). Taken together, these data implicate Hsp90 as an important determinant of tumor cell invasion and metastasis. Indeed, in the Eustace et al. study, the investigators demonstrated that GA covalently affixed to cell impermeable beads was able to significantly impair cell invasion across a Matrigel-coated membrane. These findings have been confirmed using a polar (and thus cell impermeable) derivative of 17-DMAG in place of GA-beads (Neckers et al., unpublished observations). Coincident with its inhibitory effects on cell invasiveness, cell impermeable GA also antagonized the maturation, via proteolytic self-processing, of the metalloproteinase MMP2, a cell surface enzyme whose activity has been previously demonstrated as essential to cell invasion. Further, these investigators demonstrated that Hsp90 could be found in association with MMP2 in the culture medium bathing the HT-1080 cells. It is intriguing to speculate that association with Hsp90 on the cell surface is necessary for the self-proteolysis of MMP2. Thus, a possible chaperone function for cell surface Hsp90 may be directly implicated in tumor cell invasiveness and metastasis. As such, cell surface Hsp90 may represent a novel, perhaps cancer-specific target for cell-impermeant Hsp90 inhibitors.

3
Hsp90 Inhibitors that Do not Target the Amino Terminal Nucleotide Binding Site

In the last several years, it has become apparent that Hsp90 activity can be impacted by pharmacologic attack at other sites on the protein. However, these other inhibitors are not specific for Hsp90 and thus cannot be considered to be "clean" drugs. Several years ago, we identified the coumarin antibiotics (e.g., novobiocin) as capable of binding to a C-terminal ATP-binding domain on Hsp90, with resultant inhibition of Hsp90 activity (Marcu et al. 2000a, 2000b). These results have since been confirmed and extended by others (Soti et al. 2002), but the affinity of novobiocin for Hsp90 is poor (several hundred micromolar), and it is a much better inhibitor of topoisomerase than it is of Hsp90.

Cisplatin, which exerts anti-tumor activity, at least in part, by forming DNA adducts and thus interfering in transcription, has recently been shown to also bind to the C-terminus of Hsp90, and to interfere with nucleotide binding at this site (Soti et al. 2002). However, cisplatin interacts with other thiol-containing proteins and its interaction with Hsp90 occurs at concentrations that are too high to be pharmacologically relevant. Nonetheless, the fact that both the coumarin antibiotics and cisplatin share an Hsp90 binding domain in the C-terminus of the chaperone highlights the potential importance of this domain to Hsp90 function, as well as places emphasis on the pharmacologic accessibility of this site (for review see Marcu and Neckers 2003). Nevertheless, little progress has been made in identifying high-affinity Hsp90-specific agents that target this region.

A 3rd class of agent that has been shown to affect Hsp90 are the inhibitors of histone deacetylases (HDAC inhibitors). Schrump and colleagues first demonstrated that the HDAC inhibitor depsipeptide (FR901228) was able to promote degradation of several Hsp90 client proteins, including mutated p53, c-raf-1, and ErbB2 (Yu et al. 2002). These investigators also showed that depsipeptide promoted Hsp90 acetylation and disrupted its ability to bind to ATP-Sepharose. Similar results have since been reported by Bhalla et al. (Fuino et al. 2003) using the HDAC inhibitor LAQ824, a molecule structurally different from depsipeptide. Thus, acetylation of Hsp90, which clearly occurs in response to administration of diverse HDAC inhibitors, inhibits Hsp90 function, most likely by interfering with nucleotide-dependent cycling of the chaperone complex. Certainly, Hsp90 inhibition must be considered as a component of the anti-tumor activity of these drugs, but protein hyperacetylation is now a commonly observed property of these agents, and Hsp90 is probably one of many hyperacetylated proteins. It will be of interest to determine the specific HDAC (there are several classes and multiple members of each class) or HDACs that deacetylate Hsp90 under physiologic conditions, and then to test a specific inhibitor of this HDAC, but even in the best case scenario it is unlikely that Hsp90 would be the only substrate affected. In summary, the only pure Hsp90 inhibitors identified to date are those that bind to the N-terminal nucleotide pocket of the chaperone.

4
Issues Relating to Further Development of Hsp90 Inhibitors

The toxicity, if any, produced as a result of inhibiting Hsp90 in normal tissues remains to be determined. The dose-limiting toxicity of 17-AAG (hepatic) may be due primarily to the chemical structure of the drug (e.g., presence of a quinone moiety) and not to any Hsp90-dependent effect. Ultimately, this question will be addressed when nonansamycin Hsp90 inhibitors reach the clinic. However, it is important to remember that the differential affinity of ansamycin-based Hsp90 inhibitors for tumor cell vs normal cell Hsp90 may play a role in the apparent lack of target-based toxicity of 17-AAG in normal tissues. Some of the nonansamycin Hsp90 inhibitors in development do not possess this discriminatory property, and therefore careful evaluation of their in vivo toxicities will help determine the importance of this phenomenon to overall drug efficacy. The ideal Hsp90 inhibitor should be an orally bioavailable agent that favors inhibition of tumor cell Hsp90.

Hsp90 inhibition leads to activation of the heat shock transcription factor Hsf1 and therefore promotes induction of Hsp70 and other chaperones as part of a classical heat shock response (Ali et al. 1998; Kim et al. 1999; Xiao et al. 1999; Guo et al. 2001; Lu et al. 2002; Matthews et al. 2003). Indeed, induction of Hsp70 in vivo following 17-AAG administration has proven to be the most robust pharmacodynamic indicator of drug activity in patients. However, Hsp70 is generally considered to be cytoprotective for tumor cells, and Whitesell and colleagues have reported that loss of Hsf1 expression potentiates the cytotoxicity of GA (Bagatell et al. 2000). Very recently, Gabai et al. have confirmed this hypothesis by showing that siRNA-dependent depletion of Hsp70 promotes the cytotoxicity, in vitro, of Hsp90 targeting agents (Gabai et al. 2005). Therefore, it is conceivable that the effectiveness of prolonged treatment with 17-AAG or similar Hsp90 inhibitor may be self-limiting, making selection of the appropriate schedule of drug administration a possibly critical component of its success.

Although the focus to date has been on identification of small molecule Hsp90 inhibitors, it is now clear that proper function of Hsp90 requires that it associate in a regulated manner with a host of co-chaperone proteins. Thus, investigators in the field have begun to examine whether small molecule inhibitors of specific co-chaperone–Hsp90 interactions can be designed, and if so, what effect they might have on Hsp90 function. However, this approach requires identification of inhibitors of protein–protein surface interactions, generally a more difficult undertaking than designing competitive inhibitors of the nucleotide pocket.

Another approach to pharmacologically inhibiting Hsp90 lies in understanding the role of various post-translational modifications in the function of the chaperone. For example, two studies have demonstrated that Hsp90 phosphorylation is coupled to the release of the chaperone from its client protein (Mimnaugh et al. 1995; Zhao et al. 2001). GA inhibits Hsp90 phosphorylation, suggesting that this post-translational modification is conformation-dependent, while the phosphatase inhibitor okadaic acid leads to dramatic hyperphosphorylation of Hsp90, suggesting that cycles of phosphorylation and dephosphorylation of the chaperone are continuously occurring in a regulated manner. Further, Hsp90 acetylation, as discussed earlier, has also been shown to modulate its activity, and several HDAC inhibitors have been reported to inhibit the chaperone, with consequences similar to those observed following exposure to pure Hsp90 inhibitors (Yu et al. 2002; Fuino et al. 2003).

5
Conclusion

By their very nature, cancer cells are genetically unstable. Indeed, it is this property that is the key to their etiology. Thus, in most cases, cancers display the heterogeneity and redundancy of signaling pathways that make them resistant to many environmental insults, including those imposed by chemotherapeutics and radiation. Molecular chaperones in general, and Hsp90 in particular, clearly play a major role in promoting this robustness, and therefore Hsp90 inhibitors, by targeting the very property upon which cancer cells depend for their survival, comprise a truly unique class of anti-cancer agent. Only one class of small molecule Hsp90 inhibitor has reached the clinic to date. More are certainly needed. Nonetheless, the preference of 17-AAG for tumor cell Hsp90, and the demonstration that Hsp90 can be pharmacologically modulated in vivo without severe toxicity are both very encouraging stimuli that should spur the development of second- and third-generation Hsp90 inhibitors with better pharmacologic properties. The preceding discussion of some of the latest developments in Hsp90 biology make it clear that ongoing multidisciplinary efforts of basic and clinical investigators are dramatically expanding our appreciation of the possible ways in which inhibition of Hsp90 can be exploited to mount a multifaceted attack on cancer. Further development of additional Hsp90 inhibitors is needed to most effectively take advantage of our rapidly evolving knowledge.

References

Aghajanian C, Soignet S, Dizon DS, Pien CS, Adams J, Elliott PJ, Sabbatini P, Miller V, Hensley ML, Pezzulli S, Canales C, Daud A, Spriggs DR (2002) A phase I trial of the novel proteasome inhibitor PS341 in advanced solid tumor malignancies. Clin Cancer Res 8:2505–2511

Ali A, Bharadwaj S, O'Carroll R, Ovsenek N (1998) HSP90 interacts with and regulates the activity of heat shock factor 1 in Xenopus oocytes. Mol Cell Biol 18:4949–4960

An WG, Schulte TW, Neckers LM (2000) The heat shock protein 90 antagonist geldanamycin alters chaperone association with p210bcr-abl and v-src proteins before their degradation by the proteasome. Cell Growth Differ 11:355–3360

Bagatell R, Paine-Murrieta GD, Taylor CW, Pulcini EJ, Akinaga S, Benjamin IJ, Whitesell L (2000) Induction of a heat shock factor 1-dependent stress response alters the cytotoxic activity of hsp90-binding agents. Clin Cancer Res 6:3312–3318

Bagatell R, Whitesell L (2004) Altered Hsp90 function in cancer: a unique therapeutic opportunity. Mol Cancer Ther 3:1021–1030

Basso AD, Solit DB, Chiosis G, Giri B, Tsichlis P, Rosen N (2002) Akt forms an intracellular complex with heat shock protein 90 (Hsp90) and Cdc37 and is destabilized by inhibitors of Hsp90 function. J Biol Chem 277:39858–39866

Becker B, Multhoff G, Farkas B, Wild PJ, Landthaler M, Stolz W, Vogt T (2004) Induction of Hsp90 protein expression in malignant melanomas and melanoma metastases. Exp Dermatol 13:27–32

Bisht KS, Bradbury CM, Mattson D, Kaushal A, Sowers A, Markovina S, Ortiz KL, Sieck LK, Isaacs JS, Brechbiel MW, Mitchell JB, Neckers LM, Gius D (2003) Geldanamycin and 17-allylamino-17-demethoxygeldanamycin potentiate the in vitro and in vivo radiation response of cervical tumor cells via the heat shock protein 90-mediated intracellular signaling and cytotoxicity. Cancer Res 63:8984–8995

Bonvini P, Gastaldi T, Falini B, Rosolen A (2002) Nucleophosmin-anaplastic lymphoma kinase (NPM-ALK), a novel Hsp90- client tyrosine kinase: down-regulation of NPM-ALK expression and tyrosine phosphorylation in ALK(+) CD30(+) lymphoma cells by the Hsp90 antagonist 17-allylamino,17-demethoxygeldanamycin. Cancer Res 62:1559–1566

Bottaro DP, Liotta LA (2003) Out of air is not out of action. Nature 423:593–595

Chiosis G, Huezo H, Rosen N, Mimnaugh E, Whitesell L, Neckers L (2003) 17AAG: Low target binding affinity and potent cell activity: finding an explanation. Mol Cancer Ther 2:123–129

Chiosis G, Vilenchik M, Kim J, Solit D (2004) Hsp90: the vulnerable chaperone. Drug Discov Today 9:881–888

Dias S, Shmelkov SV, Lam G, Rafii S (2002) VEGF(165) promotes survival of leukemic cells by Hsp90-mediated induction of Bcl-2 expression and apoptosis inhibition. Blood 99:2532–2540

Druker BJ, Tamura S, Buchdunger E, Ohno S, Segal GM, Fanning S, Zimmermann J, Lydon NB (1996) Effects of a selective inhibitor of the Abl tyrosine kinase on the growth of Bcr-Abl positive cells. Nat Med 2:561–566

Eiseman JL, Lan J, Lagattuta TF, Hamburger DR, Joseph E, Covey JM, Egorin MJ (2005) Pharmacokinetics and pharmacodynamics of 17-demethoxy 17-[[(2-dimethylamino)ethyl] amino]geldanamycin (17DMAG, NSC 707545) in C.B-17 SCID mice bearing MDA-MB-231 human breast cancer xenografts. Cancer Chemother Pharmacol 55:21–32

Eustace BK, Jay DG (2004) Extracellular roles for the molecular chaperone, hsp90. Cell Cycle 3:1098–1100

Eustace BK, Sakurai T, Stewart JK, Yimlamai D, Unger C, Zehetmeier C, Lain B, Torella C, Henning SW, Beste G, Scroggins BT, Neckers L, Ilag LL, Jay DG (2004) Functional proteomic screens reveal an essential extracellular role for hsp90 alpha in cancer cell invasiveness. Nat Cell Biol 6:507–514

French BA, van Leeuwen F, Riley NE, Yuan QX, Bardag-Gorce F, Gaal K, Lue YH, Marceau N, French SW (2001) Aggresome formation in liver cells in response to different toxic mechanisms: role of the ubiquitin-proteasome pathway and the frameshift mutant of ubiquitin. Exp Mol Pathol 71:241–246

Fuino L, Bali P, Wittmann S, Donapaty S, Guo F, Yamaguchi H, Wang HG, Atadja P, Bhalla K (2003) Histone deacetylase inhibitor LAQ824 down-regulates Her-2 and sensitizes human breast cancer cells to trastuzumab, taxotere, gemcitabine, and epothilone B. Mol Cancer Ther 2:971–984

Fujimoto J, Shiota M, Iwahara T, Seki N, Satoh H, Mori S, Yamamoto T (1996) Characterization of the transforming activity of p80, a hyperphosphorylated protein in a Ki-1 lymphoma cell line with chromosomal translocation t(2;5). Proc Natl Acad Sci U S A 93:4181–4186

Fumo G, Akin C, Metcalfe DD, Neckers L (2004) 17-Allylamino-17-demethoxygeldanamycin (17-AAG) is effective in down-regulating mutated, constitutively activated KIT protein in human mast cells. Blood 103:1078–1084

Gabai VL, Budagova KR, Sherman MY (2005) Increased expression of the major heat shock protein Hsp72 in human prostate carcinoma cells is dispensable for their viability but confers resistance to a variety of anticancer agents. Oncogene 24:3328–3338

Georget V, Terouanne B, Nicolas J-C, Sultan C (2002) Mechanism of antiandrogen action: key role of Hsp90 in conformational change and transcriptional activity of the androgen receptor. Biochemistry 41:11824–11831

Goetz MP, Toft DO, Ames MM, Erlichman C (2003) The Hsp90 chaperone complex as a novel target for cancer therapy. Ann Oncol 14:1169–1176

Gorre ME, Ellwood-Yen K, Chiosis G, Rosen N, Sawyers CL (2002) BCR-ABL point mutants isolated from patients with STI571-resistant chronic myeloid leukemia remain sensitive to inhibitors of the BCR-ABL chaperone heat shock protein 90. Blood 100:3041–3044

Gradin K, McGuire J, Wenger RH, Kvietikova I, Whitelaw ML, Toftgard R, Tora L, Gassmann M, Poellinger L (1996) Functional interference between hypoxia and dioxin signal transduction pathways: competition for recruitment of the Arnt transcription factor. Mol Cell Biol 16:5221–5531

Guo Y, Guettouche T, Fenna M, Boellmann F, Pratt WB, Toft DO, Smith DF, Voellmy R (2001) Evidence for a mechanism of repression of heat shock factor 1 transcriptional activity by a multichaperone complex. J Biol Chem 276:45791–45799

Hahn WC, Weinberg RA (2002) Modelling the molecular circuitry of cancer. Nat Rev Cancer 2:331–341

Hanahan D, Weinberg RA (2000) The hallmarks of cancer. Cell 100:57–70

Harris AL (2002) Hypoxia- a key regulatory factor in tumor growth. Nat Rev Cancer 2:38–47

Hur E, Kim HH, Choi SM, Kim JH, Yim S, Kwon HJ, Choi Y, Kim DK, Lee MO, Park H (2002) Reduction of hypoxia-induced transcription through the repression of hypoxia-inducible factor-1alpha/aryl hydrocarbon receptor nuclear translocator DNA binding by the 90-kDa heat-shock protein inhibitor radicicol. Mol Pharmacol 62:975–982

Isaacs JS, Jung YJ, Mimnaugh EG, Martinez A, Cuttitta F, Neckers LM (2002) Hsp90 regulates a von Hippel Lindau-independent hypoxia-inducible factor-1 alpha-degradative pathway. J Biol Chem 277:29936–29944

Isaacs JS, Xu W, Neckers L (2003) Heat shock protein 90 as a molecular target for cancer therapeutics. Cancer Cell 3:213–217

Kamal A, Thao L, Sensintaffar J, Zhang L, Boehm MF, Fritz LC, Burrows FJ (2003) A high-affinity conformation of Hsp90 confers tumour selectivity on Hsp90 inhibitors. Nature 425:407–410

Kim HR, Kang HS, Kim HD (1999) Geldanamycin induces heat shock protein expression through activation of HSF1 in K562 erythroleukemic cells. IUBMB Life 48:429–433

Kitano H (2003) Cancer robustness: tumour tactics. Nature 426:125

L'Allemain G (2002) [Update on ... the proteasome inhibitor PS341]. Bull Cancer 89:29–30

La Rosee P, O'Dwyer ME, Druker BJ (2002) Insights from pre-clinical studies for new combination treatment regimens with the Bcr-Abl kinase inhibitor imatinib mesylate (Gleevec/Glivec) in chronic myelogenous leukemia: a translational perspective. Leukemia 16:1213–1219

Lu A, Ran R, Parmentier-Batteur S, Nee A, Sharp FR (2002) Geldanamycin induces heat shock proteins in brain and protects against focal cerebral ischemia. J Neurochem 81:355–364

Mabjeesh NJ, Post DE, Willard MT, Kaur B, Van Meir EG, Simons JW, Zhong H (2002) Geldanamycin induces degradation of hypoxia-inducible factor 1α protein via the proteasome pathway in prostate cancer cells. Cancer Res 62:2478–2482

Machida H, Matsumoto Y, Shirai M, Kubota N (2003) Geldanamycin, an inhibitor of Hsp90, sensitizes human tumour cells to radiation. Int J Radiat Biol 79:973–980

Marcu MG, Chadli A, Bouhouche I, Catelli M, Neckers LM (2000a) The heat shock protein 90 antagonist novobiocin interacts with a previously unrecognized ATP-binding domain in the carboxyl terminus of the chaperone. J Biol Chem 275:37181–37186

Marcu MG, Neckers LM (2003) The C-terminal half of heat shock protein 90 represents a second site for pharmacologic intervention in chaperone function. Curr Cancer Drug Targets 3:343–347

Marcu MG, Schulte TW, Neckers L (2000b) Novobiocin and related coumarins and depletion of heat shock protein 90-dependent signaling proteins. J Natl Cancer Inst 92:242–248

Matthews RC, Rigg G, Hodgetts S, Carter T, Chapman C, Gregory C, Illidge C, Burnie J (2003) Preclinical assessment of the efficacy of mycograb, a human recombinant antibody against fungal HSP90. Antimicrob Agents Chemother 47:2208–2216

Maulik G, Kijima T, Ma PC, Ghosh SK, Lin J, Shapiro GI, Schaefer E, Tibaldi E, Johnson BE, Salgia R (2002) Modulation of the c-Met/hepatocyte growth factor pathway in small cell lung cancer. Clin Cancer Res 8:620–627

Maxwell PH, Wiesener MS, Chang G-W, Clifford SC, Vaux EC, Cockman ME, Wykoff CC, Pugh CW, Maher ER, Ratcliffe PJ (1999) The tumor suppressor protein VHL targets hypoxia-inducible factors for oxygen-dependent proteolysis. Nature 399:271–275

Mimnaugh EG, Chavany C, Neckers L (1996) Polyubiquitination and proteasomal degradation of the p185c-erbB-2 receptor protein-tyrosine kinase induced by geldanamycin. J Biol Chem 271:22796–22801

Mimnaugh EG, Worland PJ, Whitesell L, Neckers LM (1995) Possible role for serine/threonine phosphorylation in the regulation of the heteroprotein complex between the hsp90 stress protein and the pp60v-src tyrosine kinase. J Biol Chem 270:28654–28659

Mimnaugh EG, Xu W, Vos M, Yuan X, Isaacs JS, Bisht KS, Gius D, Neckers L (2004) Simultaneous inhibition of hsp 90 and the proteasome promotes protein ubiquitination, causes endoplasmic reticulum-derived cytosolic vacuolization, and enhances antitumor activity. Mol Cancer Ther 3:551–566

Minami Y, Kiyoi H, Yamamoto Y, Yamamoto K, Ueda R, Saito H, Naoe T (2002) Selective apoptosis of tandemly duplicated FLT3-transformed leukemia cells by Hsp90 inhibitors. Leukemia 16:1535–1540

Mitsiades N, Mitsiades CS, Poulaki V, Chauhan D, Fanourakis G, Gu X, Bailey C, Joseph M, Libermann TA, Treon SP, Munshi NC, Richardson PG, Hideshima T, Anderson KC (2002) Molecular sequelae of proteasome inhibition in human multiple myeloma cells. Proc Natl Acad Sci U S A 99:14374–14379

Naoe T, Kiyoe H, Yamamoto Y, Minami Y, Yamamoto K, Ueda R, Saito H (2001) FLT3 tyrosine kinase as a target molecule for selective antileukemia therapy. Cancer Chemother Pharmacol 48:S27–S30

Neckers L (2002) Hsp90 inhibitors as novel cancer chemotherapeutic agents. Trends Mol Med 8:S55–S61

Nimmanapalli R, O'Bryan E, Huang M, Bali P, Burnette PK, Loughran T, Tepperberg J, Jove R, Bhalla K (2002) Molecular characterization and sensitivity of STI-571 (imatinib mesylate, Gleevec)-resistant, Bcr-Abl-positive, human acute leukemia cells to SRC kinase inhibitor PD180970 and 17-allylamino-17-demethoxygeldanamycin. Cancer Res 62:5761–5769

Pennacchietti S, Michieli P, Galluzzo M, Mazzone M, Giordano S, Comoglio PM (2003) Hypoxia promotes invasive growth by transcriptional activation of the met protooncogene. Cancer Cell 3:347–361

Prodromou C, Pearl LH (2003) Structure and functional relationships of Hsp90. Curr Cancer Drug Targets 3:301–323

Queitsch C, Sangster TA, Lindquist S (2002) Hsp90 as a capacitor of phenotypic variation. Nature 417:618–624

Rutherford SL, Lindquist S (1998) Hsp90 as a capacitor for morphological evolution. Nature 396:336–342

Sangster TA, Queitsch C (2005) The HSP90 chaperone complex, an emerging force in plant development and phenotypic plasticity. Curr Opin Plant Biol 8:86–92

Sawyers CL, Hochhaus A, Feldman E, Goldman JM, Miller CB, Ottmann OG, Schiffer CA, Talpaz M, Guilhot F, Deininger MW, Fischer T, O'Brien SG, Stone RM, Gambacorti-Passerini CB, Russell NH, Reiffers JJ, Shea TC, Chapuis B, Coutre S, Tura S, Morra E, Larson RA, Saven A, Peschel C, Gratwohl A, Mandelli F, Ben-Am M, Gathmann I, Capdeville R, Paquette RL, Druker BJ (2002) Imatinib induces hematologic and cytogenetic responses in patients with chronic myelogenous leukemia in myeloid blast crisis: results of a phase II study. Blood 99:3530–3539

Schneider C, Sepp-Lorenzino L, Nimmesgern E, Ouerfelli O, Danishefsky S, Rosen N, Hartl FU (1996) Pharmacologic shifting of a balance between protein refolding and degradation mediated by Hsp90. Proc Natl Acad Sci U S A 93:14536–14541

Seizinger BR, Rouleau GA, Ozelius LJ, Lane AH, Farmer GE, Lamiell JM, Haines J, Yuen JW, Collins D, Majoor-Krakauer D et al. (1988) Von Hippel-Lindau disease maps to the region of chromosome 3 associated with renal cell carcinoma. Nature 332:268–269

Shah NP, Nicoll JM, Nagar B, Gorre ME, Paquette RL, Kuriyan J, Sawyers CL (2002) Multiple BCR-ABL kinase domain mutations confer polyclonal resistance to the tyrosine kinase inhibitor imatinib (STI571) in chronic phase and blast crisis chronic myeloid leukemia. Cancer Cell 2:117–125

Shiotsu Y, Neckers LM, Wortman I, An WG, Schulte TW, Soga S, Murakata C, Tamaoki T, Akinaga S (2000) Novel oxime derivatives of radicicol induce erythroid differentiation associated with preferential G(1) phase accumulation against chronic myelogenous leukemia cells through destabilization of Bcr-Abl with Hsp90 complex. Blood 96:2284–2291

Siligardi G, Hu B, Panaretou B, Piper PW, Pearl LH, Prodromou C (2004) Co-chaperone regulation of conformational switching in the Hsp90 ATPase cycle. J Biol Chem 279:51989–51998

Solit D, Zheng F, Drobnjak M, Munster P, Higgins B, Verbel D, Heller G, Tong W, Cordon-Cardo C, Agus D, Scher H, Rosen N (2002) 17-allylamino-17-demthoxygeldanamycin induces the degradation of androgen receptor and HER-2/neu and inhibits the growth of prostate cancer xenografts. Clin Cancer Res 986–993

Soti C, Racz A, Csermely P (2002) A Nucleotide-dependent molecular switch controls ATP binding at the C-terminal domain of Hsp90. N-terminal nucleotide binding unmasks a C-terminal binding pocket. J Biol Chem 277:7066–7075

Stoler DL, Chen N, Basik M, Kahlenberg MS, Rodriguez-Bigas MA, Petrelli NJ, Anderson GR (1999) The onset and extent of genomic instability in sporadic colorectal tumor progression. Proc Natl Acad Sci U S A 96:15121–15126

Tacchini L, Dansi P, Matteucci E, Desiderio MA (2001) Hepatocyte growth factor signalling stimulates hypoxia inducible factor-1 (HIF-1) activity in HepG2 hepatoma cells. Carcinogenesis 22:1363–1371

Vanaja DK, Mitchell SH, Toft DO, Young CYF (2002) Effect of geldanamycin on androgen receptor function and stability. Cell Stress Chaperones 7:55–64

Waelter S, Boeddrich A, Lurz R, Scherzinger E, Lueder G, Lehrach H, Wanker EE (2001) Accumulation of mutant huntingtin fragments in aggresome-like inclusion bodies as a result of insufficient protein degradation. Mol Biol Cell 12:1393–1407

Wegele H, Muller L, Buchner J (2004) Hsp70 and Hsp90—a relay team for protein folding. Rev Physiol Biochem Pharmacol 151:1–44

Workman P (2004) Combinatorial attack on multistep oncogenesis by inhibiting the Hsp90 molecular chaperone. Cancer Lett 206:149–157

Xiao N, Callaway CW, Lipinski CA, Hicks SD, DeFranco DB (1999) Geldanamycin provides posttreatment protection against glutamate-induced oxidative toxicity in a mouse hippocampal cell line. J Neurochem 72:95–101

Xu Y, Singer MA, Lindquist S (1999) Maturation of the tyrosine kinase c-src as a kinase and as a substrate depends on the molecular chaperone Hsp90. Proc Natl Acad Sci U S A 96:109–114

Yu X, Guo ZS, Marcu MG, Neckers L, Nguyen DM, Chen GA, Schrump DS (2002) Modulation of p53, ErbB1, ErbB2, and Raf-1 expression in lung cancer cells by depsipeptide FR901228. J Natl Cancer Inst 94:504–513

Zagzag D, Nomura M, Friedlander DR, Blanco C, Gagner JP, Nomura N, Newcomb EW (2003) Geldanamycin inhibits migration of glioma cells in vitro: a potential role for hypoxia-inducible factor (HIF-1alpha) in glioma cell invasion. J Cell Physiol 196:394–402

Zhang H, Burrows F (2004) Targeting multiple signal transduction pathways through inhibition of Hsp90. J Mol Med 82:488–499

Zhao YG, Gilmore R, Leone G, Coffey MC, Weber B, Lee PW (2001) Hsp90 phosphorylation is linked to its chaperoning function. Assembly of the reovirus cell attachment protein. J Biol Chem 276:32822–32827

Heat Shock Proteins in Immunity

G. Multhoff

Department of Hematology and Oncology, University Hospital Regensburg,
Franz-Josef Strauss Allee 11, 93053 Regensburg, Germany
gabriele.multhoff@klinik.uni-regensburg.de

1	Heat Shock Proteins: A Short Overview	280
2	The Four Paradigms: How HSPs Elicit Immune Responses	281
3	HSPs as Peptide Carriers	282
3.1	Cancer	282
3.2	Microbial Infections	287
3.2.1	*Escherichia Coli*	287
3.2.2	Human Immunodeficiency Virus	287
3.2.3	Measles	288
3.2.4	Choriomeningitis	288
3.3	Autoimmune Diseases	289
3.3.1	Rheumatoid Arthritis	289
3.3.2	Behçet's Disease	290
4	HSPs as Chaperokines	290
5	HSPs as Activatory Ligands for Natural Killer Cells	293
	References	296

Abstract This chapter focuses on immunological effects of eukaryotic and microbial heat shock proteins (HSPs), with molecular weights of about 60, 70, and 90 kDa. The search for tumor-specific antigens resulted in the identification of HSPs. They have been found to elicit a potent anti-cancer immune response mediated by the adoptive and innate immune system. Following receptor-mediated uptake of HSP (HSP70 and gp96) peptide complexes by antigen-presenting cells and representation of HSP-chaperoned peptides by MHC class I molecules, a CD8-specific T cell response is induced. Apart from chaperoning immunogenic peptides derived from tumors, bacterial and virally infected cells, they by themselves provide activatory signals for antigen-presenting cells and natural killer (NK) cells. After binding of peptide-free HSP70 to Toll-like receptors, the secretion of pro-inflammatory cytokines is initiated by antigen-presenting cells and thus results in a nonspecific stimulation of the immune system. Moreover, soluble as well as cell membrane-bound HSP70 on tumor cells can directly activate the cytolytic and migratory capacity of NK cells. Apart form cancer, HSPs of different origins, with a molecular weight of about 60, 70, and 90 kDa, also play a pivotal role in viral infections, including human and simian immunodeficiency virus (HIV, SIV), measles, and choriomeningitis. Moreover, HSPs have been found to induce tolerance against autoimmune diseases. In summary, depending on their mode of induction, intracellular/extracellular location, cellular origin (eukaryote/ prokaryote), peptide loading status, intracellular ADP/ATP content, concentration, and route of application, HSPs either exert immune activation as danger signals in cancer immunity and mediate protection against infectious diseases or exhibit regulatory activities in controlling and preventing autoimmunity.

Keywords T cells · Natural killer (NK) cells · Cancer · Microbial infections · Autoimmune diseases

1
Heat Shock Proteins: A Short Overview

Already in 1962, Ritossa established the term "heat shock response" when he discovered chromosome puffs after administration of *Drosophila* gland to elevated temperatures (Ritossa 1962). The importance of heat shock proteins (HSPs) is documented by their high degree of conservation in different species and by their enormous abundance. Under physiological conditions, HSPs can account for approximately 1%–5% of total cellular proteins (Csermely 2001).

Depending on their major molecular weights of about 20, 40, 60, 70, 90, 110, and 170 kDa (Wang et al. 2001), they are grouped into different HSP families (Lindquist and Craig 1988). Among different members of one HSP family, but not among different families, stress proteins are highly homologous in nearly all species. Even prokaryotic and eukaryotic members of the HSP70 family share more than 50% of their amino acid sequences. HSPs inhabit nearly all cellular compartments where they fulfill a broad variety of chaperoning functions, including folding of nascent polypeptides, regulation of protein import and export, assembly and disassembly of macromolecular structures, and support of antigen processing and presentation (Pierce 1994; Hartl 1996; Hartl and Hartl-Hayer 2002). With respect to their multiple functions and their mode of induction, they are also termed molecular chaperones and stress proteins.

Most chaperones are constitutively expressed under physiological conditions. However, a subgroup of them, the so-called glucose-related proteins (GRPs), with molecular weights of 34, 47, 56, 75, 78, 94, and 174 kDa, are inducible by glucose deprivation. A comparison of the composition of some major HSP families in the lysate of one million nonstressed EG7 tumor cells revealed that they contain about 10 ng of the ER-residing gp96, 70 ng HSP70, and 650 ng HSP90 (R.J. Binder, personal communication).

Following environmental stress, the synthesis of HSPs is highly upregulated to prevent cells from lethal damage induced by protein aggregation and misfolding. Except for HSPs, the synthesis of other proteins in general is down-regulated after stress. Apart from heat as the classical inducer, a variety of other stressful stimuli, including oxygen radicals, heavy metals, amino acid analog (Hightower 1980), UV-, gamma-irradiation (Gehrmann et al. 2005), cytostatic drugs (Gehrmann et al. 2002; Ciocca et al. 2003), anti-inflammatory drugs (Gehrmann et al. 2004), nutrient deprivation, bacterial and viral infections, and malignant transformation (Fuller et al. 1994), initiate the synthesis of

HSPs. Furthermore, a substantial increase in the amount of cytosolic HSPs is also detectable during cell replication, differentiation, and developmental processes (Milarski et al. 1989).

HSPs have the unique capacity to noncovalently bind to hydrophobic surfaces of unfolded proteins, but also to polypeptides and peptides. Substrate binding and release are controlled by the intracellular ADP/ATP content. The molecular mechanisms underlying their regulation by heat shock elements (HSEs) and heat shock factors (HSFs) are firmly established and have been extensively described in previous chapters (Morimoto 1993; Sarge et al. 1993).

2
The Four Paradigms: How HSPs Elicit Immune Responses

Due to their high degree of conservation and their relative broad substrate binding capacity, at first sight a specific stimulation of the immune system appeared quite unusual. However, during the last decade evidence has accumulated that HSPs are potent activators of the adoptive and innate immune system against cancer and infectious diseases (Wells and Malkovsky 2000; Wang et al. 2000). In order to shed some light into this paradoxical situation and to formally distinguish how HSPs elicit immune responses, Pramod Srivastava proposed the following four paradigms (Srivastava 1994):

1. Despite of the high degree of sequence homology within different HSP families, some variable regions exist that might function as classical species-specific, foreign antigens for the host's immune system.

2. Due to their stress inducibility and their capacity to transport proteins across membranes, HSPs might be immunogenic because they are expressed in a tissue-specific manner and only in distinct cellular and subcellular compartments.

3. An immune response might also be initiated by molecular mimicry between HSP epitopes and classical non-self antigens.

4. HSP by themselves are not immunogenic but might act as carriers for foreign antigens and thus HSP-chaperoned peptides might be responsible for the initiation of a specific immune response.

It became obvious that for cancer immunity paradigms 1, 2, and 4 are relevant, whereas paradigm 3 seems to be play a role in autoimmune and infectious diseases.

The following section deals with the role of HSPs with molecular weights of 60, 70, and 90 kDa as carriers for tumor-derived peptides. Following receptor-mediated uptake and peptide re-presentation on MHC class I molecules, a tumor-specific CD8-positive T cell response is induced.

3
HSPs as Peptide Carriers

3.1
Cancer

As mentioned earlier, homologous members of a distinct HSP family are present in nearly all cellular compartments including cytosol (Hartl 1996), nucleus, mitochondria, lysosomes (Nylandsted et al. 2004), endosomes, endoplasmic reticulum (Lammert et al. 1997; Nicchitta 1998), and on intracellular and plasma membranes (Multhoff et al. 1995a; Shin et al. 2003). Presently an association of membrane-bound HSP70 family members in detergent-resistant caveolae (Uittenbogaard et al. 1998) and in lipid rafts are discussed (Broquet et al. 2003; Bausero et al., unpublished observations). Furthermore, HSPs also have been detected in body fluids of patients suffering form cancer, infectious or autoimmune diseases, and in supernatants of tissue cultures (Barreto et al. 2003; Wang et al. 2004). This extracellular localization of HSPs with molecular weights of 60, 70, and 90 kDa is frequently associated with the appearance of HSP-specific antibodies. In addition to the induction of a humoral immune response, HSPs have the capacity to elicit cellular immunity. Depending on their intracellular, extracellular, and membranous location and their peptide loading status, a variety of different immunological functions have been established (Pockley 2003). As molecular chaperones, following nonlethal stress, elevated cytosolic and membrane-bound HSP levels are associated with protection toward a second lethal stress stimulus (Gehrmann et al. 2005; Nylandsted et al. 2000). On the other hand, membrane-bound and extracellular localized HSPs also provide danger signals for the immune system (Matzinger 2002). It was hypothesized that soluble HSPs might either originate from necrotic or apoptotic cell death or recruit from viable cells with the capacity to actively release HSPs in lipid vesicles from the endosomal compartment (Barreto et al. 2003; Hunter-Lavin et al. 2004). In any case, tumor-derived HSPs with molecular weights of 60, 70, and 90 kDa, either unloaded or peptide-loaded, have attracted significant attention from an immunological point of view. Srivastava's group was among the first who reported about their roles as adjuvant-free tumor vaccines (Srivastava et al. 1998; Menoret and Chandawarkar 1998). Injection of mice with HSP70 (the stress-inducible Hsp70 and the cognate Hsc70) or HSP90 (the glucose-related protein gp96) peptide complexes purified form the cytosol or endoplasmic reticulum (ER) of tumors generated protective immunity against subsequent tumor challenge in mice. Apart from chaperoning tumor-derived peptides, the ER-residing glycoprotein gp96 is known to interact with cholesterol esterase, fibrillin, thyroglobin, and MHC class II peptides and mediates proper folding of immunoglobulin (Ig) light chain (Nicchitta 1998; Spee et al. 1997). The cytosolic stress proteins, Hsc70 (73-kDa cognate Hsc70) and Hsp70 (72-kDa inducible Hsp70), preferentially bind early folding products, including nascent chains, and support transport of other proteins across membranes (Hartl 1996; Hartl and Hayer-Hartl 2002).

Adoptive transfer experiments of different effector cell populations convincingly demonstrated an involvement of CD8-positive cytotoxic T lymphocytes (CTLs) and of professional antigen-presenting cells, including macrophages, monocytes, and dendritic cells, in protecting mice from tumors from which the HSP preparations were derived (Menoret et al. 1995; Udono et al. 1993, 1994). HSP90 and HSP70 peptide preparations of corresponding normal tissues or of foreign tumors failed to protect mice against subsequent tumor challenge. These data provided a first hint that not HSPs by themselves but rather HSP-chaperoned tumor peptides might be of relevance for the immunostimulatory capacity. This hypothesis was further confirmed by the finding that pretreatment of HSP preparations with ATP that is known to release HSP-bound peptides completely abrogated the tumor preventive effects (Udono et al. 1993, 1994).

It was further shown that extremely small amounts in the range of pico- to nanograms of antigenic peptides, chaperoned by HSPs, were highly efficient in stimulating a CD8 T cell response in mice; in vitro concentrations of about 10 μg were found to be optimal for the stimulation of cytotoxic T lymphocytes. Characterization of gp96-bound peptides revealed that these were quite divergent in both length and sequence (Breloer et al. 1998; Lammert et al. 1997; Nieland et al. 1996). Despite these findings, binding was found to be remarkably stable and resistant to denaturing agents, although sensitive to an ATP treatment (Blachere et al. 1997).

In general, cross-presentation describes the transfer of exogenous peptides into the MHC class I pathway through an endosomal pathway. For HSP-chaperoned tumor peptides, cross-presentation on MHC class I molecules could be determined as the mode of action for stimulating a CD8-positive T cell response (Srivastava et al. 1998; Singh-Jasuja et al. 2000; Doody et al. 2004). However, the mechanism of uptake of HSP peptide complexes by antigen-presenting cells remained elusive until the molecular nature of HSP-specific receptors was identified (Binder et al. 2004). Binding studies revealed that receptor-mediated uptake of HSP peptide complexes into antigen-presenting cells was specific, saturable, and concentration-dependent (Arnold-Schild et al. 1999; Habich et al. 2002; Binder et al. 2000a). These findings provide an explanation why already small amounts of HSP peptide complexes were highly efficient in immunizing against tumors. It is well known that antigen-presenting cells rapidly internalize HSP peptide complexes by receptor-mediated endocytosis. The following paragraph summarizes some important aspects on HSP60-, HSP70-, and HSP90-specific receptors on professional and nonprofessional antigen-presenting cells and on NK cells (Table 1).

Together with the co-factor CD14, also known as the lipopolysaccharide (LPS) receptor, Toll-like receptors TLR2 and TLR4 are important players in the innate immune response against microbial infections (Medzhitov et al.

Table 1 HSP receptors and their non-HSP and HSP ligands

Receptors (cell type)	Non-HSP ligands	HSP ligands	Function
Toll-like			
TLR2/(APC)	HMGB1, LPS	HSP60, HSP70, gp96	Endocytosis
TLR4/(APC)			Chaperokine
CD14 (APC)	LPS		Signaling
TNF family			
CD40 (APC)		Mycobacteria HSP70	Signaling
		HSP70	Endocytosis
Scavenger			
CD91 (APC)	Alpha-2 macroglobulin	HSP70, DnaK, gp96, calreticulin	Adaptor endocytosis
LOX-1 (APC)	LDL	HSP70	Endocytosis
CD36 (APC)	Collagen	gp96	Signaling
SR-A (EC)	Thrombospondin Fucoidin LPS	gp96, calreticulin	Binding
C-type lectin			
CD94/ NKG2C/ A (NK)	HLA E/ HLA A, B, C Peptides		Activation/inhibition
CD94/ NKG2A (NK)	HLA E/ HSP60 Peptides	HSP70	Uptake, activation

APC, antigen-presenting cell; EC, endothelial cell; HMGB1, high mobility group box 1 protein; LDL, low-density lipoprotein; NK, natural killer cell; SR-A, scavenger receptor class A; TNF, tumor necrosis factor; if not indicated otherwise eukaryotic HSPs were used

1997). Toll-like receptors are human homologs of *Drosophila* Toll genes that are involved in the embryonic dorsoventral formation and in the immune response of the fly against fungal infections. In 1998, TLR-4 was identified as the major signal transducer for LPS, a lead component of the cell wall of gram-negative bacteria (Poltorak et al. 1998). Mutant and knock-out mouse experiments revealed that fibrinogen (Smiley et al. 2001), surfactant protein A (Guillot et al. 2002), the extra domain A of fibronectin (Okamura et al. 2001), heparan sulphate (Johnson et al. 2002), soluble hyaluronan (Termeer et al. 2002), and beta defensin 2 (Biragyn et al. 2002) function as natural ligands and activators for TLR4. For the heterodimeric TLR2–TLR4 receptor complex Hsp60 (Vabulas et al. 2001), Hsp70/Hsc70 (Asea et al. 2002), and gp96 (Vabulas et al. 2002) were characterized as interacting partners. The TLR2–

TLR4 receptor complex is frequently associated with the LPS receptor CD14 (Asea et al. 2000). Apart from LPS, high mobility goup box 1 (HMGB1) serves as a ligand for the receptor cluster (Park et al. 2004). TLR2 alone was found to be responsible for the binding of bacterial lipoproteins; a direct interaction with HSPs was not determined.

The interaction of CD154 to CD40 causes activation and differentiation of antigen-presenting cells. Various functions have been reported following contact of CD40 with HSP70s derived from different cellular sources. Interaction with mycobacterial HSP70 was found to mediate calcium-dependent cell signaling and release of CC chemokines, pro-inflammatory cytokines, and nitric oxide NO (Wang et al. 2002; Panjwani et al. 2002), whereas mammalian HSP70s were found to facilitate receptor-mediated endocytosis (Becker et al. 2002). Recently, scavenger receptors (SRs) were found to play a pivotal role in the uptake of HSP peptide complexes. LOX-1 scavenger receptor, with its natural ligand low-density lipoprotein (LDL), also mediates uptake of HSP70s (Delneste et al. 2002). The collagen and thrombospondin receptor CD36 enables signaling of gp96, and scavenger receptor class A (SR-A), the receptor for fucoidin and LPS, binding gp96 and calreticulin is being debated (Berwin et al. 2003).

More data are available on CD91, a member of the LDL family. Gp96 has been found to compete with alpha-2 macroglobulin for binding to CD91 (Binder et al. 2000b; Binder and Srivastava 2004), and CD91-deficient cell lines were unable to represent HSP-chaperoned peptides. The alpha-2 macroglobulin receptor CD91 mediates endocytosis of gp96 (Vabulas et al. 2002), HSP70 (Binder et al. 2000a); and calreticulin (Ogden et al. 2001), induces cytokine release (Asea et al. 2000) and results in NF-kappa B signaling (Basu et al. 2001).

Following uptake of gp96 peptide complexes into antigen-presenting cells, processing, and representation of HSP-bound peptides on MHC class I molecules, a CD8-positive cytotoxic T lymphocyte (CTL) response is initiated. This indicates that gp96 peptide complexes mediate pro-inflammatory signals (Doody et al. 2004; Schild and Rammensee 2000; Singh-Jasuja et al. 2000). A direct involvement of HSP in cross-presentation has recently been documented by genetic studies. Heat shock factor 1 (HSF-1) double knockout mice with a substantially decreased HSP90 and HSP70 expression also have a decreased capacity for cross-presentation of antigens to the MHC class I pathway (Zheng and Li 2004).

A schematic representation of the HSP peptide-mediated cross-priming of CTLs is summarized in Fig. 1. Following receptor-mediated uptake of HSP peptide complexes by antigen-presenting cells, the peptide becomes represented on MHC class I molecules following an endosomal/TAP-dependent pathway, and thus stimulates cytotoxic CD8-positive T cells.

In an effort to improve the efficacy of HSP peptide vaccines, several laboratories designed HSP fusion constructs with bacterial antigens. Huang's group took advantage of superantigens (SAg SEA) assisting HSP in eliciting a potent

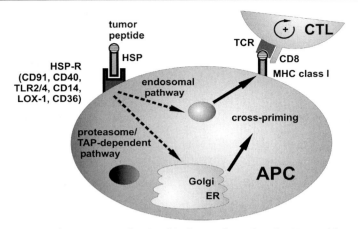

Fig. 1 HSPs as peptide carriers. Following binding and uptake of HSP-peptide complexes, tumor-derived peptides are presented on MHC class I molecules. By cross-presentation, a cytotoxic CD8-positive T cell response is initiated. In addition to Toll-like receptors 2 and 4 (*TLR2/4*), either alone or in combination with LPS receptor CD14, CD40, and alpha-2 macroglobulin receptor CD91, scavenger receptors LOX-1 and CD36 may facilitate uptake and signaling of HSP peptide complexes

anti-tumor immune response (Huang et al. 2000). HSP70-transduced tumor cells bearing SEA transmembrane fusion proteins were used successfully as a modified vaccine prolonging survival of B6 melanoma-bearing mice. The immune response against malignant melanoma was mediated through CTLs and NK cells, as demonstrated by an augmented cell proliferation of both effector cell types, in vivo.

Another approach developed primarily by StressGen Biotechnologies exploits HSP fusion constructs consisting of the viral E7 protein of human papilloma virus type 16 and Bacillus Calmette Guerin (BCG) mycobacterial Hsp65 as a vaccine (Mizzen 1998). Mice immunized with these constructs developed a strong type 1 immune response that mediates tumor regression and confers resistance to tumor challenge with the cervical cancer cell line TC-1. Again an important role for CD8-positive T lymphocytes was determined (Chu et al. 2000). Previous studies using *Mycobacterium leprae* Hsp65 also resulted in loss of tumorigenicity and conferred protection against murine reticulum sarcoma (J774) mediated through both cytotoxic CD4- and CD8-positive T cells (Lukacs et al. 1993).

Based on these encouraging preclinical data, currently, more than 150 clinical centers worldwide are testing the in vivo efficacy of autologous HSP vaccines, predominantly of gp96 and HSP70 peptide complexes, in cancer patients. Patients with highly immunogenic renal cell carcinoma and malignant melanoma are enrolled in clinical phase III trials; patients suffering form chronic myelogenous leukemia, lymphoma, and pancreatic, gastric, and colorectal cancers are presently being treated with gp96 and HSP70 peptide complexes generated from their individual tumors in phase I and II trials (Castelli et al. 2004; Hoos and Levey 2003).

In addition to antigen-presenting cells, the major stress-inducible Hsp70 stimulates natural killer (NK) cells. Binding studies and competition assays identified the C-type lectin receptor CD94 as one potential interacting partner for Hsp70 (Gross et al. 2003a, 2003b). Since CD94 lacks a cytosolic tail, signal transduction is mediated through a co-receptor. Members of the NKG2 family are forming heterodimeric receptors together with CD94. Depending on the intracellular localized long immunotyrosine-based inhibitory motif (ITIM) in NKG2A or the short activatory motif (ITAM) in NKG2C, CD94-positive NK cells become either inhibited or activated after contact with their ligands.

Under physiological conditions, HLA E molecules presenting leader peptides of HLA A, B, and C antigens serve as natural ligands for CD94/NKG2A or CD94/NKG2C (Braud et al. 1998). Following stress, HLA E appears to be associated with an HSP60 peptide derived from the mitochondrial signaling sequence (Kol et al. 2000; Michaelsson et al. 2002). Interestingly, these HLA E–Hsp60 peptide complexes are no longer recognized by the inhibitory CD94/NKG2A receptor complex. Therefore, stressed cells can be eradicated by NK cells even if they carry the inhibitory receptor complex CD94/NKG2A.

3.2
Microbial Infections

3.2.1
Escherichia Coli

Cross-presentation is not restricted to the induction of a CD8-positive T cell response against cancer. A variety of microbial infections also have the capacity to induce cross-presentation (Zugel et al. 2001). In addition to MHC class I, antigen processing and presentation of MHC class II peptides can be affected by microbial HSPs. As an example *E. coli*-derived DnaK, the human HSP70 homolog, delivers an extended ovalbumin (OVA) peptide for MHC class II presentation to CD4-positive T lymphocytes. This mechanism is highly dependent on the acidic pH in the vacuolar system but independent of TLR signaling (Tobian et al. 2004).

3.2.2
Human Immunodeficiency Virus

For human immunodeficiency virus (HIV) infection, cross-presentation of exogenous antigens was demonstrated by SenGupta's group (SenGupta et al. 2004). The clinical outcome of an HIV infection has been found to be associated with a strong and long-lasting cellular immune response (Robinson 2002). Within the HIV Gag p24 region, several peptides were identified that contain overlapping CD8 and CD4 epitopes. Gp96, an ER residing member of the HSP90 family, was found to be effective in the presentation of eight different epitopes from a single 32-mer precursor sequence for MHC class I and class II molecules. These data indicate that gp96 peptide complexes purified from HIV-infected cells might provide a useful tool to stimulate a cellular immune response against a large array of antigens. It is worth mentioning that these effects have been found to be independent of the addition of adjuvants (Suzue et al. 1997). Moreover, HSPs with molecular weights of 70 kDa have been found to provide efficient substitutes for Freund's adjuvants that cannot be used in humans.

Another study by Ahmed et al. (2005) demonstrated that in addition to CD8- and CD4-specific T cell responses, the innate immune system becomes activated by HSP peptide complexes. The role of HSPs as adjuvants in SIV vaccines has been shown to induce the production of stimulatory cytokines and chemokines for gamma/delta T cells that help to control SIV infection. Furthermore, a cell surface expression of HSP70 on HIV-infected cells has been found to stimulate antibody-dependent cellular cytotoxicity in NK cells (Di Cesare et al. 1992).

HIV infectivity is also affected by soluble HSP70 through binding of galactosyl cerebroside, by interaction with CD40, and through the production of CCR5 blocking chemokines. Epitope mapping of the HSP70 sequence revealed that the HIV inhibitory functions reside in the C-terminus (Lehner et al. 2004; Bogers et al. 2004).

3.2.3
Measles

Also in measles infections, HSPs are known to support the host's immune system. Necrosis of infected cells results in release of HSP complexed with viral proteins. After binding to antigen-presenting cells, they are forced to secrete pro-inflammatory cytokines that activate the innate immune system in a nonspecific manner. A second effect is the antigenic cargo of HSP-chaperoned peptides into the MHC presentation pathway. Cross-presentation results in the activation of naïve cytotoxic T lymphocytes against measles virus. Data from a measles mouse model provide evidence that HSPs virus complexes are also involved in the cell-mediated virus clearance from the brain (Oglesbee et al. 2002), thus indicating relevance of HSPs, not only in prevention but also in the therapeutic intervention of a persisting viral disease.

3.2.4
Choriomeningitis

Similar results have been observed in lymphocytic choriomeningitis infections. A mixture of Hsp70 proteins with bound peptides derived from lymphocytic choriomeningitis virus resulted in protection against virus mediated through CD8-positive cytotoxic T lymphocytes (Ciupitu et al. 1998). The finding that even amphibians such as *Xenopus* use an HSP-mediated cross-priming against infections points at the evolutionary conserved nature of this important immune effector mechanism (Robert 2003).

3.3
Autoimmune Diseases

3.3.1
Rheumatoid Arthritis

Due to the high degree of conservation, even prokaryotic and eukaryotic HSPs share a high sequence homology, i.e., human Hsp70 and *E. coli*-derived DnaK share more than 50% of their amino acid sequence. Cross-reactive immune responses of microbial and mammalian HSPs, as outlined in paradigm 3, have been known for a long time. With respect to the development of autoimmune diseases, the question arises whether these reactions are beneficial or harmful for the host (van Eden et al. 1988). It is well known that adjuvant arthritis (AA), a rat model for human rheumatoid arthritis (RA), can be induced by inactivated mycobacteria suspensions in incomplete Freund's adjuvant. Astonishingly, purified mycobacterial Hsp65, a component of the suspension, failed to induce AA in rats. Moreover, preimmunization of rats with full-length Hsp65 protein conferred protection against onset of the disease and was also effective in treating active adjuvant arthritis (Feige and Gasser 1994). Even a peptide (aa 180–188) derived from the mycobacterial Hsp65 sequence had comparable disease-preventive efficacy to full-length Hsp65. On the basis of these findings, it was proposed that cross-recognition of self-Hsp60 might induce peripheral tolerance and thus lead to protection against autoimmune disease (van Eden et al. 1996). Clinically, HSP-induced tolerance could be confirmed by the findings that children with juvenile chronic arthritis with a strong T cell response against Hsp60 had a milder remitting form of disease than children lacking these T cells (Graeff-Meeder et al. 1995). These HSP-activated T cells also have been found to produce anti-inflammatory cytokines, including IL-10 and IL-4. In addition to their roles as stress-inducible targets for the adaptive immune system during inflammation, they trigger factors for the innate immune system via Toll-like receptors (TLRs) (van Eden et al. 2003).

In addition to HSP60-based vaccines, mycobacterial HSP70 vaccines have been found to suppress autoimmune arthritis in rat models (Lamb et al. 1989; Tanaka et al. 1999a; Wendling et al. 2000). The mode of action has not been completely understood; however, one might speculate on low-affinity self-HSP-reactive regulatory T cells that are responsible for the secretion of IL-4, IL-10, and TGF-β. Another possibility would be that the gut environment of microbial HSPs might induce tolerance via a yet undefined mechanism. It is also hypothesized that microbial HSPs are unable to fully stimulate T cells and thus might initiate the generation of regulatory cytokines (Todryk et al. 2003). Along with paradigm 3, these data provide evidence that even slight differences in the sequence of microbial and mammalian HSPs might induce qualitatively distinct immune responses. More recent data from Srivastava's group also indicated that immunizing mice with high doses of gp96 peptide complexes has the capacity to downregulate ongoing immune responses (Chandawarkar et al. 2004). These data have been verified in autoimmune encephalomyelitis and in nonobese diabetic mice. As indicated earlier, low amounts of gp96 peptide complexes elicit efficient tumor rejection in mice, whereas high amounts failed to do so and might rather induce tolerance.

Therefore one might speculate that apart from the cellular origin, the concentration used for immunization determines whether an active immune response or tolerance is induced.

3.3.2
Behçet's Disease

A peptide within the human HSP60 sequence linked to recombinant cholera toxin B subunit prevents uveitis in rats after oral tolerization (Phipps et al. 2003). These encouraging preclinical results initiated a first clinical trial with uveitis in Behçet's disease (Tanaka et al. 1999b). Control of uveitis and extraocular manifestation was found to be associated with reduced CD4-specific T cell counts. In addition to a reduced proliferation of CD4-positive T cells, the expression of co-stimulatory molecules including CD40 and CD28 was found to be downregulated. Furthermore, a decrease in type 1 chemokines such as CCR5 and CXCR3 and in pro-inflammatory cytokines, including IFN-γ and TNF-α, was detectable in the serum of patients that responded to the HSP therapy (Stanford et al. 2004).

4
HSPs as Chaperokines

Recent studies proposed HSPs as dominant danger signals for the host's cellular immune system (Matzinger 2002). However, HSP70s, HSP60s, and HSP90s are classical cytosolic proteins lacking transmembrane domains. Although it became apparent that small amounts of HSP peptide complexes are potent activators of a cytotoxic, CD8-positive T cell response via cross-priming, the exact mechanism of how cytosolic HSPs are exported into the extracellular milieu remains to be determined. Barreto and colleagues demonstrated for the erythroleukemic cell line K562 that pro- and anti-inflammatory cytokines, including interferon gamma (IFN-γ) and IL-10, have the capacity to induce HSP70 synthesis in a manner similar to heat stress (Barreto et al. 2003; Wang et al. 2004). However, only IFN-γ, but not IL-10, is able to reduce surface expression of cognate HSP70 from the plasma membrane of tumor cells. Concomitantly, IFN-γ-pretreated tumor cells revealed an active release of Hsc70 without undergoing apoptotic or necrotic cell death. Presently, it is not absolutely clear as to whether HSP70s are released as free soluble proteins or within detergent-soluble membrane vesicles. Exosomes that have been primarily described as an export system for the transferrin receptor from reticulocytes have recently been discussed as vesicular transport vehicles for HSP70s from the endosomal compartment into the extracellular milieu. Anchorage of HSP70 in the plasma membrane was found to be associated with lipid rafts (Broquet et al. 2003).

From an immunological point of view, it becomes obvious that independent of loaded peptides, HSP70s and HSP90s also have the capacity to bind to HSP receptors, including TLR2/TLR4 and CD14 on antigen-presenting cells. A schematic representation of the role of peptide-free HSP70 on the cytokine inducing effects of antigen presenting cells is illustrated in Fig. 2. Binding of eukaryotic HSP70 to HSP-specific receptors has been found to induce the production and release of pro-inflammatory cytokines, including TNF-α, IL-1 β,

Table 2 Effects of HSP–antigen-presenting cell (APC) and HSP–natural killer (NK) cell interaction

Effect	Mediator	Cell type
Secretion of inflammatory cytokines	TNF-alpha, IL-1 beta, IL-12, IL-6, GM-CSF	Macrophages and DCs
Secretion of chemokines	MCP-1, MIP-1, RANTES	Macrophages
Secretion of immunomodulator	NO	Macrophages and DCs
Upregulation of maturation markers	MHC II, CD86, CD83, CD40	DCs
Stimulation of migration to draining lymph nodes		DCs
Signaling	Nuclear translocation of NF-kappa B	Macrophages and DCs
Cross-presentation	HSP60, HSP70, HSP90	Macrophages and DCs
Activation of migration and cytolytic function	HSP70	NK cells

DC, dendritic cell; *GM-CSF*, granulocyte monocyte-colony stimulating factor; *IL*, interleukin; *MCP-1*, monocyte chemoattractant protein 1; *MHC*, major histocompatibility complex; *MIP-1*, macrophages inflammatory protein 1; *NO*, nitric oxide; *RANTES*, regulated upon activation; normal T cell expressed and secreted; *TNF*, tumor necrosis factor

IL-12, IL-6, and GM-CSF (Lehner et al. 2004; Wang et al. 2002; Asea et al. 2000). Also, microbial HSP70s have been found to stimulate the innate immune system. Using deletion mutants, Lehner's group determined the C-terminal domain of microbial HSP70 as the stimulatory part mediating the synthesis and release of CC chemokines, IL-12, TNF-α, and nitric oxide (NO) (Lehner et al. 2004; Wang et al. 2002; Panjwani et al. 2002). HSP70 competes with CD40 ligand for binding to antigen-presenting cells as determined by the same group. Furthermore, an involvement of HSP70s in stimulation of the migration of dendritic cells to the draining lymph nodes and in maturation of dendritic cells has been determined. Following contact with HSP70, MHC class II and co-stimulatory molecules, including CD86, CD83, and CD40, were found to be upregulated on dendritic cells (Singh-Jasuja et al. 2000; Kuppner et al. 2001). Interaction of monocytes with peptide-free Hsp70 also triggers NF-kappa B translocation into the nucleus and thus initiated an important factor in the signaling transduction pathway of immune responses (Asea et al. 2000). The major effects of an interaction of different HSPs with antigen-presenting cells and NK cells are summarized in Table 2.

Although these findings have been reported independently by several groups, there is still one major concern with respect to minor amounts of endotoxin in the HSP preparations. It is hypothesized that this contamination might be responsible, at least in part, for some of the effects shown in Table 2. As indicated in Table 1, LPS, a major cell wall component of gram-negative bacteria is known to preferentially bind to HSP receptors CD14 and TLR-4 and the scavenger receptor SR-A (Tsan et al. 2004; Bausinger et al. 2002). Unless

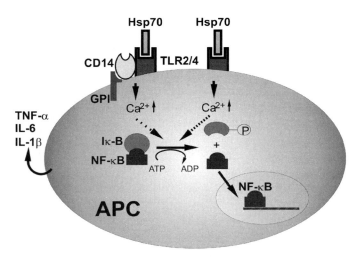

Fig. 2 HSPs as chaperokines. Interaction of peptide-free HSP70 with TLR2/4, either alone or in combination with CD14, results in release of pro-inflammatory cytokines TNF-α, IL-6, and IL-1β via NF-kappa B activation pathway

clearly distinct signal transduction pathways have been elucidated following contact of CD14, TLR4, and SR-A, either with LPS or HSPs, this topic will remain a matter of debate.

5
HSPs as Activatory Ligands for Natural Killer Cells

NK cells comprise between 5% and 20% of the peripheral blood lymphocytes and are well known players in the control of bacteria, parasites, and viruses, but also against cancer (Trinchieri 1989). In contrast to T lymphocytes, NK cells do not require primary stimulation. However, several cytokines, including IL-2, IL-15, or danger signals including LPS or HSPs, can enhance their lytic capacity. Until NK cell-specific receptors were identified, it remained unclear how NK cells distinguish between self and non-self. NK cells lack classical alpha/beta and gamma/delta T cell receptors and the complete multimeric protein complex CD3. For a long period of time, the low-affinity Fc gamma receptor CD16, responsible for antibody-dependent cellular cytotoxicity (ADCC) (Lanier et al. 1988), and the homophilic adhesion molecule CD56 have been recognized as the only NK cell-specific markers. More recently, the molecular nature of a number of killer cell inhibitory and activatory receptors was identified. These receptors either belong into the killer cell immunoglobulin-like receptor (KIR), the immunoglobulin-like transcript (ILT), C-type lectin receptor (Lanier et al. 1998), or the natural cytotoxicity receptor (NCR) (Morretta et al. 2001) families. Depending on their intracellular immunoreceptor tyrosine-based inhibitory or activatory motifs (ITAM/ITIM), these receptors mediate activating and inhibiting signals (Long 1999). A variety of different MHC class I allele groups, including HLA C, were characterized as negative regulatory ligands for NK cells. Unfortunately, less knowledge is available on activating signals. According to the "missing self" theory (Ljunggren and Kärre 1990), tumor cells with an altered or missing MHC expression pattern provide ideal targets for the cytolytic attack mediated by NK cells. However, evidence appears that in addition to "missing self," activatory ligands play important supportive roles in the activation of NK cells. It is speculated that tumor-specific ligands exist for a group of NCRs, including NKp30, NKp44, NKp46, and NKp80 (Morretta et al. 2001). Mandelboim's group (Hanna et al. 2004) provided evidence that viral hemagglutinin also plays a role for the NCRs NKp44 and NKp46.

For the homomeric C-type lectin receptor NKG2D, stress-inducible MICA and MICB, nonclassical MHC molecules, as well as the glycosyl-phosphatidylinositol (GPI-linked) UL-16-binding proteins, the retinoic acid early (RAE-1) protein and HA60, a minor histocompatibility antigen (Lanier et al. 1998; Bauer et al. 1999; Cosman et al. 2004), provide target structures.

Under physiological conditions, nonclassical HLA E molecules presenting leader peptides of HLA A, B, and C alleles serve as natural ligands for the inhibitory receptor complex CD94/NKG2A. Following stress, an HSP60-derived signaling peptide was found to gain access to HLA E. This interaction results in an upregulated surface expression of HLA E–Hsp60 complexes that are no longer recognized by the inhibitory receptor complex CD94/NKG2A (Hickman-Miller et al. 2004; Michaelsson et al. 2002). These data indicate that exogenous stress modulates immune responses of NK cells.

Our group identified the major stress-inducible Hsp70 as an activatory ligand for CD94-expressing NK cells (Gross et al. 2003a, 2003b). Mapping of the Hsp70 sequence revealed that the 14-mer peptide TKDNNLLGRFELSG derived from the C-terminal substrate binding region has similar immunostimulatory capacity on NK cells such as full-length Hsp70 or the C-terminal domain (Botzler et al. 1998; Multhoff et al. 1999). These findings are in line with the work of Colombo's group (Massa et al. 2004) who showed that tumors secreting the inducible Hsp70 displayed an increased immunogenicity against cancer. Mice injected with genetically engineered, Hsp70-secreting tumors showed an increased survival and impaired tumor take due to an activation of the adoptive and innate immune system. The genetic manipulation of tumor cells did not affect the chaperone activity of Hsp70. Tumor rejection was mediated on the one hand via an increased number of dendritic cells mediating a robust CD8-positive T cell response and on the other hand by an enhanced susceptibility of tumors toward NK cell recognition (Multhoff et al. 2000).

Binding studies revealed that similar to antigen-presenting cells, interaction of Hsp70 to NK cells is also saturable and dose-dependent (Gross et al. 2003a, 2003b). Therefore, we speculated about an Hsp70-specific receptor on NK cells. Since most HSP receptors that were found on antigen-presenting cells are only weakly expressed on NK cells, we hypothesized that different receptors might be involved in the interaction with Hsp70. Co-incubation of NK cells with soluble Hsp70 protein has been found to increase not only the cytolytic capacity of NK cells by an upregulated production and release of the pro-apoptotic serin-protease granzyme B, but it also initiates migration (Gastpar et al. 2004). Concomitantly, an increased cell surface density of the C-type lectin receptor CD94 is observed (Gross et al. 2003c). Moreover, antibody blocking studies suggested an important role of CD94 in the interaction of NK cells with Hsp70 (Gross et al. 2003b) (Fig. 3).

A variety of HSPs are found on the plasma membrane of tumor cell lines, as determined by selective cell surface protein profiling by Shin et al. (2003). A broad screening program of human tumor biopsies in our laboratory confirmed these findings. Phenotypic analysis revealed that especially Hsp70, the major stress-inducible member of the HSP70 group is frequently found on the plasma membrane of colon, lung, pancreas, mammary, head and neck tumors and metastases derived thereof (Multhoff et al. 1995a, 1995b; Chen et al. 2002). Also, bone marrow samples of patients suffering from acute and chronic myeloid leukemia are frequently Hsp70 membrane-positive (Gehrmann et al. 2003). Interestingly, the corresponding normal tissues were always Hsp70

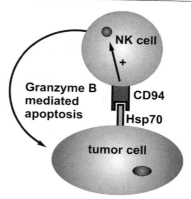

Fig. 3 HSPs as activatory ligands for NK cells. Membrane-bound Hsp70 interacts with the C-type lectin receptor CD94 on NK cells and thus causes perforin-independent, granzyme B-mediated apoptosis in tumor cells

membrane-negative. The Hsp70 density on the cell surface of tumors could be further enhanced by clinically applied reagents, including membrane-interactive alkyl-lysophospholipids (Botzler et al. 1996), cytostatic drugs including taxoids and vincristinsulfate (Gehrmann et al. 2002), cyclooxygenase (COX-1/2) inhibitors, acetyl salicyl acid, and insulin sensitizers (Gehrmann et al. 2004). This stress-inducible, increased Hsp70 surface density correlates with an increased sensitivity of these tumors toward lysis mediated by NK cells.

Hsp70 as a target structure for CD94-positive NK cells was assessed in different tumor cell systems differing profoundly in their capacity to present Hsp70 on their cell surface (Gastpar et al. 2004). Hsp70 protein-blocking studies clearly indicate that Hsp70 but no other associated co-chaperones are responsible for the interaction of NK cells with tumor cells (Gehrmann et al. 2005). The mechanism of lysis of Hsp70 membrane-positive tumor cells was characterized as granzyme B-mediated apoptosis (Gross et al. 2003c). Membrane-bound Hsp70 not only facilitates binding and uptake of granzyme B that was produced by activated NK cells, but also initiates apoptosis in Hsp70 membrane-positive tumors. Affinity chromatography reveals that full-length Hsp70 as well as the 14-mer Hsp70 peptide (aa 450–463), which is exposed to the extracellular milieu of tumor cells (Botzler et al. 1998), both have the capacity to interact with granzyme B. This finding is further supported by work of Liebermann's group (Beresfold et al. 1998), who demonstrated that granzyme B-coupled columns precipitate Hsp70 and Hsp27.

Although the immunological role of membrane-bound Hsp70 appears apparent the transport mechanism and membrane, anchorage remains elusive. Cytosolic HSPs do not contain leader peptides enabling membrane localization; however, transport of other proteins across membranes is one of the tasks of members of the HSP70 family. Nelson Arispe and Antonio DeMaio's group (Arispe et al. 2004) demonstrated a direct interaction of HSP70s with the lipid phosphatidylserine in plasma membranes of PC12 tumor cells. Earlier studies of the same group showed that Hsc70 has the capacity of inducing the ion conductance pathway in artificial bilayers that was regulated by the ATP/ADP content (Arispe and DeMaio 2000). Other laboratories found HSP70 in detergent-soluble microdomains, which are enriched with sphingolipids (Broquet et al. 2003). Work of Triantafilou et al. (2004) demonstrated that HSP70s associate with the TLR4 clusters in lipid rafts after addition of bacterial lipopolysaccharides (LPS). Confocal microscopy further suggested that upon LPS stimulus TLR4 is targeted to the Golgi apparatus along with HSPs. These data indicate that HSP not only support binding and transfer of LPS to the TLR4 complex to the cell surface, but also assist in trafficking and targeting of LPS to the Golgi compartment. The same group also demonstrated that residues 318–387, the base of the binding cleft of HSP70, are critical for LPS binding.

Along with HSP70s gp96 and ER-residing HSP90 members containing the KDEL retention sequence have been found on the plasma membrane of tumors (Altmeyer et al. 1996). It is speculated that by masking or suppressing the ER-retention sequence, gp96 might reach the plasma membrane through the ER-Golgi compartment.

In summary, additional work is required to address functional aspects of HSP70s in lipid rafts and transport mechanisms to the plasma membrane of tumor cells. However, it is apparent that membrane-bound and extracellular HSPs play important roles in stimulating the adoptive and innate immune response against cancer and infectious diseases and might even have the capacity to protect against autoimmune diseases.

References

Ahmed RK, Biberfeld G, Thorstensson R (2005) Innate immunity in experimental SIV infection and vaccination. Mol Immunol 42:251–258

Altmeyer A, Maki RG, Feldweg AM, Heike M, Protopopov VP, Masur SK, Srivastava PK (1996) Tumor-specific cell surface expression of the-KDEL containing, endoplasmic reticular heat shock protein gp96. Int J Cancer 69:340–349

Arispe N, De Maio A (2000) ATP and ADP modulate a cation channel formed by Hsc70 in acidic phospholipid membranes. J Biol Chem 275:30839–30843

Arispe N, Doh M, Simakova O, Kurganov B, De Maio A (2004) Hsc70 and Hsp70 interact with phosphatidylserine on the surface of PC12 cells resulting in a decrease of viability. FASEB J 18:1636–1645

Arnold-Schild D, Hanau D, Spehner D, Schmid C, Rammensee HG, de la Salle H, Schild H (1999) Cutting edge: receptor-mediated endocytosis of heat shock proteins by professional antigen-presenting cells. J Immunol 162:3757–3760

Asea A, Kraeft SK, Kurt-Jones EA, Stevenson MA, Chen LB, Finberg RW, Koo GC, Calderwood SK (2000) HSP70 stimulates cytokine production through a CD14-dependant pathway, demonstrating its dual role as a chaperone and cytokine. Nat Med 6:435–442

Asea A, Rehli M, Kabingu E, Boch JA, Bare O, Auron PE, Stevenson MA, Calderwood SK (2002) Novel signal transduction pathway utilized by extracellular HSP70: role of toll-like receptor (TLR) 2 and TLR4. J Biol Chem 277:15028–15034

Barreto A, Gonzalez JM, Kabingu E, Asea A, Fiorentino S (2003) Stress-induced release of HSC70 from human tumors. Cell Immunol 222:97–104

Basu S, Binder RJ, Ramalingham T, Srivastava PK (2001) CD91: a receptor for heat shock proteins gp96, hsp90, and calreticulin. Immunity 14:303–313

Bauer S, Groh V, Wu J, Steinle A, Phillips JH, Lanier LL, Spies T (1999) Activation of NK cells and T cells by NKG2D, a receptor for stress-inducible MICA. Science 285:727–729

Bausinger H, Lipsker D, Ziylan U, Manie S, Briand JP, Cazenave JP, Muller S, Haeuw JF, Ravanat C, de la SH, Hanau D (2002) Endotoxin-free heat-shock protein 70 fails to induce APC activation. Eur J Immunol 32:3708–3713

Becker T, Hartl FU, Wieland F (2002) CD40, an extracellular receptor for binding and uptake of Hsp70-peptide complexes. J Cell Biol 158:1277–1285

Beresfold PJ, Jaju M, Friedman RS, Yoon MJ, Liebermann J (1988) A role for heat shock protein 27 in CTL-mediated cell death. J Immunol 161:161–167

Berwin B, Hart JP, Pizzo SV, Nicchitta CV (2002) Cutting edge: CD91-independent cross-presentation of GRP94(gp96)-associated peptides. J Immunol 168:4282–4286

Berwin B, Hart JP, Rice S, Gass C, Pizzo SV, Post SR, Nicchitta CV (2003) Scavenger receptor-A mediates gp96/GRP94 and calreticulin internalization by antigen-presenting cells. EMBO J 22:6127–6136

Binder RJ, Harris ML, Menoret A, Srivastava PK (2000a) Saturation, competition, and specificity in interaction of heat shock proteins (hsp) gp96, hsp90, and hsp70 with CD11b+ cells. J Immunol 165:2582–2587

Binder RJ, Han DK, Srivastava PK (2000b) CD91: a receptor for heat shock protein gp96. Nat Immunol 1:151–155

Binder RJ, Vatner R, Srivastava P (2004) The heat-shock protein receptors: some answers and more questions. Tissue Antigens 64:442–451

Binder RJ, Srivastava PK (2004) Essential role of CD91 in re-presentation of gp96-chaperoned peptides. Proc Natl Acad Sci U S A 101:6128–6133

Biragyn A, Ruffini PA, Leifer CA, Klyushnenkova E, Shakhov A, Chertov O, Shirakawa AK, Farber JM, Segal DM, Oppenheim JJ, Kwak LW (2002) Toll-like receptor 4-dependent activation of dendritic cells by beta-defensin 2. Science 298:1025–1029

Blachere NE, Li Z, Chandawarkar RY, Suto R, Jaikaria NS, Basu S, Udono H, Srivastava PK (1997) Heat shock protein-peptide complexes, reconstituted in vitro, elicit peptide-specific cytotoxic T lymphocyte response and tumor immunity. J Exp Med 186:1315–1322

Bogers WM, Bergmeier LA, Oostermeijer H, ten Haaft P, Wang Y, Kelly CG, Singh M, Heeney JL, Lehner T (2004) CCR5 targeted SIV vaccination strategy preventing or inhibiting SIV infection. Vaccine 22:2974–2984

Botzler C, Kolb HJ, Issels RD, Multhoff G (1996) Noncytotoxic alkyl-lysophospholipid treatment increases sensitivity of leukemic K562 cells to lysis by natural killer (NK) cells. Int J Cancer 65:633–638

Botzler C, Li G, Issels RD, Multhoff G (1998) Definition of extracellular localized epitopes of Hsp70 involved in an NK immune response. Cell Stress Chaperones 3:6–11

Braud VM, Allan DS, O'Callaghan CA, Soderstrom K, D'Andrea A, Ogg GS, Lazetic S, Young NT, Bell JI, Phillips JH, Lanier LL, McMichael AJ (1998) HLA-E binds to natural killer cell receptors CD94/NKG2A, B and C. Nature 391:795–799

Breloer M, Marti T, Fleischer B, von Bonin A (1998) Isolation of processed, H-2Kb-binding ovalbumin-derived peptides associated with the stress proteins HSP70 and gp96. Eur J Immunol 28:1016–1021

Broquet AH, Thomas G, Masliah J, Trugnan G, Bachelet M (2003) Expression of the molecular chaperone Hsp70 in detergent-resistant microdomains correlates with its membrane delivery and release. J Biol Chem 278:21601–21606

Castelli C, Rivoltini L, Rini F, Belli F, Testori A, Maio M, Mazzaferro V, Coppa J, Srivastava PK, Parmiani G (2004) Heat shock proteins: biological functions and clinical application as personalized vaccines for human cancer. Cancer Immunol Immunother 53:227–233

Chandawarkar RY, Wagh MS, Kovalchin JT, Srivastava P (2004) Immune modulation with high-dose heat-shock protein gp96: therapy of murine autoimmune diabetes and encephalomyelitis. Int Immunol 16:615–624

Chen X, Tao Q, Yu H, Zhang L, Cao X (2002) Tumor cell membrane-bound heat shock protein 70 elicits antitumor immunity. Immunol Lett 84:81–87

Chu NR, Wu HB, Wu TC, Boux LJ, Mizzen LA, Siegel MI (2000) Immunotherapy of a human papillomavirus type 16 E7-expressing tumor by administration of fusion protein comprised of Mycobacterium bovis BCG Hsp65 and HPV16 E7. Cell Stress Chaperones 5:401–405

Ciocca DR, Rozados VR, Cuello Carrion FD, Gervasoni SI, Matar P, Scharovsky OG (2003) Hsp25 and Hsp70 in rodent tumors treated with doxorubicin and lovastatin. Cell Stress Chaperones 8:26–36

Ciupitu AM, Petersson M, O'Donnell CL, Williams K, Jindal S, Kiessling R, Welsh RM (1998) Immunization with a lymphocytic choriomeningitis virus peptide mixed with heat shock protein 70 results in protective antiviral immunity and specific cytotoxic T lymphocytes. J Exp Med 187:685–691

Cosman D, Mullberg J, Fanslow W, Armitage R, Chin W, Cassiano I (2004) The human cytomegalovirus (HCMV) glycoprotein, UL16, binds to the MHC class I-related protein, MICB/PERB11, and to two novel, MHC class I related molecules ULBP1 and ULBP2. FASEB J 14:1018–1023

Csermely P (2001) A nonconventional role of molecular chaperones: involvement in the cytoarchitecture. News Physiol Sci 16:123–126

Delneste Y, Magistrelli G, Gauchat J, Haeuw J, Aubry J, Nakamura K, Kawakami-Honda N, Goetsch L, Sawamura T, Bonnefoy J, Jeannin P (2002) Involvement of LOX-1 in dendritic cell-mediated antigen cross-presentation. Immunity 17:353–362

Di Cesare S, Poccia F, Mastino A, Colizzi V (1992) Surface expressed heat-shock proteins by stressed or human immunodeficiency virus (HIV)-infected lymphoid cells represent the target for antibody-dependent cellular cytotoxicity. Immunology 76:341–343

Doody AD, Kovalchin JT, Mihalyo MA, Hagymasi AT, Drake CG, Adler AJ (2004) Glycoprotein 96 can chaperone both MHC class I- and class II-restricted epitopes for in vivo presentation, but selectively primes CD8+ T cell effector function. J Immunol 172:6087–6092

Feige U, Gasser J (1994) Therapeutic intervention with mycobacterial 65 kDa heat shock protein peptide 180–188 in adjuvant arthritis in Lewis rats. Immunobiol 191:281–288

Fuller KJ, Issels RD, Slosman DO, Guillet JG, Soussi T, Polla BS (1994) Cancer and the heat shock response. Eur J Cancer 30:1884–1891

Gastpar R, Gross C, Rossbacher L, Ellwart J, Riegger J, Multhoff G (2004) The cell surface-localized heat shock protein 70 epitope TKD induces migration and cytolytic activity selectively in human NK cells. J Immunol 172:972–980

Gehrmann M, Pfister K, Hutzler P, Gastpar R, Margulis B, Multhoff G (2002) Effects of antineoplastic agents on cytoplasmic and membrane-bound heat shock protein 70 (Hsp70) levels. Biol Chem 383:1715–1725

Gehrmann M, Schmetzer H, Eissner G, Haferlach T, Hiddemann W, Multhoff G (2003) Membrane-bound heat shock protein 70 (Hsp70) in acute myeloid leukemia: a tumor specific recognition structure for the cytolytic activity of autologous NK cells. Haematologica 88:474–476

Gehrmann M, Brunner M, Pfister K, Reichle A, Kremmer E, Multhoff G (2004) Differential up-regulation of cytosolic and membrane-bound heat shock protein 70 in tumor cells by anti-inflammatory drugs. Clin Cancer Res 10:3354–3364

Gehrmann M, Marienhagen J, Eichholtz-Wirth H, Fritz E, Ellwart J, Jaattela M, Zilch T, Multhoff G (2005) Dual function of membrane-bound heat shock protein 70 (Hsp70), Bag-4, and Hsp40: protection against radiation-induced effects and target structure for natural killer cells. Cell Death Differ 12:38–51

Graeff-Meeder ER, van Eden W, Rijkers GT, Prakken BJ, Kuis W, Voorhorst-Ogink MM, van der ZR, Schuurman HJ, Helders PJ, Zegers BJ (1995) Juvenile chronic arthritis: T cell reactivity to human HSP60 in patients with a favorable course of arthritis. J Clin Invest 95:934–940

Gross C, Hansch D, Gastpar R, Multhoff G (2003a) Interaction of heat shock protein 70 peptide with NK cells involves the NK receptor CD94. Biol Chem 384:267–279

Gross C, Schmidt-Wolf IG, Nagaraj S, Gastpar R, Ellwart J, Kunz-Schughart LA, Multhoff G (2003b) Heat shock protein 70-reactivity is associated with increased cell surface density of CD94/CD56 on primary natural killer cells. Cell Stress Chaperones 8:348–360

Gross C, Koelch W, DeMaio A, Arispe N, Multhoff G (2003c) Cell surface-bound heat shock protein 70 (Hsp70) mediates perforin-independent apoptosis by specific binding and uptake of granzyme B. J Biol Chem 278:41173–41181

Guillot L, Balloy V, McCormack FX, Golenbock DT, Chignard M, Si-Tahar M (2002) Cutting edge: the immunostimulatory activity of the lung surfactant protein-A involves Toll-like receptor 4. J Immunol 168:5989–5992

Habich C, Baumgart K, Kolb H, Burkart V (2002) The receptor for heat shock protein 60 on macrophages is saturable, specific, and distinct from receptors for other heat shock proteins. J Immunol 168:569–576

Hanna J, Bechtel P, Zhai Y, Youssef F, McLachlan K, Mandelboim O (2004) Novel insights on human NK cells' immunological modalities revealed by gene expression profiling. J Immunol 173:6547–6563

Hartl FU (1996) Molecular chaperones in cellular protein folding. Nature 381:571–579

Hartl FU, Hayer-Hartl M (2002) Molecular chaperones in the cytosol: from nascent chain to folded protein. Science 295:1852–1858

Hickman-Miller HD, Hildebrand WH (2004) The immune response under stress: the role of HSP-derived peptides. Trends Immunol 25:427–433

Hightower LE (1980) Cultured animal cells exposed to amino acid analogues or puromycin rapidly synthesize several polypeptides. J Cell Physiol 102:407–427

Hoos A, Levey DL (2003) Vaccination with heat shock protein-peptide complexes: from basic science to clinical applications. Expert Rev Vaccines 2:369–379

Huang Q, Richmond JF, Suzue K, Eisen HN, Young RA (2000) In vivo cytotoxic T lymphocyte elicitation by mycobacterial heat shock protein 70 fusion proteins maps to a discrete domain and is CD4(+) T cell independent. J Exp Med 191:403–408

Hunter-Lavin C, Davies EL, Bacelar MM, Marshall MJ, Andrew SM, Williams JH (2004) Hsp70 release from peripheral blood mononuclear cells. Biochem Biophys Res Commun 324:511–517

Johnson GB, Brunn GJ, Kodaira Y, Platt JL (2002) Receptor-mediated monitoring of tissue well-being via detection of soluble heparan sulfate by Toll-like receptor 4. J Immunol 168:5233–5239

Kol A, Lichtman AH, Finberg RW, Libby P, Kurt-Jones EA (2000) Cutting edge: heat shock protein (HSP) 60 activates the innate immune response: CD14 is an essential receptor for HSP60 activation of mononuclear cells. J Immunol 164:13–17

Kuppner MC, Gastpar R, Gelwer S, Nossner E, Ochmann O, Scharner A, Issels RD (2001) The role of heat shock protein (hsp70) in dendritic cell maturation: hsp70 induces the maturation of immature dendritic cells but reduces DC differentiation from monocyte precursors. Eur J Immunol 31:1602–1609

Lamb JR, Bal V, Mendez-Samperio P, Mehlert A, So A, Rothbard J, Jindal S, Young RA, Young DB (1989) Stress proteins may provide a link between the immune response to infection and autoimmunity. Int Immunol 1:191–196

Lammert E, Arnold D, Nijenhuis M, Momburg F, Hammerling GJ, Brunner J, Stevanovic S, Rammensee HG, Schild H (1997) The endoplasmic reticulum-resident stress protein gp96 binds peptides translocated by TAP. Eur J Immunol 27:923–927

Lanier LL, Ruitenberg JJ, Phillips JH (1988) Functional and biochemical analysis of CD16 antigen on natural killer cells and granulocytes. J Immunol 141:3478–3485

Lanier LL, Corliss B, Wu J, Phillips JH (1998) Association of DAP12 with activating CD94/NKG2C NK cell receptors. Immunity 8:693–701

Lehner T, Wang Y, Whittall T, McGowan E, Kelly CG, Singh M (2004) Functional domains of HSP70 stimulate generation of cytokines and chemokines, maturation of dendritic cells and adjuvanticity. Biochem Soc Trans 32:629–632

Lindquist S, Craig EA (1988) The heat shock proteins. Annu Rev Genet 22:631–677

Ljunggren HG, Karre K (1990) In search of the 'missing self': MHC molecules and NK cell recognition. Immunol Today 11:237–244

Long EO (1999) Regulation of immune responses through inhibitory receptors. Annu Rev Immunol 17:875–904

Lukacs KV, Lowrie DB, Stokes RW, Colston MJ (1993) Tumor cells transfected with a bacterial heat-shock gene lose tumorigenicity and induce protection against tumors. J Exp Med 178:343–348

Massa C, Guiducci C, Arioli I, Parenza M, Colombo MP, Melani C (2004) Enhanced efficacy of tumor cell vaccines transfected with secretable hsp70. Cancer Res 64:1502–1508

Matzinger P (2002) The danger model: a renewed sense of self. Science 296:301–305

Medzhitov R, Preston-Hurlburt P, Janeway CA Jr (1997) A human homologue of the Drosophila Toll protein signals activation of adaptive immunity. Nature 388:394–397

Menoret A, Patry Y, Burg C, Le Pendu J (1995) Co-segregation of tumor immunogenicity with expression of inducible but not constitutive hsp70 in rat colon carcinomas. J Immunol 155:740–747

Menoret A, Chandawarkar R (1998) Heat-shock protein-based anticancer immunotherapy: an idea whose time has come. Semin Oncol 25:654–660

Michaelsson J, Teixeira DM, Achour A, Lanier LL, Karre K, Soderstrom K (2002) A signal peptide derived from hsp60 binds HLA-E and interferes with CD94/NKG2A recognition. J Exp Med 196:1403–1414

Milarski KL, Welch WJ, Morimoto RI (1989) Cell cycle-dependent association of HSP70 with specific cellular proteins. J Cell Biol 108:413–423

Mizzen L (1998) Immune responses to stress proteins: applications to infectious disease and cancer. Biotherapy 10:173–189

Moretta A, Bottino C, Vitale M, Pende D, Cantoni C, Mingari MC, Biassoni R, Moretta L (2001) Activating receptors and co-receptors involved in human natural killer cell-mediated cytolysis. Annu Rev Immunol 19:197–223

Morimoto RI (1993) Cells in stress: transcriptional activation of heat shock genes. Science 259:1409–1410

Multhoff G, Botzler C, Wiesnet M, Muller E, Meier T, Wilmanns W, Issels RD (1995a) A stress-inducible 72-kDa heat-shock protein (HSP72) is expressed on the surface of human tumor cells, but not on normal cells. Int J Cancer 61:272–279

Multhoff G, Botzler C, Wiesnet M, Eissner G, Issels R (1995b) CD3- large granular lymphocytes recognize a heat-inducible immunogenic determinant associated with the 72-kD heat shock protein on human sarcoma cells. Blood 86:1374–1382

Multhoff G, Botzler C, Jennen L, Ellwart J, Issels R (1997) Heat shock protein cell surface expression on colon carcinoma cells correlates with the sensitivity to lysis mediated by NK cells. J Immunol 158:4341–4350

Multhoff G, Mizzen L, Winchester CC, Milner CM, Wenk S, Eissner G, Kampinga HH, Laumbacher B, Johnson J (1999) Heat shock protein 70 (Hsp70) stimulates proliferation and cytolytic activity of natural killer cells. Exp Hematol 27:1627–1636

Multhoff G, Pfister K, Botzler C, Jordan A, Scholz R, Schmetzer H, Burgstahler R, Hiddemann W (2000) Adoptive transfer of human natural killer cells in mice with severe combined immunodeficiency inhibits growth of Hsp70-expressing tumors. Int J Cancer 88:791–797

Nicchitta CV (1998) Biochemical, cell biological and immunological issues surrounding the endoplasmic reticulum chaperone GRP94/gp96. Curr Opin Immunol 10:103–109

Nieland TJ, Tan MC, Monne-van Muijen M, Koning F, Kruisbeek AM, van Bleek GM (1996) Isolation of an immunodominant viral peptide that is endogenously bound to the stress protein GP96/GRP94. Proc Natl Acad Sci U S A 93:6135–6139

Nylandsted J, Gyrd-Hansen M, Danielewicz A, Fehrenbacher N, Lademann U, Hoyer-Hansen M, Weber E, Multhoff G, Rohde M, Jaattela M (2004) Heat shock protein 70 promotes cell survival by inhibiting lysosomal membrane permeabilization. J Exp Med 200:425–435

Nylandsted J, Rohde M, Brand K, Bastholm L, Elling F, Jaattela M (2000) Selective depletion of heat shock protein 70 (Hsp70) activates a tumor-specific death program that is independent of caspases and bypasses Bcl-2. Proc Natl Acad Sci U S A 97:7871–7876

Ogden CA, deCathelineau A, Hoffmann PR, Bratton D, Ghebrehiwet B, Fadok VA, Henson PM (2001) C1q and mannose binding lectin engagement of cell surface calreticulin and CD91 initiates macropinocytosis and uptake of apoptotic cells. J Exp Med 194:781–795

Oglesbee MJ, Pratt M, Carsillo T (2002) Role for heat shock proteins in the immune response to measles virus infection. Viral Immunol 15:399–416

Okamura Y, Watari M, Jerud ES, Young DW, Ishizaka ST, Rose J, Chow JC, Strauss JF III (2001) The extra domain A of fibronectin activates Toll-like receptor 4. J Biol Chem 276:10229–10233

Panjwani NN, Popova L, Srivastava PK (2002) Heat shock proteins gp96 and hsp70 activate the release of nitric oxide by APCs. J Immunol 168:2997–3003

Park JS, Svetkauskaite D, He Q, Kim JY, Strassheim D, Ishizaka A, Abraham E (2004) Involvement of toll-like receptors 2 and 4 in cellular activation by high mobility group box 1 protein. J Biol Chem 279:7370–7377

Phipps PA, Stanford MR, Sun JB, Xiao BG, Holmgren J, Shinnick T, Hasan A, Mizushima Y, Lehner T (2003) Prevention of mucosally induced uveitis with a HSP60-derived peptide linked to cholera toxin B subunit. Eur J Immunol 33:224–232

Pierce SK (1994) Molecular chaperones in the processing and presentation of antigen to helper T cells. Experientia 50:1026–1030

Pockley AG (2003) Heat shock proteins as regulators of the immune response. Lancet 362:469–476

Poltorak A, He X, Smirnova I, Liu MY, Van Huffel C, Du X, Birdwell D, Alejos E, Silva M, Galanos C, Freudenberg M, Ricciardi-Castagnoli P, Layton B, Beutler B (1998) Defective LPS signaling in C3H/HeJ and C57BL/10ScCr mice: mutations in Tlr4 gene. Science 282:2085–2088

Ritossa FM (1962) A new puffing pattern induced by a temperature shock and DNP in Drosophila. Experientia 18:571–573

Robert J (2003) Evolution of heat shock protein and immunity. Dev Comp Immunol 27:449–464

Robinson HL (2002) New hope for an AIDS vaccine. Nat Rev Immunol 2:239–250

Sarge KD, Murphy SP, Morimoto RI (1993) Activation of heat shock gene transcription by heat shock factor 1 involves oligomerization, acquisition of DNA-binding activity, and nuclear localization and can occur in the absence of stress. Mol Cell Biol 13:1392–1407

Schild H, Rammensee HG (2000) gp96—the immune system's Swiss army knife. Nat Immunol 1:100–101

SenGupta D, Norris PJ, Suscovich TJ, Hassan-Zahraee M, Moffett HF, Trocha A, Draenert R, Goulder PJ, Binder RJ, Levey DL, Walker BD, Srivastava PK, Brander C (2004) Heat shock protein-mediated cross-presentation of exogenous HIV antigen on HLA class I and class II. J Immunol 173:1987–1993

Shin BK, Wang H, Yim AM, Le Naour F, Brichory F, Jang JH, Zhao R, Puravs E, Tra J, Michael CW, Misek DE, Hanash SM (2003) Global profiling of the cell surface proteome of cancer cells uncovers an abundance of proteins with chaperone function. J Biol Chem 278:7607–7616

Singh-Jasuja H, Toes RE, Spee P, Munz C, Hilf N, Schoenberger SP, Ricciardi-Castagnoli P, Neefjes J, Rammensee HG, Arnold-Schild D, Schild H (2000) Cross-presentation of glycoprotein 96-associated antigens on major histocompatibility complex class I molecules requires receptor-mediated endocytosis. J Exp Med 191:1965–1974

Smiley ST, King JA, Hancock WW (2001) Fibrinogen stimulates macrophage chemokine secretion through toll-like receptor 4. J Immunol 167:2887–2894

Spee P, Neefjes J (1997) TAP-translocated peptides specifically bind proteins in the endoplasmic reticulum, including gp96, protein disulfide isomerase and calreticulin. Eur J Immunol 27:2441–2449

Srivastava PK (1994) Heat shock proteins in immune response to cancer: the fourth paradigm. Experientia 50:1054–1060

Srivastava PK, Menoret A, Basu S, Binder RJ, McQuade KL (1998) Heat shock proteins come of age: primitive functions acquire new roles in an adaptive world. Immunity 8:657–665

Stanford M, Whittall T, Bergmeier LA, Lindblad M, Lundin S, Shinnick T, Mizushima Y, Holmgren J, Lehner T (2004) Oral tolerization with peptide 336–351 linked to cholera toxin B subunit in preventing relapses of uveitis in Behçet's disease. Clin Exp Immunol 137:201–208

Suzue K, Zhou X, Eisen HN, Young RA (1997) Heat shock fusion proteins as vehicles for antigen delivery into the major histocompatibility complex class I presentation pathway. Proc Natl Acad Sci U S A 94:13146–13151

Tanaka S, Kimura Y, Mitani A, Yamamoto G, Nishimura H, Spallek R, Singh M, Noguchi T, Yoshikai Y (1999a) Activation of T cells recognizing an epitope of heat-shock protein 70 can protect against rat adjuvant arthritis. J Immunol 163:5560–5565

Tanaka T, Yamakawa N, Koike N, Suzuki J, Mizuno F, Usui M (1999b) Behçet's disease and antibody titers to various heat-shock protein 60s. Ocul Immunol Inflamm 7:69–74

Termeer C, Benedix F, Sleeman J, Fieber C, Voith U, Ahrens T, Miyake K, Freudenberg M, Galanos C, Simon JC (2002) Oligosaccharides of hyaluronan activate dendritic cells via toll-like receptor 4. J Exp Med 195:99–111

Tobian AA, Canaday DH, Boom WH, Harding CV (2004) Bacterial heat shock proteins promote CD91-dependent class I MHC cross-presentation of chaperoned peptide to CD8+ T cells by cytosolic mechanisms in dendritic cells versus vacuolar mechanisms in macrophages. J Immunol 172:5277–5286

Todryk SM, Gough MJ, Pockley AG (2003) Facets of heat shock protein 70 show immunotherapeutic potential. Immunology 110:1–9

Triantafilou M, Triantafilou K (2004) Heat-shock protein 70 and heat-shock protein 90 associate with Toll-like receptor 4 in response to bacterial lipopolysaccharide. Biochem Soc Trans 32:636–639

Trinchieri G (1989) Biology of natural killer cells. Adv Immunol 47:187–376

Tsan MF, Gao B (2004) Endogenous ligands of Toll-like receptors. J Leukoc Biol 76:514–519

Udono H, Levey DL, Srivastava PK (1994) Cellular requirements for tumor-specific immunity elicited by heat shock proteins: tumor rejection antigen gp96 primes CD8+ T cells in vivo. Proc Natl Acad Sci U S A 91:3077–3081

Udono H, Srivastava PK (1993) Heat shock protein 70-associated peptides elicit specific cancer immunity. J Exp Med 178:1391–1396

Uittenbogaard A, Ying Y, Smart EJ (1998) Characterization of a cytosolic heat shock protein-caveolin chaperone complex. Involvement in cholesterol trafficking. J Biol Chem 273:6525–6532

Vabulas RM, Ahmad-Nejad P, Da Costa C, Miethke T, Kirschning CJ, Hacker H, Wagner H (2001) Endocytosed HSP60s use toll-like receptor 2 (TLR2) and TLR4 to activate the toll/interleukin-1 receptor signaling pathway in innate immune cells. J Biol Chem 276:31332–31339

Vabulas RM, Braedel S, Hilf N, Singh-Jasuja H, Herter S, Ahmad-Nejad P, Kirschning CJ, Da Costa C, Rammensee HG, Wagner H, Schild H (2002) The endoplasmic reticulum-resident heat shock protein Gp96 activates dendritic cells via the Toll-like receptor 2/4 pathway. J Biol Chem 277:20847–20853

Van Eden W, Thole JE, van der Zee R, Noordzij A, van Embden JD, Hensen EJ, Cohen IR (1988) Cloning of the mycobacterial epitope recognized by T lymphocytes in adjuvant arthritis. Nature 331:171–173

Van Eden W, Anderton SM, van der Zee R, Prakken BJ, Broeren CP, Wauben MH (1996) (Altered) self peptides and the regulation of self reactivity in the peripheral T cell pool. Immunol Rev 149:55–73

Van Eden W, Koets A, van Kooten P, Prakken B, van der ZR (2003) Immunopotentiating heat shock proteins: negotiators between innate danger and control of autoimmunity. Vaccine 21:897–901

Wang XY, Kaneko Y, Repasky E, Subjeck JR (2000) Heat shock proteins and cancer immunotherapy. Immunol Invest 29:131–137

Wang XY, Kazim L, Repasky EA, Subjeck JR (2001) Characterization of heat shock protein 110 and glucose-regulated protein 170 as cancer vaccines and the effect of fever-range hyperthermia on vaccine activity. J Immunol 166:490–497

Wang Y, Kelly CG, Singh M, McGowan EG, Carrara AS, Bergmeier LA, Lehner T (2002) Stimulation of Th1-polarizing cytokines, CC chemokines, maturation of dendritic cells, and adjuvant function by the peptide binding fragment of heat shock protein 70. J Immunol 169:2422–2429

Wang M-H, Grossman ME, Young CYE (2004) Forced expression of Hsp70 increases the secretion of Hsp70 and provides protection against tumor growth. Br J Cancer 90:926–931

Wells AD, Malkovsky M (2000) Heat shock proteins, tumor immunogenicity and antigen presentation: an integrated view. Immunol Today 21:129–132

Wendling U, Paul L, van der Zee R, Prakken B, Singh M, van Eden W (2000) A conserved mycobacterial heat shock protein (hsp) 70 sequence prevents adjuvant arthritis upon nasal administration and induces IL-10-producing T cells that cross-react with the mammalian self-hsp70 homologue. J Immunol 164:2711–2717

Zheng H, Li Z (2004) Cutting edge: cross-presentation of cell-associated antigens to MHC class I molecule is regulated by a major transcription factor for heat shock proteins. J Immunol 173:5929–5933

Zugel U, Sponaas AM, Neckermann J, Schoel B, Kaufmann SH (2001) gp96-peptide vaccination of mice against intracellular bacteria. Infect Immun 69:4164–4167

Molecular Chaperones and Cancer Immunotherapy

X.-Y. Wang[1] (✉) · J.G. Facciponte[2] · J.R. Subjeck[1]

[1]Department of Cellular Stress Biology and Urologic Oncology, Roswell Park Cancer Institute, Buffalo NY, 14263, USA
xiang-yang.wang@roswellpark.org

[2]Department of Immunology, Roswell Park Cancer Institute, Buffalo NY, 14263, USA

1	Introduction	306
2	Stress Proteins as Molecular Chaperones	306
3	Chaperones as Tumor-Rejection Antigens and Peptide-Binding Proteins	307
4	Chaperones and Immune-Modulating Effects	309
5	Potential Use of Chaperones in Cancer Immunotherapy	310
5.1	Tumor-Derived Chaperones Vaccine	310
5.2	Reconstituted Chaperone–Peptide Complex	312
5.3	Chaperone–Antigen Protein Complex	312
5.4	Chaperone–Antigen Fusion Gene	314
5.5	Chaperone–Antigen Fusion Protein	315
5.6	Chaperone-Based Cell Vaccine	316
5.6.1	DCs Pulsed with Chaperone Vaccine	316
5.6.2	Tumor Cells Expressing Engineered Chaperones	316
6	Clinical Application and Efficacy of Chaperone-Based Immunotherapy	317
7	Concluding Remarks	320
	References	320

Abstract As one of the most abundant and evolutionarily conserved intracellular proteins, heat shock proteins, also known as stress proteins or molecular chaperones, perform critical functions in maintaining cell homeostasis under physiological as well as stress conditions. Certain chaperones in extracellular milieu are also capable of modulating innate and adaptive immunity due to their ability to chaperone polypeptides and to interact with the host's immune system, particularly professional antigen-presenting cells. The immunomodulating properties of chaperones have been exploited for cancer immunotherapy. Clinical trials using chaperone-based vaccines to treat various malignancies are ongoing.

Keywords Molecular chaperone · Heat shock proteins · Immunity · Vaccine · Cancer immunotherapy

1
Introduction

Molecular identification of many tumor-associated antigens has provided targets for the development of new immunotherapy for the treatment of cancer (Novellino et al. 2004). In the heat shock protein field, the roles of HSPs/molecular chaperones in tumor immunity and their effective applications in cancer therapy have gained increasing attention ever since they were initially identified as tumor-rejection antigens (Ullrich et al. 1986). Immunoregulatory functions of chaperones can be grouped into three tenets: The first is the ability of chaperones to bind to a myriad of tumor-associated peptides/proteins. The second is the existence of specific receptors on the surfaces of antigen-presenting cells (APCs), which allows efficient uptake of chaperones–peptide/protein complexes. This process results in cross-presentation of exogenous antigens shuttled by chaperones and subsequent priming of antigen-specific cytotoxic T cell (CTL) responses. Thirdly, chaperones interact and stimulate innate immune components (e.g., APCs or NK cells), which help initiate adaptive immune responses (i.e., activation of $CD8^+$ CTL and $CD4^+$ T helper cells) (Doody et al. 2004; Srivastava 2002b). Discovery that chaperone preparation from autologous tumor is applicable to the treatment of cancer has opened up a new perspective in immunotherapy. Human clinical trials using tumor-derived chaperones for vaccination are now underway and have shown promising results. In addition to tumor-derived chaperone vaccines, novel strategies for chaperone-based vaccination are being developed and evaluated. This review summarizes the immunomodulating functions of molecular chaperones and their potential application in cancer immunotherapy.

2
Stress Proteins as Molecular Chaperones

It has been known for the last quarter of a century that heat shock selectively induces the expression of a set of proteins called heat shock protein (HSP) (Tissieres et al. 1974). The major HSPs of mammalian cells are divided into several families based on their molecular mass. These include the hsp20–30, hsp50–60, hsp70, hsp90, and hsp110 families. There is little obvious amino acid homology between HSP families, although members of individual families are closely related. These proteins are also called stress proteins since they are strongly induced by other stresses, i.e., ethanol, amino acid analogs, oxidative reagents, recovery from anoxia, heavy metals, and inflammation (Black and Subjeck 1991). While HSPs reside primarily in the cytosol, nucleus, and mitochondria, a second and distinct family of stress proteins known as glucose-regulated proteins (GRPs) resides in the endoplasmic reticulum (ER). These proteins differ from HSPs in their cellular compartmentalization and regulation. GRPs are induced by glucose starvation, reducing agents, anoxia, heavy metals, amino acid analogs, reagents interfering with calcium homeostasis and inhibitors of glycosylation (Easton et al. 2000; Subjeck and Shyy 1986). The primary GRPs include grp78/BiP (a hsp70 homolog), grp94 (a hsp90 homolog), and grp170 (a hsp110 homolog). Therefore, GRPs are basically HSPs with regard to function and sequence homology. The cytosol is in a reductive state while the ER is in an oxidative state. The redox balance across the ER membrane appears to play a key role in a reciprocal regulation of these two sets of stress proteins (Sciandra and Subjeck 1983; Sciandra et al. 1984; Whelan and Hightower 1985). As the most abundant and ubiquitous intracellular proteins, stress proteins are considered to have essential roles in the survival of cells due to their diverse functions.

The functional state of a newly synthesized protein is strictly associated with the acquisition of a unique three-dimensional structure. However, it is now recognized that the primary amino acid sequence is not sufficient to determine tertiary protein structure with physiological functions (Johnson and Craig 1997). Stress proteins, as molecular chaperones, act in a coordinated fashion to assist the folding and translocation of newly synthesized proteins (Chirico et al. 1988; Deshaies et al. 1988; Gething and Sambrook 1992). Under stress conditions, chaperones bind to hydrophobic regions of denatured proteins, thereby preventing their aggregation (Kiang and Tsokos 1998). Chaperones can initiate the refolding of their bound substrate or target protein for proteolytic degradation if the unfolding is irreversible (Agarraberes and Dice 2001; Agarraberes et al. 1997). In addition, chaperones are involved in the assembly and disassembly of protein multiunit complexes (Lindquist and Craig 1988; Skowronek et al. 1998; Solheim 1999), the regulation of native proteins including steroid receptors and protein kinases (Craig et al. 1993; Lund 1995; Pratt and Toft 1997; Stirling et al. 2003), thermotolerance (Lindquist and Craig 1988; Oh et al. 1997; Parsell et al. 1993; Subjeck and Shyy 1986), and buffering the expression of mutations (Rutherford and Lindquist 1998).

3
Chaperones as Tumor-Rejection Antigens and Peptide-Binding Proteins

In the 1940s, inbred mice vaccinated with irradiated syngeneic tumor cells were found to be resistant to subsequent challenge with the same tumor cells for vaccination (Gross 1943). This important discovery led immunologists to search for the immunogenic components within tumor cells that could confer immunity. Surprisingly, chaperone proteins were identified as tumor rejection antigens when isolated from chemically induced mouse tumors (Srivastava et al. 1986; Udono and Srivastava 1993; Ullrich et al. 1986). Vaccination with tumor-derived chaperones elicited a strong immune response against the syngeneic tumor, but not against antigenically distinct tumors. Although chaperone genes from tumors and normal tissues reveal no differences in the amino acid sequences, chaperones derived from normal tissues do not significantly induce tumor immunity (Udono and Srivastava 1994). The immunogenicity of these tumor-derived chaperones was subsequently attributed to the individually distinct array of antigenic peptides associated with chaperones (Udono and Srivastava 1993). This is consistent with the well-recognized capacity of chaperones to bind polypeptide chains in response to physiological stress (Welch 1993).

Despite many questions concerning the mode in which peptides interact with chaperones in vivo, numerous experimental observations suggest that chaperones display intrinsic peptide-binding activity in vitro (Blachere et al. 1993; Blond-Elguindi et al. 1993; Flynn et al. 1989; Flynn et al. 1991; Linderoth et al. 2001; Wearsch and Nicchitta 1996). Using an ex vivo system, peptides have been shown to interact with the chaperones localized in the ER (Arnold et al. 1997; Lammert et al. 1997; Spee and Neefjes 1997; Spee et al. 1999). To date, not much is known regarding the structural requirements for peptides bound to chaperones. Hsp70 seems to have a broad binding specificity that only requires the presence of hydrophobic regions composed of six or more amino acid residues (Blond-Elguindi et al. 1993; Flynn et al. 1991; Fourie et al. 1994; Gragerov et al. 1994; Rudiger et al. 1997; Takenaka et al. 1995). Different chaperones also demonstrate different affinities for a single peptide, suggesting that they may have distinct substrate specificity (Fourie et al. 1994; Gragerov et al. 1994). The hsp70 family members use their hydrophobic binding pocket to bind to hydrophobic segments of peptide in the extended conformation (Zhu et al. 1996). The structure of the gp96-binding pocket has been mapped to the C-terminal domain near the dimerization site (Linderoth et al. 2000), but has also been identified in the N-terminal nucleotide-binding domain as well (Vogen et al. 2002). Hsp70-peptide interactions are regulated by cycle of ATP binding and hydrolysis (Flynn et al. 1989). However, the regulation of the peptide-binding activity of gp96 is not well understood (Wearsch and Nicchitta 1997; Wearsch et al. 1998). The primary sequence of several peptides eluted from hsp70 and gp96 is now available, but information is still too limited to permit formulation of general rules to determine the binding capacity of a given peptide (Arnold et al. 1995; Breloer et al. 1998; Demine and Walden 2005; Grossmann et al. 2004; Ishii et al. 1999; Meng et al. 2002; Nieland et al. 1996).

In addition to the structural studies reported above, considerable immunological evidence also supports the notion that the peptides associated with these chaperone proteins contribute significantly to the immunogenicity of stress proteins. Studies in murine systems have shown that peptides carried by chaperone proteins can be derived from self-proteins, tumor, bacterial, viral, or minor histocompatibility antigens (Srivastava 2002a; Srivastava et al. 1998). More recent data have shown that hsp70 and gp96 may chaperone tumor-derived peptides (e.g., the differentiation antigens Mart-1, gp100, TRP-2, and tyrosinase) in different human tumors (Castelli et al. 2001; Noessner et al. 2002;

Rivoltini et al. 2003). Although the biochemical basis and biological relevance for peptide interactions with chaperones needs to be further explored, there is no doubt that chaperone–peptide complexes can be used as vaccines for immunotherapeutic applications.

4
Chaperones and Immune-Modulating Effects

The immunological function of chaperones that has received the most attention thus far is the ability to shuttle peptides into the endogenous presentation pathway of professional APCs. Traditionally, proteins acquired from extracellular spaces through receptor-mediated endocytosis are processed within endocytic compartments and presented on major histocompatibility complex (MHC) class II molecules. In contrast, endogenous proteins are degraded via the proteasome and are transported into the ER by TAP, where they are further processed and complexed with MHC I molecules in the presence of a macromolecular peptide-loading complex (Gromme and Neefjes 2002). Thus, the presentation of exogenous chaperone-associated peptides on MHC I molecules represents an archetype of a process referred to as cross-presentation or cross-priming. This processing pathway, in overview, is comprised of APC receptor recognition of the chaperone–peptide complex, internalization, and a subcellular trafficking and processing pathway that yields the presentation of chaperone-associated antigens on MHC I molecules. Srivastava first postulated the existence of a receptor specifically mediating the uptake of chaperones several years ago (Srivastava et al. 1994). Recent studies have documented that specific receptor-mediated endocytosis by APCs is critical for the cross-presentation of stress protein-chaperoned peptides (Arnold-Schild et al. 1999; Castellino et al. 2000; Singh-Jasuja et al. 2000b; Wassenberg et al. 1999). Indeed, several receptors, e.g., CD91, LOX1, scavenger receptor class-A (SR-A), and SREC have been identified or suggested to be involved in the cross-priming event (Basu et al. 2001; Berwin et al. 2003, 2004; Binder et al. 2000a; Delneste et al. 2002).

In addition to promoting cross-priming events, hsp70 and gp96 upon specific interaction with their cognate receptors can promote phenotypic and functional maturation of APCs such as dendritic cells (DCs) or monocytes (Asea et al. 2000; Kuppner et al. 2001; Singh-Jasuja et al. 2000a). Independent of the chaperoned peptides, APC–chaperone interaction leads to upregulation of co-stimulatory molecules and secretion of pro-inflammatory cytokines or chemokines (Asea et al. 2002; Wang et al. 2002b). Toll-like receptors (TLR) 2/4 and the downstream MyD88/NF-κB pathway have been proposed to mediate the hsp70- and gp96-triggered DC activation (Asea et al. 2002; Vabulas et al. 2002a, 2002b). Other cell surface receptors, including CD14 (Asea et al. 2000) and CD40 (Becker et al. 2002; Wang et al. 2002b), are also potentially involved in transducing activation signals of hsp70 to APCs. Notably, internalization of gp96 by active endocytosis is required in order to achieve activation of the signaling cascade and consequently the maturation of DCs. It has been postulated that chaperones transported in endocytic vesicles by CD91-mediated internalization might be able to trigger signaling through TLR2/4 present in these vesicles (Vabulas et al. 2002b). Endocytic receptors and signaling receptors could cooperate and achieve the cross-presentation of antigenic peptides and activation of APCs (Delneste et al. 2002). Identification of chaperone receptors is still in its infancy, but it appears that different families of chaperones may share common receptors (Binder et al. 2000b).

A confounding factor in these studies to determine innate stimulatory activities of chaperone proteins includes the potential contamination of endotoxin, which could be responsible for, or at least contribute to the activation of APCs by chaperones. Therefore, data derived from studies of the role of chaperones in innate immunity have to be interpreted with caution (Manjili et al. 2004; Tsan and Gao 2004). Nonetheless, emerging evidence strongly supports the idea that chaperones are capable of activating innate immune responses, which play an important role in the generation of CTLs and tumor rejection (Baker-LePain et al. 2002, 2003; Strbo et al. 2003). In addition, hsp70 has been shown to be a cell surface immune mediator acting as a target of NK cells, resulting in proliferation and cytotoxic activity of NK cells (Multhoff 2002; Multhoff et al. 1999). A specific 14-mer peptide of the inducible hsp70 has been identified as the ligand interacting with CD94 on NK cells (Gross et al. 2003; Multhoff et al. 2001). Therefore, chaperones are capable of integrating both innate and adaptive immunity.

5
Potential Use of Chaperones in Cancer Immunotherapy

There are two main features that make chaperones appealing candidates for cancer immunotherapy. First, receptor-mediated uptake of chaperones by professional APCs ensures specificity and sensitivity of antigen targeting. Second, chaperones serve as "danger signals" and activate innate immune components, which is crucial in development of an active immune response. Thus, these unique immunostimulatory properties enable chaperones to be utilized as physiological adjuvants to develop different immunotherapeutic approaches.

5.1
Tumor-Derived Chaperones Vaccine

Potent anti-tumor immune responses elicited by tumor-derived chaperone–peptide complexes have been extensively demonstrated in both prophylactic and therapeutic settings using numerous animal models (Srivastava 2002b).

To date, anti-tumor immunity elicited by chaperones has been shown against a variety of tumors of different histologic origins such as fibrosarcomas, lung carcinomas, melanomas, colon cancers, B cell lymphoma, and prostate cancer (Graner et al. 2000; Janetzki et al. 2000; Srivastava et al. 1986; Tamura et al. 1997; Vanaja et al. 2000; Wang et al. 2001; Yedavelli et al. 1999). Tumor-derived chaperone proteins that have been shown to be effective cancer vaccines include the cytosolic heat shock proteins hsp70, hsp90, hsp110, or the ER resident grp94/gp96, grp170, and calreticulin. Not all proteins with chaperone capacity display immunological functions. The ER residents PDI, BiP/grp78, and ERp72, although exerting peptide-binding activity, were unable to elicit a specific CTL response (Nair et al. 1999). Recently, studies show that multiple chaperones enriched from tumor lysate (referred to as chaperone-rich cell lysate) also exhibit anti-tumor efficacy, which would simplify the procedure necessary to prepare individual chaperones (Graner et al. 2000, 2003; Zeng et al. 2003, 2004).

Purification of chaperones from a cancer is believed to co-purify an antigenic peptide fingerprint of the cell of origin. Thus, vaccination with chaperone–peptide complexes derived from tumor circumvents the need to identify CTL epitopes from individual cancers. This unique advantage extends the use of chaperone-based immunotherapy to cancers where specific tumor antigens have not yet been characterized. Indeed, chaperone–peptide complexes from autologous tumor represent a patient-specific polyvalent vaccine, thus generating specific immunity only to the cancer from which the chaperones are isolated. This approach drastically reduces the possibility of tumor escape from immunotherapy due to antigen loss, since immunization with chaperones derived from tumor cells is directed against a diverse antigenic repertoire. The possibility of eliciting autoimmunity may exist because a significant number of self-peptides may be chaperoned by chaperones. However, no autoimmune reactions or severe side effects of the vaccine have so far been observed in any immunized patients (Janetzki et al. 2000). Interestingly, chaperones do not elicit an immune response to themselves, which makes chaperone vaccines effective and less toxic, thus safe for human use.

While potentially powerful, the use of this approach clinically has several limitations. It is very likely that only a small percentage of chaperone-associated peptides would be capable of eliciting tumor-specific CTLs. A major limitation of this approach is that vaccine preparation is time-consuming and requires a patient specimen. In general, a minimum of 1–5 g of tumor tissue for low-dose vaccination and up to 10 g of tumor tissue for high-dose vaccination regimens is required to produce a sufficient amount of gp96, a highly abundant cellular chaperone (Oki and Younes 2004). The whole process for vaccine preparation usually takes 6–8 weeks (Gordon and Clark 2004). As a result, some patients are unable to participate in these autologous vaccine trials (Belli et al. 2002). The lack of information on targeted antigens also limits the ability to monitor immune responses generated by this autologous vaccine approach. Beyond this conventional approach, other novel antigen-specific vaccine strategies using the unique adjuvant properties of chaperones have been developed and tested in preclinical studies.

5.2
Reconstituted Chaperone–Peptide Complex

Srivastava and others have shown that HSP/chaperone-peptide complexes can be generated in vitro and can elicit a peptide-specific CTL response without any additional adjuvant. Remarkably, the quantity of antigenic peptides associated with an immunogenic dose of HSP/chaperone-peptide complexes was found to be extremely small (~1–2 ng) (Blachere et al. 1997; Ciupitu et al. 2002; Moroi et al. 2000; Roman and Moreno 1996). The mechanism of peptide binding to HSPs may or may not be related to actual chaperoning property (Demine and Walden 2005; Nicchitta et al. 2004). This vaccine formulation directed against a tumor-specific immunogenic peptide epitope has advantages such as relative ease of construction and production and chemical stability. Moreover, the ability to specifically select peptides (dominant, subdominant, and T helper) allows the design of highly tailored chaperone vaccines. In order to minimize antigen escape, multiple peptides from different tumor antigens can be selected to create a polyvalent vaccine to generate a more vigorous and diverse immune response. Due to HLA restrictions, peptides have to be carefully chosen to match the HLA phenotype of the patient.

5.3
Chaperone–Antigen Protein Complex

Chaperone proteins have long been recognized to be involved in numerous cellular processes via interaction with different client protein substrates. This polypeptide-binding property can be used to form a chaperone–protein antigen complex by heat shock, which mimics a natural complex that occurs in vivo (Fig. 1a). The immunoadjuvant activities of hsp110, a large cytosolic chaperone in mammalian cells, can significantly enhance the immunogenicity of targeted antigens (e.g., L523S, a lung carcinoma antigen), resulting in cell-mediated as well as humoral immune responses (Fig. 1b–d). This recombinant chaperone vaccine approach has several unique and significant advantages:

- Chaperones efficiently bind substrate proteins. Therefore, a highly concentrated vaccine would be presented to the immune system.
- The whole protein antigen employed in this approach contains a large reservoir of potential peptides that allow the individual's own MHC alleles to select the appropriate epitope for presentation. Such a chaperone complex vaccine not only increases the chance of polyepitope-directed T and B cell responses, but also circumvents HLA restriction and allows most patients to be eligible for this form of vaccine treatment.

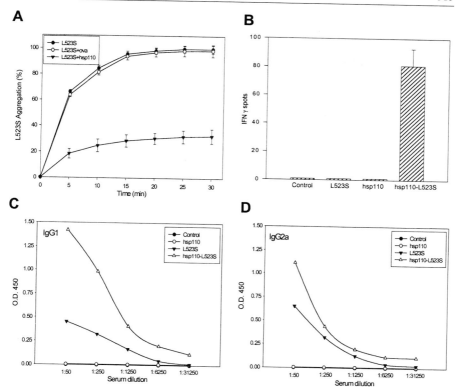

Fig. 1a–d Immunization with the hsp110–L523S chaperone complex elicits antigen-specific immune responses. **a** hsp110 protects L523S from heat shock-induced aggregation by forming chaperone complexes. Recombinant hsp110 and L523S protein (1:1 molar ratio) were incubated at 50 °C and optical density changes were measured at 320 nm using a spectrophotometer. **b** C57BL/6 mice (6/group) were immunized i.p. with 30 μg of the hsp110–L523S complex, hsp110 alone, L523S alone, or left untreated. The vaccinations were repeated 2 weeks later. Two weeks after the second immunization. Splenocytes (5×10^5 cells/well) were cultured in vitro with L523S (20 μg/ml) overnight, and IFN-γ secretion was detected using an ELISPOT assay. *. $p<0.005$ compared with splenocytes from naïve mice by Student's t test. Additionally, serum samples were collected and pooled 2 weeks after the last immunization. The presence of L523S-specific IgG1 (**c**) and IgG2a (**d**) were measured using an ELISA assay. The results from a serial dilution of sera are presented showing mean absorbance (OD) at 450 nm. The data shown here are representative of three independent experiments

- Since a tumor specimen is not required for vaccine production, patients with no measurable disease or inaccessible tumor can still be treated using this approach. This makes it an ideal adjuvant therapy for patients with completely resected disease. Furthermore, this approach is theoretically applicable as a preventive therapy for patients at high risk for cancer or for recurrence of cancer.

- Production of the recombinant chaperone complex is less time-consuming and far less expensive than the production of allogeneic vaccines, autologous vaccines, or tumor-derived chaperone vaccines. In addition, vaccine can be generated in considerable quantities with significant uniformity from batch to batch.
- Using well-defined antigens as opposed to whole tumor cells or lysate allows better characterization of the resultant immune responses.
- This synthetic approach of combining antigen with chaperone can serve as a model to develop and evaluate many different antigen targets, either alone or in combination vaccines.

Our studies have demonstrated that large chaperones, i.e., hsp110 and grp170, are excellent candidates for this natural chaperone–antigen protein complex formulation, since they are much more potent in holding protein antigens compared to other chaperones such as hsp70 (Oh et al. 1997; Park et al. 2003). Furthermore, vaccination with hsp110-Her-2/neu or gp100 chaperoned by hsp110 elicits strong antigen-specific anti-tumor immunity (Manjili et al. 2002, 2003; Wang et al. 2003). Most strikingly, mouse hsp110 (in mouse) is more effective as an adjuvant than is complete Fruend's adjuvant (CFA), suggesting potentially significant clinical applications (Wang et al. 2003).

5.4
Chaperone–Antigen Fusion Gene

This approach uses chaperones in the form of chimeric DNA through linkage of a known antigen gene (Chen et al. 2000; Hsu et al. 2001). DNA vaccines are usually administered intramuscularly or intradermally using a hypodermic needle or by bombarding plasmid DNA coated onto colloidal gold microparticles into the dermis or muscle using a gene gun. The injected DNA could transfect somatic cells (myocytes, keratinocytes) or DCs that infiltrate muscle or skin as a part of the inflammatory response to vaccination, resulting in a subsequent cross-priming (Ulmer et al. 1996a, 1996b) or direct presentation (Casares et al. 1997). One major advantage of DNA vaccines is the ability to endogenously generate a CTL response against the antigen, since it is difficult to induce a CTL response using a protein-based vaccine (not inclusive of the chaperone vaccines discussed here, which can induce strong CTL responses). DNA vaccines containing human papillomavirus (HPV) type 16 E7 and mycobacterial hsp70 fusion genes demonstrated preventive and therapeutic anti-tumor effects via a CD8-dependent pathway (Chen et al. 2000; Hauser et al. 2004). *Mycobacterium tuberculosis* hsp70 and MAGE-1 fusion gene enhanced the frequency of MAGE-1-specific CTLs, resulting in a potent anti-tumor effect against established tumors (Ye et al. 2004). E7-calreticulin fusion gene in the context of plasmid or vaccinia vector exhibits a strong anti-tumor response as well as antiangiogenesis effects (Cheng et al. 2001; Hsieh et al. 2004). In addition, direct intratumor injection of a hsp70 plasmid, oncolytic adenovirus expressing the hsp70 protein or replication-defective adenovirus that co-expresses hsp70, and a thymidine kinase suicide gene elicited potent systemic anti-tumor activities leading to eradication of tumors (Huang et al. 2003; Rafiee et al. 2001; Ren et al. 2004). Because of its capacity to induce individual tumor-specific immune responses, chaperone-mediated oncolytic tumor vaccine might become a universally applicable, personalized vaccine against any type of solid tumor. However, it was reported that chaperones fused to tumor-specific antigens (e.g., PSA) did not result in the enhancement of antigen-specific CTL responses (Pavlenko et al. 2004). The safety issues such as the risk of integration of DNA into the host genome and the development of autoimmune disease due to the induction of responses directed against transfected cells need to be considered carefully (Smith and Klinman 2001). However, chaperone fusion gene vaccines hold promise for future clinical use.

5.5
Chaperone–Antigen Fusion Protein

Another approach utilizes recombinant chaperone-antigen fusion proteins for immunotherapy. It has been demonstrated that a recombinant protein consisting of the HIV-1 p24 antigen or ovalbumin fused to the N-terminus of mycobacterial hsp70 elicited a potent antigen-specific CTL response against tumor (Suzue and Young 1996; Suzue et al. 1997). In a separate study, microbial hsp65/E7 fusion protein protected mice against challenge with an E7-expressing tumor cell line (TC-1 cells), a model of cervical carcinoma (Chu et al. 2000). Therapeutic immunization with the vaccine completely eradicated the established tumors (Chu et al. 2000). It has been proposed that the immunostimulatory effects of certain chaperone fusion proteins may be due to the bacterial origin of the chaperone moiety. However, OVA protein or its CTL epitope fused to murine hsp70 elicited CTL responses equivalent to those generated by the bacterial hsp70 fusion protein (Huang et al. 2000).

The mechanisms by which these fusion proteins act to stimulate CTL responses are unknown, but may be a result of strong chaperone-specific $CD4^+$ helper cell responses that enhance what might otherwise be a minimal response to the soluble proteins (Barrios et al. 1992; Konen-Waisman et al. 1999; Suzue and Young 1996). A recent study revealed that heat shock cognate protein (hsc70) contains two antigenic epitopes (i.e., 106–114aa and 233–241aa) recognized by HLA-B46-restricted and tumor-reactive CTLs (Azuma et al. 2003). Another possibility is that chaperones can deliver the fusion protein to the intracellular compartments of APCs for processing (Arnold-Schild et al. 1999). It was shown that hsp70 fusion proteins elicit a CTL response in the absence of $CD4^+$ T lymphocytes and the immunogenicity of hsp70 resides in a 200-amino acid (161–370 aa) segment of ATP-binding region of hsp70. This would support a model in which hsp70 bypasses the need for CD4$^+$ help by directly or indirectly activating APCs (i.e., DCs) independent of its peptide-binding property (Huang et al. 2000; Udono et al. 2001). Studies in our laboratory examined the fusion construct of hsp110 and E7. We found that antigen covalently fused to hsp110 interferes with its chaperoning function and the hsp110-E7 fusion protein cannot bind to receptors on APCs in vitro (Li et al., unpublished data). Therefore, these data suggest that chaperone fusion proteins may trigger immune responses via a different mechanism.

5.6
Chaperone-Based Cell Vaccine

5.6.1
DCs Pulsed with Chaperone Vaccine

DCs have been recognized as the most efficient APCs that have the capacity to initiate naïve T cell response in vitro and in vivo (Banchereau et al. 2003). Advances in DC generation, loading, and maturation methodologies have made it possible to generate clinical grade vaccines for various human trials (Cranmer et al. 2004). Given that chaperones can promote functional cross-priming, adoptive transfer of ex vivo expanded DCs pulsed with autologous chaperones or recombinant chaperone vaccines may be used as a cell vaccine strategy for cancer treatment. It has been shown that immunization with tumor-derived chaperone-pulsed DCs induces a tumor-specific CTL response in a transporter-associated antigen processing (TAP)-dependent manner (Ueda et al. 2004). Human DCs pulsed with mycobacterial hsp70–peptide complex is capable of activating peptide-specific T cells (MacAry et al. 2004). Interestingly, tumor-derived chaperone proteins were observed to be more potent than tumor lysate when pulsed onto DCs in the form of a cellular vaccine (Graner et al. 2003; Zeng et al. 2003). DC-based immunotherapy is labor- and resource-intensive. However, this approach would be beneficial to the patients with suppressed immune systems, since impaired APC activity could be partially responsible for the defective immune responses observed in tumor-bearing hosts (Restifo et al. 1991).

5.6.2
Tumor Cells Expressing Engineered Chaperones

Another cell vaccine approach is the use of tumor cells engineered to highly express chaperones in different cellular locations (i.e., cytosolic, membrane-anchored, or extracellular). It is postulated that chaperone overexpression or exposure of chaperones to the host immune system may enhance tumor immunogenicity. It has been shown that murine tumor cell lines stably transfected with mycobacterial hsp65, autologous hsp70, or hsp110 gene abrogated its tumorigenicity in rodent models and significantly enhanced the immunogenicity of the tumor in vivo (Lukacs et al. 1993; Melcher et al. 1998; Wang et al. 2002a). Dual-modified B16 cells with transfected hsp70 and superantigen as a vaccine induced a potent tumor-specific immune response as well as nonspecific immunity (Huang et al. 2004). Tumor cells with chaperones genetically engineered to be expressed on the cell surface (Chen et al. 2002; Zheng et al. 2001) or secreted into the extracellular environment (Massa et al. 2004; Yamazaki et al. 1999) displayed dramatically reduced tumorigenicity and were effective vaccines against subsequent tumor challenge. More recently, Baker-LePain et al. demonstrated that syngeneic murine fibroblasts transfected with secretable gp96 lacking the ER retention signals or a truncated gp96 mutant lacking the presumptive peptide-binding domain was able to significantly delay tumor growth, when used as prophylactic vaccines (Baker-LePain et al. 2002). Tumor antigen independent effects observed in this study highlight the role of chaperone-mediated innate immunity in tumor control. Regardless of the immunological mechanism involved, these findings suggest that certain chaperones are associated with tumor immunogenicity and that the induction or manipulation of certain chaperones may provide a novel approach to boost the immune response and help break tolerance to tumor antigens that otherwise remain immunologically silent in the progressively growing tumor. A significant disadvantage with the use of genetically modified autologous tumor cells is the substantial efforts of tumor cell harvest, propagation, gene engineering, and validation. Whole cell vaccines could also introduce transforming DNA or potentially immunosuppressive factors into the host. Nonetheless, the concept of exposing tumor-associated chaperones to the immune system appears to be an effective means of mounting an immune response against cancer. In order to bypass the complicated genetic modification of the tumor cell, attempts have been made to achieve biologic targeting of chaperones to the cell surface. Cell surface expression of chaperones induced by heat stress was found to strongly enhance the immunogenicity of apoptotic cells (Feng et al. 2002, 2003).

6
Clinical Application and Efficacy of Chaperone-Based Immunotherapy

The approaches to the use of chaperones in cancer immunotherapy, described in the preceding sections, are still in various stages of development. Only tumor-derived chaperone preparations are being actively examined in clinical trials. Successful preclinical studies in animal models have led to several phase I/II clinical trials using tumor-derived gp96 or hsp70 for cancer immunotherapy (Castelli et al. 2004; Hoos and Levey 2003; Oki and Younes 2004; Parmiani et al. 2004). Janetzki et al. first conducted a pilot study examining autologous tumor-derived chaperone gp96 vaccine (HSPPC-96, Oncophage, Antigenics Inc., New York, NY, USA) in 16 patients with advanced cancer who had become refractory to other therapies (Janetzki et al. 2000). Autologous gp96 vaccine was individually prepared for each patient and four weekly vaccinations were performed. Immunological monitoring consisted of an ELISPOT assay for IFN-γ following peripheral blood T cell stimulation with the autologous tumor and evaluation of NK cell activation. There was no significant treatment-related toxicity or evidence of autoimmune reactions. Six patients demonstrated a tumor-specific $CD8^+$ T cell response and eight patients displayed a significant increase in the number of NK cells. Clinically, one patient experienced a greater than 50% reduction in overall tumor burden and four patients developed stable disease for at least 3 months.

In another phase I/II clinical trial, Belli et al. specifically examined autologous tumor-derived gp96-peptide complex vaccine in 64 patients with stage IV melanoma (Belli et al. 2002). Given that resection of a portion of tumor was required for vaccine generation, only 28 of the 64 patients had residual measurable disease after surgery and were assessable for a clinical response. Eleven out of 23 patients (47.8%) showed an increased number of tumor-specific T cells after vaccination as evaluated by IFN-γ ELISPOT. There were two complete responses (7%) involving regression of both cutaneous and visceral metastasis and three cases of disease stabilization (11%) that lasted for more than 5 months in the 28 patients. Patients were randomly assigned to receive 5 or 50 μg of gp96 vaccine at weekly intervals. No correlation between clinical responses and dose of vaccine could be found. Despite the very promising results of this vaccine approach, 19% of patients were unable to participate in the trial due to lack of adequate autologous tumor or an insufficient amount of produced vaccine.

Cancer vaccines are generally expected to be more effective in an adjuvant setting where all clinical detectable disease has been removed. The immunotherapeutic effects of chaperones were mostly observed in animal with minimal residual disease (Srivastava 2000). A recent phase I/II study was conducted using tumor-derived gp96 to treat colorectal cancer patients rendered disease-free by complete resection of liver metastasis (Mazzaferro et al. 2003). Patients responding (i.e., showing an increased frequency of tumor-specific T cells after vaccination) had statistically significant survival advantage over nonresponding patients (2-year overall survival: 100% vs 50%; disease-free survival: 51% vs 8%). Although the results obtained were promising, the number of patients was too small for any definitive conclusion on possible clinical advantages of adjuvant vaccination with gp96. In the setting of minimal residual disease, tumor-derived hsp70 is currently being investigated in patients with chronica phase CML who did not achieve a complete response to imatinib mesylate (Bleevec, Novartis, East Hanover, NJ, USA) (Hoos and Levey 2003; Oki and Younes 2004).

At least ten phase I/II clinical trials have been conducted or are ongoing to examine the safety profile, the potential efficacy, and clinical benefit of autologous tumor-derived gp96/grp94 or hsp70 in different cancer types (Hoos and Levey 2003; Oki and Younes 2004). Clinical activity has been observed in melanoma, renal cell carcinoma, colorectal cancer, pancreatic cancer, non-Hodgkin's lymphoma, and chronic myelogenous leukemia (CML). Autologous gp96 was used in all these trials except CML, where hsp70 was used. Among all those patients immunized with autologous chaperones, overall toxicity due to gp96 is reported to be minimal and no signs of autoimmunity were observed. Only mild and transient side effects, such as injection site reactions or low-grade fevers were seen (Cohen et al. 2002). T cell responses in cancer patients have been detected after vaccination with tumor-derived gp96, as measured by ELISPOT or Tetramer assays. Measurable immune responses were correlated with clinical outcome in phase II trails in both metastatic melanoma and colorectal cancer (Belli et al. 2002; Mazzaferro et al. 2003). Objective clinical responses (complete responses) were seen in gp96-treated patients with metastatic melanoma and renal cell carcinoma that had failed several other therapies (Belli et al. 2002; Cohen et al. 2002). Well-controlled, randomized phase III trials are now underway in both adjuvant (renal cell carcinoma at high-risk of recurrence after nephrectomy) and metastatic disease settings (metastatic melanoma). This will help determine the durability of immune response and concomitant effect on the course of disease in patients, which will ultimately define the magnitude of clinical efficacy of this autologous chaperone vaccine.

In addition to autologous tumor-derived chaperone vaccine trials, a phase I trial using ex vivo hsp70 peptide (TKD) stimulation and reinfusion of autologous NK cells have been recently completed (Krause et al. 2004). Patients with metastatic colorectal cancer ($n = 11$) and non-small cell lung cancer ($n = 1$) who had failed standard therapies were enrolled. The cytolytic activity of NK cells against hsp70 membrane-positive colon carcinoma cells was enhanced after TKD/IL-2 stimulation in 10 of 12 patients. An excellent toxicity profile and some clinical responses were observed in these patients.

Other than its application in cancer immunotherapy, chaperones derived from virus or pathogen-infected cells or reconstituted chaperone–virus peptide complex was found to generate an antigen-specific CTL response (Blachere et al. 1997; Heikema et al. 1997; Suto and Srivastava 1995). In some cases, this immunity provided in vivo protection from challenge with live virus or bacteria (Ciupitu et al. 2002; Kumaraguru et al. 2003; Rapp and Kaufmann 2004; Zugel et al. 2001). Therefore, chaperone-based vaccines also have applications in the prevention and treatment of infectious diseases. Two chaperone-based vaccines targeting chronic viral infections are currently in clinical development. These are HSP-E7 for treatment of anogenital warts and cervical dysplasia (Stressgen Biotechnologies, BC, Canada) and AG-702 for treatment of genital herpes (Antigenics). Both approaches use recombinant chaperone as carrier of defined pathogen-specific antigens, and have been shown to be safe and well tolerated (Hoos and Levey 2003).

7
Concluding Remarks

Highly conserved chaperone proteins have long been recognized as essential components in numerous functions within cells from prokaryotes to man. It is now evident that chaperones also interact with the immune system in different and fundamental ways. The use of this information in formulation of new approaches to immunotherapy has now attracted the attention of numerous research groups. However, as a basic area of biomedical research, this field remains in its infancy and many new, unexpected, and exciting discoveries can be expected. Further elucidation of the basic mechanisms of immune responses induced by chaperone (i.e., characterization of chaperone receptors and their role in the generation of innate and adaptive immunity) will facilitate the design of novel chaperone-based vaccine strategies.

Chaperone-based immunotherapy (individual tumor-specific and tumor-associated antigen-specific) represents a novel approach with a promising role in cancer therapy or management. It can be applicable in a broad range of patients and may provide significant benefits as an adjuvant to conventional cancer therapies. It is well accepted that the immune system consists of different effector cells functioning together in a highly integrated fashion. New insights in molecular immunology related to mechanisms of T cell activation, regulation and effector function, and tumor invasion emphasize that antigen-driven responses such as vaccines need to be combined with manipulation of other immune parameters to achieve sufficient therapeutic potency (Pardoll and Allison 2004). Chaperone-based cancer vaccine administered in conjunction other immunomodulators such as agonists for TLR receptors or T cell costimulatory pathways and inhibitors for CTLA-4 or $CD4^+CD25^+$ regulatory cells could overcome tolerance and generate optimal anti-tumor responses.

In the next few years, data from ongoing phase III studies testing autologous chaperones for treatment of cancer will answer the question of how efficacious this vaccination strategy will be in humans. In view of the limitations associated with autologous tumor-derived chaperones as a cancer vaccine (i.e., requirement of specimen, costly and time-consuming vaccine preparation), testing of other chaperone-based vaccine approaches, especially recombinant chaperone DNA or protein formulations, at preclinical and clinical levels is strongly warranted. It is expected that other chaperone-based vaccine approaches will be translated into clinical practice for immunotherapy of cancer as well as infectious disease in the near future.

References

Agarraberes FA, Dice JF (2001) A molecular chaperone complex at the lysosomal membrane is required for protein translocation. J Cell Sci 114:2491–2499

Agarraberes FA, Terlecky SR, Dice JF (1997) An intralysosomal hsp70 is required for a selective pathway of lysosomal protein degradation. J Cell Biol 137:825–834

Arnold D, Faath S, Rammensee H, Schild H (1995) Cross-priming of minor histocompatibility antigen-specific cytotoxic T cells upon immunization with the heat shock protein gp96. J Exp Med 182:885–889

Arnold D, Wahl C, Faath S, Rammensee HG, Schild H (1997) Influences of transporter associated with antigen processing (TAP) on the repertoire of peptides associated with the endoplasmic reticulum-resident stress protein gp96. J Exp Med 186:461–466

Arnold-Schild D, Hanau D, Spehner D, Schmid C, Rammensee HG, de la Salle H, Schild H (1999) Cutting edge: receptor-mediated endocytosis of heat shock proteins by professional antigen-presenting cells. J Immunol 162:3757–3760

Asea A, Kraeft SK, Kurt-Jones EA, Stevenson MA, Chen LB, Finberg RW, Koo GC, Calderwood SK (2000) HSP70 stimulates cytokine production through a CD14-dependant pathway, demonstrating its dual role as a chaperone and cytokine. Nat Med 6:435–442

Asea A, Rehli M, Kabingu E, Boch JA, Bare O, Auron PE, Stevenson MA, Calderwood SK (2002) Novel signal transduction pathway utilized by extracellular HSP70: role of toll-like receptor (TLR) 2 and TLR4. J Biol Chem 277:15028–15034

Azuma K, Shichijo S, Takedatsu H, Komatsu N, Sawamizu H, Itoh K (2003) Heat shock cognate protein 70 encodes antigenic epitopes recognised by HLA-B4601-restricted cytotoxic T lymphocytes from cancer patients. Br J Cancer 89:1079–1085

Baker-LePain JC, Reed RC, Nicchitta CV (2003) ISO: a critical evaluation of the role of peptides in heat shock/chaperone protein-mediated tumor rejection. Curr Opin Immunol 15:89–94

Baker-LePain JC, Sarzotti M, Fields TA, Li CY, Nicchitta CV (2002) GRP94 (gp96) and GRP94 N-terminal geldanamycin binding domain elicit tissue nonrestricted tumor suppression. J Exp Med 196:1447–1459

Banchereau J, Fay J, Pascual V, Palucka AK (2003) Dendritic cells: controllers of the immune system and a new promise for immunotherapy. Novartis Found Symp 252:226–235; discussion 235–238, 257–267

Barrios C, Lussow AR, Van Embden J, Van der Zee R, Rappuoli R, Costantino P, Louis JA, Lambert PH, Del Giudice G (1992) Mycobacterial heat-shock proteins as carrier molecules. II: the use of the 70-kDa mycobacterial heat-shock protein as carrier for conjugated vaccines can circumvent the need for adjuvants and Bacillus Calmette Guerin priming. Eur J Immunol 22:1365–1372

Basu S, Binder RJ, Ramalingam T, Srivastava PK (2001) CD91 is a common receptor for heat shock proteins gp96, hsp90, hsp70, and calreticulin. Immunity 14:303–313

Becker T, Hartl FU, Wieland F (2002) CD40, an extracellular receptor for binding and uptake of hsp70-peptide complexes. J Cell Biol 158:1277–1285

Belli F, Testori A, Rivoltini L, Maio M, Andreola G, Sertoli MR, Gallino G, Piris A, Cattelan A, Lazzari I, Carrabba M, Scita G, Santantonio C, Pilla L, Tragni G, Lombardo C, Arienti F, Marchiano A, Queirolo P, Bertolini F, Cova A, Lamaj E, Ascani L, Camerini R, Corsi M, Cascinelli N, Lewis JJ, Srivastava P, Parmiani G (2002) Vaccination of metastatic melanoma patients with autologous tumor-derived heat shock protein gp96-peptide complexes: clinical and immunologic findings. J Clin Oncol 20:4169–4180

Berwin B, Delneste Y, Lovingood RV, Post SR, Pizzo SV (2004) SREC-I, a type F scavenger receptor, is an endocytic receptor for calreticulin. J Biol Chem 279:51250–51257

Berwin B, Hart JP, Rice S, Gass C, Pizzo SV, Post SR, Nicchitta CV (2003) Scavenger receptor-A mediates gp96/GRP94 and calreticulin internalization by antigen-presenting cells. EMBO J 22:6127–6136

Binder RJ, Han DK, Srivastava PK (2000a) CD91: a receptor for heat shock protein gp96. Nat Immunol 1:151–155

Binder RJ, Harris ML, Menoret A, Srivastava PK (2000b) Saturation, competition, and specificity in interaction of heat shock proteins (HSP) gp96, hsp90, and hsp70 with CD11b+ cells. J Immunol 165:2582–2587

Blachere NE, Li Z, Chandawarkar RY, Suto R, Jaikaria NS, Basu S, Udono H, Srivastava PK (1997) Heat shock protein-peptide complexes, reconstituted in vitro, elicit peptide-specific cytotoxic T lymphocyte response and tumor immunity. J Exp Med 186:1315–1322

Blachere NE, Udono H, Janetzki S, Li Z, Heike M, Srivastava PK (1993) Heat shock protein vaccines against cancer. J Immunother 14:352–356

Black AR, Subjeck JR (1991) The biology and physiology of the heat shock and glucose-regulated stress protein systems. Methods Achiev Exp Pathol 15:126–166

Blond-Elguindi S, Cwirla SE, Dower WJ, Lipshutz RJ, Sprang SR, Sambrook JF, Gething MJ (1993) Affinity panning of a library of peptides displayed on bacteriophages reveals the binding specificity of BiP. Cell 75:717–728

Breloer M, Marti T, Fleischer B, von Bonin A (1998) Isolation of processed, H-2Kb-binding ovalbumin-derived peptides associated with the stress proteins HSP70 and gp96. Eur J Immunol 28:1016–1021

Casares S, Inaba K, Brumeanu TD, Steinman RM, Bona CA (1997) Antigen presentation by dendritic cells after immunization with DNA encoding a major histocompatibility complex class II-restricted viral epitope. J Exp Med 186:1481–1486

Castelli C, Ciupitu AM, Rini F, Rivoltini L, Mazzocchi A, Kiessling R, Parmiani G (2001) Human heat shock protein 70 peptide complexes specifically activate antimelanoma T cells. Cancer Res 61:222–227

Castelli C, Rivoltini L, Rini F, Belli F, Testori A, Maio M, Mazzaferro V, Coppa J, Srivastava PK, Parmiani G (2004) Heat shock proteins: biological functions and clinical application as personalized vaccines for human cancer. Cancer Immunol Immunother 53:227–233

Castellino F, Boucher PE, Eichelberg K, Mayhew M, Rothman JE, Houghton AN, Germain RN (2000) Receptor-mediated uptake of antigen/heat shock protein complexes results in major histocompatibility complex class I antigen presentation via two distinct processing pathways. J Exp Med 191:1957–1964

Chen CH, Wang TL, Hung CF, Yang Y, Young RA, Pardoll DM, Wu TC (2000) Enhancement of DNA vaccine potency by linkage of antigen gene to an HSP70 gene. Cancer Res 60:1035–1042

Chen X, Tao Q, Yu H, Zhang L, Cao X (2002) Tumor cell membrane-bound heat shock protein 70 elicits antitumor immunity. Immunol Lett 84:81–87

Cheng WF, Hung CF, Chai CY, Hsu KF, He L, Ling M, Wu TC (2001) Tumor-specific immunity and antiangiogenesis generated by a DNA vaccine encoding calreticulin linked to a tumor antigen. J Clin Invest 108:669–678

Chirico WJ, Waters MG, Blobel G (1988) 70 K heat shock related proteins stimulate protein translocation into microsomes. Nature 332:805–810

Chu NR, Wu HB, Wu TC, Boux LJ, Mizzen LA, Siegel MI (2000) Immunotherapy of a human papillomavirus type 16 E7-expressing tumor by administration of fusion protein comprised of Mycobacterium bovis BCG hsp65 and HPV16 E7. Cell Stress Chaperones 5:401–405

Ciupitu AM, Petersson M, Kono K, Charo J, Kiessling R (2002) Immunization with heat shock protein 70 from methylcholanthrene-induced sarcomas induces tumor protection correlating with in vitro T cell responses. Cancer Immunol Immunother 51:163–170

Cohen L, de Moor C, Parker PA, Amato RJ (2002) Quality of life in patients with metastatic renal cell carcinoma participating in a phase I trial of an autologous tumor-derived vaccine. Urol Oncol 7:119–124

Craig EA, Gambill BD, Nelson RJ (1993) Heat shock proteins: molecular chaperones of protein biogenesis. Microbiol Rev 57:402–414

Cranmer LD, Trevor KT, Hersh EM (2004) Clinical applications of dendritic cell vaccination in the treatment of cancer. Cancer Immunol Immunother 53:275–306

Delneste Y, Magistrelli G, Gauchat J, Haeuw J, Aubry J, Nakamura K, Kawakami-Honda N, Goetsch L, Sawamura T, Bonnefoy J, Jeannin P (2002) Involvement of LOX-1 in dendritic cell-mediated antigen cross-presentation. Immunity 17:353–362

Demine R, Walden P (2005) Testing the role of gp96 as peptide chaperone in antigen processing. J Biol Chem 280:17573–17578

Deshaies RJ, Koch BD, Werner-Washburne M, Craig EA, Schekman R (1988) A subfamily of stress proteins facilitates translocation of secretory and mitochondrial precursor polypeptides. Nature 332:800–805

Doody AD, Kovalchin JT, Mihalyo MA, Hagymasi AT, Drake CG, Adler AJ (2004) Glycoprotein 96 can chaperone both MHC class I- and class II-restricted epitopes for in vivo presentation, but selectively primes CD8+ T cell effector function. J Immunol 172:6087–6092

Easton DP, Kaneko Y, Subjeck JR (2000) The hsp110 and Grp170 stress proteins: newly recognized relatives of the hsp70s. Cell Stress Chaperones 5:276–2790

Feng H, Zeng Y, Graner MW, Katsanis E (2002) Stressed apoptotic tumor cells stimulate dendritic cells and induce specific cytotoxic T cells. Blood 100:4108–4115

Feng H, Zeng Y, Graner MW, Likhacheva A, Katsanis E (2003) Exogenous stress proteins enhance the immunogenicity of apoptotic tumor cells and stimulate antitumor immunity. Blood 101:245–252

Flynn GC, Chappell TG, Rothman JE (1989) Peptide binding and release by proteins implicated as catalysts of protein assembly. Science 245:385–390

Flynn GC, Pohl J, Flocco MT, Rothman JE (1991) Peptide-binding specificity of the molecular chaperone BiP. Nature 353:726–730

Fourie AM, Sambrook JF, Gething MJ (1994) Common and divergent peptide binding specificities of hsp70 molecular chaperones. J Biol Chem 269:30470–30478

Gething MJ, Sambrook J (1992) Protein folding in the cell. Nature 355:33–45

Gordon NF, Clark BL (2004) The challenges of bringing autologous HSP-based vaccines to commercial reality. Methods 32:63–69

Gragerov A, Zeng L, Zhao X, Burkholder W, Gottesman ME (1994) Specificity of DnaK-peptide binding. J Mol Biol 235:848–854

Graner M, Raymond A, Romney D, He L, Whitesell L, Katsanis E (2000) Immunoprotective activities of multiple chaperone proteins isolated from murine B-cell leukemia/lymphoma. Clin Cancer Res 6:909–915

Graner MW, Zeng Y, Feng H, Katsanis E (2003) Tumor-derived chaperone-rich cell lysates are effective therapeutic vaccines against a variety of cancers. Cancer Immunol Immunother 52:226–234

Gromme M, Neefjes J (2002) Antigen degradation or presentation by MHC class I molecules via classical and non-classical pathways. Mol Immunol 39:181–202

Gross C, Hansch D, Gastpar R, Multhoff G (2003) Interaction of heat shock protein 70 peptide with NK cells involves the NK receptor CD94. Biol Chem 384:267–279

Gross L (1943) Intradermal immunization of C3H mice against a sarcoma that originated in an animal of the same line. Cancer Res 3:323–326

Grossmann ME, Madden BJ, Gao F, Pang YP, Carpenter JE, McCormick D, Young CY (2004) Proteomics shows hsp70 does not bind peptide sequences indiscriminately in vivo. Exp Cell Res 297:108–117

Hauser H, Shen L, Gu QL, Krueger S, Chen SY (2004) Secretory heat-shock protein as a dendritic cell-targeting molecule: a new strategy to enhance the potency of genetic vaccines. Gene Ther 11:924–932

Heikema A, Agsteribbe E, Wilschut J, Huckriede A (1997) Generation of heat shock protein-based vaccines by intracellular loading of gp96 with antigenic peptides. Immunol Lett 57:69–74

Hoos A, Levey DL (2003) Vaccination with heat shock protein-peptide complexes: from basic science to clinical applications. Expert Rev Vaccines 2:369–379

Hsieh CJ, Kim TW, Hung CF, Juang J, Moniz M, Boyd DA, He L, Chen PJ, Chen CH, Wu TC (2004) Enhancement of vaccinia vaccine potency by linkage of tumor antigen gene to gene encoding calreticulin. Vaccine 22:3993–4001

Hsu KF, Hung CF, Cheng WF, He L, Slater LA, Ling M, Wu TC (2001) Enhancement of suicidal DNA vaccine potency by linking Mycobacterium tuberculosis heat shock protein 70 to an antigen. Gene Ther 8:376–383

Huang C, Yu H, Wang Q, Ma W, Xia D, Yi P, Zhang L, Cao X (2004) Potent antitumor effect elicited by superantigen-linked tumor cells transduced with heat shock protein 70 gene. Cancer Sci 95:160–167

Huang Q, Richmond JF, Suzue K, Eisen HN, Young RA (2000) In vivo cytotoxic T lymphocyte elicitation by mycobacterial heat shock protein 70 fusion proteins maps to a discrete domain and is CD4(+) T cell independent. J Exp Med 191:403–408

Huang XF, Ren W, Rollins L, Pittman P, Shah M, Shen L, Gu Q, Strube R, Hu F, Chen SY (2003) A broadly applicable, personalized heat shock protein-mediated oncolytic tumor vaccine. Cancer Res 63:7321–7329

Ishii T, Udono H, Yamano T, Ohta H, Uenaka A, Ono T, Hizuta A, Tanaka N, Srivastava PK, Nakayama E (1999) Isolation of MHC class I-restricted tumor antigen peptide and its precursors associated with heat shock proteins hsp70, hsp90, and gp96. J Immunol 162:1303–1309

Janetzki S, Palla D, Rosenhauer V, Lochs H, Lewis JJ, Srivastava PK (2000) Immunization of cancer patients with autologous cancer-derived heat shock protein gp96 preparations: a pilot study. Int J Cancer 88:232–238

Johnson JL, Craig EA (1997) Protein folding in vivo: unraveling complex pathways. Cell 90:201–204

Kiang JG, Tsokos GC (1998) Heat shock protein 70 kDa: molecular biology, biochemistry, and physiology. Pharmacol Ther 80:183–201

Konen-Waisman S, Cohen A, Fridkin M, Cohen IR (1999) Self heat-shock protein (hsp60) peptide serves in a conjugate vaccine against a lethal pneumococcal infection. J Infect Dis 179:403–413

Krause SW, Gastpar R, Andreesen R, Gross C, Ullrich H, Thonigs G, Pfister K, Multhoff G (2004) Treatment of colon and lung cancer patients with ex vivo heat shock protein 70-peptide-activated, autologous natural killer cells: a clinical phase I trial. Clin Cancer Res 10:3699–3707

Kumaraguru U, Gouffon CA Jr, Ivey RA 3rd, Rouse BT, Bruce BD (2003) Antigenic peptides complexed to phylogenically diverse hsp70s induce differential immune responses. Cell Stress Chaperones 8:134–143

Kuppner MC, Gastpar R, Gelwer S, Nossner E, Ochmann O, Scharner A, Issels RD (2001) The role of heat shock protein (hsp70) in dendritic cell maturation: hsp70 induces the maturation of immature dendritic cells but reduces DC differentiation from monocyte precursors. Eur J Immunol 31:1602–1609

Lammert E, Arnold D, Nijenhuis M, Momburg F, Hammerling GJ, Brunner J, Stevanovic S, Rammensee HG, Schild H (1997) The endoplasmic reticulum-resident stress protein gp96 binds peptides translocated by TAP. Eur J Immunol 27:923–927

Linderoth NA, Popowicz A, Sastry S (2000) Identification of the peptide-binding site in the heat shock chaperone/tumor rejection antigen gp96 (Grp94). J Biol Chem 275:5472–5477

Linderoth NA, Simon MN, Hainfeld JF, Sastry S (2001) Binding of antigenic peptide to the endoplasmic reticulum-resident protein gp96/GRP94 heat shock chaperone occurs in higher order complexes. Essential role of some aromatic amino acid residues in the peptide-binding site. J Biol Chem 276:11049–11054

Lindquist S, Craig EA (1988) The heat-shock proteins. Annu Rev Genet 22:631–677

Lukacs KV, Lowrie DB, Stokes RW, Colston MJ (1993) Tumor cells transfected with a bacterial heat-shock gene lose tumorigenicity and induce protection against tumors. J Exp Med 178:343–348

Lund PA (1995) The roles of molecular chaperones in vivo. Essays Biochem 29:113–123

MacAry PA, Javid B, Floto RA, Smith KG, Oehlmann W, Singh M, Lehner PJ (2004) HSP70 peptide binding mutants separate antigen delivery from dendritic cell stimulation. Immunity 20:95–106

Manjili MH, Henderson R, Wang XY, Chen X, Li Y, Repasky E, Kazim L, Subjeck JR (2002) Development of a recombinant HSP110-HER-2/neu vaccine using the chaperoning properties of HSP110. Cancer Res 62:1737–1742

Manjili MH, Wang XY, Chen X, Martin T, Repasky EA, Henderson R, Subjeck JR (2003) HSP110-HER2/neu chaperone complex vaccine induces protective immunity against spontaneous mammary tumors in HER-2/neu transgenic mice. J Immunol 171:4054–4061

Manjili MH, Wang XY, MacDonald IJ, Arnouk H, Yang GY, Pritchard MT, Subjeck JR (2004) Cancer immunotherapy and heat-shock proteins: promises and challenges. Expert Opin Biol Ther 4:363–373

Massa C, Guiducci C, Arioli I, Parenza M, Colombo MP, Melani C (2004) Enhanced efficacy of tumor cell vaccines transfected with secretable hsp70. Cancer Res 64:1502–1508

Mazzaferro V, Coppa J, Carrabba MG, Rivoltini L, Schiavo M, Regalia E, Mariani L, Camerini T, Marchiano A, Andreola S, Camerini R, Corsi M, Lewis JJ, Srivastava PK, Parmiani G (2003) Vaccination with autologous tumor-derived heat-shock protein gp96 after liver resection for metastatic colorectal cancer. Clin Cancer Res 9:3235–3245

Melcher A, Todryk S, Hardwick N, Ford M, Jacobson M, Vile RG (1998) Tumor immunogenicity is determined by the mechanism of cell death via induction of heat shock protein expression. Nat Med 4:581–587

Meng SD, Song J, Rao Z, Tien P, Gao GF (2002) Three-step purification of gp96 from human liver tumor tissues suitable for isolation of gp96-bound peptides. J Immunol Methods 264:29–35

Moroi Y, Mayhew M, Trcka J, Hoe MH, Takechi Y, Hartl FU, Rothman JE, Houghton AN (2000) Induction of cellular immunity by immunization with novel hybrid peptides complexed to heat shock protein 70. Proc Natl Acad Sci U S A 97:3485–3490

Multhoff G (2002) Activation of natural killer cells by heat shock protein 70. Int J Hyperthermia 18:576–585

Multhoff G, Mizzen L, Winchester CC, Milner CM, Wenk S, Eissner G, Kampinga HH, Laumbacher B, Johnson J (1999) Heat shock protein 70 (hsp70) stimulates proliferation and cytolytic activity of natural killer cells. Exp Hematol 27:1627–1636

Multhoff G, Pfister K, Gehrmann M, Hantschel M, Gross C, Hafner M, Hiddemann W (2001) A 14-mer hsp70 peptide stimulates natural killer (NK) cell activity. Cell Stress Chaperones 6:337–344

Nair S, Wearsch PA, Mitchell DA, Wassenberg JJ, Gilboa E, Nicchitta CV (1999) Calreticulin displays in vivo peptide-binding activity and can elicit CTL responses against bound peptides. J Immunol 162:6426–6432

Nicchitta CV, Carrick DM, Baker-Lepain JC (2004) The messenger and the message: gp96 (GRP94)-peptide interactions in cellular immunity. Cell Stress Chaperones 9:325–331

Nieland TJ, Tan MC, Monne-van Muijen M, Koning F, Kruisbeek AM, van Bleek GM (1996) Isolation of an immunodominant viral peptide that is endogenously bound to the stress protein GP96/GRP94. Proc Natl Acad Sci U S A 93:6135–6139

Noessner E, Gastpar R, Milani V, Brandl A, Hutzler PJ, Kuppner MC, Roos M, Kremmer E, Asea A, Calderwood SK, Issels RD (2002) Tumor-derived heat shock protein 70 peptide complexes are cross-presented by human dendritic cells. J Immunol 169:5424–5432

Novellino L, Castelli C, Parmiani G (2004) A listing of human tumor antigens recognized by T cells: March 2004 update. Cancer Immunol Immunother 54:187–207

Oh HJ, Chen X, Subjeck JR (1997) hsp110 protects heat-denatured proteins and confers cellular thermoresistance. J Biol Chem 272:31636–31640

Oki Y, Younes A (2004) Heat shock protein-based cancer vaccines. Expert Rev Vaccines 3:403–411

Pardoll D, Allison J (2004) Cancer immunotherapy: breaking the barriers to harvest the crop. Nat Med 10:887–892

Park J, Easton DP, Chen X, MacDonald IJ, Wang XY, Subjeck JR (2003) The chaperoning properties of mouse grp170, a member of the third family of hsp70 related proteins. Biochemistry 42:14893–14902

Parmiani G, Testori A, Maio M, Castelli C, Rivoltini L, Pilla L, Belli F, Mazzaferro V, Coppa J, Patuzzo R, Sertoli MR, Hoos A, Srivastava PK, Santinami M (2004) Heat shock proteins and their use as anticancer vaccines. Clin Cancer Res 10:8142–8146

Parsell DA, Taulien J, Lindquist S (1993) The role of heat-shock proteins in thermotolerance. Philos Trans R Soc Lond B Biol Sci 339:279–285; discussion 285–286

Pavlenko M, Roos AK, Lundqvist A, Palmborg A, Miller AM, Ozenci V, Bergman B, Egevad L, Hellstrom M, Kiessling R, Masucci G, Wersall P, Nilsson S, Pisa P (2004) A phase I trial of DNA vaccination with a plasmid expressing prostate-specific antigen in patients with hormone-refractory prostate cancer. Br J Cancer 91:688–694

Pratt WB, Toft DO (1997) Steroid receptor interactions with heat shock protein and immunophilin chaperones. Endocr Rev 18:306–360

Rafiee M, Kanwar JR, Berg RW, Lehnert K, Lisowska K, Krissansen GW (2001) Induction of systemic antitumor immunity by gene transfer of mammalian heat shock protein 70.1 into tumors in situ. Cancer Gene Ther 8:974–981

Rapp UK, Kaufmann SH (2004) DNA vaccination with gp96-peptide fusion proteins induces protection against an intracellular bacterial pathogen. Int Immunol 16:597–605

Ren W, Strube R, Zhang X, Chen SY, Huang XF (2004) Potent tumor-specific immunity induced by an in vivo heat shock protein-suicide gene-based tumor vaccine. Cancer Res 64:6645–6651

Restifo NP, Esquivel F, Asher AL, Stotter H, Barth RJ, Bennink JR, Mule JJ, Yewdell JW, Rosenberg SA (1991) Defective presentation of endogenous antigens by a murine sarcoma. Implications for the failure of an anti-tumor immune response. J Immunol 147:1453–1459

Rivoltini L, Castelli C, Carrabba M, Mazzaferro V, Pilla L, Huber V, Coppa J, Gallino G, Scheibenbogen C, Squarcina P, Cova A, Camerini R, Lewis JJ, Srivastava PK, Parmiani G (2003) Human tumor-derived heat shock protein 96 mediates in vitro activation and in vivo expansion of melanoma- and colon carcinoma-specific T cells. J Immunol 171:3467–3474

Roman E, Moreno C (1996) Synthetic peptides non-covalently bound to bacterial hsp 70 elicit peptide-specific T-cell responses in vivo. Immunology 88:487–492

Rudiger S, Buchberger A, Bukau B (1997) Interaction of hsp70 chaperones with substrates. Nat Struct Biol 4:342–349

Rutherford SL, Lindquist S (1998) hsp90 as a capacitor for morphological evolution. Nature 396:336–342

Sciandra JJ, Subjeck JR (1983) The effects of glucose on protein synthesis and thermosensitivity in Chinese hamster ovary cells. J Biol Chem 258:12091–12093

Sciandra JJ, Subjeck JR, Hughes CS (1984) Induction of glucose-regulated proteins during anaerobic exposure and of heat-shock proteins after reoxygenation. Proc Natl Acad Sci U S A 81:4843–4847

Singh-Jasuja H, Scherer HU, Hilf N, Arnold-Schild D, Rammensee HG, Toes RE, Schild H (2000a) The heat shock protein gp96 induces maturation of dendritic cells and down-regulation of its receptor. Eur J Immunol 30:2211–2215

Singh-Jasuja H, Toes RE, Spee P, Munz C, Hilf N, Schoenberger SP, Ricciardi-Castagnoli P, Neefjes J, Rammensee HG, Arnold-Schild D, Schild H (2000b) Cross-presentation of glycoprotein 96-associated antigens on major histocompatibility complex class I molecules requires receptor-mediated endocytosis. J Exp Med 191:1965–1974

Skowronek MH, Hendershot LM, Haas IG (1998) The variable domain of nonassembled Ig light chains determines both their half-life and binding to the chaperone BiP. Proc Natl Acad Sci U S A 95:1574–1578

Smith HA, Klinman DM (2001) The regulation of DNA vaccines. Curr Opin Biotechnol 12:299–303

Solheim JC (1999) Class I MHC molecules: assembly and antigen presentation. Immunol Rev 172:11–19

Spee P, Neefjes J (1997) TAP-translocated peptides specifically bind proteins in the endoplasmic reticulum, including gp96, protein disulfide isomerase and calreticulin. Eur J Immunol 27:2441–2449

Spee P, Subjeck J, Neefjes J (1999) Identification of novel peptide binding proteins in the endoplasmic reticulum: ERp72, calnexin, and grp170. Biochemistry 38:10559–10566

Srivastava P (2002a) Interaction of heat shock proteins with peptides and antigen presenting cells: chaperoning of the innate and adaptive immune responses. Annu Rev Immunol 20:395–425

Srivastava P (2002b) Roles of heat-shock proteins in innate and adaptive immunity. Nat Rev Immunol 2:185–194

Srivastava PK (2000) Immunotherapy of human cancer: lessons from mice. Nat Immunol 1:363–366

Srivastava PK, DeLeo AB, Old LJ (1986) Tumor rejection antigens of chemically induced sarcomas of inbred mice. Proc Natl Acad Sci U S A 83:3407–3411

Srivastava PK, Menoret A, Basu S, Binder RJ, McQuade KL (1998) Heat shock proteins come of age: primitive functions acquire new roles in an adaptive world. Immunity 8:657–665

Srivastava PK, Udono H, Blachere NE, Li Z (1994) Heat shock proteins transfer peptides during antigen processing and CTL priming. Immunogenetics 39:93–98

Stirling PC, Lundin VF, Leroux MR (2003) Getting a grip on non-native proteins. EMBO Rep 4:565–570

Strbo N, Oizumi S, Sotosek-Tokmadzic V, Podack ER (2003) Perforin is required for innate and adaptive immunity induced by heat shock protein gp96. Immunity 18:381–390

Subjeck JR, Shyy TT (1986) Stress protein systems of mammalian cells. Am J Physiol 250:C1–C17

Suto R, Srivastava PK (1995) A mechanism for the specific immunogenicity of heat shock protein-chaperoned peptides. Science 269:1585–1588

Suzue K, Young RA (1996) Adjuvant-free hsp70 fusion protein system elicits humoral and cellular immune responses to HIV-1 p24. J Immunol 156:873–879

Suzue K, Zhou X, Eisen HN, Young RA (1997) Heat shock fusion proteins as vehicles for antigen delivery into the major histocompatibility complex class I presentation pathway. Proc Natl Acad Sci U S A 94:13146–13151

Takenaka IM, Leung SM, McAndrew SJ, Brown JP, Hightower LE (1995) Hsc70-binding peptides selected from a phage display peptide library that resemble organellar targeting sequences. J Biol Chem 270:19839–19844

Tamura Y, Peng P, Liu K, Daou M, Srivastava PK (1997) Immunotherapy of tumors with autologous tumor-derived heat shock protein preparations. Science 278:117–120

Tissieres A, Mitchell HK, Tracy UM (1974) Protein synthesis in salivary glands of Drosophila melanogaster: relation to chromosome puffs. J Mol Biol 84:389–398

Tsan MF, Gao B (2004) Endogenous ligands of Toll-like receptors. J Leukoc Biol 76:514–519

Udono H, Srivastava PK (1993) Heat shock protein 70-associated peptides elicit specific cancer immunity. J Exp Med 178:1391–1396

Udono H, Srivastava PK (1994) Comparison of tumor-specific immunogenicities of stress-induced proteins gp96, hsp90, and hsp70. J Immunol 152:5398–5403

Udono H, Yamano T, Kawabata Y, Ueda M, Yui K (2001) Generation of cytotoxic T lymphocytes by MHC class I ligands fused to heat shock cognate protein 70. Int Immunol 13:1233–1242

Ueda G, Tamura Y, Hirai I, Kamiguchi K, Ichimiya S, Torigoe T, Hiratsuka H, Sunakawa H, Sato N (2004) Tumor-derived heat shock protein 70-pulsed dendritic cells elicit tumor-specific cytotoxic T lymphocytes (CTLs) and tumor immunity. Cancer Sci 95:248–253

Ullrich SJ, Robinson EA, Law LW, Willingham M, Appella E (1986) A mouse tumor-specific transplantation antigen is a heat shock-related protein. Proc Natl Acad Sci U S A 83:3121–3125

Ulmer JB, Deck RR, Dewitt CM, Donnhly JI, Liu MA (1996a) Generation of MHC class I-restricted cytotoxic T lymphocytes by expression of a viral protein in muscle cells: antigen presentation by non-muscle cells. Immunology 89:59–67

Ulmer JB, Sadoff JC, Liu MA (1996b) DNA vaccines. Curr Opin Immunol 8:531–536

Vabulas RM, Ahmad-Nejad P, Ghose S, Kirschning CJ, Issels RD, Wagner H (2002a) HSP70 as endogenous stimulus of the Toll/interleukin-1 receptor signal pathway. J Biol Chem 277:15107–15112

Vabulas RM, Braedel S, Hilf N, Singh-Jasuja H, Herter S, Ahmad-Nejad P, Kirschning CJ, Da Costa C, Rammensee HG, Wagner H, Schild H (2002b) The endoplasmic reticulum-resident heat shock protein Gp96 activates dendritic cells via the Toll-like receptor 2/4 pathway. J Biol Chem 277:20847–20853

Vanaja DK, Grossmann ME, Celis E, Young CY (2000) Tumor prevention and antitumor immunity with heat shock protein 70 induced by 15-deoxy-delta12,14-prostaglandin J2 in transgenic adenocarcinoma of mouse prostate cells. Cancer Res 60:4714–4718

Vogen S, Gidalevitz T, Biswas C, Simen BB, Stein E, Gulmen F, Argon Y (2002) Radicicol-sensitive peptide binding to the N-terminal portion of GRP94. J Biol Chem 277:40742–40750

Wang XY, Chen X, Manjili MH, Repasky E, Henderson R, Subjeck JR (2003) Targeted immunotherapy using reconstituted chaperone complexes of heat shock protein 110 and melanoma-associated antigen gp100. Cancer Res 63:2553–2560

Wang XY, Kazim L, Repasky EA, Subjeck JR (2001) Characterization of heat shock protein 110 and glucose-regulated protein 170 as cancer vaccines and the effect of fever-range hyperthermia on vaccine activity. J Immunol 166:490–497

Wang XY, Li Y, Manjili MH, Repasky EA, Pardoll DM, Subjeck JR (2002a) hsp110 overexpression increases the immunogenicity of the murine CT26 colon tumor. Cancer Immunol Immunother 51:311–319

Wang Y, Kelly CG, Singh M, McGowan EG, Carrara AS, Bergmeier LA, Lehner T (2002b) Stimulation of Th1-polarizing cytokines, C-C chemokines, maturation of dendritic cells, and adjuvant function by the peptide binding fragment of heat shock protein 70. J Immunol 169:2422–2429

Wassenberg JJ, Dezfulian C, Nicchitta CV (1999) Receptor mediated and fluid phase pathways for internalization of the ER hsp90 chaperone GRP94 in murine macrophages. J Cell Sci 112:2167–175

Wearsch PA, Nicchitta CV (1996) Endoplasmic reticulum chaperone GRP94 subunit assembly is regulated through a defined oligomerization domain. Biochemistry 35:16760–16769

Wearsch PA, Nicchitta CV (1997) Interaction of endoplasmic reticulum chaperone GRP94 with peptide substrates is adenine nucleotide-independent. J Biol Chem 272:5152–5156

Wearsch PA, Voglino L, Nicchitta CV (1998) Structural transitions accompanying the activation of peptide binding to the endoplasmic reticulum hsp90 chaperone GRP94. Biochemistry 37:5709–5719

Welch WJ (1993) Heat shock proteins functioning as molecular chaperones: their roles in normal and stressed cells. Philos Trans R Soc Lond B Biol Sci 339:327–333

Whelan SA, Hightower LE (1985) Differential induction of glucose-regulated and heat shock proteins: effects of pH and sulfhydryl-reducing agents on chicken embryo cells. J Cell Physiol 125:251–258

Yamazaki K, Nguyen T, Podack ER (1999) Cutting edge: tumor secreted heat shock-fusion protein elicits CD8 cells for rejection. J Immunol 163:5178–5182

Ye J, Chen GS, Song HP, Li ZS, Huang YY, Qu P, Sun YJ, Zhang XM, Sui YF (2004) Heat shock protein 70 / MAGE-1 tumor vaccine can enhance the potency of MAGE-1-specific cellular immune responses in vivo. Cancer Immunol Immunother 53:825–834

Yedavelli SP, Guo L, Daou ME, Srivastava PK, Mittelman A, Tiwari RK (1999) Preventive and therapeutic effect of tumor derived heat shock protein, gp96, in an experimental prostate cancer model. Int J Mol Med 4:243–248

Zeng Y, Feng H, Graner MW, Katsanis E (2003) Tumor-derived, chaperone-rich cell lysate activates dendritic cells and elicits potent antitumor immunity. Blood 101:4485–4491

Zeng Y, Graner MW, Thompson S, Marron M, Katsanis E (2005) Induction of BCR-ABL specific immunity following vaccination with chaperone rich cell lysates (CRCL) derived from BCR-ABL+ tumor cells. Blood 105:2016–2022

Zheng H, Dai J, Stoilova D, Li Z (2001) Cell surface targeting of heat shock protein gp96 induces dendritic cell maturation and antitumor immunity. J Immunol 167:6731–6735

Zhu X, Zhao X, Burkholder WF, Gragerov A, Ogata CM, Gottesman ME, Hendrickson WA (1996) Structural analysis of substrate binding by the molecular chaperone DnaK. Science 272:1606–1614

Zugel U, Sponaas AM, Neckermann J, Schoel B, Kaufmann SH (2001) gp96-Peptide vaccination of mice against intracellular bacteria. Infect Immun 69:4164–4167

Hsp90 Inhibitors in the Clinic

S. Pacey[1] · U. Banerji[2] · I. Judson[1] · P. Workman[1] (✉)

[1]Cancer Research UK Centre for Cancer Therapeutics, The Institute of Cancer Research, Haddow Laboratories, Sutton, Surrey SM2 5NG, UK
paul.workman@icr.ac.uk

[2]The Royal Marsden Foundation NHS Trust, Downs Road, Sutton, Surrey SM2 5PT, UK

1	Introduction	332
1.1	Therapeutic Use of Hsp90 Inhibitors in Humans	332
1.2	Cancer: Summary of the Problem	334
1.3	Hsp90 as an Anti-cancer Target	335
2	Development of Hsp90 Inhibitors	336
2.1	Benzoquinone Ansamycins	336
2.2	Radicicol Derivatives	338
2.3	Synthetic Small Molecule Inhibitors	338
2.3.1	Purines	338
2.3.2	Diaryl Pyrazole Resorcinols	338
2.4	Other Compounds with Inhibitory Effects on Hsp90	338
2.4.1	Novobiocin	338
2.4.2	Histone Deacetylase Inhibitors	339
2.4.3	GRP94 Inhibitors	339
2.4.4	Mycograb	339
2.5	Miscellaneous	340
2.5.1	Cisplatin	340
2.5.2	Sodium Cromoglycate and Amlexanox	340
3	Hsp90 Inhibition in Preclinical Modells	340
4	Clinical Trials in Oncology	342
4.1	Phase I Clinical Trials with 17-AAG	342
4.2	Phase I Combination Studies with 17-AAG	345
4.3	Phase II Trials	345
4.3.1	Prostate Cancer	346
4.3.2	Malignant Melanoma	347
4.3.3	Multiple Myeloma	347
4.3.4	Mastocytosis	348
4.4	Phase I Trials of 17-DMAG	348
5	Summary and Lessons Learnt to Date	349
6	Concluding Remarks	350
References		351

Abstract Specific inhibitors of Hsp90 have recently entered human clinical trials. At the time of writing, trials have been initiated only in metastatic cancer, although a rationale exists for using these agents in a variety of human diseases where protein (mis)folding is involved in the disease pathophysiology. Hsp90 inhibitors offer a unique anti-cancer opportunity because they provide simultaneous combinatorial blockade of multiple oncogenic pathways. The first compound in this class, 17-AAG, has completed phase I trials and phase II trials are in progress. The toxicity has been manageable and evidence of possible clinical activity has been seen in metastatic melanoma, prostate cancer and multiple myeloma. Other inhibitors with improved properties are approaching clinical trials. This chapter presents an update of the current clinical trials using Hsp90 inhibitors, focussing on the areas that will be increasingly relevant in the next 5 years.

Keywords Hsp90 · Inhibitor · Disease · Clinical · Trial

1
Introduction

There were initial reservations that Hsp90 inhibition was an achievable clinical goal. Hsp90 accounts for 1%–2% of a cell's total protein and is involved in chaperoning many important proteins. Nevertheless, drugs that produce specific Hsp90 inhibition have progressed through clinical trials without the excess toxicity widely predicted.

This chapter reviews the progress of Hsp90 inhibitors through clinical development and a presents a summary of the lessons that have been learnt to date, along with potential future directions and challenges.

1.1
Therapeutic Use of Hsp90 Inhibitors in Humans

A rationale exists for the therapeutic use of Hsp90 inhibitors in several human diseases, as summarized in Table 1 and discussed further elsewhere in this book.

Specific Hsp90 inhibitors entered clinical trial in 1999 in patients with advanced, metastatic cancer. The first in class Hsp90 inhibitor 17-allylamino, 17-demethoxy geldanamycin (17-AAG) has completed dose-finding, phase I testing as a single agent. A variety of tumour-specific phase II and combination phase I trials are ongoing.

The geldanamycin analogue 17-demethoxy, dimethylaminoethylamino geldanamycin (17-DMAG) has recently entered phase I trials in cancer patients. A range of other Hsp90 inhibitors are currently undergoing preclinical development (Chiosis et al. 2004; Dymock et al. 2004).

Mycograb is a genetically recombinant antibody against yeast Hsp90, which increases the efficacy of amphotericin B against *Candida neoformans* infection (Nooney et al. 2005). It is currently undergoing phase II evaluation in patients with systemic candiasis (Matthews and Burnie 2004).

Table 1 Examples of diseases where Hsp90 inhibitors might be used clinically

Indication	Rationale and current position
Malignancy	Specific Hsp90 inhibitors in early clinical trials. Hsp90 inhibitors provide the potential for simultaneous, combinatorial blockade of multiple oncogenic pathways, as well as inhibition of stress responses in cancer cells. For more details see text.
Central Nervous System	Several chronic, neurodegenerative diseases are attractive targets for drugs that interfere with abnormal protein folding or function. A hallmark of such conditions is the deposition of abnormal protein aggregates (Barral et al. 2004; Cohen and Kelly 2003). Debate remains as to whether these are pathogenic or an incidental (protective) response (Orr 2004), which would have implications for Hsp targeting in therapeutics.
Parkinson's disease	Geldanamycin shown to suppress neurotoxicity in alpha-synuclein model in *Drosophila* (Auluck et al. 2002)
Prion disease (Creutzfeldt Jacob disease)	Caused by accumulation of misfolded PrP^c protein (PrP^{sc}) (Weissmann 2004). Novobiocin depletes PrP^c in a cell line model. In part this effect is postulated to be via Hsp90 inhibition, and treatment with geldanamycin partially antagonizes this process (Ochel and Gademann 2004). Less substrate is then available for conversion to PrP^{sc} (Daude et al. 2003; Mallucci et al. 2003).
Huntington's disease	Treatment of mammalian cells overexpressing mutant huntingtin protein with geldanamycin prevents deposition of polyQ protein in amyloid-like aggregates (Sittler et al. 2001).
Alzheimer's disease	Deposition of abnormal amyloid (Aβ) protein is thought to underlie development of progressive neurotoxicity, manifest as dementia.
Respiratory	
Cystic fibrosis	Caused by a mutant chloride channel (ΔF508 CFTR). Studies in rabbit reticulocyte lysates have shown defective intracellular trafficking so that mutant protein fails to reach the cell membrane. It retains cAMP activated Cl-channel activity (potentially enough to ameliorate the disease). Geldanamycin reduces the ubiquitination and degradation of the mutant protein (Fuller and Cuthbert 2000). Additionally, Hsp70 induction promotes ΔF508 CFTR trafficking and maturation (Choo-Kang and Zeitlin 2001).
Immunological	Several Hsp (including Hsp90) have been shown to activate the immune system by stimulating production of pro-inflammatory cytokines. Debate is ongoing as to the validity of these model systems (Tsan and Gao 2004).
Musculoskeletal	Models have shown immunosuppressant effects of geldanamycin (modulating MAPK and NF-κ-B pathways), including reduced progression of adjuvant-induced arthritis (Sugita et al. 1999). Hsp90 regulates post-transcriptional control of the pro-inflammatory cytokines TNFα and IL-6. Geldanamycin treatment reduces levels of TNFα and IL-6 (Wax et al. 2003).

Table 1 (continued)

Indication	Rationale and current position
Other	
Anti-microbial	*Plasmodium* growth inhibited in culture by geldanamycin with an IC_{50} comparable to current treatment (chloroquine), at similar concentrations to those achieved with 17-AAG in anti-cancer human trials. Synergism noted by combining geldanamycin and chloroquine in chloroquine-resistant and -sensitive plasmodium strains (Kumar et al. 2003). Mycograb (genetically recombinant antibody against yeast Hsp90) increases efficacy of amphotericin B against *Candida neoformans* (Nooney et al. 2005) and is currently undergoing clinical trial in patients with systemic candidiasis (Matthews et al. 2004).
Modulation of other Hsp	For example Hsp70 induction, as follows Hsp90 inhibition, shown to be protective in a variety of situations. One example is limitation of ischaemic damage, such as that seen in haemorrhagic shock and subsequent cardiovascular resuscitation injury damage (Kiang et al. 2004). Another would be post myocardial infarction or protecting donor hearts prior to transplant (Latchman, 2001).

Hsp, heat shock protein

If these drugs are to be used outside the setting of advanced cancer there are significant challenges to address. For example, the acceptable toxicity profile varies depending on a patient's prognosis or specifically with respect to neurodegenerative disease, the current compounds show minimal penetration of the blood–brain barrier.

To what extent, if at all, of the therapeutic potential is realized will only become clear if and when Hsp90 inhibitors progress into clinical trial for these indications. As the use of Hsp90 inhibitors in advanced, metastatic cancer remains the most mature indication at the time of writing, the remainder of this chapter will focus on this area.

1.2
Cancer: Summary of the Problem

Cancer represents a diverse and complex therapeutic problem. The burden on the world's healthcare systems remains vast despite advances in treatment. For example, in the United States, 1.2 million people are diagnosed with cancer each year, leading to over 500,000 deaths (DeVita et al. 2001). The outcome varies greatly depending on many factors, not least the origin of the tumour. This can be seen from the range of 5-year survival rates, examples being testicular (95%), breast (>70%) and pancreatic (2%) cancers (Toms 2004). However, these figures hide a heterogeneous set of outcomes. In advanced, metastatic cancer, treatment options become limited; cytotoxic chemotherapy is employed in this setting, which exposes patients to potentially severe side effects. There have been advances in treatment of human cancer with chemotherapy since the first use of nitrogen mustards and anti-folate drugs in the 1940s (Chabner and Roberts 2005). While we should be optimistic, the statistics remind us that there is still much room for improvement.

Therapeutic strategies targeted specifically to the causes of disease pathophysiology are being developed as the molecular basis of cancer is defined. In the majority of advanced solid tumours, several genes are likely to exist that are aberrant in copy number, structure or expression (Balmain et al. 2003).

The new molecular therapeutic agents have enjoyed clinical success where single genes or pathways drive tumour progression, such as imatinib (Gleevec) in chronic myeloid leukaemia and gastrointestinal stromal tumours (Druker 2003; van Oosterom et al. 2002) or with trastuzumab (Herceptin) in breast cancer (Baselga et al. 2001). Early optimism has been blunted by tumours acquiring resistance to these new agents, for example by further mutations in the kinase domain giving rise to imatinib resistance (Chen et al. 2004; Gorre et al. 2001; Shah et al. 2002). Such evidence has led investigators to suggest that inhibiting a single target will not be adequate to produce a cure in most human tumours (Workman 2005).

1.3
Hsp90 as an Anti-cancer Target

The functions of Hsp90 make it a unique cancer target. Many current cancer drug targets are contained within the ever-growing list of proteins that Hsp90 interacts with (for up to date list, visit http://www.picard.ch). At a time when the limitations in knowledge of the molecular biology of cancer are evident as much as the advances, inhibition of Hsp90 may produce the best of both worlds – specific inhibition of key individual targets such as BCR-ABL (An et al.

Table 2 Examples of Hsp90 client proteins that influence the six hallmark traits of cancer

Hall Mark Trait	Hsp 90 client protein
Self-sufficiency in growth signals	RAF (Schulte et al. 1996)
	ERB-B2 (Xu et al. 2001)
	BCR-ABL (An et al. 2000)
Insensitivity to anti-growth signals	PLK-1 (de Carcer et al. 2001)
Evasion of apoptosis	AKT (Sato et al. 2000)
Limitless replicative potential	h-TERT (Forsythe et al. 2001)
Sustained angiogenesis	AKT (Sato et al. 2000)
	HIF-1 (Minet et al. 1999)
Tissue invasion and metastasis	MET (Kawano et al. 2004)

Table 3 List of advantages for drugs that target Hsp90 (adapted from Dymock et al. 2004)

Property	Detail	Reference
Mutant protein are more dependent on Hsp90	Greater requirement for Hsp90 to stabilize mutant protein compared to their wild type counterpart.	(Whitesell et al. 1998)
Broad range of anti-cancer effects	Potential to inhibit all six of Hannahan and Weinberg's hallmark cancer traits.	(Maloney et al. 2002)
Trigger cell death by lethal protein mutations	The Hsp90 inhibitors may unmask lethal mutations in cancer cells.	(Rutherford and Lindquist 1998)
Intra-tumoural drug accumulation	17-AAG accumulates in cancer cells and human tumour xenograft models.	(Banerji et al. 2001; Chiosis et al. 2003; Egorin et al. 2001)
Tumour Hsp90 is more sensitive than normal cell Hsp90 to inhibitors	Differential expression of a super-chaperone complex in cancer cells, which is more sensitive to Hsp90 inhibitors.	(Kamal et al. 2003)
Synergism with other therapy	Cytotoxic drugs such as paclitaxol.	(Solit et al. 2003a)
Exploiting multiple oncogene addiction	Cancer cells develop "addiction to" or dependence upon multiple oncogenic pathways. Hsp90 inhibitors provide combinatorial effects on multiple oncoproteins.	(Workman 2004)
Overcoming drug resistance	Drug resistance to single molecular therapeutics may arise by activation of alternative pathways or mutation of the target. Hsp90 inhibitors can retain activity in both settings.	(Gorre et al. 2001)

2000) or ERB-B2 (Xu et al. 2001) and modulation of multiple pathways and the compensatory cross-talk that may otherwise prevent response or generate resistance.

Cancer cells differ from normal cells by six hall mark traits (Hanahan and Weinberg 2000). Multiple client proteins of Hsp90 are known to be deregulated in cancer. Some examples are shown in Table 2.

Drugs targeting Hsp90 gain several advantages, which are listed in Table 3 (see also Dymock et al. 2004).

2
Development of Hsp90 Inhibitors

The structures of selected Hsp90 inhibitors are shown in Fig. 1. The most relevant classes of compound are discussed in more detail in the next section.

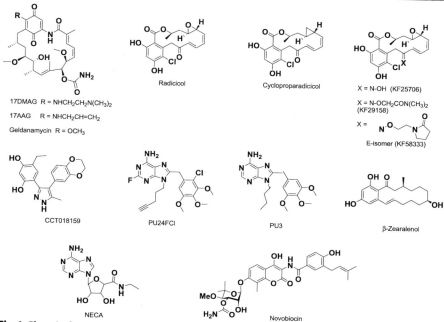

Fig. 1 Chemical structures of selected Hsp90 inhibitors

2.1
Benzoquinone Ansamycins

First-generation benzoquinone ansamycins such as herbimycin were thought to be tyrosine kinase inhibitors (Okabe et al. 1992). Study of the mechanism of action of ansamysins, such as geldanamycin and its analogue 17-AAG, have demonstrated that these compounds compete with ATP at the nucleotide binding site in the N-terminal domain of Hsp90 (Grenert et al. 1997; Prodromou et al. 1997; Stebbins et al. 1997).

Geldanamycin was found to be too hepatotoxic in preclinical models (Supko et al. 1995). Further analogues were developed for clinical use, including 17-AAG (Schnur et al. 1995) and 17-DMAG (Eiseman et al. 2005); 17-AAG caused cell cycle arrest and apoptosis in cancer cells (Hostein et al.). It also showed activity in human tumour xenograft models (Kelland et al. 1999; Solit et al. 2002), leading to phase I evaluation (discussed later). The cumbersome formulation has limited true evaluation of its maximum tolerated dose (MTD) (Banerji et al. 2003; Erlichman et al. 2001; Goetz et al. 2002; Sausville et al. 2003).

The analogue 17-DMAG is more water soluble and is under evaluation in preclinical and clinical settings (Bagatell and Whitesell, 2004; Hollingshead et al. 2005). Attempts to reformulate 17-AAG have resulted in clinical trials commencing with CF1010 (www.conformacorp.com) and KOS 953 (www.Kosan.com).

2.2
Radicicol Derivatives

Radicicol is another natural Hsp90 inhibitor with potent in vitro activity. It has poor in vivo activity, possibly due to its expoxide moiety. Oxime derivatives of radicicol have been shown to be active in human tumour xenograft models (KF58333), while others have directed their efforts to replacing the reactive epoxide moiety to improve the properties of the molecule, as in cycloproparadicicol (Dymock et al. 2004; Soga et al. 2003). A clinical candidate has yet to be developed.

2.3
Synthetic Small Molecule Inhibitors

2.3.1
Purines

Rational drug design has led to a series of drug-like small molecules that inhibit Hsp90 (Chiosis et al. 2002). The scaffold is amenable to further manipulation, providing the scope to improve on the properties of the current lead compound PU24FCl (Wright et al. 2004). More potent analogues have been discovered recently (Llauger et al. 2005).

2.3.2
Diaryl Pyrazole Resorcinols

The diaryl pyrazole resorcinol CCT018159 was discovered by high-throughput screening (Rowlands et al. 2004; Cheung et al. 2005). CCT018159 has a growth inhibitory IC_{50} of 4.5 μM in HCT116 cells (Cheung et al. 2005) and the corresponding value for inhibiting ATPase activity is 7.1 μM. Optimization of the CCT018159 lead is being carried out with the aid of X-ray crystallography (Cheung et al. 2005). This is exemplified by the 5-amide analogue VER-49009, a nanomolar inhibitor that compares favourably with 17-AAG (Dymock et al. 2005).

2.4
Other Compounds with Inhibitory Effects on Hsp90

2.4.1
Novobiocin

Novobiocin (an antibiotic which targets eukaryotic topoisomerase II) may inhibit Hsp90 by binding the C-terminal rather than the N-terminal domain

(Marcu et al. 2000). Poor pharmacokinetic properties limit its use; however, it may be a template for developing alternate strategies of Hsp90 inhibition (Marcu and Neckers 2003). In oncology, chemomodulatory clinical trials have been conducted with novobiocin and alkylating agents, but results were disappointing (Eder et al. 1991; Ellis et al. 1991).

2.4.2
Histone Deacetylase Inhibitors

This class of compounds has been developed to target epigenetic modification of gene transcription; inhibition leads to hyperacetylation of cellular proteins. By inhibiting deacetylation of Hsp90, these compounds inhibit Hsp90, as shown using depsipeptide in cancer cell line models (Yu et al. 2002) and LAQ824 during human clinical trials (Kristeleit et al. 2004). These drugs are at varying stages of clinical development, with hints of clinical activity. For example, in a phase I trial of MS-275 a partial response was observed in a patient with melanoma and stabilization of disease was seen in melanoma, Ewing's sarcoma and colorectal carcinoma (Gore et al. 2004). Combination with specific Hsp90 inhibitors may be beneficial, as suggested by cancer cell line data in chronic and acute myeloid leukaemia models, where co-treatment using LBH589 and 17-AAG is synergistic (George et al. 2005).

2.4.3
GRP94 Inhibitors

GRP94 is an endoplasmic reticulum homologue of Hsp90 and is structurally similar to Hsp90. It is known to chaperone client proteins important to cancer cells. Both radicicol and geldanamycins are active against both GRP94 and Hsp90; however, it has been possible develop inhibitors that act on GRP94 and not Hsp90, as exemplified by $5'$-N-ethylcarboxamideadenosine (NECA) (Dymock et al. 2004; Soldano et al. 2003).

2.4.4
Mycograb

Mycograb is a human genetically recombinant antibody against fungal Hsp90. Clinical data shows that the development of antibodies to Hsp90 is closely related to patient recovery from systemic candidal infection (Matthews et al. 2003). Multinational clinical trials have commenced in patients with invasive candidiasis giving Mycograb in combination with Amphotericin B (Matthews et al. 2004). This agent clearly has potential utility in cancer.

2.5
Miscellaneous

2.5.1
Cisplatin

Cisplatin binds Hsp90 and produces a conformation change (Itoh et al. 1999). Inhibition of steroid hormone receptor function has been demonstrated, but effects on kinases were not (Rosenhagen et al. 2003). Others have observed that the oxidation state of cysteine residues are important for Hsp90 function (Nardai et al. 2000), leading to the suggestion that these nucleophilic residues are the target for cisplatin, and to the speculation that these observations may reveal how Hsp90 differentially regulates protein kinases and steroid hormone receptors (Chiosis et al. 2004)

2.5.2
Sodium Cromoglycate and Amlexanox

These compounds are used to treat allergic reactions. Recent evidence has suggested a pharmacological effect on Hsp90 at micromolar concentrations (Okada et al. 2003). The relevance of this has yet to be determined.

3
Hsp90 Inhibition in Preclinical Modells

Preclinical models established a basis upon which to initiate clinical trials with specific Hsp90 inhibitors. They provided estimates of maximally tolerated dose (MTD), a pharmacokinetic (PK) profile, evidence of target inhibition in the tumour, i.e. pharmacodynamic (PD) effect and demonstration of tumour growth inhibition. Table 4 details some of the evidence obtained for 17-AAG in mouse models.

Early clinical trials now routinely incorporate PD studies developed during the preclinical phase to better understand the mechanism of action, dose and scheduling of new drugs (Gelmon et al. 1999). In the case of targeted agents where traditional MTD may be less relevant than with cytotoxic therapy, these studies are increasingly valuable to define the relationship between clinical response, PK and PD. This process is summarized in Fig. 2.

Hsp90 inhibition produces a characteristic molecular signature of protein expression. These include the reduction in expression of client proteins such as ERB-B2, C-RAF and CDK-4 combined with increased expression of Hsp70 (Maloney and Workman 2002; Workman 2004). Depletion of LCK in peripheral blood lymphocytes can also be a useful marker. Gene expression profiling by cDNA microarray has been used to provide further details of the molecular signature of Hsp90 inhibition (Clarke et al. 2000). PK-PD relationships have

Table 4 Human tumour xenograft studies for evaluating growth inhibition, PK and PD profile of 17-AAG

Xenograft Model	Growth inhibition	PK studies	PD studies in vivo	Reference
Melanoma MEXF276L	+	Peak plasma levels 5–10 µM	Hsp70 induction and Hsp90 depletion	Burger et al. 2004
Ovarian A2780, CH1	+	Peak plasma levels 27–34 µM. Half life longer in tumour than in plasma	C-RAF and CDK4 depletion with Hsp70 induction in tumour and PBLs	Banerji et al. 2001; Banerji et al. 2005
Colon HT29	+	ND	Hsp70 induction and increase in phospho-choline levels on MRI	Chung et al. 2003
Colon HT29, BE	+	ND	Hsp70 induction, depletion of C-RAF, mutant p53	Kelland et al. 1999
Breast TAMR-1	+	ND	C-RAF, AKT and ER depletion and Hsp70 induction	Beliakoff et al. 2003
Prostate CWR22, CWRSA6	+	ND	Androgen receptor, HER2, HER3, AKT, depletion and Hsp70 induction	Solit et al. 2002
Breast MDA-MB-473	ND	Peak plasma level 19.3 µg/ml	C-RAF-1, ERB-B2, depletion and Hsp70 induction	Xu et al. 2003

ND, not determined

been discussed and validated in a human ovarian cancer xenograft model (Banerji et al. 2005).

Induction of Hsp70, while a useful marker of target inhibition, may be a concern. Hsp70 has been shown to be anti-apoptotic, reviewed by Jaattela (1999). Induction of Hsp70 represents a compensatory effect which antagonizes the anti-cancer effects of Hsp90 inhibition. This is a good example of how PD measurements not only provide information useful for altering drug dose and scheduling, but also lead to increased understanding of molecular pharmacology. Work has shown that inhibition of Hsp90, with 17-AAG, in cells that are manipulated so that they are not able to induce Hsp70 (by using siRNA to Hsp70) leads to increased apoptotic effects (Sheriff et al. 2004). Further efforts to demonstrate the clinical relevance of these observations are ongoing.

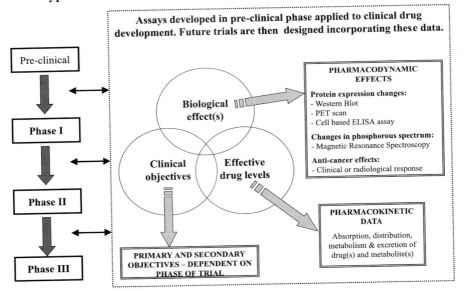

Fig. 2 Summary of PK/PD-driven drug development

4
Clinical Trials in Oncology

Drug development in oncology follows a defined logical sequence. The objectives of each phase of the process are summarized in Fig. 3.

4.1
Phase I Clinical Trials with 17-AAG

Multiple phase I trials have explored different dose and scheduling strategies. These are summarized in Fig. 4.

Goetz and colleagues administered 17-AAG on days 1, 8 and 15 of a 28-day cycle: 21 patients were evaluated, and the maximum dose administered was 431 mg/m^2. At this dose, two patients experienced dose-limiting toxicity (DLT). One suffered hyperbilirubinaemia and transaminitis, while the other patient required hospital treatment for nausea and vomiting. At the recommended dose of 308 mg/m^2 diarrhoea, nausea, vomiting, fatigue, anorexia and anaemia were most commonly noted. One death from respiratory failure occurred 24 h after receiving 17-AAG in a patient who had past oxaliplatin exposure and was known to have pulmonary infiltrates. At the MTD, biologically active concentrations of 17-AG and 17-AAG were detected. Nonlinear PK behaviour was noted, the interpretation being limited by the single-patient

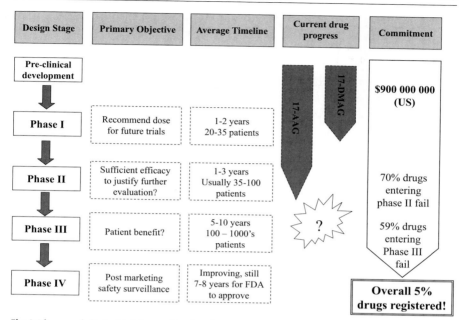

Fig. 3 Phases of clinical trials used in developing new drugs (Kola and Landis, 2004; Nottage and Siu 2002; Scherer 2004)

SCHEDULE (Days marked with arrow)	SUMMARY
Continuous weekly schedules. 1 8 15 22 29	Weekly, with day 22 infusion omitted. Maximum dose reached 431mg/m². Dose limiting toxicities were liver dysfunction, nausea and vomiting. Recommended Phase II dose 308 mg/m² (Goetz, M., et al. 2005). Weekly, maximum and recommended Phase II doses were 450 mg/m². Further escalation limited by drug formulation. Dose limiting toxicity was diarrhoea and transaminitis. Linear relationship between dose and AUC of 17-AAG (Banerji, U., et al. 2005).
Weekly x3, week off and repeat. 1 8 15 29	Forty five patients evaluated in 11 dose levels. Maximum dose reached 395 mg/m²/wk. Dose limiting toxicities were pancreatitis and fatigue. Linear relationship between dose and AUC of 17-AAG. Recommended phase II dose 295 mg/m² (Ramanathan, R., et al 2004)
Twice weekly, week off and repeat. 1 4 8 11 22	Twelve patients treated in 6 dose levels. Maximum dose reached 308 mg/kg. Dose limiting toxicity included diarrhea, dehydration and liver enzyme abnormalities in two out of two patients treated. Plasma half life of 17-AAG 2.9 hrs and clearance 18.5 L/hr/m². PD markers in PBLs showed HSP70 and HSF-1 induction at 24 hrs and integrin linked kinase depleted at 6 hrs. One patient with metastatic melanoma had stable disease. Recommended phase II dose 220mg/m² (Erlichman, C., et al., 2004)
Daily x5, repeated every 3 weeks. 12345 22	The highest dose reached 56 mg/m². Dose limiting toxicities included diarrhoea, thrombocytopenia and elevated liver enzymes. Maximum plasma concentrations reported 2080nM. Recommended phase II dose 40 mg/m² (Grem, J.L., et al., 2005)

Fig. 4 Summary of phase I trials with 17-AAG in advanced cancer

cohorts treated at lower drug doses (a common method used to expedite trial progress). *CYP3A4-* and *A5-*dependent metabolism was determined to affect circulating drug levels. It was suggested that *NQO1* polymorphisms did not affect toxicity (although only two patients were *1/*3 genotype). PD results showed an increase in Hsp70 expression on Western blot analysis of proteins extracted from peripheral blood lymphocytes (PBL) (Goetz et al. 2005).

A parallel study by Banerji and colleagues (2005) administered 17-AAG weekly with no breaks in treatment. The maximum dose achieved was 450 mg/m^2, with two out of nine patients experiencing DLT (diarrhoea and transaminitis) at this dose. The relationship between dose and plasma area under curve (AUC) for 17-AAG was noted to be linear across the dose cohorts. Effective plasma drug levels compared to preclinical models were seen up to 24 h after infusion. PD data (Hsp70, CDK-4 and C-RAF expression analysis by Western blotting) showed target inhibition with doses of 320 and 450 mg/m^2 per week. In both surrogate (PBL) and tumour tissue target, inhibition persisted for between 1 and 5 days after infusion. Although phase I trials do not look for tumour response as a trial endpoint, interesting clinical activity was seen. Two patients (out of 12 with metastatic melanoma treated in the course of the study) experienced disease stabilization for 15 and 46 months. This is noteworthy due to the short life-expectancy of such patients, with a median survival around 6.4 months (Manola et al. 2000). The investigators hypothesized a cytostatic action of 17-AAG in these patients (Banerji et al. 2005), consistent with preclinical data.

The difference in MTD between the two studies (of Banerji and Goetz) highlights a weakness of current phase I design. The small sample size, combined with the practical difficulty in equating very different toxicities can lead to apparent inconsistency in results. Goetz and colleagues declared MTD due to transaminitis (in a patient with already compromised liver function) and uncontrolled vomiting (where the patient's previous experience and expectations of therapy can have a large effect). The progress of the trial by Ramanathan and colleagues (2004) was halted by patient fatigue (which can be hard to interpret in a patient with advanced cancer) and pancreatitis. In contrast, Banerji and colleagues (2005) were prevented from further escalation by drug formulation. Given the level of toxicity, the next cohort may have represented the MTD, especially when taken in conjunction with data from Erlichman and colleagues, who were able to deliver 220 mg/m^2 twice weekly but not 308 mg/m^2 (Erlichman et al. 2004).

To date no advantage has been demonstrated for more frequent schedules, such as that of Grem and colleagues (2005). The convenience and tolerability of the weekly schedule, combined with convincing PK, PD and clinical data, have led to this schedule being utilized in several phase II trial designs. The formulation of 17-AAG dissolved in 0.4% DMSO has already been improved. DMSO at this concentration is known to cause nausea, vomiting and diarrhoea (Santos et al. 2003). How this affects the toxicity profile and whether such formulations present an opportunity to fully realize the clinical potential of 17-AAG will become clear in the near future.

4.2
Phase I Combination Studies with 17-AAG

The completion of single-agent phase I trials allows further trials to be initiated investigating combinations of 17-AAG and other therapeutic modalities. Most often this will involve a combination of two or more drugs but effects of a drug combined with radiotherapy, for example, may be tested.

Combinations are chosen on the basis of preclinical evidence and practical issues. Practical issues might include combining 17-AAG with agent(s) that are currently standard therapy by predicting indications where 17-AAG may be tested in the future, or combining 17-AAG with drugs with nonoverlapping toxicities.

The involvement of the U.S. National Cancer Institute in the clinical development of 17-AAG has allowed combinations to be tested that would otherwise not be feasible, due to commercial interest or sponsorship issues when combining more than one investigational drug.

Multiple phase I studies are currently in progress to define the MTD in combination with a range of cytotoxic and molecularly targeted agents, as summarized in Table 5.

Trials are also ongoing using alternative formulations of 17-AAG, such as KOS-953 in multiple myeloma patients combined with bortezomib (www.kosan.com).

4.3
Phase II Trials

Preclinical evidence and data from phase I trials has guided the phase II programme for 17-AAG (Table 6). Completion of phase II studies over the

Table 5 Ongoing combination phase I trials with 17-AAG (www.nci.nih.gov/clinicaltrials)

17-AAG combined with	Population on study	Protocol ID
Pactlitaxel	Advanced, metastatic cancer	PCI-03-152
Docetaxel	Hormone refractory prostate cancer	MSKCC-03006
Cisplatin and gemcitabine	Advanced, metastatic cancer	MAYO-MC0111
Imatinib	Gleevec refractory CML	UCLA 0408048-01
Bortezomib	Advanced, metastatic cancer	MAYO-MC0214
	Refractory NHL/AML	OSU-2004C0084
Cytarabine	Refractory leukaemia	MAYO-MC0313
Fludarabine and rituximab	Refractory leukaemia	OSU-0429

Table 6 Current phase II trials of 17-AAG (www.nci.nih.gov/clinicaltrials; Bagatell et al. 2004)

Tumour type	Example of rationale or relevant client protein	Protocol ID
Metastatic melanoma	B-RAF (Davies et al. 2002)	MSKCC-04056
Metastatic melanoma		CRUK-PH2/049
Renal cancer:	HIF-1 (Chung et al. 2004)	
+ Von-Hippel-Lindau disease		NCI-04-C-0238
OR papillary/clear cell		MSKCC-04082
Breast (adenocarcinoma)	ERBB2 (Slamon and Pegram 2001) h-TERT(Bearss et al. 2000)	WSU-C-2803
Mastocytosis	c-KIT (Longley et al. 1999)	–
Thyroid	17-AAG showed activity in a panel of thyroid cancer cell lines, correlating with Hsp90 levels, not histological tumour subtype. (Braga-Basaria et al. 2004)	–
Ovary	Cell line panel (ovarian, cervical, breast and endometrial cancer lines) all responsive to 17AAG and 17DMAG (Gossett et al. 2005)	–
Hormone refractory prostate	AR and HER2 (Solit et al. 2002)	–

next few years should provide answers as to whether 17-AAG is an effective anti-cancer agent. At the present time, early indications of clinical activity has been seen in metastatic melanoma, prostate cancer and multiple myeloma.

4.3.1
Prostate Cancer

Hormone refractory disease represents a prime opportunity to test new agents. Only one drug (docetaxel) to date has demonstrated a survival benefit, and this is just a few months in duration (Petrylak et al. 2004; Tannock et al. 2004). Drugs can therefore be tested in patients who are either chemotherapy-naïve or have minimal previous exposure to chemotherapy.

Wild type androgen receptor, ERBB2 and AKT are all depleted by inhibition of Hsp90 (with 17-AAG) and are key components of continued cell proliferation and survival (Solit et al. 2003b). Alongside pronounced effects in preclinical models, patients have experienced reduction in tumour marker (prostate-specific antigen) levels within the phase I combination study with docetaxel (Solit et al. 2004).

4.3.2
Malignant Melanoma

The incidence of melanoma continues to increase in the developed world and affects a younger population than other solid tumours (Armstrong and Kricker 1994). In 2002 melanoma accounted for 53,600 new cases and 7,400 deaths in the United States (Jemal et al. 2002). Current therapy for metastatic disease produces response rates in the order of 5%–35% whether single, combined or introducing biological agents (Eigentler et al. 2003), but there is minimal effect on patient survival and great scope for improvement.

Preclinical data suggest that melanoma might be sensitive to Hsp90 inhibition (Burger et al. 2004). While no responses have been seen, prolonged disease stabilization has been noted when giving weekly 17-AAG (Banerji et al. 2005). Investigators have hypothesized a cytostatic action of Hsp90 inhibition; phase II trials have been designed with this in mind. Within the UK, under the auspices of Cancer Research UK, the Royal Marsden Hospital/Institute of Cancer Research and the Royal Free Hospital have commenced a trial where disease stabilization after 6 months of therapy will be the primary endpoint. This represents a departure from the more traditional endpoint of tumour shrinkage.

Other phase II trials are asking whether 17-AAG can produce tumour shrinkage in patients with melanoma, led by the Sloane Kettering Memorial Medical Centre. The B-RAF mutation has been shown to be common and important in melanoma (Davies et al. 2002) and this trial is stratifying patients according to their B-RAF mutation status. Studies in pre-clinical models have shown that mutant B-RAF, including the most common V600E mutant, is hypersensitive to depletion by 17-AAG (daRocha Dias et al. 2005; Neil Rosen personal communication).

4.3.3
Multiple Myeloma

Multiple myeloma is the second commonest haematological malignancy diagnosed in the Western world. This neoplasm of plasma cells is characterized by their presence in the bone marrow and overproduction of (monoclonal) immunoglobulin. Until recently, attempts to prolong survival were limited to patients fit enough to undergo high-dose chemotherapy with stem cell rescue (Child et al. 2003). Less fit patients were offered cytotoxic therapy (such as melphalan and prednisolone), with minimal impact on survival. Median survival for symptomatic patients ranges from 3 to 5 years (Weber 2003).

Molecular targeted agents have made an impact recently. Thalidomide has shown efficacy (Singhal et al. 1999) and most recently the proteasome inhibitor bortezomib (PS-341) has shown clinical activity (Richardson et al. 2003). Preclinically, co-treatment using 17-AAG and bortezomib is potentially synergistic (Mitsiades et al. 2002). A phase I trial, using the cremaphor formulation of 17-AAG in patients with myeloma is currently recruiting patients clinical activity has been noted. The company expect to initiate a phase II trial once a dose has been recommended (Mitsiade et al. 2005).

4.3.4
Mastocytosis

This rare disease is due to a proliferation of mast cells. Malignant transformation then occurs in 7%–30% patients, dependent on patient age at disease onset. Mastocytosis presents a valuable clinical model in which to study Hsp90 inhibition. The disease is associated with activating mutations in *c-KIT* (Longley et al. 1999). In-vivo models have shown 17-AAG to reduce KIT protein and activity in both sensitive and insensitive (to imatinib) cell lines. Even in imatinib-resistant cell lines, 17-AAG is able to induce apoptosis (Gorre et al. 2001). This trial allows data to be collected in humans where the molecular basis of the disease is well understood – comparable to development of imatinib in CML and GIST, as well as looking for therapeutic benefit.

4.4
Phase I Trials of 17-DMAG

Clinical trials with 17-AAG have established that an Hsp90 inhibitor can be given to patients with tolerable side effects (see above). 17-AAG has a variety

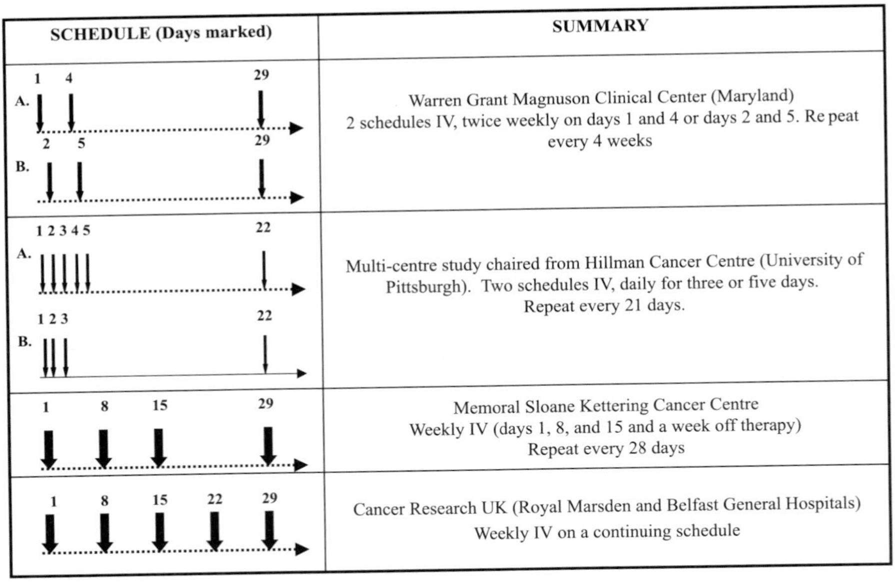

SCHEDULE (Days marked)	SUMMARY
A. 1 4 29 / B. 2 5 29	Warren Grant Magnuson Clinical Center (Maryland) 2 schedules IV, twice weekly on days 1 and 4 or days 2 and 5. Repeat every 4 weeks
A. 1 2 3 4 5 22 / B. 1 2 3 22	Multi-centre study chaired from Hillman Cancer Centre (University of Pittsburgh). Two schedules IV, daily for three or five days. Repeat every 21 days.
1 8 15 29	Memoral Sloane Kettering Cancer Centre Weekly IV (days 1, 8, and 15 and a week off therapy) Repeat every 28 days
1 8 15 22 29	Cancer Research UK (Royal Marsden and Belfast General Hospitals) Weekly IV on a continuing schedule

Fig. 5 Current phase I trials using 17-DMAG in advanced cancer (www.nci.nih.gov/clinicaltrials)

of limitations such as complex formulation, poor solubility; odour and lack of oral bioavailability.

Therefore, 17-DMAG has been developed to overcome some of these limitations. Evidence from preclinical models suggests that 17-DMAG and 17-AAG share inhibition of Hsp90 as the mechanism of action and that they have a similar spectrum of toxicity. It appears that 17-DMAG is more potent as an Hsp90 inhibitor but also more toxic (Eiseman et al. 2005; Hollingshead et al. 2005).

Phase I trials investigating a variety of schedules have been initiated. Each of these trials is seeking to build on the initial PK/PD work that has proven beneficial in the development of 17-AAG to date (Fig. 5).

Although 17-AAG has been well tolerated on a weekly schedule and less so on more frequent schedules, the relative contribution of drug and class effect vs the formulation vehicle (DMSO) is obscure. The phase I trials of 17-DMAG have been designed pragmatically to recommend doses on a variety of schedules.

5
Summary and Lessons Learnt to Date

Hsp90 is a highly conserved chaperone protein and its role in cellular biology has been extensively discussed in other chapters of this book.

The clinical use to date of Hsp90 inhibitors has been focused within the field of cancer medicine. There are many other areas where roles of Hsp90 in protein maturation, folding and trafficking make it an attractive therapeutic target, such as neurodegenerative and immune-mediated diseases.

The most successful attempts to date in developing an Hsp90 drug have been seen with small molecule inhibitors against the ATPase activity of the chaperone's N-terminus. The benzoquinone ansamycins have been studied in detail and the first drug in clinical trials (17-AAG) has been shown to be tolerated at doses which possess acceptable PK properties, achieve target inhibition and show possible anti-tumour activity. Toxicity, whilst troublesome, is manageable, although room for improvement remains. The DMSO-containing formulation of 17-AAG has been improved and how this affects the toxicity profile will become clear in the near future. Although care must be applied interpreting efficacy data from phase I trials, there have been several cases of prolonged disease stabilization in patients with metastatic melanoma (Banerji et al. 2005; Erlichman et al. 2004), prompting the instigation of phase II trials. Activity has also been described in hormone-refractory prostate cancer (Solit et al. 2004).

Investigation of PK and PD effects have shown 17-AAG to be biologically active against Hsp90 at doses used within the clinical trials. Improved understanding has led to the concern that one of the most robust PD markers, Hsp70 induction, may represent a compensatory mechanism by which cancer cells avoid death due to Hsp90 inhibition. It is possible this will limit the use of Hsp90 inhibitors as single agents.

The study of further geldanamycin analogues such as 17-DMAG is underway. Development programmes with non-ansamycin drugs such as radicicol, purine and diaryl pyrazole resorcinol compounds are aiming to produce clinical candidates.

6
Concluding Remarks

Anti-cancer drug discovery has shifted focus from the development of DNA-damaging agents to treatments that target the underlying mechanisms of oncogenesis.

Signal transduction inhibitors have generated considerable interest, several of which have been licensed. Concern remains that such drugs will be vulnerable to redundancy between pathways, allowing cells to compensate for inhibition of the initial target. Hsp90 inhibitors deplete multiple client proteins, leading to a combinatorial attack on multiple oncogenic targets. This may prevent cross-talk effects and provides for a broad spectrum of activity against many tumour types, as well as reducing the opportunity for resistance to develop (Workman 2004). While we wait for technology to advance so that personalized therapy becomes reality, drugs such as Hsp90 inhibitors that deliver combinatorial blockade of multiple oncogenic pathways and hallmark traits of cancer offer exciting potential.

These are exciting times in cancer drug development and Hsp90 inhibitors represent a unique opportunity. A foundation exists on which to build treatment regimens containing Hsp90 inhibitors, whether as a single agent or in combination with other therapies. Excellent pre-clinical research rational drug trial design, international cooperation and an emphasis on translational medicine has guided this field to date and must be allowed to direct progress going forward. Without such efforts, there is a risk of failing to realize the potential of new agents (not just Hsp90 inhibitors), whether by setting incorrect objectives, failing to gather the correct information, or simply not understanding fully what has been achieved to date.

This chapter has highlighted the current state of clinical trials using inhibitors of the Hsp90 molecular chaperone protein. We predict that, as the field of cancer chemotherapy matures, chaperone targets (in particular Hsp90) will play a role for some time to come. In other areas of medicine, it will be fascinating to watch for the advances that Hsp90 inhibitors may present.

Acknowledgements We are grateful to Andrew Kalusa for preparing Fig. 1. Work by the authors on Hsp90 is funded by Cancer Research UK (CR-UK) grant number 309/A2187 and Paul Workman is a Cancer Research UK life fellow.

References

An WG, Schulte TW, Neckers LM (2000) The heat shock protein 90 antagonist geldanamycin alters chaperone association with p210bcr-abl and v-src proteins before their degradation by the proteasome. Cell Growth Differ. 11:355–360

Armstrong BK, Kricker A (1994) Cutaneous melanoma. Cancer Surv. 19–20:219–240

Auluck PK, Chan HY, Trojanowski JQ, Lee VM, Bonini NM (2002) Chaperone suppression of alpha-synuclein toxicity in a Drosophila model for Parkinson's disease. Science 295:865–868

Bagatell R, Whitesell L (2004) Altered Hsp90 function in cancer: a unique therapeutic opportunity. Mol Cancer Ther 3:1021–1030

Balmain A, Gray J, Ponder B (2003) The genetics and genomics of cancer. Nat Genet 33 Suppl:238–244

Banerji U, Walton M, Raynaud F, Kelland LR, Judson I, Workman P (2001) Validation of pharmacodynamic endpoints for the Hsp90 molecular chaperone inhibitor 17-allylamino 17-demethoxygeldanamycin (17-AAG) in a human tumour xenograft model. Proc Am Asso Cancer Res 42:833

Banerji U, O'Donnell A, Scurr M, Benson C, Stapleton S, Raynaud F, Clarke PA, Turner A, Workman P, Judson I (2003) A pharmacokinetically (PK) – pharmacodynamically (PD) guided phase I trial of the heat shock protein (Hsp90) inhibitor 17-Allylamino, 17-demethoxygeldanamycin (17AAG). Proc Am Soc Clin Oncol 22:199

Banerji U, O'Donnell A, Scurr M, Pacey S, Stapleton S, Asad Y, Simmons L, Maloney A, Raynaud F, Campbell M, Walton M, Lakhani S, Kaye S, Workman P, Judson I (2005) A Phase I pharmacokinetic (PK) and pharmacodynamic (PD) study of 17-allylamino, 17-demethoxygeldanamycin (17-AAG) in patients with advanced malignancies. J Clin Oncol 23:4152–4161)

Banerji U, Walton M, Raynaud F, Grimshaw R, Kelland LR, Valentini M, Judson I, Workman P (2005b) Pharmacokinetic-Pharmacodynamic relationships for the HSP90 molecular chaperone inhbitors 17-Allylamino, 17-demethoxygeldanamycin (17-AAG) in human ovarian cancer models. Clin Cancer Res (in press)

Barral JM, Broadley SA, Schaffar G, Hartl FU (2004) Roles of molecular chaperones in protein misfolding diseases. Semin Cell Dev Biol 15:17–29

Baselga J, Albanell J, Molina MA, Arribas J (2001) Mechanism of action of trastuzumab and scientific update. Semin Oncol 28:4–11

Bearss DJ, Hurley LH, Von Hoff DD (2000) Telomere maintenance mechanisms as a target for drug development. Oncogene 19:6632–6641

Beliakoff J, Bagatell R, Paine-Murrieta G, Taylor CW, Lykkesfeldt AE, Whitesell L (2003) Hormone-refractory breast cancer remains sensitive to the antitumor activity of heat shock protein 90 inhibitors. Clin Cancer Res 9:4961–4971

Braga-Basaria M, Hardy E, Gottfried R, Burman KD, Saji M, Ringel MD (2004) 17-Allylamino-17-demethoxygeldanamycin activity against thyroid cancer cell lines correlates with heat shock protein 90 levels. J Clin Endocrinol Metab 89:2982–2988

Burger AM, Fiebig HH, Stinson SF, Sausville EA (2004) 17-(Allylamino)-17-demethoxygeldanamycin activity in human melanoma models. Anticancer Drugs 15:377–387

Chabner BA, Roberts TG (2005) Timeline: chemotherapy and the war on cancer. Nat Rev Cancer 5:65–72

Chen LL, Trent JC, Wu EF, Fuller GN, Ramdas L, Zhang W, Raymond AK, Prieto VG, Oyedeji CO, Hunt KK, Pollock RE, Feig BW, Hayes KJ, Choi H, Macapinlac HA, Hittelman W, Velasco MA, Patel S, Burgess MA, Benjamin RS, Frazier ML (2004) A missense mutation in KIT kinase domain 1 correlates with imatinib resistance in gastrointestinal stromal tumors. Cancer Res 64:5913–5919

Cheung KM, Matthews T, James K, Aherne W, Rowlands M, Boxall K, Sharp S, Prodromou C, Pearl L, McDonald E, Workman P (2005) The identification, synthesis and in vitro biochemical evaluation of a new class of Hsp90 inhibitors. J Med Chem Bioorg Med Chem Lett 15:3338–3343

Child JA, Morgan GJ, Davies FE, Owen RG, Bell SE, Hawkins K, Brown J, Drayson MT, Selby PJ (2003) High-dose chemotherapy with hematopoietic stem-cell rescue for multiple myeloma. N Engl J Med 348:1875–1883

Chiosis G, Lucas B, Shtil A, Huezo H, Rosen N (2002) Development of a purine-scaffold novel class of Hsp90 binders that inhibit the proliferation of cancer cells and induce the degradation of Her2 tyrosine kinase. Bioorg Med Chem 10:3555–3564

Chiosis G, Huezo H, Rosen N, Mimnaugh E, Whitesell L, Neckers L (2003) 1AAG: low target binding affinity and potent cell activity—finding an explanation. Mol Cancer Ther 2:123–129

Chiosis G, Vilenchik M, Kim J, Solit D (2004) Hsp90: the vulnerable chaperone. Drug Discov Today 9:881–888

Choo-Kang LR, Zeitlin PL (2001) Induction of Hsp70 promotes DeltaF508 CFTR trafficking. Am J Physiol Lung Cell Mol Physiol 281:L58–L68

Chung J, Yoon S, Datta K, Bachelder RE, Mercurio AM (2004) Hypoxia-induced vascular endothelial growth factor transcription and protection from apoptosis are dependent on alpha6beta1 integrin in breast carcinoma cells. Cancer Res 64:4711–4716

Chung YL, Troy H, Banerji U, Jackson LE, Walton M, Stubbs M, Griffiths JR, Judson I, Leach MO, Workman P, Ronen SM (2003) Magnetic resonance spectroscopic pharmacodynamic markers of the heat shock protien inhibitor 17-allyamino, 17-demethoxygeldanamycin (17-AAG) in human colon cancer models. J Natl Cancer Inst 95:1624–1633

Clarke PA, Hostein I, Banerji U, Stefano FD, Maloney A, Walton M, Judson I, Workman P (2000) Gene expression profiling of human colon cancer cells following inhibition of signal transduction by 17-allylamino-17-demethoxygeldanamycin, an inhibitor of the hsp90 molecular chaperone. Oncogene 19: 4125–4133

Cohen FE, Kelly JW (2003) Therapeutic approaches to protein-misfolding diseases. Nature 426:905–909

Daude N, Marella M, Chabry J (2003) Specific inhibition of pathological prion protein accumulation by small interfering RNAs. J Cell Sci 116:2775–2779

daRocha Dias S, Light Y, Friedlos F, Springer C, Workman P, Marais R (2005) Oncogenic B-RAF is an Hsp90 client protein that is targetted by the anti-cancer drug 17-AAG. Cancer Res (in press)

Davies H, Bignell GR, Cox C, Stephens P, Edkins S, Clegg S, Teague J, Woffendin H, Garnett MJ, Bottomley W, Davis N, Dicks E, Ewing R, Floyd Y, Gray K, Hall S, Hawes R, Hughes J, Kosmidou V, Menzies A, Mould C, Parker A, Stevens C, Watt S, Hooper S, Wilson R, Jayatilake H, Gusterson BA, Cooper C, Shipley J, Hargrave D, Pritchard-Jones K, Maitland N, Chenevix-Trench G, Riggins GJ, Bigner DD, Palmieri G, Cossu A, Flanagan A, Nicholson A, Ho JW, Leung SY, Yuen ST, Weber BL, Seigler HF, Darrow TL, Paterson H, Marais R, Marshall CJ, Wooster R, Stratton MR, Futreal PA (2002) Mutations of the BRAF gene in human cancer. Nature 417:949–954

De Carcer G, do Carmo Avides M, Lallena MJ, Glover DM, Gonzalez C (2001) Requirement of Hsp90 for centrosomal function reflects its regulation of Polo kinase stability. EMBO J 20:2878–2884

DeVita VT Jr, Hellman S, Rosenberg SA (2001) Cancer : principles and practise of oncology, 6 edn. Lippincott, Williams and Wilkins, Philadelphia

Druker BJ (2003) David A. Karnofsky Award lecture. Imatinib as a paradigm of targeted therapies. J Clin Oncol 21:239S–245S

Dymock B, Drysdale M, McDonald E, Workman P (2004) Inhibitors of Hsp90 and other chaperones for the treatment of cancer. Expert Opin Ther Patents 14:837–847

Dymock BW, Barril X, Brough PA, Cansfield JE, Massey A, McDonald E, Hubbard RE, Surgenor A, Roughley SD, Webb P, Workman P, Wright L, Drysdale MJ (2005) Novel, potent small-molecule inhibitors of the molecular chaperone Hsp90 discovered through structure-based design. J Med Chem 48:4212–4215

Eder JP, Wheeler CA, Teicher BA, Schnipper LE (1991) A phase I clinical trial of novobiocin, a modulator of alkylating agent cytotoxicity. Cancer Res 51:510–513

Egorin MJ, Zuhowski EG, Rosen DM, Sentz DL, Covey JM, Eiseman JL (2001) Plasma pharmacokinetics and tissue distribution of 17-(allylamino)-17-demethoxygeldanamycin (NSC 330507) in CDF1 mice. Cancer Chemother Pharmacol 47:291–302

Eigentler TK, Caroli UM, Radny P, Garbe C (2003) Palliative therapy of disseminated malignant melanoma: a systematic review of 41 randomised clinical trials. Lancet Oncol 4:748–759

Eiseman JL, Lan J, Lagattuta TF, Hamburger DR, Joseph E, Covey JM, Egorin MJ (2005) Pharmacokinetics and pharmacodynamics of 17-demethoxy 17-[[(2-dimethylamino)ethyl] amino]geldanamycin (1DMAG, NSC 707545) in C.B-17 SCID mice bearing MDA-MB-231 human breast cancer xenografts. Cancer Chemother Pharmacol 55:21–32

Ellis GK, Crowley J, Livingston RB, Goodwin JW, Hutchins L, Allen A (1991) Cisplatin and novobiocin in the treatment of non-small cell lung cancer. A Southwest Oncology Group study. Cancer 67:2969–2973

Erlichman C, Toft D, Reid J, Sloan J, Atherton P, Adjei A, Ames M, Croghan G (2001) A phase I trial of 17 allylamino geldanamycin in patients with advanced cancer. Proc Am Asso Cancer Res 42:833

Erlichman C, Toft D, Reid J, Goetz M, Ames M, Mandrekar A, Ajei A, McCollum A, Ivy P (2004) A Phase I trial of 17-allylamino-geldanamycin) (1AAG) in patients with advanced cancer (abstract). Proc Am Assoc Clin Oncol 23:3030

Forsythe HL, Jarvis JL, Turner JW, Elmore LW, Holt SE (2001) Stable association of hsp90 and p23, but Not hsp70, with active human telomerase. J Biol Chem 276:15571–15574

Fuller W, Cuthbert AW (2000) Post-translational disruption of the delta F508 cystic fibrosis transmembrane conductance regulator (CFTR)-molecular chaperone complex with geldanamycin stabilizes delta F508 CFTR in the rabbit reticulocyte lysate. J Biol Chem 275:37462–37468

Gelmon KA, Eisenhauer EA, Harris AL, Ratain MJ, Workman P (1999) Anticancer agents targeting signaling molecules and cancer cell environment: challenges for drug development? J Natl Cancer Inst 91:1281–1287

George P, Bali P, Annavarapu S, Scuto A, Fiskus W, Guo F, Sigua C, Sondarva G, Moscinski L, Atadja P, Bhalla K (2005) Combination of the histone deacetylase inhibitor LBH589 and the hsp90 inhibitor 17-AAG is highly active against human CML-BC cells and AML cells with activating mutation of FLT. Blood 105:1768–1776

Goetz M, Toft D, Reid J, Sloan J, Atherton P, Adjei A, Croghan G, Weinshilboum R, Erlichman C, Ames M (2002) A phase I trial of 17-allyaminodemethoxygeldanamycin (17-AAG) in patients with advanced cancer. Proceedings EORTC-NCI-AACR Symposium of Molecular Cancer Therapeutics. Eur J Cancer 38 [Suppl]:S54

Goetz MP, Toft D, Reid J, Ames M, Stensgard B, Safgren S, Adjei AA, Sloan J, Atherton P, Vasile V, Salazaar S, Adjei A, Croghan G, Erlichman C (2005) Phase I trial of 17-allylamino-17-demethoxygeldanamycin in patients with advanced cancer. J Clin Oncol 23:1078–1087

Gore L, Holden SN, Basche M, Raj SKS, Arnold I, O'Bryant C, Witta S, Rohde B, McCoy C, Eckhardt SG (2004) Updated results from a phase I trial of the histone deacetylase (HDAC) inhibitor MS-275 in patients with refractory solid tumours. Proc Am Soc Clin Oncol 22:3026

Gorre ME, Mohammed M, Ellwood K, Hsu N, Paquette R, Rao PN, Sawyers CL (2001) Clinical resistance to STI-571 cancer therapy caused by BCR-ABL gene mutation or amplification. Science 293:876–880

Gossett DR, Bradley MS, Jin X, Lin J (2005) 17-Allyamino-17-demethoxygeldanamycin and 17-NN-dimethyl ethylene diamine-geldanamycin have cytotoxic activity against multiple gynecologic cancer cell types. Gynecol Oncol 96:381–388

Grem JL, Morrison G, Guo XD, Agnew E, Takimoto CH, Thomas R, Szabo E, Grochow L, Grollman F, Hamilton JM, Neckers L, Wilson RH (2005) Phase I and pharmacologic study of 17-(allylamino)-17-demethoxygeldanamycin in adult patients with solid tumors. J Clin Oncol 23:1885–1893

Grenert JP, Sullivan WP, Fadden P, Haystead TA, Clark J, Mimnaugh E, Krutzsch H, Ochel HJ, Schulte TW, Sausville E, Neckers LM, Toft DO (1997) The amino-terminal domain of heat shock protein 90 (hsp90) that binds geldanamycin is an ATP/ADP switch domain that regulates hsp90 conformation. J Biol Chem 272:23843–23850

Hanahan D, Weinberg RA (2000) The hallmarks of cancer. Cell 100:57–70

Hollingshead M, Alley M, Burger AM, Borgel S, Pacula-Cox C, Fiebig HH, Sausville EA (2005) In vivo antitumor efficacy of 17-DMAG (17-dimethylaminoethylamino-17-demethoxygeldanamycin hydrochloride), a water-soluble geldanamycin derivative. Cancer Chemother Pharmacol. Cancer Chemother Pharmacol 56:115–125

Hostein I, Robertson D, DiStefano F, Workman P, Clarke PA (2001) Inhibtion of signal transduction by Hsp90 inhibitor 17-allylamino, 17-demethoxygeldanamycin results in cytostasis and apoptosis. Cancer Res 61:4003–4009

Itoh H, Ogura M, Komatsuda A, Wakui H, Miura AB, Tashima Y (1999) A novel chaperone-activity-reducing mechanism of the 90-kDa molecular chaperone Hsp90. Biochem J 343:697–703

Jaattela M (1999) Escaping cell death: survival proteins in cancer. Exp Cell Res 248:30–43

Jemal A, Thomas A, Murray T, Thun M (2002) Cancer statistics, 20CA. Cancer J Clin 52:23–47

Kamal A, Thao L, Sensintaffar J, Zhang L, Boehm MF, Fritz LC, Burrows FJ (2003) A high-affinity conformation of Hsp90 confers tumour selectivity on Hsp90 inhibitors. Nature 425:407–410

Kawano R, Ohshima K, Karube K, Yamaguchi T, Kohno S, Suzumiya J, Kikuchi M, Tamura K (2004) Prognostic significance of hepatocyte growth factor and c-MET expression in patients with diffuse large B-cell lymphoma. Br J Haematol 127:305–307

Kelland LR, Sharp SY, Rogers PM, Myers TG, Workman P (1999) DT-Diaphorase expression and tumor cell sensitivity to 17-allylamino, 17-demethoxygeldanamycin, an inhibitor of heat shock protein 90. J Natl Cancer Inst 91:1940–1949

Kiang JG, Bowman PD, Wu BW, Hampton N, Kiang AG, Zhao B, Juang YT, Atkins JL, Tsokos GC (2004) Geldanamycin treatment inhibits hemorrhage-induced increases in KLF6 and iNOS expression in unresuscitated mouse organs: role of inducible Hsp70. J Appl Physiol 97:564–569

Kola I, Landis J (2004) Can the pharmaceutical industry reduce attrition rates? Nat Rev Drug Discov 3:711–715

Kristeleit RS, Tandy D, Atadja P, Patnaik A, Scott J, De Bono JS, Judson I, Kaye SB, Workman P, Aherne W (2004) Effects of the histone deacetylase inhibitor (HDACi) LAQ824 on histone acetylation, Hsp70 and c-Raf in peripheral blood lymphocytes from patients with advanced solid tumours enrolled in a phase I clinical trial. Proc Am Soc Clin Oncol 22:3032

Kumar R, Musiyenko A, Barik S (2003) The heat shock protein 90 of Plasmodium falciparum and antimalarial activity of its inhibitor, geldanamycin. Malar J 2:30

Latchman DS (2001) Heat shock proteins and cardiac protection. Cardiovasc Res 51:637–646

Llauger L, He H, Kim J, Aguirre J, Rosen N, Peters U, Davies P, Chiosis G (2005) Evaluation of 8-arylsulfanyl, 8-arylsulfoxyl, and 8-arylsulfonyl adenine derivatives as inhibitors of the heat shock protein 90. J Med Chem 48:2892–2905

Longley BJ Jr, Metcalfe DD, Tharp M, Wang X, Tyrrell L, Lu SZ, Heitjan D, Ma Y (1999) Activating and dominant inactivating c-KIT catalytic domain mutations in distinct clinical forms of human mastocytosis. Proc Natl Acad Sci U S A 96:1609–1614

Mallucci G, Dickinson A, Linehan J, Klohn PC, Brandner S, Collinge J (2003) Depleting neuronal PrP in prion infection prevents disease and reverses spongiosis. Science 302:871–874

Maloney A, Workman P (2002) Hsp90 as a new therapeutic target for cancer therapy: the story unfolds. Expert Opin Biol Ther 2:3–24

Manola J, Atkins M, Ibrahim J, Kirkwood J (2000) Prognostic factors in metastatic melanoma: a pooled analysis of Eastern Cooperative Oncology Group trials. J Clin Oncol 18:3782–3793

Marcu M, Neckers L (2003) The C-Terminal half of heat shock protein 90 represents a second site for pharmacologic intervention in chaperone function. Curr Cancer Drug Targets 3:343–347

Marcu MG, Chadli A, Bouhouche I, Catelli M, Neckers LM (2000) The heat shock protein 90 antagonist novobiocin interacts with a previously unrecognized ATP-binding domain in the carboxyl terminus of the chaperone. J Biol Chem 275:37181–37186

Matthews RC, Burnie JP (2004) Recombinant antibodies: a natural partner in combinatorial antifungal therapy. Vaccine 22:865–871

Matthews RC, Rigg G, Hodgetts S, Carter T, Chapman C, Gregory C, Illidge C, Burnie J (2003) Preclinical assessment of the efficacy of mycograb, a human recombinant antibody against fungal HspAntimicrob Agents Chemother 47:2208–2216

Minet E, Mottet D, Michel G, Roland I, Raes M, Remacle J, Michiels C (1999) Hypoxia-induced activation of HIF-1: role of HIF-1alpha-Hsp90 interaction. FEBS Lett 460:251–256

Mitsiades C, Chanan-Khan A, Alsina M, Dass D, Landrigan D, Keitner M, Albitar GF, Hannah AL, Richardson P (2005) Phase I trial of 17-AAG in patients with relapsed and refractory multiple myeloma (MM). Proc American Assoc Clin Onc 23:3056

Mitsiades N, Mitsiades CS, Poulaki V, Chauhan D, Fanourakis G, Gu X, Bailey C, Joseph M, Libermann TA, Treon SP, Munshi NC, Richardson PG, Hideshima T, Anderson KC (2002) Molecular sequelae of proteasome inhibition in human multiple myeloma cells. Proc Natl Acad Sci U S A 99:14374–14379

Nardai G, Sass B, Eber J, Orosz G, Csermely P (2000) Reactive cysteines of the 90-kDa heat shock protein, Hsp90. Arch Biochem Biophys 384:59–67

Neckers L, Ivy P (2003). Heat Shock Protein 90. Curr Opin Oncol 15:419–24

Nooney L, Matthews RC, Burnie JP (2005) Evaluation of mycograb, amphotericin B, caspofungin, and fluconazole in combination against Cryptococcus neoformans by checkerboard and time-kill methodologies. Diagn Microbiol Infect Dis. 51:19–29

Nottage M, Siu LL (2002) Principles of clinical trial design. J Clin Oncol 20:42S–46S

Ochel HJ, Gademann G (2004) Destabilization of the non-pathogenic, cellular prion-protein by a small molecular drug. Antivir Ther 9:441–445

Okabe M, Uehara Y, Miyagishima T, Itaya T, Tanaka M, Kuni-Eda Y, Kurosawa M, Miyazaki T (1992) Effect of herbimycin A, an antagonist of tyrosine kinase, on bcr/abl oncoprotein-associated cell proliferations: abrogative effect on the transformation of murine hematopoietic cells by transfection of a retroviral vector expressing oncoprotein P210bcr/abl and preferential inhibition on Ph1-positive leukemia cell growth. Blood 80:1330–1338

Okada M, Itoh H, Hatakeyama T, Tokumitsu H, Kobayashi R (2003) Hsp90 is a direct target of the anti-allergic drugs disodium cromoglycate and amlexanox. Biochem J 374:433–441

Orr HT (2004) Neurodegenerative disease: neuron protection agency. Nature 431:747–748

Petrylak DP, Tangen CM, Hussain MH, Lara PN Jr, Jones JA, Taplin ME, Burch PA, Berry D, Moinpour C, Kohli M, Benson MC, Small EJ, Raghavan D, Crawford ED (2004) Docetaxel and estramustine compared with mitoxantrone and prednisone for advanced refractory prostate cancer. N Engl J Med 351:1513–1520

Prodromou C, Roe SM, O'Brien R, Ladbury JE, Piper PW, Pearl LH (1997) Identification and structural characterization of the ATP/ADP-binding site in the Hsp90 molecular chaperone. Cell 90:65–75

Ramanathan RK, Trump DL, Eiseman JL, Belani P, Agarwala S, Zuhowski EG, Lan J, Ivy P (2004) A phase I pharmacokinetic (PK) and pharmacodynamic (PD) trial of weekly 17-allylamino-17 demethoxygeldanamcyin (17-AAG, NSC-704057) in patients with advanced tumors (abstract). Proc Am Assoc Clin Oncol 23:3031

Richardson PG, Barlogie B, Berenson J, Singhal S, Jagannath S, Irwin D, Rajkumar SV, Srkalovic G, Alsina M, Alexanian R, Siegel D, Orlowski RZ, Kuter D, Limentani SA, Lee S, Hideshima T, Esseltine DL, Kauffman M, Adams J, Schenkein DP, Anderson KC (2003) A phase 2 study of bortezomib in relapsed, refractory myeloma. N Engl J Med 348:2609–2617

Rosenhagen MC, Soti C, Schmidt U, Wochnik GM, Hartl FU, Holsboer F, Young JC, Rein T (2003) The heat shock protein 90-targeting drug cisplatin selectively inhibits steroid receptor activation. Mol Endocrinol 17:1991–2001

Rowlands MG, Newbatt YM, Prodromou C, Pearl LH, Workman P, Aherne W (2004) High-throughput screening assay for inhibitors of heat-shock protein 90 ATPase activity. Anal Biochem 327:176–183

Rutherford SL, Lindquist S (1998) Hsp90 as a capacitor for morphological evolution. Nature 396:336–342

Santos NC, Figueria-Coelho J, Martin-Silva J (2003) Multidisciplinary utilization of dimethyl sulfoxide: pharmacological, cellular and molecular aspects. Biochem Pharmacol 65:1035–1041

Sato S, Fujita N, Tsuruo T (2000) Modulation of Akt kinase activity by binding to HspProc Natl Acad Sci U S A 97:10832–10837

Sausville EA, Tomaszewski JE, Ivy P (2003) Clinical development of 17-allylamino, 17-demethoxygeldanamycin. Curr Cancer Drug Targets 3:377–383

Scherer FM (2004) The pharmaceutical industry—prices and progress. N Engl J Med 351:927–932

Schnur RC, Corman ML, Gallaschun RJ, Cooper BA, Dee MF, Doty JL, Muzzi ML, DiOrio CI, Barbacci EG, Miller PE et al. (1995) erbB-2 oncogene inhibition by geldanamycin derivatives: synthesis, mechanism of action, and structure-activity relationships. J Med Chem 38:3813–3820

Schulte TW, Blagosklonny MV, Romanova L, Mushinski JF, Monia BP, Johnston JF, Nguyen P, Trepel J, Neckers LM (1996) Destabilization of Raf-1 by geldanamycin leads to disruption of the Raf-1-MEK-mitogen-activated protein kinase signalling pathway. Mol Cell Biol 16:5839–5845

Shah NP, Nicoll JM, Nagar B, Gorre ME, Paquette RL, Kuriyan J, Sawyers CL (2002) Multiple BCR-ABL kinase domain mutations confer polyclonal resistance to the tyrosine kinase inhibitor imatinib (STI571) in chronic phase and blast crisis chronic myeloid leukemia. Cancer Cell 2:117–125

Sheriff M, Clarke PA, Workman P (2004) Factors that govern the cell death response induced by inhibition of the molecular chaperone heat shock protein 90. Eur J Cancer 2:97

Singhal S, Mehta J, Desikan R, Ayers D, Roberson P, Eddlemon P, Munshi N, Anaissie E, Wilson C, Dhodapkar M, Zeddis J, Barlogie B (1999) Antitumor activity of thalidomide in refractory multiple myeloma. N Engl J Med 341:1565–1571

Sittler A, Lurz R, Lueder G, Priller J, Lehrach H, Hayer-Hartl MK, Hartl FU, Wanker EE (2001) Geldanamycin activates a heat shock response and inhibits huntingtin aggregation in a cell culture model of Huntington's disease. Hum Mol Genet 10:1307–1315

Slamon D, Pegram M (2001) Rationale for trastuzumab (Herceptin) in adjuvant breast cancer trials. Semin Oncol 28:13–19

Soga S, Shiotsu Y, Akinaga S, Sharma SV (2003) Development of radicicol analogues. Curr Cancer Drug Targets. 3:359–369

Soldano KL, Jivan A, Nicchitta CV, Gewirth DT (2003) Stucture of the N-terminal domain of GRP94: basis for ligand specificity and regulation. J Biol Chem 48:48330–48338

Solit DB, Zheng FF, Drobnjak M, Munster PN, Higgins B, Verbel D, Heller G, Tong W, Cordon-Cardo C, Agus DB, Scher HI, Rosen N (2002) 17-Allylamino-17-demethoxygeldanamycin induces the degradation of androgen receptor and HER-2/neu and inhibits the growth of prostate cancer xenografts. Clin Cancer Res 8:986–993

Solit DB, Basso AD, Olshen AB, Scher HI, Rosen N (2003a) Inhibition of heat shock protein 90 function down-regulates Akt kinase and sensitizes tumors to Taxol. Cancer Res 63:2139–2144

Solit DB, Scher HI, Rosen N (2003b) Hsp90 as a therapeutic target in prostate cancer. Semin Oncol 30:709–716

Solit DB, Egorin M, Valetin G, Delacruz QYe, Schwartz L, Larson N, Rosen N, Scher HI (2004) A Phase 1 pharmacokinetic and pharmacodynamic trial of decetaxol and 1AAG (17-allylamin-17-demethoxygeldanamycin). Proc Am Assoc Clin Oncol 22:3032

Stebbins CE, Russo AA, Schneider C, Rosen N, Hartl FU, Pavletich NP (1997) Crystal structure of an Hsp90-geldanamycin complex: targeting of a protein chaperone by an antitumor agent. Cell 89:239–250

Sugita T, Tanaka S, Murakami T, Miyoshi H, Ohnuki T (1999) Immunosuppressive effects of the heat shock protein 90-binding antibiotic geldanamycin. Biochem Mol Biol Int 47:587–595

Supko JG, Hickman RL, Grever MR, Malspeis L (1995) Preclinical pharmacologic evaluation of geldanamycin as an antitumor agent. Cancer Chemother Pharmacol 36:305–515

Tannock IF, de Wit R, Berry WR, Horti J, Pluzanska A, Chi KN, Oudard S, Theodore C, James ND, Turesson I, Rosenthal MA, Eisenberger MA (2004) Docetaxel plus prednisone or mitoxantrone plus prednisone for advanced prostate cancer. N Engl J Med 351:1502–1512

Toms JR (2004) CancerStats Monograph. Cancer Research UK, London

Tsan MF, Gao B (2004) Cytokine function of heat shock proteins. Am J Physiol Cell Physiol 286:C739–C744

Van Oosterom AT, Judson IR, Verweij J, Stroobants S, Dumez H, Donato dP, Sciot R, van Glabbeke M, Dimitrijevic S, Nielsen OS (2002) Update of phase I study of imatinib (STI571) in advanced soft tissue sarcomas and gastrointestinal stromal tumors: a report of the EORTC Soft Tissue and Bone Sarcoma Group. Eur J Cancer 38 [Suppl] 5:S83-S87

Wax S, Piecyk M, Maritim B, Anderson P (2003) Geldanamycin inhibits the production of inflammatory cytokines in activated macrophages by reducing the stability and translation of cytokine transcripts. Arthritis Rheum 48:541–550

Weber D (2003) Thalidomide and its derivatives: new promise for multiple myeloma. Cancer Control 10:375–383

Weissmann C (2004) The state of the prion. Nat Rev Microbiol. 2:861–871

Whitesell L, Sutphin PD, Pulcini EJ, Martinez JD, Cook PH (1998) The physical association of multiple molecular chaperone proteins with mutant p53 is altered by geldanamycin, an hsp90-binding agent. Mol Cell Biol 18:1517–1524

Workman P (2005) Genomics and the second golden era of cancer drug development. Molecular BioSystems in press

Workman P (2004) Combinatorial attack on multistep oncogenesis by inhibiting the Hsp90 molecular chaperone. Cancer Lett 206:149–157

Wright L, Barril X, Dymock B, Sheridan L, Surgenor A, Beswick M, Drysdale M, Collier A, Massey A, Davies N, Fink A, Fromont C, Aherne W, Boxall K, Sharp S, Workman P, Hubbard RE (2004) Structure-activity relationships in purine-based inhibitor binding to Hsp90 isoforms. Chem Biol 11:775–785

Xu L, Eiseman JL, Egorin MJ, D'Argenio DZ (2003) Physiologically-based pharmacokinetics and molecular pharmacodynamics of 17-(allylamino)-17-demethoxygeldanamycin and its active metabolite in tumor-bearing mice. J Pharmacokinet Pharmacodyn 30:185–219

Xu W, Mimnaugh E, Rosser MF, Nicchitta C, Marcu M, Yarden Y, Neckers L (2001) Sensitivity of mature Erbb2 to geldanamycin is conferred by its kinase domain and is mediated by the chaperone protein Hsp90. J Biol Chem 276:3702–3708

Yu X, Guo ZS, Marcu MG, Neckers L, Nguyen DM, Chen GA, Schrump DS (2002) Modulation of p53, ErbB1, ErbB2, and Raf-1 expression in lung cancer cells by depsipeptide FR901228. J Natl Cancer Inst 94:504–513

Pharmacological Targeting of Catalyzed Protein Folding: The Example of Peptide Bond *cis/trans* Isomerases

F. Edlich · G. Fischer (✉)

Max-Planck Research Unit for Enzymology of Protein Folding, Weinbergweg 22, 06120 Halle/Saale, Germany
fischer@enzyme-halle.mpg.de

1	Introduction	360
2	Cyclophilins	364
2.1	Immunosuppression	364
2.2	Viral Infections	366
2.3	Parasitic Infections	369
2.4	Malignancies	369
2.5	Ischemia/Reperfusion Injury	370
2.6	Amyotrophic Lateral Sclerosis	371
2.7	Inflammation	372
2.8	Allergy	373
3	FKBPs	373
3.1	Immune Response	374
3.2	Skin Disorders	376
3.3	Neuropathies	376
3.4	Inherited Disease	378
3.5	Pathogenic Microorganisms	378
3.6	Steroid Responsiveness	379
3.7	Malignancies	381
3.8	Arylhydrocarbon (Dioxin) Responsiveness	383
3.9	Cardiovascular Disorders	384
4	Parvulins	386
4.1	Malignancies	386
4.2	Neuropathies	388
References		389

Abstract Peptide bond isomerases are involved in important physiological processes that can be targeted in order to treat neurodegenerative disease, cancer, diseases of the immune system, allergies, and many others. The folding helper enzyme class of Peptidyl-Prolyl-*cis/trans* Isomerases (PPIases) contains the three enzyme families of cyclophilins (Cyps), FK506 binding proteins (FKBPs), and parvulins (Pars). Although they are structurally unrelated, all PPIases catalyze the *cis/trans* isomerization of the peptide bond preceding the proline in a polypeptide chain. This process not only plays an important role in de novo protein folding, but also in isomerization of native proteins. The native state isomerization plays a role in physiological processes by influencing receptor ligand recognition or isomer-specific enzyme reaction or by regulating protein function by catalyzing the switch between native isomers differing in their activity, e.g., ion channel regulation. Therefore elucidating PPIase involvement in physiological processes and development of specific inhibitors will be a suitable attempt to design therapies for fatal and deadly diseases.

Keywords Peptide bond isomerases · PPIases · FKBP · Cyclophilin · Parvulin

1
Introduction

In principle, the native conformation of a protein is determined by its amino acid sequence since many isolated proteins can be denatured and refolded in vitro in the absence of other cellular components. However, it is now clear that assembly and folding of polypeptides in vivo involves helper proteins that assist in protein folding. The spatiotemporal characteristics of the various conformational states or folding states of a protein dictate important events in cell life under normal and pathophysiological conditions. Physically, protein folding is a complex process whereby a polypeptide chain can adopt a huge number of conformations that may differ greatly from its native folding state, but represent, as well, a transiently stable form of protein. These different folding intermediates can be seen as a polypeptide chain on an energy landscape containing many energy minima. Obviously, the connection of different energy minima requires folding pathways in order to allow adoption of the native protein fold (Dobson et al. 1998). Notably, the same rules apply for de novo protein folding and conformational changes in native proteins. Therefore chain reshuffling by protein–protein and protein–ligand interactions, de novo protein folding, and refolding of denatured proteins are based on similar molecular principles (Tsai et al. 1999).

Protein-mediated folding assistance was found to exploit many biochemical mechanisms resulting in different modes of action. It includes folding in cavities, enzymatically catalyzed folding and coupled chain holding-chain folding cycles. The pathways of assisted folding play a key role in cell response to physiological signals, and effectors of the helper proteins involved represent a potential for therapeutic intervention for many human diseases. In the last two decades, different families of folding helper proteins have been characterized that proved to be highly conserved during evolution. Among them, the folding helper enzymes act catalytically to control the folding dynamics and the products of folding. Both protein disulfide isomerases and peptide bond *cis/trans* isomerases form the sole examples of folding helper enzymes known to date. Particular attention is given to peptide bond *cis/trans* isomerases because peptide bond *cis/trans* isomerization plays an obligatory role for protein folding, whereas disulfide bond formation does not.

Notably, most rotations about covalent bonds, events that dictate the progress of the folding reaction, proved intrinsically to be very fast and do not need further rate acceleration by either external factors or intramolecular assistance. Potentially, external influences on rotational rates could be mediated by a wide variety of physical and chemical means, such as enzyme catalysis, catalysis by low-molecular-mass compounds, heat, mechanical forces, and supportive microenvironments. Acid-base catalysis, torsional strain or proximity, and field effects are major forces with the potential to act in intramolecular assistance or enzyme catalysis. Among the folding helper enzymes only peptide bond *cis/trans* isomerases are able to catalyze conformational interconversions in unfolded, partially folded, and native states of proteins (Fischer and Aumuller 2003). In all cases, enzymatic rate enhancement requires a particular substrate structure usually characterized by a combination of primary and secondary binding sites. In fact, on the basis of accelerated bond rotation, peptide bond *cis/trans* isomerases can be characterized as conformases. Currently known enzymes show specificity for the peptide bond preceding either a proline residue (prolyl peptide bond) or a variety of secondary amide peptide bonds. These enzymes termed peptidyl prolyl *cis/trans* isomerases (PPIases) and the secondary amide peptide bond *cis/trans* isomerases (APIases), can be divided into subfamilies, some of which are characterized by their ability to recruit secondary binding sites of the substrate polypeptide chain for catalysis. Current knowledge on PPIases goes back to first observations in 1984 (Fischer et al. 1984), whereas the discovery of APIases is a more recent event (Schiene-Fischer et al. 2002).

Extensive studies of PPIases have provided reasonable knowledge of the mechanisms determining the relationship between prolyl isomerization in proteins and cell signaling.

It was shown that PPIases not only increase the rate of the slow kinetic phases in the refolding of denatured proteins, but also accelerate efficiently the interconversion between native state isomers of proteins. Regardless of their sequential context, the *cis* and *trans* prolyl bond isomers in unstructured, partially structured, or even native proteins exhibit energetic differences, which tend to be small, thus leading to comparable levels of isomers in solution.

Given that one isomer is not reactive, coupling of this equilibrium to a fast subsequent reaction step must cause a transient discrimination of an isomer in the overall reaction (isomer-specificity, Fig. 1a). Isomer-specific reactivity differences have been demonstrated in many bioreactions, including enzyme catalyzed phosphate transfer, proteolysis, protein folding, and receptor/ligand recognition (Fischer and Aumuller 2003). In addition, peptide bond stiffness, which is reduced under PPIase catalysis, may be a reactivity determining molecular parameter in polypeptides (Fig. 1b). Furthermore, the switch-like character of peptide bond *cis/trans* isomerization allows for chemomechanical coupling in proteins (Fig. 1c) (Fischer 1994; Tchaicheeyan 2004).

In the years following the discovery of the PPIase in pig kidney, cyclophilins, FK506-binding proteins, and parvulins have been characterized, which now collectively form the enzyme class of PPIases (EC 5.2.1.8.). Prolyl bonds usually

A) Isomer-specificity of polypeptide recognition

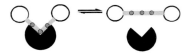

B) Segmental flexibilization in the context of the rigid backbone

C) Mechanical force generator for structural distortion

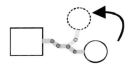

D) One-bond folding energy trap in native proteins

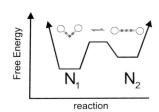

Fig. 1a–d Biochemical mechanisms underlying the physiological role of peptidyl prolyl bond *cis/trans* isomerizations of proteins. **a** Proteins (*dark gray*) capable of isomer-specific recognition of a polypeptide ligand (*gray*). An extreme manifestation of isomer specificity is seen if one prolyl isomer releases free binding energy (*left*), while the other isomer is not able to do so (*right*). Examples have been found in protein folding, enzyme catalysis, and receptor/ligand interactions. **b** Local control of chain flexibility in the context of the rigid polypeptide backbone. A decrease in energy barrier to rotation of a prolyl bond has the conformational consequence of promoting nearly unconstrained motion of adjacent covalent bonds similar to those present in alkanes. This process might play a role in protein folding. **c** Mechanical force generator for structural distortion. The *cis/trans* isomerization of a single peptidyl-prolyl bond may confer to global structural rearrangements changing the relative positions of a moving domain (*circle*) to a static domain (*rectangle*). This process is thought to play a role in allosteric regulation of proteins. In addition, peptidyl prolyl bond *cis/trans* isomerization addresses, on the submolecular level, a broad spectrum of actomyosin's functional characteristics (Tchaicheeyan 2004). **d** One-bond folding energy trap is important in conferring protein metastability. The native protein adopts *cis/trans* isomerism of prolyl bonds, which must lead to conformational polymorphism. Assuming that protein molecules populate the N_2 state, the higher energy implies kinetic instability. Conversion to the stable folding state may proceed very slowly for prolyl isomers

confer uncoupling of a bond rotation from the framework movement of the backbone and side chains by virtue of an exceptionally high torsion barrier. As a result, the time scale of the uncatalyzed interconversion between the prolyl *cis/trans* isomers ranges from several hundred milliseconds to a few hours under physiological conditions.

The singularity of prolyl bonds in peptide chains results from the N-alkylated amino acid proline forming an imidic peptide bond at its N-terminus. This property is exceptional since all other gene-coded amino acids form secondary amide peptide bonds. In both types of peptide bond, the nitrogen atom is able to delocalize its lone electron pair over the whole functional group. This results in a planar framework of electronic organization and the consequent partial double-bond character of the C-N bond, restricting the energy minima in peptide bond torsion. The free energy dependence of the peptide bond *cis/trans* isomerization in a chain fragment indicates just two minimum-energy states: the geometric isomers *cis* and *trans*. Both isomers are separated by the above-mentioned rotational barrier corresponding to the perpendicular high-energy intermediate state of rotation. While three-dimensional protein structures achieved by X-ray crystallography reveal prolyl bonds that adopt a conformational homogeneous state, either *cis* or *trans* conformation, solution NMR spectroscopy reveals a more complex picture. Many native proteins show limited conformational polymorphism, termed native state isomerization (Fig. 1d). These conformational states are mainly caused by the structural properties of a particular proline residue in the polypeptide chain. Protein conformational polymorphism often indicates the existence of peptide bond *cis/trans* isomers of different biological activity (Andreotti 2003).

The enzyme class of PPIases presently comprises three enzyme families that represent structurally and functionally distinct proteins. Enzyme families were named according to the affinity of prototypic enzymes for the immunosuppressive drugs cyclosporin A (CsA) and FK506, cyclophilins (Cyps), and FK506-binding proteins (FKBPs). "Immunophilin" is a term that collectively comprises members of both enzyme families. The parvulin family of PPIases has not been reported thus far to exhibit affinity for these immunosuppressive drugs. Heat shock proteins of the hsp70 family comprise the first examples of APIases (Schiene-Fischer et al. 2002).

Numerous biochemical investigations have led to the elucidation of four different modes of action, some of which were shown to represent auxiliary features of PPIases in cells (Fischer and Aumuller 2003). However, rate acceleration of prolyl isomerization plays a central role in their physiological function. Consequently, these enzymes may contribute to de novo protein folding as well as polypeptide chain rearrangements in native proteins. Logically, PPIases involved in de novo protein folding must show distinct characteristics as compared to enzymes involved in the interconversion between native states of proteins. To effectively perform this broad range of tasks, PPIase families, in particular FKBPs and cyclophilins, consist of many individual members of different molecular masses that encompass one or more PPIase domains complemented with other functional polypeptide segments (Galat 2004a, 2004b). In addition, PPIase activity might be controlled by accessory cellular factors, such as second messengers or interacting proteins. Such molecular characteristics are expected to contribute to the regulatory function of the PPIase, a situation reminiscent of the regulation of protein kinases and protein phosphatases.

Among the plethora of human PPIases, several members seem to be specifically involved in the interconversion of native protein conformers, which differ in their biological activity. However, it is reasonable to assume functional redundancy among paralogous PPIases. It should also be noted that complete depletion of the catalytic activity in human cells, which may be possible for lower eukaryotes, proved to be difficult (Dolinski et al. 1997; Wang et al. 2001).

Functional analyses of PPIases in cells, performed with PPIase variants, attenuated in their catalytic activity, protein knock-down by genetic means, or small cell-permeable inhibitors, allowed unequivocal identification of many signaling pathways that are based on catalyzed prolyl isomerization. Consequently, specific inhibition of PPIases might cause pathway blockade, which can have beneficial effects in pathologically altered cells, tissues and organs, by interfering with catalyzed folding processes.

However, it remains a challenge to identify differential inhibitors for the individual members of the PPIase families.

2
Cyclophilins

Members of this PPIase family are ubiquitously expressed throughout the human organism (Fischer and Aumuller 2003; Galat 2004a). In their catalytic domain, cyclophilins share high sequence similarity to the prototypic cyclophilin 18 (Cyp18, CypA), which has a molecular mass of 18 kDa. The immunosuppressive drug cyclosporine A (CsA) represents a tight-binding inhibitor for the PPIase activity of most human cyclophilins. For many human cyclophilins, additional polypeptide segments complete the full-length proteins. They were found to be located N-terminally and C-terminally to the catalytic core domain. Functionally, the extra domains and segments are coupled with intracellular targeting, RNA recognition, tetratricopeptide repeat (TPR)-mediated protein–protein interactions and macro-complex assembly of Ran-binding proteins.

2.1
Immunosuppression

Low-molecular-mass inhibitors of cyclophilins are utilized in prevention of allograft rejection in transplantation medicine, for treatment of autoimmune diseases, and for prophylaxis of graft-versus-host diseases (Arai and Vogelsang 2000; Gremese and Ferraccioli 2004). Prior binding of CsA derivatives to Cyp18 is one requirement for inhibition of T cell activation after antigenic stimulation and immunosuppressive activity (Fischer et al. 1989; Takahashi et al. 1989). A causal relationship between inhibition of the Cyp18 PPIase activity and T cell replication does not exist (Navia 1996; Schreier et al. 1993). Instead, the protein phosphatase calcineurin appears to be involved in Ca^{2+}-dependent signal transduction pathways in, among others, T cells and mast cells. Cyp18 and FKBP12 probably mediate the immunosuppressive actions of CsA and FK506, respectively, by a gain-of-function effect of the immunosuppressive drug/immunophilin complex with and inhibiting the protein phosphatase activity of calcineurin (Friedman and Weissman 1991; Liu et al. 1991). Mechanistically, Cyp18 forms a binding platform for CsA, with a portion of the drug molecule remaining presented to calcineurin by the resulting complex. Probably the most important reason for the occurrence of gain-of-function effects is alteration in macrocyclic conformation of CsA when bound to its presenter PPIase. Downregulation of the NF-AT, NFκB, and the c-jun N-terminal kinase (JNK) signaling pathways that determine expression of IL-2 and other cytokines is thought to play a key role in conferring susceptibility to CsA and FK506 effects on the growth of T cells involved in cellular immune response (Vogel et al. 2001). It has been reported that continuous treatment of transplant patients with CsA induces numerous side effects such as hypertension, encephalopathy, malignancy, neurotoxicity, hyperglycemia and chronic nephropathy. The most prominent CsA side effects are nephrotoxicity and oxidative stress causing a block in muscle differentiation that is mediated by ROS generation (Hong et al. 2002). In addition, CsA application causes nephrotoxicity by initiation of vacuolization and fatty change in tubular epithelial and endothelial cells, inhibition of cell growth, detachment, and cell death in a time- and dose-dependent manner (Ryffel et al. 1988). Besides its ability to interfere with calcineurin-mediated signaling, cyclosporins inhibit PPIase activity of many cellular cyclophilins in drug-treated patients. A variety of physiological processes exist where cyclophilins were shown to be involved (Bennett et al. 1998; Waldmeier et al. 2002, 2003; Wei et al. 2004; Yurchenko et al. 2002). Thus, inhibition of the PPIase activity of cyclophilins may be associated with the occurrence of adverse effects in long-term treatment with cyclosporins. In fact, calcineurin-independent pathways affect peripheral T cell deletion induced by either superantigens or anti-TCR αβ mAb, and allergy-induced by superantigens, and this may be relevant to the development of immune tolerance (Prudhomme et al. 1995).

To add to the puzzle, the estimation of calcineurin activity in CsA-treated renal transplant patients raised doubts about the direct relationship between calcineurin inhibition and immunosuppression. In circulating lymphocytes of the immunosuppressed patients, calcineurin activity is only partially reduced (50%–85% of the control) (Batiuk et al. 1995). The calcineurin fraction resistant to inhibition is even larger for FK506 administration in several tissues. The serum CsA level can approximately predict clinical parameters of immunosuppression in individuals but is poorly correlated with the calcineurin activity of lymphocytes. It was hypothesized that partial calcineurin inhibition might account for both the immunosuppression and the immunocompetence of CsA-treated patients (Batiuk et al. 1997). Sanglifehrin macrolides, which are structurally distinct from cyclosporins, offer a new approach for preventing clonal T cell expansion via tight binding to cyclophilins. Sanglifehrin A blocks T cell proliferation in response to IL-2 by inhibiting the appearance of activity in the cell cycle kinase cyclinE-Cdk2 (Zhang et al. 2001). Interestingly, a Cyp18-sanglifehrin complex does not inhibit the protein phosphatase calcineurin but is completely inactive in PPIase assays (Zenke et al. 2001). Sanglifehrin A was shown to inhibit mitochondrial cyclophilin, and is thus an inactivating agent of the mitochondrial permeability transition pore, which is highly involved in the promotion of apoptosis and necrosis (Clarke et al. 2002). In addition, Sanglifehrin A interferes with the differentiation of dendric cells and promotes endocytosis of antigens (Steinschulte et al. 2003; Woltman et al. 2004).

Cyp18 can influence T cell activation on its own utilizing a pathway that does not require the application of cyclosporins. Knock-down of Cyp18 in mice promotes development of allergic diseases with elevated IgE and tissue infiltration by mast cells and eosinophils, which is driven by CD4(+) T helper type II (Th2) cytokines (Colgan et al. 2004). Cyp18 inhibits the interleukin-2 tyrosine kinase (Itk) that is a nonreceptor kinase, promoting T cell activation (Brazin et al. 2002; Mallis et al. 2002). Mechanistically, the PPIase interacts with the Src homology 2 (SH2) domain of the kinase known to control the activity of the neighboring catalytic domain. This interaction mediates a native state prolyl isomerization, causing a conformational change in the structure of Itk. Consequently, Cyp18 catalytically accelerates switching between the active and the inactive isomer of Itk. This process prevents T cell activation by cellular factors acting downstream of the Itk. Application of CsA prevents the interaction of Cyp18 and Itk.

2.2
Viral Infections

Cyp18 plays an important functional role in virus replication among the host cell proteins of human immunodeficiency virus type-1 (HIV-1) virions because it enhances viral infectivity (Luban et al. 1993). Other folding helper enzymes are also found in complex with viral proteins (Brenner and Wainberg 1999). Virions released from infected cells show a ratio of ten copies of p24^{gag} to one copy of Cyp18, which is a unique feature among the virions of retroviruses (Thali et al. 1994). The results of some studies underscore the importance of the Cyp18 population of the target cell vs Cyp18 packaged in virions (Ikeda et al. 2004). During early phases of replication, Cyp18 was found to associate with proline-rich segments of both the retroviral capsid protein p24^{gag} and the viral protein R (Vpr) (Zander et al. 2003).

Consequently, blocking the host cell Cyp18–provirus interaction by cyclophilin inhibitors, such as immunosuppressive and nonimmunosuppressive CsA derivatives or reducing the cellular concentration of Cyp18 by antisense

U7 snRNAs and siRNAs targeting Cyp18 was beneficial for limiting the virus load (Bartz et al. 1995; Liu et al. 2004; Thali et al. 1994). Indeed, previous analyses already suggested the beneficial effect of CsA treatment on the progression of disease and outcome of AIDS-related mortality (Huss et al. 1995). In another study, CsA restored normal CD4(+) T cell levels, both in terms of percentage and absolute numbers in AIDS patients (Rizzardi et al. 2002).

Recent data show that three isoforms of Cyp18 assemble with HIV-1 particles (Misumi et al. 2002). One of the three isoforms was found outside of the viral membrane. It was suggested that this Cyp18 isoform might play a role in the attachment of the virons to the surface of the target cell. It has been discussed that Cyp18 mediates HIV-1 attachment to a cell by targeting heparans on the cellular surface (Saphire et al. 1999). This observation is supported by data from groups investigating the role of Cyp18 and Cyp23 in response to inflammation (Yurchenko et al. 2002).

The two remaining isoforms were isolated as components of the viral membrane. They most likely play a role in the regulation of the HIV-1 p24^{gag} conformation (Misumi et al. 2002). They are included in the viral membrane as a result of interactions with Gag polyprotein when the virion is assembled and released from the host cell (Franke et al. 1994). Cyp18 binds p24^{gag} at three distinct sites. One is located in the N-terminal part of the protein around Gly$_{89}$-Pro$_{90}$ (Gamble et al. 1996; Howard et al. 2003) (Fig. 2). The interaction at this site certainly influences virion packaging.

In the C-terminal part of p24^{gag}, the other two interaction sites are at Gly$_{156}$-Pro$_{157}$ and Gly$_{223}$-Pro$_{224}$ and may play a role in destabilizing the capsid cone (Endrich et al. 1999). Viruses assembled in cells with low Cyp18 concentrations are less infectious than particles assembled at high Cyp18 concentrations (Liu et al. 2004).

Another aspect of the Cyp18-p24^{gag} interaction is the Cyp18-mediated evasion of the antiviral action of a human restriction factor that targets p24^{gag} soon after virus entry into the cell. This situation is revealed in owl monkey cells where restriction is released by capsid mutants or CsA that disrupt capsid interaction with Cyp18. The structures of viral capsid proteins largely determines restriction (Perron et al. 2004). It was suggested that HIV-1 co-opted Cyp18 to counteract restriction factors and that this adaptation can confer sensitivity to restriction in unnatural hosts (Towers et al. 2003). A chimerical protein containing a restriction factor and Cyp18 expressed in owl monkey was shown to account for postentry restriction of HIV-1 and block HIV-1 infection when transferred to otherwise infectable human or rat cells (Sayah et al. 2004).

Cyp18-catalyzed prolyl isomerization has been detected in capsid-derived oligopeptides and the N-terminal fragment of HIV-1 capsid at the Gly89-Pro90 site (Bosco et al. 2002). On the other hand, a Vpr-Cyp18 fusion protein variant, which lacks PPIase activity, rescues HIV-1 replication in a *trans* complementation assay (Saphire et al. 2002).

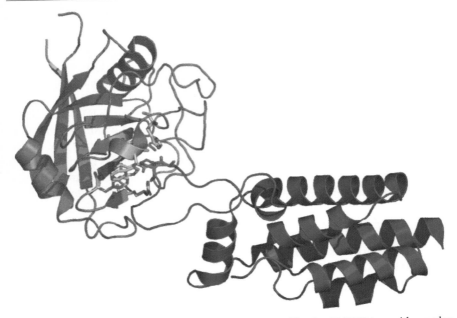

Fig. 2 Representation of the three-dimensional structure of the Cyp18/HIV-1 capsid complex (Howard et al. 2003). The prototypic cyclophilin folds a β-barrel structure containing eight antiparallel β-sheets and two amphipatic α-helices depicted in *red*. The PPIase-active site of Cyp18 is assembled by the amino acid side chains of R55, F60, Q63, F113, W121, and H126 (*orange*). The active site of Cyp18 binds to P90 and its preceding glycine residue (*green*) within the exposed loop of the capsid protein, displayed in *blue*

Interestingly, application of CsA and a nonimmunosuppressive CsA derivate to virus-infected quiescent cells preloaded with anti-CD25 immunotoxin diminishes virus production and thus suppress infectivity (Borvak et al. 1996).

In addition to data showing the importance of Cyp18 for HIV-1 replication, there is evidence that the cyclophilins are utilized by other viruses, as well. Because CsA dose-dependently inhibits herpes simplex virus production in resting monkey kidney cells, the involvement of cyclophilins in virus replication is obvious (Vahlne et al. 1992). A considerable variation in CsA sensitivity of replication has been obtained for serologically distinct types of vesicular stomatitis virus (VSV) in baby hamster kidney cells (Bose et al. 2003). Overexpression of a catalytically inactive Cyp18 variant parallels the CsA effects. It was shown that Cyp18 interacts with the nucleocapsid protein and is incorporated into VSV particles. These data imply that the processes that Cyp18 mediates in the VSV life cycle differ from its function in HIV-1. A calcineurin-independent effect of CsA has been reported for the replication of the hepatitis C virus and data from bone marrow transplantation patients and nontransplant patient populations confirm that CsA inhibits HCV replication (Pollard 2004; Watashi et al. 2003).

2.3
Parasitic Infections

CsA is an antiparasitic drug for many different parasitic infections whose mode of action might involve both parasite and host cell cyclophilins. Response to CsA treatment has already been observed for malaria, leishmaniasis, trypanosomiasis, schistosomiasis, and filariasis (Bell et al. 1996). In fact, *Brugia malayi* is resistant to the antiparasitic activity of cyclosporin A (CsA), in accordance with the relatively low CsA affinity of parasitic cyclophilins (Ellis et al. 2000). On the other hand, the antihelmintic action of different cyclosporins against *Hymenolepis microstoma* does not correspond with the degree of cyclophilin binding and implies that a parasite surface component is the drug target (McLauchlan et al. 2000). Typically, survival to infection with *Trypanosoma cruzi* was 50% higher for CsA-treated mice than in non-treated animals (Calabrese et al. 2000). Immunosuppression and antimalarial activity of cyclosporin derivatives does not correlate in parasites. Therefore calcineurin is not involved in cyclosporin-sensitive pathogenesis (Bell et al. 1996). This study investigates whether complexing of CsA with parasite Cyp may account for its antihelmintic action.

2.4
Malignancies

Using differential display techniques, cDNA microarrays and proteome projects frequently revealed upregulation of Cyp18 in various types of cancer cells (Campa et al. 2003; Grzmil et al. 2004; Lim et al. 2002; Rey et al. 1999). The mitochondrial Cyp22 is specifically upregulated in human tumors of the breast, ovary, and uterus (Schubert and Grimm 2004). Among the cyclophilins, Cyp40 is unique for its ability to sequester heat shock proteins and its involvement in the control of mitogenic signaling mediated by glucocorticoids (Renoir et al. 1995; Ward et al. 1999). Consequently, deletion of the Cyp40 gene promotes breast tumor progression in late developmental phases (Ward et al. 2001). Treatment of breast cancer cells with estradiol increases steady state concentrations of Cyp40 mRNA through both transcriptional and post-transcriptional mechanisms (Kumar et al. 2001). The function of Cyp40 in the steroid hormone receptor complex seems to be related to the role of FKBP51 and FKBP52 in that the unactivated receptor complex is stabilized, and assistance is provided for subcellular translocation processes upon steroid binding. In addition, Cyp40 seems required to form the functional peroxisome proliferator-activated receptor alpha (PPARα) as well. This complex might play a role in the development of liver cancer after chronic exposure to peroxisome proliferators, including certain industrial and pharmaceutical chemicals, and in mitogenic or apoptotic regulation of growing tumors (Miller et al. 2000).

Both Cyp40 and FKBPs have been found in complex with Hsp90 and p53 via their tetratricopeptide repeat (TPR) domains and dynein with their PPIase domains. This heterocomplex forms a functional link between the PPIases and p53 transport into the nucleus in colon carcinoma cells (Galigniana et al. 2004).

A Cyp18 isoform sharing about 84% sequence identity with Cyp18 were identified to be encoded in the q21-23 region of chromosome 1, which has been recognized as being extensively transcribed during the development of metastases in liver, breast, and bladder cancer (Meza-Zepeda et al. 2002; Nilsson et al. 2004). The gene product was designated chromosome one amplified sequence 2 (COAS2), but has not yet been characterized enzymatically.

Immunosuppressive drugs could promote tumor progression due to calcineurin inhibition, but immune response-independent mechanisms have also been suggested (London et al. 1995; Van de Vrie et al. 1997). In contrast, rapamycin, another immunosuppressive agent acting through a calcineurin/ cyclophilin-independent mechanism, exhibits potent anti-tumor activity. It was shown that CsA could induce cancer cell-typical phenotypic changes in normal cells. The host immune defense does not account for this effect. The CsA-mediated enhancement of tumor growth in immunodeficient SCID-beige mice can be prevented by simultaneous treatment with mAb directed against transforming growths factor-β (TGF-ß) (Hojo et al. 1999).

A 166-kDa cyclophilin (NKTR), which harbors a N-terminal cyclophilin domain followed by a long C-terminal extension, is an important determinant of natural killer (NK) cells. A signaling pathway has been described that utilizes this cyclophilin in MHC unrestricted killing of tumor cells (Giardina et al. 1995). This protein belongs to the large subfamily of SR cyclophilins thought to be involved in cell cycle regulation (Dubourg et al. 2004).

Cyclosporin A and its nonimmunosuppressive derivative $[3'\text{-Keto-Bmt}^1]$-$[\text{Val}^2]$-cyclosporin (PSC 833) has been shown to increase the sensitivity of multidrug-resistant (MDR) cells to chemotherapeutic agents used in anti-tumor treatment (Boesch et al. 1991). Labeling of the cyclosporin derivatives demonstrated the binding to and blocking of the human MDR1 P-glycoprotein, a transmembranous adenosine 5'-triphosphate binding cassette (ABC) transporter (Loor et al. 2002).

2.5
Ischemia/Reperfusion Injury

Mitochondria play a major role in ischemic cell death in peripheral organs and the central nervous system, because activation of the mitochondrial permeability transition pore (mtPTP) causes apoptosis and necrosis in injured tissue (Li et al. 2004). During pore opening, the mitochondrial inner membrane becomes freely permeable to solutes of less than 1.5 kDa. The mtPTP is a protein complex consisting of a cyclophilin (Cyp22, CypD), the voltage-dependent anion channel, members of the pro- and anti-apoptotic Bax/Bcl-2 protein family, and the adenine nucleotide (ADP/ATP) translocator. Calcium-triggered functional switching of the mtPTP from a specific transporter to a nonspecific pore is facilitated by the binding of Cyp22 to the adenine nucleotide (ADP/ATP) translocator (probably on proline residue 61) (Halestrap and Brenner 2003). Overexpression of Cyp22 had opposite effects on apoptosis and necrosis, obtaining suppression of apoptosis at high Cyp22 levels (Li et al. 2004; Schubert and Grimm 2004).

On the other hand, permeability transition induced by calcium ions was powerfully inhibited by CsA and nonimmunosuppressive CsA derivatives at nanomolar concentrations, indicating a potential anti-apoptotic drug effect due to inhibition of Cyp22 enzyme activity (Griffiths and Halestrap 1995; Hansson et al. 2004). Consequently, studies suggest preservation of mitochondrial integrity in animal stroke models, in injured rat liver and heart as well as neuroprotective effects of CsA in animal models (LeDucq et al. 1998; Waldmeier et al. 2003; Yu et al. 2004). However, different cyclophilins are thought to be involved in the beneficial CsA effects (Li et al. 2004).

Using the calcineurin-inactive drug sanglifehrin new light was shed on Cyp22 function in the mtPTP complex (Clarke et al. 2002). Sanglifehrin inhibits mtPTP activation but not formation of an adenine nucleotide (ADP/ATP) translocator–Cyp22 complex. Therefore, proteins assembling the mtPTP probably undergo structural changes catalyzed by Cyp22 that do not correlate with complex formation but with the pore function. Despite similar K_i values for inhibition of the PPIase activity of Cyp22 in vitro, sanglifehrin binds more tightly to mitochondria than CsA. The sigmoidal dose-response curves obtained for the pore inhibition by sanglifehrin allowed assessment of pore complex composition and detection of cooperative interactions between the pore components during opening.

2.6
Amyotrophic Lateral Sclerosis

Hyperactive, mutant Cu/Zn superoxide dismutase-1 (SOD), which can be linked to familial amyotrophic lateral sclerosis (familial ALS), induces apoptosis of neuronal cells in culture because of increase in reactive oxygen species (Lee et al. 1999). Overexpressed wild type Cyp18, but not a PPIase-inactive Cyp18 variant, protected cells from death after hyperactive SOD variant expression. This result indicates involvement of misfolding reactions of unknown client proteins in the loss of motor neurons. Application of CsA enhanced neuronal cell death. Silencing of SOD hyperactivity by RNAi protects against CsA effects. Consequently, CsA was not shown to be beneficial in patients with allergic contact dermatitis, multiple sclerosis, or amyotrophic lateral sclerosis (Faulds et al. 1993). It is somewhat surprising that CsA treatment is also able to attenuate degeneration and cell death of injured neurons in mouse brain tissue (Karlsson et al. 2004). Results showing that CsA in doses of (10.0 mg kg^{-1}) can prevent axotomized neonatal motor neuron death are counter to the expectations drawn from the SOD variant-based approach (Iwasaki et al. 2002). Interestingly, the downregulation of another PPIase, the multidomain enzyme FKBP52, was also found to be important for pathogenesis of ALS (Manabe et al. 2002).

2.7
Inflammation

Dependent on its concentration level, prototypic Cyp18 acts as proinflammatory molecule in many diseases. The protein mediates endothelial cell proliferation, migration, invasive capacity, and tubulogenesis at low concentrations. Opposite effects have been observed at high Cyp18 concentrations (Kim et al. 2004). Increased concentrations of Cyp18 are found in the blood stream of patients with inflammatory diseases. Cyp18 acts chemotactically to leukocytes that are found in increased levels, as well (Xu et al. 1992). Cyp18 and the highly homologous Cyp23 are also found in high concentrations in the serum of patients with rheumatic arthritis and sepsis (Billich et al. 1997; Tegeder et al. 1997). The chemotactic action of PPIases depends on the free accessibility of the active site of the cyclophilins, because application of PPIase inhibitors can completely suppress the effect. Most likely, the cyclophilins are exocytosed to the plasma during inflammation and act as a chemotaxis factor. Once in the blood stream, the cyclophilins might bind with high affinity to heparin sulfate proteoglycans and thus induce chemotaxis for neutrophils and T cells (De Ceuninck et al. 2003).

Upon stimulation of arterial endothelia cells by lipopolysaccharides, Cyp18 is expressed and secreted (Coppinger et al. 2004; Kim et al. 2004). Smooth muscle cells and macrophages show similar effects in response to stimulation by lipopolysaccharides and oxidative stress, suggesting a role in the pathogenesis of inflammatory diseases, such as arteriosclerosis (Jin et al. 2000, 2004).

Heat and chronic hypoxia mediate overexpression of Cyp18 up to threefold in myogenic cells (Andreeva et al. 1997). Furthermore, it was found that hypoxic cardiac muscle cells overexpress the cell surface receptor CD147, considered to be the Cyp18 receptor (Seko et al. 2004). Cyp18 signaling via CD147 involves the Cyp18 active site that probably binds to the Pro_{180}-Gly_{181} moiety of CD147 (Yurchenko et al. 2002). CD147 signaling culminates in ERK activation. PPIase-inactive Cyp18 mutants failed to initiate the signaling cascade. Cyp23 is able to promote CD147 signaling, as well (Yurchenko et al. 2001). However, the mechanism by which cyclophilins protect cells against oxidative stress remains to be discovered. After chemical modifications, the systemically active cyclosporin A can be converted to a drug that after topical application is beneficial in cutaneous inflammation (Rothbard et al. 2000).

Characteristic features of asthma are allergic bronchial inflammation and airway hyperresponsiveness, with increased numbers of eosinophils and activated T cells in the airways. The efficacy and the pharmacological profile of cyclosporin A treatment in patients with chronic severe asthma argues for T cell involvement, but a direct effect on the pro-inflammatory function of cyclophilins can also be assumed (Kon and Kay 1999). For the allergen-induced late asthmatic reaction, inhibitory effects on eosinophil-associated cytokines and chemokines have been discussed. The beneficial effect of CsA may also be the result of a reduced accumulation of eosinophils (Khan et al. 2000). Blocking the chemotactic action of Cyp18 might contribute to the latter effect.

2.8
Allergy

FK506, CsA, and CsA derivatives attenuate IgE-mediated histamine release from human basophils according to their relative immunosuppressive potency (Sperr et al. 1997). Exposure of cyclophilins, which form a pan-allergen family, to sensitized individuals could lead to the release of anaphylactogenic mediators due to cross-linking of IgE bound to the high-affinity surface protein Fc(ε)RI (Fluckiger et al. 2002). FK506 effects on Fc(ε)RI-mediated exocytosis of preformed mediators directly correlates with the amount of calcineurin B subunits in the cytosol of mast cells and basophils (Hultsch et al. 1998a, 1998b). Prototypic cyclophilins of carrots, birch, *A. fumigatus*, *E. granulosus*, and *basidiospores* belong to the allergens that share a high degree of sequence identity with the corresponding human protein. In fact, serum of individuals sensitized against mold proteins shows autoreactivity against human Cyp18 (Appenzeller et al. 1999).

3
FKBPs

Members of the FKBP family have rather different molecular masses in humans. Overall, catalytic domains in FKBPs are much less conserved than the catalytic domains within the cyclophilin family of PPIases. Both PPIase families exhibit significant differences in their catalytic mechanisms and substrate specificities (Fanghanel and Fischer 2004). The domain composition is also more diverse, ranging from a multiplicity of catalytic domains, EF hands, calmodulin binding sites to TPR motifs (Fischer and Aumuller 2003). Among the 16 different FKBPs in humans, there are members that are crucial in apoptosis (Edlich et al. 2005), receptor signaling (Pratt et al. 1999), calcium homeostasis (Schiene-Fischer and Yu 2001), and spermatogenesis (Crackower et al. 2003). Most human FKBPs, if not all, bind to and become inhibited by FK506, with K_i values ranging from high picomolar to high nanomolar levels. The three-dimensional structure of the inhibitory complex with FBBP12-FK506 complex might identify the catalytic site of FKBP12 at the FK506 position in the complex (Fig. 3). As in the case of cyclophilins, differ-

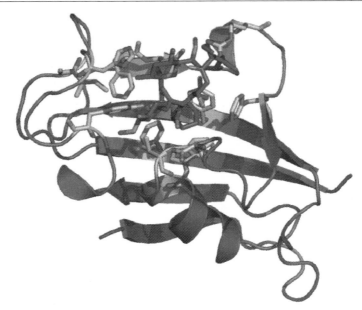

Fig. 3 Representation of the three-dimensional structure of the inhibitory FK506/FKBP12 complex (Van Duyne et al. 1991). The global fold of the prototypic FKBP12 (*red*) is assembled by five antiparallel β-sheets and an α-helix. The peptidomacrolide FK506 (*green*) binds to the putative active site of FKBP12. The interaction of FKBP12 to FK506 is mediated by the residues Y26, F36, D37, R42, F46, V55, J56, W59, Y82, H87, I91, and F99, depicted in *orange*, which are thought to assemble the active site of this PPIase. The pipecolinyl moiety of FK506 penetrates the prolyl binding pocket of FKBP12

ential low-molecular-mass inhibitors for the human members of the FKBPs family are still lacking.

3.1
Immune Response

Unlike Cyp18, FKBP12 utilizes two different modes of action to contribute to drug-mediated immunosuppression. First, human FKBPs, like cyclophilins, belong to the immunophilins because the endogenous FKBP12 of T cells binds the microbial drug FK506 (also known as tacrolimus), with subsequent block of antigen-stimulated T cell replication due to calcineurin targeting. Inhibition of the protein phosphatase activity of calcineurin by the FKBP12–FK506 complex is thought to form the biochemical basis of immunosuppression (Liu et al. 1991). Complex formation between calcineurin and FKBP12-FK506 does not directly interfere with the catalytic site of calcineurin (Griffith et al. 1995). Instead, complex formation hinders proteinaceous calcineurin substrates from entering the active site by applying steric constraints to binding of protein substrates (Fig. 4). Notably, the active FK506 conformer competitively inhibits the PPIase activity of FKBP12 with a K_i value of 0.5 nM (Zarnt et al. 1995). The dissociation constants of the ternary CaN/FKBP/FK506 complex range from 88 nM to 27 μM when different FKBPs are allowed to share the ternary complex (Edlich, unpublished observations).

Secondly, formation of a tight complex of the peptidomacrolide rapamycin (also known as sirolimus) with endogenous FKBP12 inhibits T cell proliferation by blocking protein synthesis and arresting the cell cycle in the G1 phase. Similar to the Cyp18-CsA combination, FKBP12-attached rapamycin experiences gain-of-function that leads to inhibition of mTOR kinase, a downstream effector of the phosphatidylinositol 3-kinase (PI3 K)/Akt (protein kinase B) signaling pathway. mTOR plays an important role in RNA stability and transcription, and controls the translation machinery, in response to amino acids and growth factors, via activation of p70 ribosomal S6 kinase, and inhibition of the eIF-4E binding protein (Fingar and Blenis 2004; Kahan 2004).

Given the common inhibition of many PPIases in the cell, CsA, FK506, and rapamycin may share similar side effects in patients (Hong et al. 2002; Mihatsch et al. 1998).

The multidomain enzyme FKBP52 was shown to interact with the peroxisomal phytanoyl-CoA α-hydroxylase, a protein that has homology to the LN1 sup-

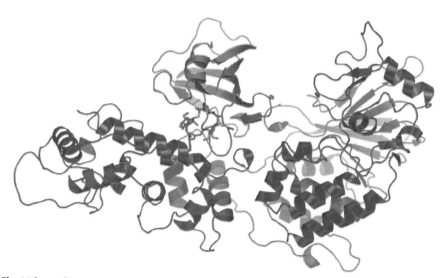

Fig. 4 Three-dimensional structure of the CaN/FK506/FKBP12 complex (Griffith et al. 1995). The protein phosphatase CaN is assembled from two subunits (*blue*), the catalytic subunit CaNA and the regulatory subunit CaNB. FKBP12 is displayed in *red*. After FKBP12 binding, FK506 (*green*) experiences gain of function enabling the resulting complex to form a inhibitory CaN complex. Both FKBP12 and FK506 participate in the interaction with calcineurin. Neither FK506 nor FKBP12 alone are able to inhibit CaN

pressor of progression of the human autoimmune disease lupus erythematosus (Chambraud et al. 1999). It was shown that the N-terminal PPIase-domain of FKBP52 mediates the interaction. A physiological role of this interaction has not yet been elucidated. FKBP12-directed autoantibodies, which prevent the formation of the inhibitory FKBP12–FK506–calcineurin complex, have been frequently found in serum of patients with autoimmune diseases (Shinkura et al. 1999).

3.2
Skin Disorders

The therapeutic effects of FKBP active site ligands to inflammatory skin disorders suggest a major role of FKBP in functional modulations of epidermal cells. Cyclosporin A, rapamycin and FK506 were found to exhibit differential effects on T cell and keratinocyte proliferation. Recent reports link between FK506 application in basal epidermal keratinocytes and therapeutic success in skin disease such as psoriasis (Al-Daraji et al. 2002; Panhans-Gross et al. 2001). Ongoing autoreactive Th-1 response of psoriatic epidermis has genetically determined immunogenic and inflammatory components. It is thought that the calcineurin inhibition in T cells may account in epidermal keratinocytes for the block of cell division causing psoriasis or atopic dermatitis (Hultsch et al. 1998b). A topically active FK506 derivative (pimecrolimus) with reduced system exposure and thus increased immunological safety has been launched for therapeutic application in atopic dermatitis, psoriasis, and allergic contact dermatitis (Marsland and Griffiths, 2004).

3.3
Neuropathies

Analyses of the neurotrophic and neuroprotective properties of FK506 and CsA show convincingly that inhibition of enzyme catalyzed prolyl isomerizations is a major factor in neuronal cell signaling under physiological conditions, but the nature of the PPIase substrates is still unknown (Brecht et al. 2003; Christner et al. 2001; Gold 1997; Hamilton et al. 1997; Steiner et al. 1997). Besides protection and regeneration following nerve fiber injury, FK506 and derivatives have further beneficial effects including alteration of neurotransmitter release (Steiner et al. 1996), protection against ischemic brain injury (Sharkey and Butcher 1994; Shichinohe et al. 2004), attenuation of glutamate neurotoxicity in vitro (Dawson et al. 1993), prevention of N-methyl-D-aspartate (NMDA)-receptor desensitization (Tong et al. 1995), modulation of long-term potentiation (LTP) (Terashima et al. 2000), and the blockage/prevention of long-term depression (LTD) in the rat hippocampus (Hodgkiss and Kelly 1995) as well as prevention of LTP and LTD in the visual cortex (Funauchi et al. 1994).

Multiple beneficial factors might contribute to the neuroprotective effects of PPIase-inhibitory FKBP ligands, including anti-apoptotic properties and activation of neurotrophic factors (Avramut et al. 2001; Lyons et al. 1994).

Selective enrichment in neurons of the peripheral and central nervous system is the hallmark of the neuronal FKBP, occasionally termed neuroimmunophilins (Steiner et al. 1992). The involvement of at least one FKBP in the regulation of neuronal cell death was identified by in vivo experiments studying the effect of FK506 application (Lyons et al. 1994). It has been observed, as well, that FK506 binding of neuronal cell protein and FKBP expression levels are significantly increased following the time course of neuronal-damaging processes, such as ischemia (Araki et al. 1998; Avramut and Achim 2003). Nonimmunosuppressive FKBP inhibitors, such as JNJ460, GPI1046, V10,367, GPI1048, and GPI1485, to name the most prominent candidates, have already been used in preclinical or clinical studies for treatment of Parkinson's disease (Birge et al. 2004; Gold et al. 2004; Steiner et al. 1997). In a mouse model, systemic administration of GPI1046 resulted in sparing of dopaminergic neurons and fibers. Schwann cells were found to be crucial components for the neurotrophic effects of the FKBP inhibitors (Birge et al. 2004). However, GPI1046 did not provide convincing evidence supporting neuroprotection in methyl-4-phenyl-1,2,3,6 tetrahydropyridine (MPTP)-treated rhesus monkeys after oral administration probably resulting from pharmacokinetics and pharmacodynamics (Emborg et al. 2001). Treatment with FKBP inhibitors resulted in dramatic recovery of penile erectile function after unilateral and bilateral cavernous nerve injury in a rat model. Therefore, FKBP inhibitors offer a new potential for patients who suffer nerve damage as a result of or during prostate surgery (Burnett and Becker 2004; Sezen et al. 2001).

Despite its abundance in neuronal tissues, inhibition of FKBP12 is unlikely to be functionally significant (Costantini et al. 2001; Tanaka et al. 2002).

Identification of the drug-targeted neuroimmunophilin from the whole collection of human FKBPs has advanced slowly. Unlike CsA and FK506, rapamycin was rarely shown to promote neuroregeneration and neuroprotection (Alemdar et al. 2004; Parker et al. 2000). In many instances, it proved to be inactive on its own but was shown to antagonize the neurotrophic effects of FK506 in some models (Costantini and Isacson 2000; Dawson et al. 1993; Sharkey and Butcher 1994). Thus, FKBPs responsible for the interaction with neurotrophic FKBP ligands can be identified among the rapamycin-insensitive PPIases.

Inhibition constants of rapamycin for several FKBPs have been collected (Christner et al. 2001). Because the K_i values for FKBP12, FKBP13, FKBP25, and FKBP52 were found to be in the low nanomolar range, the targeted FKBP must reside among the remaining members of the human FKBPs. FKBP52, which has already been discussed as mediating neurotrophic actions of neuroimmunophilin ligands (Gold et al. 1999) must be excluded using the K_i criterion.

However, there is one promising candidate FKBP that might regulate neuronal cell death. The multidomain protein FKBP38 is predominantly expressed in brain tissue and shows a secondary messenger-regulated PPIase activity that can be efficiently inhibited by FK506 and its nonimmunosuppressive derivatives, but shows 1,000-fold lower affinity for rapamycin when compared to the rapamycin–FKBP12 complex (Edlich et al., 2005). FKBP38 has been reported to be involved in cell size control, In addition, it shows considerable antimetastatic activity in mice tumor models (Fong et al. 2003; Rosner et al. 1997). Furthermore, the protein interacts functionally with the anti-apoptotic Bcl-2 (Edlich et al., 2005). Interestingly FKBP38 is required in hedgehog signal transduction during central nervous system development by antagonizing the secreted sonic hedgehog morphogen in neural tissues (Bulgakov et al. 2004).

3.4
Inherited Disease

There are several FKBP genes reported to have a defective function in inherited diseases, such as the Williams Beuren syndrome and Lebers congenital amarosis for FKBP36 (FKBP6) and FKBP44 (AIPL1) mutations, respectively (Meng et al. 1998; Ramamurthy et al. 2003; Sohocki et al. 2000) Lebers congenital amaurosis is diagnosed early in life with severely impaired vision or blindness, nystagmus, and an abnormal or flat electroretinogram. The onset of disease might be linked to the regulation of cell cycle progression during photoreceptor maturation via a FKBP44–NUB1 (NEDD8 ultimate buster 1) interaction (Akey et al. 2002).

The observed overlap of clinical phenotypes of autosomal dominant supravalvar aortic stenosis (SVAS) and Williams syndrome do not correlate with a common loss of FKBP6 gene integrity (Morris et al. 2003). Similar to wheat FKBP73, FKBP36 deletion results in defect spermatogenesis, suppressed testis development, and male-sterile organisms (Crackower et al. 2003; Kurek et al. 2002). Fertility and meiosis are normal in mutant females devoid of FKBP36. These findings suggest a novel strategy for the development contraceptive drugs based on the inhibition of FKBP36.

3.5
Pathogenic Microorganisms

Multidomain FKBPs of the Mip-type have been identified as virulence factors of many human pathogens including *Legionella pneumophila*, *Neisseria meningitidis*, *Chlamydia trachomatis*, *Coxiella burnetii*, *Trypanosoma cruzi*, *Aeromonas hydrophila*, and *Salmonella typhimurium*. For example, *Legionella pneumophila*, the causative agent of Legionnaires disease, is able to parasitize human lung macrophages and to cause this severe form of pneumonia.

The three-dimensional structures of Mip proteins from *Legionella pneumophila* (*Lp*FKBP25) and *Trypanosoma cruzi* (*Tc*FKBP18.8) reveal that each monomer has a FKBP-like catalytic core domain attached to a very long α-helical rod N-terminal to the core domain (Pereira et al. 2002; Riboldi-Tunnicliffe et al. 2001).

Monoclonal antibodies raised against (*Lp*FKBP25) significantly inhibit the early establishment and initiation of an intracellular infection of the bacteria in *Acanthamoeba castellanii*, the natural host, and in the human U937 macrophages (Helbig et al. 2003). Utilizing its PPIase site, *Lp*FKBP25 greatly enhances bacterial infectivity in animal models. Consequently, FK506 has been found to attenuate intracellular infections with *Chlamydia trachomatis* (Lundemose et al. 1993) and *Trypanosoma cruzi* (Moro et al. 1995). The mechanism by which Mip-catalyzed prolyl isomerization facilitates intracellular infections is still unknown. Typically, deletion of a single FKBP gene in *Flavobacterium johnsoniae* gave rise to phenotypic alterations, such as abolishing cell motility and inability to digest biopolymers that might have implications for understanding Mip-mediated bacterial virulence (McBride and Braun, 2004).

3.6
Steroid Responsiveness

Glucocorticoid treatment affects many diseases through a spectrum of anti-inflammatory and immunosuppressive effects, interferences in the growth factor-mediated pathways, as well as responses in the hematopoietic system and the calcium phosphate turnover. In addition to intrinsic functions such as steroid and DNA binding, proteins of the glucocorticoid receptor superfamily assemble nonreceptor proteins associated with the unactivated forms of the receptor, and among them, PPIases appear to play a crucial role in receptor function. Unactivated glucocorticoid receptors form hetero-oligomeric complexes of different composition that are located in the peripheral cytoplasm in order to detect their ligands. Upon ligand binding, the receptor molecule is translocated into the nucleus forming homodimers. The receptor homodimers induce transcription of several target genes (Fig. 5).

Two large human FKBPs, FKBP51 (also designated FKBP5, FKBP54) and FKBP52 (also designated FKBP4, FKBP56, FKBP59, p59), are intimately linked to steroid hormone signaling.

Both proteins possess two FKBP-like domains, and two putative calmodulin-binding sites adjacent to three TPR (tetratricopeptide repeat) domains. One molecule of FKBP51 or FKBP52 is involved in the formation of unactivated receptor complex in the cytosol that is capable of steroid hormone recognition and promotion of downstream signaling. FKBP52 develops affinity to client proteins in the presence of Hsp90 that assembles via the TPR motifs of the FKBP. Besides the steroid receptors Hsp90 and FKBP52, further components serve to assemble the unactivated receptor complex such as FKBP51, p23, Hsp70,

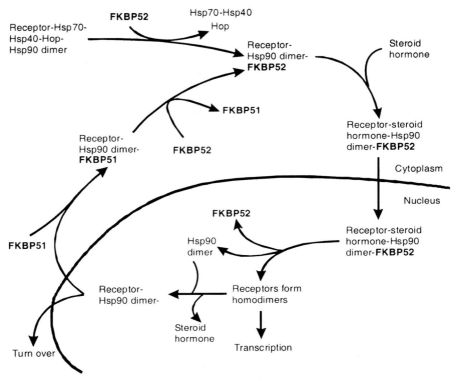

Fig. 5 Schematic representation of the involvement of multidomain FKBP in steroid hormone signaling. FKBP52 assembles with the receptor complex in order to stabilize the unactivated receptor complex in the periphery of the cell. It takes part in the receptor maturation altering proteins, assembling the hormone receptor complex. Upon steroid binding, the receptor is translocated into the nucleus, where it disassembles in order to allow dimerization of the ligand-bound receptor molecules that induce transcription. FKBP52 is believed to participate in receptor complex translocation. Once the receptor dimer disassembles, the receptor complex with Hsp90 translocates to the cytosol, where FKBP51 might regulate the transport of the receptor complex to the periphery of the cytosol

Hsp40, protein phosphatase 5, cyclophilin 40, Hip, and Hop. A conserved functional element of most protein components is their sequence similarity to already established peptide bond *cis/trans* isomerases (Fischer and Aumuller 2003; Lee et al. 2004; Schiene-Fischer et al. 2002; Silverstein et al. 1997). Among them, FKBP52 maintains the glucocorticoid receptor in a high-affinity state while FKBP51 decreases steroid affinity if it is present in the unactivated receptor complex (Denny et al. 2000). An increase in steroid affinity requires both the Hsp90-binding ability and the prolyl isomerase activity of FKBP52 (Riggs et al. 2003). An early step in receptor response to steroids might be associated with an exchange process between FKBP51 and FKBP52, which subsequently releases affinity to dynein and receptor trafficking only in the FKBP52-bound state (Davies et al. 2002).

In addition, the expression of FKBP51 is specifically enhanced by glucocorticoids, progestins, and androgens. This effect might attenuate the cellular response to the steroid hormones (Hubler et al. 2003).

A role for FKBP52 in subcellular trafficking has also been considered, because dynein co-purifies and microtubules co-localizes with FKBP52 and anchors the receptor in the cytosol (Czar et al. 1994; Galigniana et al. 2004). The PPIase site of FKBP52 is responsible for interaction with dynein.

Recent research shows that both receptor-associated FKBPs can play a role in other disorders based on steroid hormone signaling, such as the androgen insensitivity syndrome (AIS) or hypospadius. For example, polymorphisms in the FKBP5 gene, encoding FKBP51, were identified as causing more frequent depression periods, rapidly responding to antidepressant treatment (Binder et al. 2004).

Cross-talk between Hsp90 and FKBP52 has been observed for the interaction of FKBP52 with the single-stranded D-sequence binding protein of the adeno-associated virus type 2 that is important for the prevention of viral replication (Zhong et al. 2004). Adeno-associated virus 2 (AAV) vectors are currently in use in phase I/II clinical trials for gene therapy of cystic fibrosis and hemophilia B.

Differential gene expression following androgen-deprivation of an androgen-dependent prostate tumor xenograft revealed decreased expression of FKBP51. When the tumor was grown to an androgen-independent state, FKBP51 increased to its original level (Amler et al. 2000).

3.7
Malignancies

Biochemical investigations have led to the elucidation of two different molecular mechanisms by which FKBP might contribute to the control of malignant growth and proliferation.

First, several studies have suggested a regulatory role for FKBPs in the cell cycle and tumor progression (Aghdasi et al. 2001; Fong et al. 2003). In yeast, the Hmo1 protein, which is likely to be the homolog of HMG1/2 in higher organisms, plays an important role in genome maintenance, because it binds single-stranded DNA and unwinds DNA in the presence of eukaryotic DNA topoisomerase I (Dolinski and Heitman 1999). FKBP52 has been shown to be a direct interaction partner of the gene product of the *c-Myc* protooncogene by Northern blotting (Coller et al. 2000). The overexpression of two PPIases, Cyp40 and FKBP52, coupled with relative differences in their expression pattern among individual tumors, may have important functional implications for steroid response in breast cancer (Ward et al. 1999). Similarly, the function of the aryl hydrocarbon receptor-interacting protein-like 1 (AIPL1, FKBP44) might conceivably involve cell cycle control, because it participates in cell cycle progression via the NUB1 (NEDD8 ultimate buster 1)/NEDD8 pathway (Akey et al. 2002).

Promotion of malignant astrocytoma-induced angiogenesis characterizes the differential expression patterns of angiogeneses related genes in malignant high-grade astrocytomas. In a microarray of 133 angiogenesis-related genes, FKBP12 is among the few significantly overexpressed genes, and 9 of 21 (43%) genes overexpressed by high-grade astrocytomas were genes associated with either FKBP or hypoxia-inducible transcription factor 2α (Khatua et al. 2003). Among the proteins with anti-invasive and antimetastatic functions, FKBP38 and FKBP12 have been identified using systemic gene transfer in tumor-bearing mice. Syndecan and matrix metalloproteinase 9 were thought to mediate the antitumor activities of these enzymes (Fong et al. 2003).

Reduced cell growth and cyclin D1 level in fibroblasts from FKBP12-deficient (FKBP12$^{-/-}$) mice corresponds to cell cycle arrest in G_1 phase and to the fact that these cells can be rescued by FKBP12 transfection (Aghdasi et al. 2001). Generally, proteins associated with G1 regulation have been shown to play a key role in proliferation, differentiation, and oncogenic transformation as well as apoptosis, and represent promising targets for cancer treatment.

The mechanism suggested for the FKBP12-driven cell cycle is based on the regulation of TGF-beta receptor signaling. The prototypic FKBP12 interacts with the transforming growth factor β (TGFβ) (Wang et al. 1994), acting as a negative regulator of TGFβ receptor endocytosis (Aghdasi et al. 2001; Yao et al. 2000). The immunophilin binds the unphosphorylated form of the TGFβ receptor type I (Huse et al. 1999). Within the complex FKBP12 caps the phosphorylation site of the receptor and further stabilizes the inactive conformation of the receptor (Huse et al. 1999). Even though FKBP12 binds directly to the GS2 helix in the receptor molecule, the conformation of the GS2 region is also stable in the absence of FKBP12 and not induced by its binding (Huse et al. 1999). GS loop phosphorylation induces FKBP12 dissociation. Recent data points to a function of FKBP12 as buffer in TGFβ signaling, stabilizing and prolonging the lifetime of the dephosphorylated receptor (Wang and Donahoe 2004). This function might become important in preventing ligand-independent activation of the receptor at very high concentrations of the receptor molecules.

Similar effects were observed with the epidermal growth factor (EGF) receptor that utilizes the PPIase activity of cytosolic FKBP12 to downregulate receptor tyrosine autophosphorylation (Lopez-Ilasaca et al. 1998). Inhibition of FKBP12 through FK506 and rapamycin leads to stimulation of EGF receptor autophosphorylation.

Secondly, the peptidomacrolide rapamycin is a major player in antitumor activities mediated by FKBPs. Both rapamycin and CCI-779, an ester analog of rapamycin with improved pharmaceutical properties and solubility in water, have demonstrated impressive activity against a broad range of human cancers growing in tissue culture and in human tumor xenograft models (Hidalgo and Rowinsky 2000). Problems might arise from genetic mutations or compensatory changes in tumor cells that could enable cells to escape rapamycin action (Huang and Houghton 2001). Wyeth Research (USA) is developing CCI-779 as a anticancer drug. By November 2001, phase III trials had been initiated. In October 2001, filing was predicted for 2003, with a potential launch in 2005 (Elit 2002).

The FKBP12/rapamycin (CCI-779) complexes are probably the biologically relevant species because these drugs experience gain-of-function when bound to the FKBP. The drug-targeted protein is the serine/threonine protein kinase TOR (target of rapamycin) (Sabers et al. 1995).

The protein kinase TOR is an essential protein in eukaryotes that acts as a gatekeeper for the progression from G1 to S phase in cell cycle and is therefore a central regulator of cell growth and proliferation (Brunn et al. 1997). In fact, rapamycin-induced apoptosis in tumor cells is a consequence of continued G_1 progression during mTOR inhibition (Huang et al. 2001).

The protein controls cap-dependent translation initiation by inactivating eukaryotic initiation factor 4E binding proteins in response to mitogen, nutrient, and energy levels (Bjornsti and Houghton 2004; Chen and Fang 2002).

Under normal conditions, TOR is precisely regulated interacting with a protein, called raptor (regulatory associated protein of TOR). The raptor–TOR interaction changes TOR affinity to its substrates in dependence to amino acid and ATP levels (Oshiro et al. 2004). It is now controversially discussed whether the FKBP12–rapamycin complex changes the conformation of TOR, masks the active site for TOR substrates or mimics directly a substrate of the protein kinase activity, but there is growing evidence that the FKBP12–rapamycin complex inhibits the raptor–TOR interaction. The disruption of this interaction may uncouple TOR from its substrates rather than inhibiting the intrinsic kinase activity (Oshiro et al. 2004).

Importantly, application of rapamycin, a product of the bacteria *Streptomyces hygroscopicus*, as well as its synthetic analogous to early mouse embryos resulted in the same "flat top" phenotype that is developed by TOR mutant mice, causing death after less than 2 weeks (Bjornsti and Houghton 2004). These results imply that application of rapamycin leads to complete inhibition of TOR and therefore affects cell cycle progression. Other results suggest either that rapamycin reversibly acts in living cells or that the cellular effects of rapamycin are not mediated through global inhibition of mTOR kinase activity (Edinger et al. 2003).

3.8
Arylhydrocarbon (Dioxin) Responsiveness

The arylhydrocarbon (dioxin) receptor (AhR) is a ligand-inducible transcriptional activator that exhibits structural and functional similarities to steroid hormone receptors. It is a cytosolic receptor complex that can detect 2,3,7,8-tetrachlorodibenzo-p-dioxin in order to start transcription of xenobiotic response genes encoding drug-metabolizing enzymes. Induced enzymes include those required for carcinogen activation. Furthermore, AhR ligands are involved in cardiovascular diseases through lipid peroxidation and endothelium dysfunction.

The cytosolic AhR assembles the ligand-binding subunit, a dimer of Hsp90 and FKBP37.7 (XAP2, AIP, ARA9) that dissociate subsequent to 2,3,7,8-tetrachlorodibenzo-p-dioxin treatment (Carver and Bradfield 1997; Ma and Whitlock 1997). To molecularly characterize FKBP37.7, three TPR (tetratricopeptide repeat) motifs follow a single FKBB-like domain (Petrulis and Perdew 2002). In a dioxin-responsive element reporter gene assay the presence of FKBP37.7 in endogenous AhR complexes causes twofold higher signal for the luciferase activity signal (Meyer et al. 1998) and FKBP37.7 stabilizes the ternary receptor complex molecule. In addition, it diminishes receptor degradation by protection against ubiquitination (Kazlauskas et al. 2000). Furthermore, interactions between FKBP37.7 and Hsp90 with the AhR subunit proved to be similar to those observed within the glucocorticoid receptor complexes, pointing to a common mode of action. However, the number of peptide bond *cis/trans* isomerases involved in receptor signaling is quite different in both cases. It raises the question whether there is a yet unidentified folding helper enzyme that adds functionally and physically to FKBP37.7 in the cytosolic AhR. Recently, an anchoring function of FKBP37.7 for the ligand-free receptor complex to cytoskeletal structures of the actin network has been demonstrated (Berg and Pongratz 2002). In addition, FKBP37.7 inhibits the ligand-independent shuttling of the receptor into the nucleus by the a conformational alteration of the nuclear localization signal of the AhR that prevents binding of importin-β (Petrulis et al. 2003). For a nuclear receptor, termed peroxisome proliferator-activated receptor alpha (PPARα), the presence of FKBP37.7 in a ternary complex with Hsp90 and the receptor protein was also shown. In humans, this member of the nuclear receptor superfamily regulates energy homeostasis by controlling the lipid metabolism (Sumanasekera et al. 2003).

3.9
Cardiovascular Disorders

Prototypic FKBPs, such as FKBP12 and FKBP12.6, belong to the receptor-associated folding helper enzymes (Schiene-Fischer and Yu 2001). For example, they co-purify and physically interact with intracellular Ca^{2+} release channels, the ryanodine receptors (RyR), and contribute to the regulation of the release of intracellular Ca^{2+} stores (Jayaraman et al. 1992). FKBP12 and RyR1 assemble in the skeletal receptor in a 4:1 molar ratio of FKBP12-tetrameric receptor protein; the cardiac RyR2 contains FKBP12.6 (Marks et al. 2002). Receptor-binding affinities of FKBPs have been found in the high nanomolar range (Jeyakumar et al. 2001). The heterocomplex dissociation by FK506 titration reveals that the active site of FKBP12 is involved in the interaction. Binding of FKBPs to RyRs are thought to stabilize the close state of the Ca^{2+} channel, removal of FKBP makes the channel leaky. However, there are conflicting reports about the involvement of a catalyzed prolyl isomerization in the FKBP-mediated regulation of the open probability and mean open time of terminal cisternae (Marks 1996; Timerman et al. 1995). Additional control of RyR1 channel function is provided by protein kinase A-mediated phosphorylation, preventing FKBP12 binding to the hyperphosphorylated receptor. Under these conditions, impaired sarcoplasmic Ca^{2+} release and early fatigue in the skeletal muscle results in heart failure among test animals (Reiken et al. 2003a). Generally, prototypic FKBP deficiency and failure of RyR mutants to bind prototypic FKBP have similar phenotypic responses, such as arrhythmogenic right ventricular dysplasia/cardiomyopathy type 2 and stress-induced polymorphic ventricular tachycardia (Tiso et al. 2002). The phenotype of FKBP12.6-deficient mice is consistent with this model (Wehrens et al. 2003).

In patients with heart failure, improved cardiac muscle function under treatment with blockers of β-adrenergic receptor was associated with restoration of normal FKBP12.6 levels in the RyR2 heterocomplex and RyR2 channel function (Reiken et al. 2003b).

FKBP12.6 stabilizes RyR2, preventing aberrant activation of the channel during the resting phase of the cardiac cycle. $FKBP12.6^{-/-}$ mice consistently exhibited exercise-induced cardiac ventricular arrhythmias that caused sudden cardiac death. Cyclic ADP ribose (cADPR) was reported to be a ligand for FKBP12.6 in islet RyR, and the binding of cADPR to FKBP12.6 dissociates it from the channel, causing Ca^{2+} release (Noguchi et al. 1997) However, cADPR neither binds to nor inhibits recombinant FKBP12.6 (Edlich and Fanghänel, unpublished results).

In conclusion, FKBP12.6 plays a role in the regulation of calcium homeostasis, which is of importance for heart muscle contraction. When RyR2 is hyperphosphorylated, FKBP12.6 is depleted from the receptor. Similarly, mutations in RyR2 that lead to FKBP12.6-deficient receptors cause intracellular calcium leakage triggering fatal cardiac arrhythmia failure and exercise-induced cardiac death (Lehnart et al. 2003, 2004; Wehrens et al. 2004). Several attempts show that the application of numerous drugs can improve FKBP-mediated regulation of ryanodyne receptors (Doi et al. 2002; Kohno et al. 2003).

It appears that a different function of FKBP12 plays a role in the cardiac growth and chamber maturation. Transcripts of the cardiac restricted cytokine bone morphogenic protein 10 (BMP10) are upregulated in hypertrabeculated hearts of $FKBP12^{-/-}$ mouse embryos. BMP10 upregulation may be causative of congenital heart disease, such as ventricular septal defect (VSD), myocardium noncompaction, and ventricular hypertrabeculation (Chen et al. 2004).

4
Parvulins

Parvulins constitute the third PPIase family that, unlike enzymes of the other two families, cyclophilins and FKBPs, do not express affinity for immunosuppressive drugs (Rahfeld et al. 1994). In humans, Par14 and Par18 (also known as Pin1) represent the only members of the parvulin family. Among PPIases Pin1 provides a rare example of relatively high substrate specificity, because polypeptide chains require a pSer(pThr)-Pro-moiety (where p denotes phosphoesterification) if orderly Pin1 catalysis is to occur (Ranganathan et al. 1997; Zhou et al. 2000). Pin1 contains an N-terminal group IV WW domain and a C-terminal parvulin-like catalytic domain connected by a flexible linker (Ranganathan et al. 1997). The mechanism by which Pin1 exerted its critical role in the cell cycle became obvious by combining in vitro studies on isomer-specific dephosphorylation by protein phosphatase 2a (PP2a) with dephosphorylation studies in *Xenopus* mitotic extracts, and rescue experiments in yeast (Lu et al. 2002; Zhou et al. 2000). For example, the cell cycle regulatory protein phosphatase Cdc25C contains the pThr48-Pro and pThr67-Pro moieties that represent critical regulatory phosphorylation sites. The prolyl bond conformation can be either *cis* or *trans* or a mixture of both. For PP2a the inability to dephosphorylate *cis* pThr(Ser)-Pro moieties can be detected in vivo by a reciprocal genetic interactions in temperature-sensitive PP2a-deficient *PPH* and the Pin1-homolog-deficient *ESS1/PTF1* mutant strains of budding yeast and in vitro using oligopeptide dephosphorylation experiments. Despite the presence of sufficient activity of PP2a a fraction of Cdc25C containing *cis* pThr-Pro is left over until it slowly isomerizes to the *trans* isomer. This isomerization is accelerated markedly in the presence of Pin1. Transgenic expression of point-mutated variants of Pin1 and isolated Pin1 domains in yeast indicates that the enzymatically active PPIase domain is necessary and sufficient to carry out conformational tuning of a PP2a substrate to become dephosphorylated in time. The group IV WW domain alone, although exhibiting affinity to the pThr(Ser)-Pro containing polypeptide chains, cannot assist in this function (Fig. 6).

In contrast, Par14 has some specificity for arginine preceding the proline residue (Uchida et al. 1999). This enzyme seems to localize to the nucleus and the N-terminal extension of the catalytic domain might mediate interactions with the preribosomal nucleoprotein complex (Fujiyama et al. 2002).

4.1
Malignancies

Pin1 was identified in its involvement in the *Aspergillus* mitotic kinase NIMA pathway and thus prevention of NIMA-induced mitotic cell death (Lu et al. 1996). Depletion of Pin1 induces mitotic arrest, whereas HeLa cells overexpressing Pin1 arrest in the G2 phase of the cell cycle. While expression is at

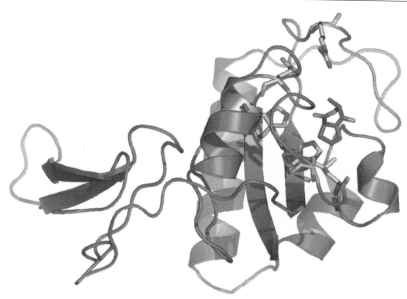

Fig. 6 Representation of the global fold of Pin1 (*red*) harboring a parvulin-like PPIase domain (*right*), which consists of two antiparallel β-sheets and four α-helices, and a WW domain (*left*); both domains are connected by a flexible hinge region (Ranganathan et al. 1997). The structure includes the dipeptide Ala-Pro bound to the active site of Pin1. The parvulin-type PPIase site is formed by the amino acid residue side chains of H59, L122, F125, M130, Q131, F134, and H157 (*orange*). The R68 residue (*orange*) completes the substrate binding pocket of Pin1. The WW domain is described to mediate phosphorylation-dependent protein–protein interactions

low level in most normal tissues, Pin1 is highly expressed in many different human cancers including prostate, lung, ovary, cervical, and brain tumors and melanoma (Bao et al. 2004; Miyashita et al. 2003). In breast cancer cells, early transformed properties correlate with the Pin1 content of the cells, because they are effectively suppressed by Pin1 deletion (Wulf et al. 2004). In conclusion, depletion of pThr(pSer)-Pro-specific PPIase activity point mutations, gene deletion, or expression of antisense RNA induce mitotic arrest and apoptosis in budding yeast and human tumor cell lines (Hani et al. 1999; Rippmann et al. 2000; Wu et al. 2000).

The probably best studied Pin1 substrate in the regulation of mitosis is Cdc25C (Crenshaw et al. 1998; Patra et al. 1999; Yaffe and Cantley 1999). Pin1 induces a structural change in Cdc25C and directly affects the phosphatase activity of Cdc25C by triggering the isomer-specific Cdc25C dephosphorylation by PP2A. This protein phosphatase dephosphorylates Cdc25C, which probably exhibits native state conformational polymorphism with regard to pSer(pThr)-Pro moieties, only if the protein conformer has *trans* pSer(pThr)-Pro moieties.

Pin1 facilitates dephosphorylation through catalyzed *cis/trans* isomerizations. Furthermore, Pin1 can antagonize the Cdc2-associated proteins Suc1/Cks1 by binding the same site in Cdc25C (Landrieu et al. 2001). Pin1 has also been reported to bind to Cdc25C motifs that play a crucial role in Cdc2 activation and subsequently for G2-M transition in the cell cycle.

Among the many tumor-growth-related substrates of Pin1 (Lu, 2004), p53 interacts with Pin1 in response to p53 phosphorylation on a number of Ser-Pro and Thr-Pro moieties. On p53 targeting, Pin1 generates conformational changes in p53, enhancing its transactivation activity. Genotoxicity-induced phosphorylation of p53 causes Ser(Thr) phosphoester sites, the prolyl bond *cis/trans* interconversion of which requires Pin1 catalyzed conformational interconversions in order for the p53-mediated tumor suppression to occur (Wulf et al. 2002; Zacchi et al. 2002). Consequently, Pin1 depletion reduces the stress response of p53 function. Interestingly, another p53 family member, the c Abl-linked p73, responds to Pin1 depletion by protein destabilization in the presence and the absence of DNA (Mantovani et al. 2004). Anticancer drugs, such as cisplatin and adriamycin, exert their cytotoxicity in p73-dependent manner characterized by activating the mismatch-repair-dependent apoptosis pathway. These data converge on prolyl bond isomerization since cAbl activates proline-directed p38 MAP kinase to phosphorylate Thr-Pro sites within p73. The p38-mediated protein phosphorylation depends on the isomeric state of the Thr-Pro, and could be catalyzed by PPIases (Weiwad et al. 2004). The mammalian germ cell development and spermatogenesis is discussed in the context of the interplay between Pin1, and the Ras/MEK/MAPK pathway may also explain the effects of Pin1 deletion observed in primordial germ cell proliferation during embryonic development, along with degenerative or proliferative defects in the adult testis, retina, mammary gland, and brain in mice (Atchison and Means 2004).

Another cancer-relevant finding shows that the dephosphorylation of the proto-oncogenic transcription factor c-Myc by protein phosphatase 2a is assisted by Pin1, and Pin1 facilitates c-My protein degradation (Yeh et al. 2004).

Summarizing these studies demonstrates that Pin1-specific PPIase inhibitors may represent potential lead structures for therapeutic intervention in malignancies.

4.2
Neuropathies

Among the many Pin1-targeted phosphoproteins potentially present in neurons, microtubule-associated tau-protein is a prominent member, whose dephosphorylation restores its ability to stabilize microtubules. Mice with Pin1 knock-out suffer from many age-dependent neurological deficits, such as motor and behavioral problems, tau hyperphosphorylation, tau filament formation and neuronal degeneration (Liou et al. 2002). Neuronal apoptosis following nuclear depletion of Pin1 was reported to be a contributing factor to frontotemporal dementias (Thorpe et al. 2004). The involvement of Pin1 in the G0/G1 transition in neurons makes its presence crucial to neuronal apoptosis (Hamdane et al. 2002). The fraction of soluble Pin1 is reduced in the brains of Alzheimer's disease patients (Lu et al. 1999). Rather, it consistently associates with the pThr231 residues of the various hyperphosphorylated tau proteins that characterize degenerating neurons. Surprisingly, Pin1 may restore the tubulin polymerization function of these hyperphosphorylated tau species (Lu et al. 2002). According to its biochemical role, Pin1 can restore tau function by facilitating its enzymatic dephosphorylation and the subsequent recovery of microtubule assembly (Lu et al. 1999; Zhou et al. 2000). In fact, neurons containing Pin1 granules were devoid of neurofibrillary tangles in Alzheimer patients' brains (Holzer et al. 2002).

References

Aghdasi B, Ye KQ, Resnick A, Huang A, Ha HC, Guo X, Dawson TM, Dawson VL, Snyder SH (2001) FKBP12, the 12-kDa FK506-binding protein, is a physiologic regulator of the cell cycle. Proc Natl Acad Sci U S A 98:2425–2430

Akey DT, Zhu X, Dyer M, Li A, Sorensen A, Blackshaw S, Fukuda-Kamitani T, Daiger SP, Craft CM, Kamitani T, Sohocki MM (2002) The inherited blindness associated protein AIPL1 interacts with the cell cycle regulator protein NUB1. Hum Mol Genet 11:2723–2733

Al-Daraji WI, Grant KR, Ryan K, Saxton A, Reynolds NJ (2002) Localization of calcineurin/NFAT in human skin and psoriasis and inhibition of calcineurin/NFAT activation in human keratinocytes by cyclosporin A. J Invest Dermatol 118:779–788

Alemdar AY, Sadi D, McAlister VC, Mendez I (2004) Liposomal formulations of tacrolimus and rapamycin increase graft survival and fiber outgrowth of dopaminergic grafts. Cell Transplant 13:263–271

Amler LC, Agus DB, LeDuc C, Sapinoso ML, Fox WD, Kern S, Lee D, Wang V, Leysens M, Higgins B, Martin J, Gerald W, Dracopoli N, Cordon-Cardo C, Scher HI, Hampton GM (2000) Dysregulated expression of androgen-responsive and nonresponsive genes in the androgen-independent prostate cancer xenograft model CWR22-R1. Cancer Res 60:6134–6141

Andreeva L, Motterlini R, Green CJ (1997) Cyclophilins are induced by hypoxia and heat stress in myogenic cells. Biochem Biophys Res Commun 237:6–9

Andreotti AH (2003) Native state proline isomerization: an intrinsic molecular switch. Biochemistry 42:9515–9524

Appenzeller U, Meyer C, Menz G, Blaser K, Crameri R (1999) IgE-mediated reactions to autoantigens in allergic diseases. Int Arch Allergy Immunol 118:193–196

Arai S, Vogelsang GB (2000) Management of graft-versus-host disease. Blood Rev 14:190–204

Araki T, Kato H, Shuto K, Itoyama Y (1998) Post-ischemic alterations in [3H]FK506 binding in the gerbil and rat brains. Metab Brain Dis 13:9–19

Atchison FW, Means AR (2004) A role for Pin1 in mammalian germ cell development and spermatogenesis. Front Biosci 9:3248–3256

Avramut M, Achim CL (2003) Immunophilins in nervous system degeneration and regeneration. Current Topics Med Chem 3:1376–1382

Avramut M, Zeevi A, Achim CL (2001) The immunosuppressant drug FK506 is a potent trophic agent for human fetal neurons. Brain Res Develop Brain Res 132:151–157

Bao L, Kimzey A, Sauter G, Sowadski JM, Lu KP, Wang DG (2004) Prevalent overexpression of prolyl isomerase Pin1 in human cancers. Am J Pathol 164:1727–1737

Bartz SR, Hohenwalter E, Hu MK, Rich DH, Malkovsky M (1995) Inhibition of human immunodeficiency virus replication by nonimmunosuppressive analogs of cyclosporin A. Proc Natl Acad Sci U S A 92:5381–5385

Batiuk TD, Kung LN, Halloran PF (1997) Evidence that calcineurin is rate-limiting for primary human lymphocyte activation. J Clin Invest 100:1894–1901

Batiuk TD, Pazderka F, Halloran PF (1995) Cyclosporine-treated renal transplant patients have only partial inhibition of calcineurin phosphatase activity. Transplant Proc 27:840–841

Bell A, Roberts HC, Chappell LH (1996) The antiparasite effects of cyclosporin A—possible drug targets and clinical applications. Gen Pharmacol 27:963–971

Bennett PC, Singaretnam LG, Zhao WQ, Lawen A, Ng KT (1998) Peptidyl-prolyl-cis/trans-isomerase activity may be necessary for memory formation. FEBS Lett 431:386–390

Berg P, Pongratz I (2002) Two parallel pathways mediate cytoplasmic localization of the dioxin (Aryl hydrocarbon) receptor. J Biol Chem 277:32310–32319

Billich A, Winkler G, Aschauer H, Rot A, Peichl P (1997) Presence of cyclophilin a in synovial fluids of patients with rheumatoid arthritis. J Exp Med 185:975–980

Binder EB, Salyakina D, Lichtner P, Wochnik GM, Ising M, Putz B, Papiol S, Seaman S, Lucae S, Kohli MA, Nickel T, Kunzel HE, Fuchs B, Majer M, Pfennig A, Kern N, Brunner J, Modell S, Baghai T, Deiml T, Zill P, Bondy B, Rupprecht R, Messer T, Kohnlein O, Dabitz H, Bruckl T, Muller N, Pfister H, Lieb R, Mueller JC, Lohmussaar E, Strom TM, Bettecken T, Meitinger T, Uhr M, Rein T, Holsboer F, Muller-Myhsok B (2004) Polymorphisms in FKBP5 are associated with increased recurrence of depressive episodes and rapid response to antidepressant treatment. Nat Genet 36:1319–1325

Birge RB, Wadsworth S, Akakura R, Abeysinghe H, Kanojia R, MacIelag M, Desbarats J, Escalante M, Singh K, Sundarababu S, Parris K, Childs G, August A, Siekierka J, Weinstein DE (2004) A role for Schwann cells in the neuroregenerative effects of a non-immunosuppressive FK506 derivative, JNJ460. Neuroscience 124:351–366

Bjornsti MA, Houghton PJ (2004) The TOR pathway: a target for cancer therapy. Nat Rev Cancer 4:335–348

Boesch D, Muller K, Pourtier-Manzanedo A, Loor F (1991) Restoration of daunomycin retention in multidrug-resistant P388 cells by submicromolar concentrations of SDZ PSC 833, a nonimmunosuppressive cyclosporin derivative. Exp Cell Res 196:26–32

Borvak J, Chou CS, Van Dyke G, Rosenwirth B, Vitetta ES, Ramilo O (1996) The use of cyclosporine, FK506, and SDZ NIM811 to prevent CD25-quiescent peripheral blood mononuclear cells from producing human immunodeficiency virus. J Infect Dis 174:850–853

Bosco DA, Eisenmesser EZ, Pochapsky S, Sundquist WI, Kern D (2002) Catalysis of cis/trans isomerization in native HIV-1 capsid by human cyclophilin A. Proc Natl Acad Sci U S A 99:5247–5252

Bose S, Mathur M, Bates P, Joshi N, Banerjee AK (2003) Requirement for cyclophilin A for the replication of vesicular stomatitis virus New Jersey serotype. J Gen Virol 84:1687–1699

Brazin KN, Mallis RJ, Fulton DB, Andreotti AH (2002) Regulation of the tyrosine kinase Itk by the peptidyl-prolyl isomerase cyclophilin A. Proc Natl Acad Sci U S A 99:1899–1904

Brecht S, Schwarze K, Waetzig V, Christner C, Heiland S, Fischer G, Sartor K, Herdegen T (2003) Changes in peptidyl-prolyl cis/trans isomerase activity and FK506 binding protein expression following neuroprotection by FK506 in the ischemic rat brain. Neuroscience 120:1037–1048

Brenner BG, Wainberg MA (1999) Heat shock protein-based therapeutic strategies against human immunodeficiency virus type 1 infection. Infect Dis Obstet Gynecol 7:80–90

Brunn GJ, Fadden P, Haystead TAJ, Lawrence JC (1997) The mammalian target of rapamycin phosphorylates sites having a (Ser/Thr)-pro motif and is activated by antibodies to a region near its cooh terminus. J Biol Chem 272:32547–32550

Bulgakov OV, Eggenschwiler JT, Hong DH, Anderson KV, Li TS (2004) FKBP8 is a negative regulator of mouse sonic hedgehog signaling in neural tissues. Development 131:2149–2159

Burnett AL, Becker RE (2004) Immunophilin ligands promote penile neurogenesis and erection recovery after cavernous nerve injury. J Urol 171:495–500

Calabrese KS, Paradela A, do Valle TZ, Tedesco RC, Silva S, Mortara RA, da Costa SCG (2000) Study of acute chagasic mice under immunosuppressive therapy by cyclosporin A: modulation and confocal analysis of inflammatory reaction. Immunopharmacology 47:1–11

Campa MJ, Wang MZ, Howard B, Fitzgerald MC, Patz EF (2003) Protein expression profiling identifies macrophage migration inhibitory factor and cyclophilin A as potential molecular targets in non-small cell lung cancer. Cancer Res 63:1652–1656

Carver LA, Bradfield CA (1997) Ligand-dependent interaction of the aryl hydrocarbon receptor with a novel immunophilin homolog in vivo. J Biol Chem 272:11452–11456

Chambraud B, Radanyi C, Camonis JH, Rajkowski K, Schumacher M, Baulieu EE (1999) Immunophilins, refsum disease, and lupus nephritis: The peroxisomal enzyme phytanoyl-COA alpha-hydroxylase is a new FKBP-associated protein. Proc Natl Acad Sci U S A 96:2104–2109

Chen HY, Shi S, Acosta L, Li WM, Lu J, Bao SD, Chen ZA, Yang ZC, Schneider MD, Chien KR, Conway SJ, Yoder MC, Haneline LS, Franco D, Shou WN (2004) BMP10 is essential for maintaining cardiac growth during murine cardiogenesis. Development 131:2219–2231

Chen J, Fang YM (2002) A novel pathway regulating the mammalian target of rapamycin (mTOR) signaling. Biochem Pharmacol 64:1071–1077

Christner C, Herdegen T, Fischer G (2001) FKBP ligands as novel therapeutics for neurological disorders. Mini-Rev Med Chem 1:377–397

Clarke SJ, McStay GP, Halestrap AP (2002) Sanglifehrin A acts as a potent inhibitor of the mitochondrial permeability transition and reperfusion injury of the heart by binding to cyclophilin-D at a different site from cyclosporin A. J Biol Chem 277:34793–34799

Colgan J, Asmal M, Neagu M, Yu B, Schneidkraut J, Lee Y, Sokolskaja E, Andreotti A, Luban J (2004) Cyclophilin A regulates TCR signal strength in CD4(+) T cells via a proline-directed conformational switch in Itk. Immunity 21:189–201

Coller HA, Grandori C, Tamayo P, Colbert T, Lander ES, Eisenman RN, Golub TR (2000) Expression analysis with oligonucleotide microarrays reveals that MYC regulates genes involved in growth, cell cycle, signaling, and adhesion. Proc Natl Acad Sci U S A 97:3260–3265

Coppinger JA, Cagney G, Toomey S, Kislinger T, Belton O, McRedmond JP, Cahill DJ, Emili A, Fitzgerald DJ, Maguire PB (2004) Characterization of the proteins released from activated platelets leads to localization of novel platelet proteins in human atherosclerotic lesions. Blood 103:2096–2104

Costantini LC, Cole D, Chaturvedi P, Isacson O (2001) Immunophilin ligands can prevent progressive dopaminergic degeneration in animal models of Parkinson's disease. Eur J Neurosci 13:1085–1092

Costantini LC, Isacson O (2000) Immunophilin ligands and GDNF enhance neurite branching or elongation from developing dopamine neurons in culture. Exp Neurol 164:60–70

Crackower MA, Kolas NK, Noguchi J, Sarao R, Kikuchi K, Kaneko H, Kobayashi E, Kawai Y, Kozieradzki I, Landers R, Mo R, Hui CC, Nieves E, Cohen PE, Osborne LR, Wada T, Kunieda T, Moens PB, Penninger JM (2003) Essential role of Fkbp6 in male fertility and homologous chromosome pairing in meiosis. Science 300:1291–1295

Crenshaw DG, Yang J, Means AR, Kornbluth S (1998) The mitotic peptidyl-prolyl isomerase, Pin1, interacts with Cdc25 and Plx1. EMBO J 17:1315–1327

Czar MJ, Owensgrillo JK, Yem AW, Leach KL, Deibel MR, Welsh MJ, Pratt WB (1994) The Hsp56 immunophilin component of untransformed steroid receptor complexes is localized both to microtubules in the cytoplasm and to the same nonrandom regions within the nucleus as the steroid receptor. Mol Endocrinol 8:1731–1741

Davies TH, Ning YM, Sanchez ER (2002) A new first step in activation of steroid receptors—hormone-induced switching of FKBP51 and FKBP52 immunophilins. J Biol Chem 277:4597–4600

Dawson TM, Steiner JP, Dawson VL, Dinerman JL, Uhl GR, Snyder SH (1993) Immunosuppressant Fk506 enhances phosphorylation of nitric oxide synthase and protects against glutamate neurotoxicity. Proc Natl Acad Sci U S A 90:9808–9812

De Ceuninck F, Allain F, Caliez A, Spik G, Vanhoutte PM (2003) High binding capacity of cyclophilin B to chondrocyte heparan sulfate proteoglycans and its release from the cell surface by matrix metalloproteinases: possible role as a proinflammatory mediator in arthritis. Arthritis Rheumatism 48:2197–2206

Denny WB, Valentine DL, Reynolds PD, Smith DF, Scammell JG (2000) Squirrel monkey immunophilin FKBP51 is a potent inhibitor of glucocorticoid receptor binding. Endocrinology 141:4107–4113

Dobson CM, Sali A, Karplus M (1998) Protein folding—a perspective from theory and experiment. Angewandte Chemie International Edition in English 37:868–893

Doi M, Yano M, Kobayashi S, Kohno M, Tokuhisa T, Okuda S, Suetsugu M, Hisamatsu Y, Ohkusa T, Matsuzaki M (2002) Propranolol prevents the development of heart failure by restoring FKBP12.6-mediated stabilization of ryanodine receptor. Circulation 105:1374–1379

Dolinski K, Muir S, Cardenas M, Heitman J (1997) All cyclophilins and Fk506 binding proteins are, individually and collectively, dispensable for viability in Saccharomyces cerevisiae. Proc Natl Acad Sci U S A 94:13093–13098

Dolinski KJ, Heitman J (1999) Hmo1p, a high mobility group 1/2 homolog, genetically and physically interacts with the yeast FKBP12 prolyl isomerase. Genetics 151:935–944

Dubourg B, Kamphausen T, Weiwad M, Jahreis G, Feunteun J, Fischer G, Modjtahedi N (2004) The human nuclear SRcyp is a cell cycle-regulated cyclophilin. J Biol Chem 279:22322–22330

Edinger AL, Linardic CM, Chiang GG, Thompson CB, Abraham RT (2003) Differential effects of rapamycin on mammalian target of rapamycin signaling functions in mammalian cells. Cancer Res 63:8451–8460

Edlich F, Weiwad M, Erdmann F, Fanghanel J, Jarczowski F, Rahfeld JU, Fischer G (2005) Bcl-2 regulator FKBP38 is activated by Ca^{2+}/calmodulin. EMBO J (in press)

Elit L (2002) CCI-779 Wyeth. Curr Opin Invest Drugs 3:1249–1253

Ellis PJ, Carlow CK, Ma D, Kuhn P (2000) Crystal structure of the complex of brugia malayi cyclophilin and cyclosporin A. Biochemistry 39:592–598

Emborg ME, Shin P, Roitberg B, Sramek JG, Chu Y, Stebbins GT, Hamilton JS, Suzdak PD, Steiner JP, Kordower JH (2001) Systemic administration of the immunophilin ligand GPI 1046 in MPTP-treated monkeys. Exp Neurol 168:171–182

Endrich MM, Gehrig P, Gehring H (1999) Maturation-induced conformational changes of HIV-1 capsid protein and identification of two high affinity sites for cyclophilins in the C-terminal domain. J Biol Chem 274:5326–5332

Fanghanel J, Fischer G (2004) Insights into the catalytic mechanism of peptidyl prolyl cis/trans isomerases. Front Biosci 9:3453–3478

Faulds D, Goa KL, Benfield P (1993) Cyclosporin. A review of its pharmacodynamic and pharmacokinetic properties, and therapeutic use in immunoregulatory disorders. Drugs 45:953–1040

Fingar DC, Blenis J (2004) Target of rapamycin (TOR): an integrator of nutrient and growth factor signals and coordinator of cell growth and cell cycle progression. Oncogene 23:3151–3171

Fischer G (1994) Peptidyl-prolyl cis/trans isomerases and their effectors. Angew Chem International Edition in English. 33:1415–1436

Fischer G, Aumuller T (2003) Regulation of peptide bond cis/trans isomerization by enzyme catalysis and its implication in physiological processes. Rev Physiol Biochem Pharmacol 148:105–150

Fischer G, Bang H, Mech C (1984) Nachweis einer Enzymkatalyse für die cis-trans-Isomerisierung der Peptidbindung in prolinhaltigen Peptiden. Biomed Biochim Acta 43:1101–1111

Fischer G, Wittmann-Liebold B, Lang K, Kiefhaber T, Schmid FX (1989) Cyclophilin and peptidyl-prolyl cis-trans isomerase are probably identical proteins. Nature 337:476–478

Fluckiger S, Fijten H, Whitley P, Blaser K, Crameri R (2002) Cyclophilins, a new family of cross-reactive allergens. Eur J Immunol 32:10–17

Fong S, Mounkes L, Liu Y, Maibaum M, Alonzo E, Desprez PY, Thor AD, Kashani-Sabet M, Debs RJ (2003) Functional identification of distinct sets of antitumor activities mediated by the FKBP gene family. Proc Natl Acad Sci U S A 100:14253–14258

Franke EK, Yuan HEH, Luban J (1994) Specific incorporation of cyclophilin A into HIV-1 virions. Nature 372:359–362

Friedman J, Weissman I (1991) Two cytoplasmic candidates for immunophilin action are revealed by affinity for a new cyclophilin: one in the presence and one in the absence of CsA. Cell 66:799–806

Fujiyama S, Yanagida M, Hayano T, Miura Y, Isobe T, Fujimori F, Uchida T, Takahashi N (2002) Isolation and proteomic characterization of human Parvulin-associating preribosomal ribonucleoprotein complexes. J Biol Chem 277:23773–23780

Funauchi M, Haruta H, Tsumoto T (1994) Effects of an inhibitor for calcium/calmodulin-dependent protein phosphatase, calcineurin, on induction of long-term potentiation in rat visual cortex. Neurosci Res 19:269–278

Galat A (2004a) Function-dependent clustering of orthologues and paralogues of cyclophilins. Proteins 56:808–820

Galat A (2004b) A note on clustering the functionally-related paralogues and orthologues of proteins: a case of the FK506-binding proteins (FKBPs). Comput Biol Chem 28:129–140

Galigniana MD, Harrell JM, O'Hagen HM, Ljungman M, Pratt WB (2004) Hsp90-binding immunophilins link p53 to dynein during p53 transport to the nucleus. J Biol Chem 279:22483–22489

Gamble TR, Vajdos FF, Yoo SH, Worthylake DK, Houseweart M, Sundquist WI, Hill CP (1996) Crystal structure of human cyclophilin A bound to the amino-terminal domain of HIV-1 capsid. Cell 87:1285–1294

Giardina SL, Anderson SK, Sayers TJ, Chambers WH, Palumbo GA, Young HA, Ortaldo JR (1995) Selective loss of NK cytotoxicity in antisense Nk-Tr1 rat Lgl cell lines—abrogation of antibody-independent tumor and virus-infected target cell killing. J Immunol 154:80–87

Gold BG (1997) FK506 and the role of immunophilins in nerve regeneration. Mol Neurobiol 15:285–306

Gold BG, Densmore V, Shou W, Matzuk MM, Gordon HS (1999) Immunophilin FK506-binding protein 52 (not FK506-binding protein 12) mediates the neurotrophic action of FK506. J Pharmacol Exp Therap 289:1202–1210

Gold BG, Voda J, Yu XL, McKeon G, Bourdette DN (2004) FK506 and a nonimmunosuppressant derivative reduce axonal and myelin damage in experimental autoimmune encephalomyelitis: neuroimmunophilin ligand-mediated neuroprotection in a model of multiple sclerosis. J Neurosci Res 77:367–377

Gremese E, Ferraccioli GF (2004) Benefit/risk of cyclosporine in rheumatoid arthritis. Clin Exp Rheumatol 22:S101–S107

Griffith JP, Kim JL, Kim EE, Sintchak MD, Thomson JA, Fitzgibbon MJ, Fleming MA, Caron PR, Hsiao K, Navia MA (1995) X-ray structure of calcineurin inhibited by the immunophilin-immunosuppressant FKBP12-FK506 complex. Cell 82:507–522

Griffiths EJ, Halestrap AP (1995) Mitochondrial non-specific pores remain closed during cardiac ischaemia, but open upon reperfusion. Biochem J 307:93–98

Grzmil M, Voigt S, Thelen P, Hemmerlein B, Helmke K, Burfeind P (2004) Up-regulated expression of the MAT-8 gene in prostate cancer and its siRNA-mediated inhibition of expression induces a decrease in proliferation of human prostate carcinoma cells. Int J Oncol 24:97–105

Halestrap AP, Brenner C (2003) The adenine nucleotide translocase: a central component of the mitochondrial permeability transition pore and key player in cell death. Curr Med Chem 10:1507–1525

Hamdane M, Smet C, Sambo AV, Leroy A, Wieruszeski JM, Delobel P, Maurage CA, Ghestem A, Wintjens R, Begard S, Sergeant N, Delacourte A, Horvath D, Landrieu I, Lippens G, Buee L (2002) Pin1: a therapeutic target in Alzheimer neurodegeneration. J Mol Neurosci 19:275–287

Hamilton GS, Huang W, Connolly MA, Ross DT, Guo H, Valentine HL, Suzdak PD, Steiner JP (1997) Fkbp12-binding domain analogues of Fk506 are potent, nonimmunosuppressive neurotrophic agents in vitro and promote recovery in a mouse model of Parkinson's disease. Bioorg Med Chem Lett 7:1785–1790

Hani J, Schelbert B, Bernhardt A, Domdey H, Fischer G, Wiebauer K, Rahfeld JU (1999) Mutations in a peptidylprolyl-cis/trans-isomerase gene lead to a defect in 3'-end formation of a pre-mRNA in Saccharomyces cerevisiae. J Biol Chem 274:108–116

Hansson MJ, Mattiasson G, Mansson R, Karlsson J, Keep MF, Waldmeier P, Ruegg UT, Dumont JM, Besseghir K, Elmer E (2004) The nonimmunosuppressive cyclosporin analogs NIM811 and UNIL025 display nanomolar potencies on permeability transition in brain-derived mitochondria. J Bioenergetics Biomembranes 36:407–413

Helbig JH, Konig B, Knospe H, Bubert B, Yu C, Luck CP, Riboldi-Tunnicliffe A, Hilgenfeld R, Jacobs E, Hacker J, Fischer G (2003) The PPIase active site of Legionella pneumophila Mip protein is involved in the infection of eukaryotic host cells. Biol Chem 384:125–137

Hidalgo M, Rowinsky EK (2000) The rapamycin-sensitive signal transduction pathway as a target for cancer therapy. Oncogene 19:6680–6686

Hodgkiss JP, Kelly JS (1995) Only 'de novo' long-term depression (LTD) in the rat hippocampus in vitro is blocked by the same low concentration of FK506 that blocks LTD in the visual cortex. Brain Res 705:241–246

Hojo M, Morimoto T, Maluccio M, Asano T, Morimoto K, Lagman M, Shimbo T, Suthanthiran M (1999) Cyclosporine induces cancer progression by a cell-autonomous mechanism. Nature 397:530–534

Holzer M, Gartner U, Stobe A, Hartig W, Gruschka H, Bruckner MK, Arendt T (2002) Inverse association of Pin1 and tau accumulation in Alzheimer's disease hippocampus. Acta Neuropathol (Berl) 104:471–481

Hong F, Lee J, Song JW, Lee SJ, Ahn H, Cho JJ, Ha J, Kim SS (2002) Cyclosporin A blocks muscle differentiation by inducing oxidative stress and inhibiting the peptidyl-prolyl-cis-trans isomerase activity of cyclophilin A: cyclophilin A protects myoblasts from cyclosporin A-induced cytotoxicity. FASEB J 16:1633–1635

Howard BR, Vajdos FF, Li S, Sundquist WI, Hill CP (2003) Structural insights into the catalytic mechanism of cyclophilin A. Nat Struct Biol 10:475–481

Huang S, Houghton PJ (2001) Mechanisms of resistance to rapamycins. Drug Resistance Updates 4:378–391

Huang S, Liu LN, Hosoi H, Dilling MB, Shikata T, Houghton PJ (2001) p53/p21(CIP1) cooperate in enforcing rapamycin-induced G(1) arrest and determine the cellular response to rapamycin. Cancer Res 61:3373–3381

Hubler TR, Denny WB, Valentine DL, Cheung-Flynn J, Smith DF, Scammell JG (2003) The FK506-binding immunophilin FKBP51 is transcriptionally regulated by progestin and attenuates progestin responsiveness. Endocrinology 144:2380–2387

Hultsch T, Brand P, Lohmann S, Saloga J, Kincaid RL, Knop J (1998a) Direct evidence that FK506 inhibition of FcepsilonRI-mediated exocytosis from RBL mast cells involves calcineurin. Arch Dermatol Res 290:258–263

Hultsch T, Muller KD, Meingassner JG, Grassberger M, Schopf RE, Knop J (1998b) Ascomycin macrolactam derivative SDZ ASM 981 inhibits the release of granule-associated mediators and of newly synthesized cytokines in RBL 2H3 mast cells in an immunophilin-dependent manner. Arch Dermatol Res 290:501–507

Huse M, Chen YG, Massague J, Kuriyan J (1999) Crystal structure of the cytoplasmic domain of the type I TGF beta receptor in complex with FKBP12. Cell 96:425–436

Huss R, Hoy CA, Ottinger H, Grosse-Wilde H, Deeg HJ (1995) Cyclosporine-induced apoptosis in CD4+ T lymphocytes and computer-simulated analysis: modeling a treatment scenario for HIV infection. Res Immunol 146:101–108

Ikeda Y, Ylinen LM, Kahar-Bador M, Towers GJ (2004) Influence of gag on human immunodeficiency virus type 1 species-specific tropism. J Virol 78:11816–11822

Iwasaki Y, Ichikawa Y, Igarashi O, Iwamoto K, Kinoshitata M, Ikeda K (2002) Neuroprotective actions of FK506 and cyclosporin A on motor neuron survival following neonatal axotomy. Neurol Res 24:573–576

Jayaraman T, Brillantes AM, Timerman AP, Fleischer S, Erdjument-Bromage H, Tempst P, Marks AR (1992) FK506 binding protein associated with the calcium release channel (ryanodine receptor). J Biol Chem 267:9474–9477

Jeyakumar LH, Ballester L, Cheng DS, McIntyre JO, Chang P, Olivey HE, Rollins-Smith L, Barnett JV, Murray K, Xin HB, Fleischer S (2001) FKBP binding characteristics of cardiac microsomes from diverse vertebrates. Biochem Biophys Res Commun 281:979–986

Jin ZG, Lungu AO, Xie L, Wang M, Wong C, Berk BC (2004) Cyclophilin A is a proinflammatory cytokine that activates endothelial cells. Arterioscler, Thromb Vasc Biol 24:1186–1191

Jin ZG, Melaragno MG, Liao DF, Yan C, Haendeler J, Suh YA, Lambeth JD, Berk BC (2000) Cyclophilin A is a secreted growth factor induced by oxidative stress. Circ Res 87:789–796

Kahan BD (2004) Sirolimus: a ten-year perspective. Transplant Proc 36:71–75

Karlsson J, Fong KS, Hansson MJ, Elmer E, Csiszar K, Keep MF (2004) Life span extension and reduced neuronal death after weekly intraventricular cyclosporin injections in the G93A transgenic mouse model of amyotrophic lateral sclerosis. J Neurosurg 101:128–137

Kazlauskas A, Poellinger L, Pongratz I (2000) The immunophilin-like protein XAP2 regulates ubiquitination and subcellular localization of the dioxin receptor. J Biol Chem 275:41317–41324

Khan LN, Kon OM, Macfarlane AJ, Meng Q, Ying S, Barnes NC, Kay AB (2000) Attenuation of the allergen-induced late asthmatic reaction by cyclosporin A is associated with inhibition of bronchial eosinophils, interleukin-5, granulocyte macrophage colony-stimulating factor, and eotaxin. Am J Resp Crit Care Med 162:1377–1382

Khatua S, Peterson KM, Brown KM, Lawlor C, Santi MR, LaFleur B, Dressman D, Stephan DA, MacDonald TJ (2003) Overexpression of the EGFR/FKBP12/HIF-2alpha pathway identified in childhood astrocytomas by angiogenesis gene profiling. Cancer Res 63:1865–1870

Kim SH, Lessner SM, Sakurai Y, Galis ZS (2004) Cyclophilin A as a novel biphasic mediator of endothelial activation and dysfunction. Am J Pathol 164:1567–1574

Kohno M, Yano M, Kobayashi S, Doi M, Oda T, Tokuhisa T, Okuda S, Ohkusa T, Matsuzaki M (2003) A new cardioprotective agent, JTV519, improves defective channel gating of ryanodine receptor in heart failure. Am J Physiol Heart Circ Physiol 284:H1035–H1042

Kon OM, Kay AB (1999) Anti-T cell strategies in asthma. Inflamm Res 48:516–523

Kumar P, Mark PJ, Ward BK, Minchin RF, Ratajczak T (2001) Estradiol-regulated expression of the immunophilins cyclophilin 40 and FKBP52 in MCF-7 breast cancer cells. Biochem Biophys Res Commun 284:219–225

Kurek I, Dulberger R, Azem A, Tzvi BB, Sudhakar D, Christou P, Breiman A (2002) Deletion of the C-terminal 138 amino acids of the wheat FKBP73 abrogates calmodulin binding, dimerization and male fertility in transgenic rice. Plant Mol Biol 48:369–381

Landrieu I, Odaert B, Wieruszeski JM, Drobecq H, Rousselot-Pailley P, Inze D, Lippens G (2001) p13(SUC1) and the WW domain of PIN1 bind to the same phosphothreonine-proline epitope. J Biol Chem 276:1434–1438

LeDucq N, Delmas-Beauvieux MC, Bourdel-Marchasson I, Dufour S, Gallis JL, Canioni P, Diolez P (1998) Mitochondrial permeability transition during hypothermic to normothermic reperfusion in rat liver demonstrated by the protective effect of cyclosporin A. Biochem J 336:501–506

Lee JP, Palfrey HC, Bindokas VP, Ghadge GD, Ma L, Miller RJ, Roos RP (1999) The role of immunophilins in mutant superoxide dismutase-1linked familial amyotrophic lateral sclerosis. Proc Natl Acad Sci U S A 96:3251–3256

Lee YS, Marcu MG, Neckers L (2004) Quantum chemical calculations and mutational analysis suggest heat shock protein 90 catalyzes trans-cis isomerization of geldanamycin. Chem Biol 11:991–998

Lehnart SE, Huang F, Marx SO, Marks AR (2003) Immunophilins and coupled gating of ryanodine receptors. Curr Topics Med Chem 3:1383–1391

Lehnart SE, Wehrens XH, Kushnir A, Marks AR (2004) Cardiac ryanodine receptor function and regulation in heart disease. Ann N Y Acad Sci 1015:144–159

Li YM, Johnson N, Capano M, Edwards M, Crompton M (2004) Cyclophilin-D promotes the mitochondrial permeability transition but has opposite effects on apoptosis and necrosis. Biochem J 383:101–109

Lim SO, Park SJ, Kim W, Park SG, Kim HJ, Kim YI, Sohn TS, Noh JH, Jung G (2002) Proteome analysis of hepatocellular carcinoma. Biochem Biophys Res Commun 291:1031–1037

Liou YC, Ryo A, Huang HK, Lu PJ, Bronson R, Fujimori F, Uchida T, Hunter T, Lu KP (2002) Loss of Pin1 function in the mouse causes phenotypes resembling cyclin D1-null phenotypes. Proc Natl Acad Sci U S A 99:1335–1340

Liu J, Farmer JD Jr, Lane WS, Friedman J, Weissman I, Schreiber SL (1991) Calcineurin is a common target of cyclophilin-cyclosporin A and FKBP-FK506 complexes. Cell 66:807–815

Liu SK, Asparuhova M, Brondani V, Ziekau I, Klimkait T, Schumperli D (2004) Inhibition of HIV-1 multiplication by antisense U7 snRNAs and siRNAs targeting cyclophilin A. Nucleic Acids Res 32:3752–3759

London NJ, Farmery SM, Will EJ, Davison AM, Lodge JPA (1995) Risk of neoplasia in renal transplant patients. Lancet 346:403–406

Loor F, Tiberghien F, Wenandy T, Didier A, Traber R (2002) Cyclosporins: structure-activity relationships for the inhibition of the human MDR1 P-glycoprotein ABC transporter. J Med Chem 45:4598–4612

Lopez-Ilasaca M, Schiene C, Kullertz G, Tradler T, Fischer G, Wetzker R (1998) Effects of Fk506-binding protein 12 and Fk506 on autophosphorylation of epidermal growth factor receptor. J Biol Chem 273:9430–9434

Lu KP (2004) Pinning down cell signaling, cancer and Alzheimer's disease. Trends Biochem Sci 29:200–209

Lu KP, Hanes SD, Hunter T (1996) A human peptidyl-prolyl isomerase essential for regulation of mitosis. Nature 380:544–547

Lu KP, Liou YC, Zhou XZ (2002) Pinning down proline-directed phosphorylation signaling. Trends Cell Biol 12:164–172

Lu PJ, Zhou XZ, Shen M, Lu KP (1999) Function of WW domains as phosphoserine- or phosphothreonine-binding modules. Science 283:1325–1328

Luban J, Bossolt KL, Franke EK, Kalpana GV, Goff SP (1993) Human immunodeficiency virus type-1 gag protein binds to cyclophilin-a and cyclophilin-B. Cell 73:1067–1078

Lundemose AG, Rouch DA, Penn CW, Pearce JH (1993) The chlamydia-trachomatis Mip-like protein is a lipoprotein. J Bacteriol 175:3669–3671

Lyons WE, George EB, Dawson TM, Steiner JP, Snyder SH (1994) Immunosuppressant FK506 promotes neurite outgrowth in cultures of PC12 cells and sensory ganglia. Proc Natl Acad Sci U S A 91:3191–3195

Ma Q, Whitlock JP (1997) A novel cytoplasmic protein that interacts with the Ah Receptor, contains tetratricopeptide repeat motifs, and augments the transcriptional response to 2,3,7,8-tetrachlorodibenzo-P-dioxin. J Biol Chem 272:8878–8884

Mallis RJ, Brazin KN, Fulton DB, Andreotti AH (2002) Structural characterization of a proline-driven conformational switch within the Itk SH2 domain. Nat Struct Biol 9:900–905

Manabe Y, Warita H, Murakami T, Shiote M, Hayashi T, Omori N, Nagano I, Shoji M, Abe K (2002) Early decrease of the immunophilin FKBP52 in the spinal cord of a transgenic model for amyotrophic lateral sclerosis. Brain Res 935:124–128

Mantovani F, Piazza S, Gostissa M, Strano S, Zacchi P, Mantovani R, Blandino G, Del Sal G (2004) Pin1 links the activities of c-Abl and p300 in regulating p73 function. Mol Cell 14:625–636

Marks AR (1996) Cellular function of immunophilins. Physiol Rev 76:631–649

Marks AR, Marx SO, Reiken S (2002) Regulation of ryanodine receptors via macromolecular complexes: a novel role for leucine/isoleucine zippers. Trends Cardiovasc Med 12:166–170

Marsland AM, Griffiths CEM (2004) Therapeutic potential of macrolicle immunosuppressants in dermatology. Expert Opin Invest Drugs 13:125–137

McBride MJ, Braun TF (2004) GldI is a lipoprotein that is required for Flavobacterium johnsoniae gliding motility and chitin utilization. J Bacteriol 186:2295–2302

McLauchlan PE, Roberts HC, Chappell LH (2000) Mode of action of cyclosporin A against Hymenolepis microstoma (Cestoda): relationship between cyclophilin binding and drug-induced damage. Parasitology 121:661–670

Meng X, Lu X, Morris CA, Keating MT (1998) A novel human gene FKBP6 is deleted in Williams syndrome. Genomics 52:130–137

Meyer BK, Praygrant MG, Vandenheuvel JP, Perdew GH (1998) Hepatitis B virus X-associated protein 2 is a subunit of the unliganded aryl hydrocarbon receptor core complex and exhibits transcriptional enhancer activity. Mol Cell Biol 18:978–988

Meza-Zepeda LA, Forus A, Lygren B, Dahlberg AB, Godager LH, South AP, Marenholz I, Lioumi M, Florenes VA, Maelandsmo GM, Serra M, Mischke D, Nizetic D, Ragoussis J, Tarkkanen M, Nesland JM, Knuutila S, Myklebost O (2002) Positional cloning identifies a novel cyclophilin as a candidate amplified oncogene in 1q21. Oncogene 21:2261–2269

Mihatsch MJ, Kyo M, Morozumi K, Yamaguchi Y, Nickeleit V, Ryffel B (1998) The side-effects of cyclosporine-A and tacrolimus. Clin Nephrol 49:356–363

Miller RT, Anderson SP, Corton JC, Cattley RC (2000) Apoptosis, mitosis and cyclophilin-40 expression in regressing peroxisome proliferator-induced adenomas. Carcinogenesis 21:647–652

Misumi S, Fuchigami T, Takamune N, Takahashi I, Takama M, Shoji S (2002) Three isoforms of cyclophilin A associated with human immunodeficiency virus type 1 were found by proteomics by using two-dimensional gel electrophoresis and matrix-assisted laser desorption ionization-time of flight mass spectrometry. J Virol 76:10000–10008

Miyashita H, Mori S, Motegi K, Fukumoto M, Uchida T (2003) Pin1 is overexpressed in oral squamous cell carcinoma and its levels correlate with cyclin D1 overexpression. Oncol Rep 10:455–461

Moro A, Ruizcabello F, Fernandezcano A, Stock RP, Gonzalez A (1995) Secretion by Trypanosoma cruzi of a peptidyl-prolyl cis-trans isomerase involved in cell infection. EMBO J 14:2483–2490

Morris CA, Mervis CB, Hobart HH, Gregg RG, Bertrand J, Ensing GJ, Sommer A, Moore CA, Hopkin RJ, Spallone PA, Keating MT, Osborne L, Kimberley KW, Stock AD (2003) GTF2I hemizygosity implicated in mental retardation in Williams syndrome: genotype-phenotype analysis of five families with deletions in the Williams syndrome region. Am J Med Genet 123A:45–59

Navia MA (1996) Protein-drug complexes important for immunoregulation and organ transplantation. Curr Opin Struct Biol 6:838–847

Nilsson M, Meza-Zepeda LA, Mertens F, Forus A, Myklebost O, Mandahl N (2004) Amplification of chromosome 1 sequences in lipomatous tumors and other sarcomas. Int J Cancer 109:363–369

Noguchi N, Takasawa S, Nata K, Tohgo A, Kato I, Ikehata F, Yonekura H, Okamoto H (1997) Cyclic Adp-ribose binds to Fk506-binding protein 12.6 to release Ca^{2+} from islet microsomes. J Biol Chem 272:3133–3136

Oshiro N, Yoshino K, Hidayat S, Tokunaga C, Hara K, Eguchi S, Avruch J, Yonezawa K (2004) Dissociation of raptor from mTOR is a mechanism of rapamycin-induced inhibition of mTOR function. Genes Cells 9:359–366

Panhans-Gross A, Novak N, Kraft S, Bieber T (2001) Human epidermal Langerhans' cells are targets for the immunosuppressive macrolide tacrolimus (FK506). J Allergy Clin Immunol 107:345–352

Parker EM, Monopoli A, Ongini E, Lozza G, Babij CM (2000) Rapamycin, but not FK506 and GPI-1046, increases neurite outgrowth in PC12 cells by inhibiting cell cycle progression. Neuropharmacology 39:1913–1919

Patra D, Wang SX, Kumagai A, Dunphy WG (1999) The xenopus Suc1/Cks protein promotes the phosphorylation of G(2)/M regulators. J Biol Chem 274:36839–36842

Pereira PJB, Vega MC, Gonzalez-Rey E, Fernandez-Carazo R, Macedo-Ribeiro S, Gomis-Ruth FX, Gonzalez A, Coll M (2002) Trypanosoma cruzi macrophage infectivity potentiator has a rotamase core and a highly exposed alpha-helix. EMBO Rep 3:88–94

Perron MJ, Stremlau M, Song B, Ulm W, Mulligan RC, Sodroski J (2004) TRIM5 alpha mediates the postentry block to N-tropic murine leukemia viruses in human cells. Proc Natl Acad Sci U S A 101:11827–11832

Petrulis JR, Kusnadi A, Ramadoss P, Hollingshead B, Perdew GH (2003) The hsp90 co-chaperone XAP2 alters importin beta recognition of the bipartite nuclear localization signal of the Ah receptor and represses transcriptional activity. J Biol Chem 278:2677–2685

Petrulis JR, Perdew GH (2002) The role of chaperone proteins in the aryl hydrocarbon receptor core complex. Chem Biol Interact 141:25–40

Pollard S (2004) Calcineurin inhibition and disease recurrence in the hepatitis C virus-positive liver transplant recipient. Liver Int 24:402–406

Pratt WB, Silverstein AM, Galigniana MD (1999) A model for the cytoplasmic trafficking of signalling proteins involving the hsp90-binding immunophilins and p50(cdc37). Cell Signal 11:839–851

Prudhomme GJ, Vanier LE, Bocarro DC, Stecroix H (1995) Effects of cyclosporin A, rapamycin, and Fk520 on peripheral T-cell deletion and anergy. Cell Immunol 164:47–56

Rahfeld JU, Rucknagel KP, Schelbert B, Ludwig B, Hacker J, Mann K, Fischer G (1994) Confirmation of the existence of a third family among peptidyl-prolyl cis/trans isomerases. Amino acid sequence and recombinant production of parvulin. FEBS Lett 352:180–184

Ramamurthy V, Roberts M, van den Akker F, Niemi G, Reh TA, Hurley JB (2003) AIPL1, a protein implicated in Leber's congenital amaurosis, interacts with and aids in processing of farnesylated proteins. Proc Natl Acad Sci U S A 100:12630–12635

Ranganathan R, Lu KP, Hunter T, Noel JP (1997) Structural and functional analysis of the mitotic rotamase Pin1 suggests substrate recognition is phosphorylation dependent. Cell 89:875–886

Reiken S, Lacampagne A, Zhou H, Kherani A, Lehnart SE, Ward C, Huang F, Gaburjakova M, Gaburjakova J, Rosemblit N, Warren MS, He KL, Yi GH, Wang J, Burkhoff D, Vassort G, Marks AR (2003a) PKA phosphorylation activates the calcium release channel (ryanodine receptor) in skeletal muscle: defective regulation in heart failure. J Cell Biol 160:919–928

Reiken S, Wehrens XHT, Vest JA, Barbone A, Klotz S, Mancini D, Burkhoff D, Marks AR (2003b) beta-blockers restore calcium release channel function and improve cardiac muscle performance in human heart failure. Circulation 107:2459–2466

Renoir JM, Mercierbodard C, Hoffmann K, Lebihan S, Ning YM, Sanchez ER, Handschumacher RE, Baulieu EE (1995) Cyclosporin a potentiates the dexamethasone-induced mouse mammary tumor virus-chloramphenicol acetyltransferase activity in Lmcat cells—a possible role for different heat shock protein-binding immunophilins in glucocorticosteroid receptor-mediated gene expression. Proc Natl Acad Sci U S A 92:4977–4981

Rey O, Baluda MA, Park NH (1999) Differential gene expression in neoplastic and human papillomavirus-immortalized oral keratinocytes. Oncogene 18:827–831

Riboldi-Tunnicliffe A, Konig B, Jessen S, Weiss MS, Rahfeld J, Hacker J, Fischer G, Hilgenfeld R (2001) Crystal structure of Mip, a prolylisomerase from Legionella pneumophila. Nat Struct Biol 8:779–783

Riggs DL, Roberts PJ, Chirillo SC, Cheung-Flynn J, Prapapanich V, Ratajczak T, Gaber R, Picard D, Smith DF (2003) The Hsp90-binding peptidylprolyl isomerase FKBP52 potentiates glucocorticoid signaling in vivo. EMBO J 22:1158–1167

Rippmann JF, Hobbie S, Daiber C, Guilliard B, Bauer M, Birk J, Nar H, Garin-Chesa P, Rettig WJ, Schnapp A (2000) Phosphorylation-dependent proline isomerization catalyzed by Pin1 is essential for tumor cell survival and entry into mitosis. Cell Growth Differ 11:409–416

Rizzardi GP, Harari A, Capiluppi B, Tambussi G, Ellefsen K, Ciuffreda D, Champagne P, Bart PA, Chave JP, Lazzarin A, Pantaleo G (2002) Treatment of primary HIV-1 infection with cyclosporin A coupled with highly active antiretroviral therapy. J Clin Invest 109:681–688

Rosner M, Solberg Y, Turetz J, Belkin M (1997) Neuroprotective therapy for Argon-LASER induced retinal injury. Exp Eye Res 65:485–495

Rothbard JB, Garlington S, Lin Q, Kirschberg T, Kreider E, McGrane PL, Wender PA, Khavari PA (2000) Conjugation of arginine oligomers to cyclosporin A facilitates topical delivery and inhibition of inflammation. Nat Med 6:1253–1257

Ryffel B, Foxwell BM, Gee A, Greiner B, Woerly G, Mihatsch MJ (1988) Cyclosporine—relationship of side effects to mode of action. Transplantation 46:90S–96S

Sabers CJ, Martin MM, Brunn GJ, Williams JM, Dumont FJ, Wiederrecht G, Abraham RT (1995) Isolation of a protein target of the Fkbp12-rapamycin complex in mammalian cells. J Biol Chem 270:815–822

Saphire ACS, Bobardt MD, Gallay PA (1999) Host cyclophilin A mediates HIV-1 attachment to target cells via heparans. EMBO J 18:6771–6785

Saphire ACS, Bobardt MD, Gallay PA (2002) A trans-complementation rescue of cyclophilin A-deficient viruses reveals that the requirement for cyclophilin A in human immunodeficiency virus type 1 replication is independent of its isomerase activity. J Virol 76:2255–2262

Sayah DM, Sokolskaja E, Berthoux L, Luban J (2004) Cyclophilin A retrotransposition into TRIM5 explains owl monkey resistance to HIV-1. Nature 430:569–573

Schiene-Fischer C, Habazettl J, Schmid FX, Fischer G (2002) The hsp70 chaperone DnaK is a secondary amide peptide bond cis-trans isomerase. Nat Struct Biol 9:419–424

Schiene-Fischer C, Yu C (2001) Receptor accessory folding helper enzymes: the functional role of peptidyl prolyl cis/trans isomerases. FEBS Lett 495:1–6

Schreier MH, Baumann G, Zenke G (1993) Inhibition of T-cell signaling pathways by immunophilin drug complexes: are side effects inherent to immunosuppressive properties? Transplant Proc 25:502–507

Schubert A, Grimm S (2004) Cyclophilin D, a component of the permeability transition-pore, is an apoptosis repressor. Cancer Res 64:85–93

Seko Y, Fujimura T, Taka H, Mineki R, Murayama K, Nagai R (2004) Hypoxia followed by reoxygenation induces secretion of cyclophilin A from cultured rat cardiac myocytes. Biochem Biophys Res Commun 317:162–168

Sezen SF, Hoke A, Burnett AL, Snyder SH (2001) Immunophilin ligand FK506 is neuroprotective for penile innervation. Nat Med 7:1073–1074

Sharkey J, Butcher SP (1994) Immunophilins mediate the neuroprotective effects of Fk506 in focal cerebral ischaemia. Nature 371:336–339

Shichinohe H, Kuroda S, Abumiya T, Ikeda J, Kobayashi T, Yoshimoto T, Iwasaki Y (2004) FK506 reduces infarct volume due to permanent focal cerebral ischemia by maintaining BAD turnover and inhibiting cytochrome c release. Brain Res 1001:51–59

Shinkura N, Ikai I, Yamauchi A, Hirose T, Kawai Y, Inamoto T, Ozaki S, Iwai M, Bona C, Yamaoka Y (1999) Autoantibodies to FK506 binding protein 12 (FKBP12) in autoimmune diseases. Autoimmunity 29:159–170

Shirane M, Nakayama KI (2003) Inherent calcineurin inhibitor FKBP38 targets Bcl-2 to mitochondria and inhibits apoptosis. Nat Cell Biol 5:28–37

Silverstein AM, Galigniana MD, Chen MS, Owens-Grillo JK, Chinkers M, Pratt WB (1997) Protein phosphatase 5 is a major component of glucocorticoid receptor.hsp90 complexes with properties of an FK506-binding immunophilin. J Biol Chem 272:16224–16230

Sohocki MM, Bowne SJ, Sullivan LS, Blackshaw S, Cepko CL, Payne AM, Bhattacharya SS, Khaliq S, Qasim Mehdi S, Birch DG, Harrison WR, Elder FF, Heckenlively JR, Daiger SP (2000) Mutations in a new photoreceptor-pineal gene on 17p cause Leber congenital amaurosis. Nat Genet 24:79–83

Sperr WR, Agis H, Semper H, Valenta R, Susani M, Sperr M, Willheim M, Scheiner O, Liehl E, Lechner K, Valent P (1997) Inhibition of allergen-induced histamine release from human basophils by cyclosporine A and FK506. Int Arch Allergy Immunol 114:68–73

Steiner JP, Connolly MA, Valentine HI, Hamilton GS, Dawson TM, Hester L, Snyder SH (1997) Neurotrophic Actions of nonimmunosuppressive analogues of immunosuppressive drugs Fk506, rapamycin and cyclosporin A. Nat Med 3:421–428

Steiner JP, Dawson TM, Fotuhi M, Glatt CE, Snowman AM, Cohen N, Snyder SH (1992) High brain densities of the immunophilin FKBP colocalized with calcineurin. Nature 358:584–587

Steiner JP, Dawson TM, Fotuhi M, Snyder SH (1996) Immunophilin regulation of neurotransmitter release. Mol Med 2:325–333

Steinschulte C, Taner T, Thomson AW, Bein G, Hackstein H (2003) Cutting edge: sanglifehrin A, a novel cyclophilin-binding immunosuppressant blocks bioactive IL-12 production by human dendritic cells. J Immunol 171:542–546

Sumanasekera WK, Tien ES, Turpey R, Vanden Heuvel JP, Perdew GH (2003) Evidence that peroxisome proliferator-activated receptor alpha is complexed with the 90-kDa heat shock protein and the hepatitis virus B X-associated protein 2. J Biol Chem 278:4467–4473

Takahashi N, Hayano T, Suzuki M (1989) Peptidyl-prolyl cis-trans isomerase is the cyclosporin A-binding protein cyclophilin. Nature 337:473–475

Tanaka K, Fujita N, Higashi Y, Ogawa N (2002) Neuroprotective and antioxidant properties of FKBP-binding immunophilin ligands are independent on the FKBP12 pathway in human cells. Neurosci Lett 330:147–150

Tchaicheeyan O (2004) Is peptide bond cis/trans isomerization a key stage in the chemomechanical cycle of motor proteins? FASEB J 18:783–789

Tegeder I, Schumacher A, John S, Geiger H, Geisslinger G, Bang H, Brune K (1997) Elevated serum cyclophilin levels in patients with severe sepsis. J Clin Immunol 17:380–386

Terashima A, Taniguchi T, Nakai M, Yasuda M, Kawamata T, Tanaka C (2000) Rapamycin and FK506 induce long-term potentiation by pairing stimulation via an intracellular Ca^{2+} signaling mechanism in rat hippocampal CA1 neurons. Neuropharmacology 39:1920–1928

Thali M, Bukovsky A, Kondo E, Rosenwirth B, Walsh CT, Sodroski J, Gottlinger HG (1994) Functional association of cyclophilin a with HIV-1 virions. Nature 372:363–365

Thorpe JR, Mosaheb S, Hashemzadeh-Bonehi L, Cairns NJ, Kay JE, Morley SJ, Rulten SL (2004) Shortfalls in the peptidyl-prolyl cis-trans isomerase protein Pin1 in neurons are associated with frontotemporal dementias. Neurobiol Dis 17:237–249

Timerman AP, Wiederrecht G, Marcy A, Fleischer S (1995) Characterization of an exchange reaction between soluble FKBP-12 and the FKBP–ryanodine receptor complex. Modulation by FKBP mutants deficient in peptidyl-prolyl isomerase activity. J Biol Chem 270:2451–2459

Tiso N, Salamon M, Bagattin A, Danieli GA, Argenton F, Bortolussi M (2002) The binding of the RyR2 calcium channel to its gating protein FKBP12.6 is oppositely affected by ARVD2 and VTSIP mutations. Biochem Biophys Res Commun299:594–598

Tong G, Shepherd D, Jahr CE (1995) Synaptic desensitization of NMDA receptors by calcineurin. Science 267:1510–1512

Towers GJ, Hatziioannou T, Cowan S, Goff SP, Luban J, Bieniasz PD (2003) Cyclophilin A modulates the sensitivity of HIV-1 to host restriction factors. Nat Med 9:1138–1143

Tsai CJ, Ma BY, Nussinov R (1999) Folding and binding cascades: shifts in energy landscapes. Proc Natl Acad Sci U S A 96:9970–9972

Uchida T, Fujimori F, Tradler T, Fischer G, Rahfeld JU (1999) Identification and characterization of a 14 kDa human protein as a novel parvulin-like peptidyl prolyl cis/trans isomerase. FEBS Lett 446:278–282

Vahlne A, Larsson PA, Horal P, Ahlmen J, Svennerholm B, Gronowitz JS, Olofsson S (1992) Inhibition of herpes simplex virus production in vitro by cyclosporin A. Arch Virol 122:61–75

Van de Vrie W, Marquet RL, Eggermont AM (1997) Cyclosporin A enhances locoregional metastasis of the CC531 rat colon tumour. J Cancer Res Clin Oncol 123:21–24

Van Duyne GD, Standaert RF, Karplus PA, Schreiber SL, Clardy J (1991) Atomic structure of FKBP-FK506, an immunophilin-immunosuppressant complex. Science 252:839–842

Vogel KW, Briesewitz R, Wandless TJ, Crabtree GR (2001) Calcineurin inhibitors and the generalization of the presenting protein strategy. Adv Protein Chem 56:253–291

Waldmeier PC, Feldtrauer JJ, Qian T, Lemasters JJ (2002) Inhibition of the mitochondrial permeability transition by the nonimmunosuppressive cyclosporin derivative NIM811. Mol Pharmacol 62:22–29

Waldmeier PC, Zimmermann K, Qian T, Tintelnot-Blomley M, Lemasters JJ (2003) Cyclophilin D as a drug target. Curr Med Chem 10:1485–1506

Wang P, Cardenas ME, Cox GM, Perfect JR, Heitman J (2001) Two cyclophilin A homologs with shared and distinct functions important for growth and virulence of Cryptococcus neoformans. EMBO Rep 2:511–518

Wang TW, Donahoe PK (2004) The immunophilin FKBP12: a molecular guardian of the TGF-beta family type I receptors. Front Biosci 9:619–631

Wang TW, Donahoe PK, Zervos AS (1994) Specific interaction of type I receptors of the Tgf-beta family with the immunophilin Fkbp-12. Science 265:674–676

Ward BK, Kumar P, Turbett GR, Edmondston JE, Papadimitriou JM, Laing NG, Ingram DM, Minchin RF, Ratajczak T (2001) Allelic loss of cyclophilin 40, an estrogen receptor-associated immunophilin, in breast carcinomas. J Cancer Res Clinical Oncol 127:109–115

Ward BK, Mark PJ, Ingram DM, Minchin RF, Ratajczak T (1999) Expression of the estrogen receptor-associated immunophilins, cyclophilin 40 and FKBP52, in breast cancer. Breast Cancer Res Treat 58:267–280

Watashi K, Hijikata M, Hosaka M, Yamaji M, Shimotohno K (2003) Cyclosporin A suppresses replication of hepatitis C virus genome in cultured hepatocytes. Hepatology 38:1282–1288

Wehrens XH, Lehnart SE, Reiken SR, Deng SX, Vest JA, Cervantes D, Coromilas J, Landry DW, Marks AR (2004) Protection from cardiac arrhythmia through ryanodine receptor-stabilizing protein calstabin2. Science 304:292–296

Wehrens XHT, Lehnart SE, Huang F, Vest JA, Reiken SR, Mohler PJ, Sun J, Guatimosim S, Song LS, Rosemblit N, D'Armiento JM, Napolitano C, Memmi M, Priori SG, Lederer WJ, Marks AR (2003) FKBP12.6 deficiency and defective calcium release channel (ryanodine receptor) function linked to exercise-induced sudden cardiac death. Cell 113:829–840

Wei L, Steiner JP, Hamilton GS, Wu YQ (2004) Synthesis and neurotrophic activity of nonimmunosuppressant cyclosporin A derivatives. Bioorg Med Chem Lett 14:4549–4551

Weiwad M, Werner A, Rucknagel P, Schierhorn A, Kullertz G, Fischer G (2004) Catalysis of proline-directed protein phosphorylation by peptidyl-prolyl cis/trans isomerases. J Mol Biol 339:635–646

Woltman AM, Schlagwein N, van der Kooij SW, van Kooten C (2004) The novel cyclophilin-binding drug sanglifehrin A specifically affects antigen uptake receptor expression and endocytic capacity of human dendritic cells. J Immunol 172:6482–6489

Wu X, Wilcox CB, Devasahayam G, Hackett RL, Arevalo-Rodriguez M, Cardenas ME, Heitman J, Hanes SD (2000) The Ess1 prolyl isomerase is linked to chromatin remodeling complexes and the general transcription machinery. EMBO J 19:3727–3738

Wulf G, Garg P, Liou YC, Iglehart D, Lu KP (2004) Modeling breast cancer in vivo and ex vivo reveals an essential role of Pin1 in tumorigenesis. EMBO J 23:3397–3407

Wulf GM, Liou YC, Ryo A, Lee SW, Lu KP (2002) Role of Pin1 in the regulation of p53 stability and p21 transactivation, and cell cycle checkpoints in response to DNA damage. J Biol Chem 277:47976–47979

Xu Q, Leiva MC, Fischkoff SA, Handschumacher RE, Lyttle CR (1992) Leukocyte chemotactic activity of cyclophilin. J Biol Chem 267:11968–11971

Yaffe MB, Cantley LC (1999) Signal transduction. Grabbing phosphoproteins. Nature 402:30–31

Yao DY, Dore JJE, Leof EB (2000) FKBP12 is a negative regulator of transforming growth factor-beta receptor internalization. J Biol Chem 275:13149–13154

Yeh E, Cunningham M, Arnold H, Chasse D, Monteith T, Ivaldi G, Hahn WC, Stukenberg PT, Shenolikar S, Uchida T, Counter CM, Nevins JR, Means AR, Sears R (2004) A signalling pathway controlling c-Myc degradation that impacts oncogenic transformation of human cells. Nat Cell Biol 6:308–318

Yu G, Hess DC, Borlongan CV (2004) Combined cyclosporine-A and methylprednisolone treatment exerts partial and transient neuroprotection against ischemic stroke. Brain Res 1018:32–37

Yurchenko V, O'Connor M, Dai WW, Guo HM, Toole B, Sherry B, Bukrinsky M (2001) CD147 is a signaling receptor for cyclophilin B. Biochem Biophys Res Commun 288:786–788

Yurchenko V, Zybarth G, O'Connor M, Dai WW, Franchin G, Hao T, Guo HM, Hung HC, Toole B, Gallay P, Sherry B, Bukrinsky M (2002) Active site residues of cyclophilin A are crucial for its signaling activity via CD147. J Biol Chem 277:22959–22965

Zacchi P, Gostissa M, Uchida T, Salvagno C, Avolio F, Volinia S, Ronai Z, Blandino G, Schneider C, Del Sal G (2002) The prolyl isomerase Pin1 reveals a mechanism to control p53 functions after genotoxic insults. Nature 419:853–857

Zander K, Sherman MP, Tessmer U, Bruns K, Wray V, Prechtel AT, Schubert E, Henklein P, Luban J, Neidleman J, Greene WC, Schubert U (2003) Cyclophilin A interacts with HIV-1 Vpr and is required for its functional expression. J Biol Chem 278:43202–43213

Zarnt T, Lang K, Burtscher H, Fischer G (1995) Time-dependent inhibition of peptidylprolyl cis-trans-isomerases by Fk506 is probably due to cis-trans isomerization of the inhibitors imide bond. Biochem J 305:159–164

Zenke G, Strittmatter U, Fuchs S, Quesniaux VF, Brinkmann V, Schuler W, Zurini M, Enz A, Billich A, Sanglier JJ, Fehr T (2001) Sanglifehrin A, a novel cyclophilin-binding compound showing immunosuppressive activity with a new mechanism of action. J Immunol 166:7165–7171

Zhang LH, Youn HD, Liu JO (2001) Inhibition of cell cycle progression by the novel cyclophilin ligand sanglifehrin A is mediated through the NF kappa B-dependent activation of p53. J Biol Chem 276:43534–43540

Zhong L, Qing KY, Si Y, Chen LY, Tan MQ, Srivastava A (2004) Heat-shock treatment-mediated increase in transduction by recombinant adeno-associated virus 2 vectors is independent of the cellular heat-shock protein 90. J Biol Chem 279:12714–12723

Zhou XZ, Kops O, Werner A, Lu PJ, Shen M, Stoller G, Kullertz G, Stark M, Fischer G, Lu KP (2000) Pin1-dependent prolyl isomerization regulates dephosphorylation of Cdc25C and tau proteins. Mol Cell 6:873–883

Chemical Chaperones: Mechanisms of Action and Potential Use

E. Papp · P. Csermely (✉)

Department of Medical Chemistry, Semmelweis University, Budapest, Hungary
csermely@puskin.sote.hu

1	**Chemical Chaperones**	405
1.1	Osmolytes	406
1.2	Hydrophobic Compounds	409
1.2.1	PBA	409
1.2.2	Lipids and Detergents	410
2	**Pharmacological Chaperones**	410
2.1	Enzyme Antagonists	410
2.2	Folding Agonists	412
	References	413

Abstract An increasing number of studies indicate that low-molecular-weight compounds can help correct conformational diseases by inhibiting the aggregation or enable the mutant proteins to escape the quality control systems, and thus their function can be rescued. The small molecules were named chemical chaperones and it is thought that they nonselectively stabilize the mutant proteins and facilitate their folding. Chemical chaperones are usually osmotically active, such as DMSO, glycerol, or deuterated water, but other compounds, such as 4-phenylbutiric acid, are also members of the chemical chaperone group. More recently, compounds such as receptor ligands or enzyme inhibitors, which selectively recognize the mutant proteins, were also found to rescue conformational mutants and were termed pharmacological chaperones. An increasing amount of evidence suggests that the action of pharmacological chaperones could be generalized to a large number of misfolded proteins, representing new therapeutic possibilities for the treatment of conformational diseases. A new and exciting strategy has recently been developed, leading to the new chemical group called folding agonist. These small molecules are designed to bind proteins and thus restore their native conformation.

Keywords Chemical chaperones · Pharmacological chaperones · Conformational diseases · Protein misfolding · Quality control machinery

1
Chemical Chaperones

Chemical chaperones are small molecules with a common feature mimicking the chaperone function of molecular chaperones. Many osmolytes as well as compounds with the ability to bind to hydrophobic surfaces can rescue mutant proteins from aggregation or can help them to escape from quality control and subsequent degradation (Sitia and Braakman 2003; Conn et al. 2002). This feature can be very helpful in the so-called folding diseases (Selkoe 2003). In this group of diseases, the disorder results from a mutation of one specific protein (Kopito and Ron 2000). The symptoms can result from the aggregation of the protein, such as the beta-amyloid plaques in Alzheimer's disease (Soto et al. 2003), as well as problems resulting from the loss-of-function of the protein in question, such as Cl^- ion transport deficiency in cystic fibrosis, where a point mutation in the gene of CFTR (cystic fibrosis transmembrane conductance regulator) protein results in its capturing by the quality control system within the endoplasmic reticulum and its quick degradation by the proteasomal machinery (Denning et al. 1992). Absence of this channel results in an imbalance of ion concentrations across the cell membrane, which leads to the onset of many severe symptoms, such as chronic inflammation and fibrosis of the lung, problems with all the secretory glands, especially with the pancreas, resulting in the devastation of beta cells, and promotes diabetes. Diseases in which the combination of both types of the above-mentioned disturbance occur are also known. In alpha-1 antitrypsin deficiency a serum elastase inhibitor, alpha-1 antitrypsin (AAT) is mutated. The specific point mutation often referred to as Z type AAT results in fatal folding deficiency of the protein. The uncovered hydrophobic surface of the protein leads to its prompt aggregation inside the lumen of the endoplasmic reticulum, forming large insoluble aggregates in the organelle. The loss-of-function of AAT results in emphysema in human patients and, additionally, the protein deposits in the endoplasmic reticulum of the liver cells generates cirrhosis, hepatitis, and elevated sensitivity to hepatocellular carcinoma (Needham and Stockley 2004). Chemical chaperones are widely used in experimental systems (Smith et al. 1998); however, their use in human patients is limited due to their general impact on the whole organism. The main groups of classical chemical chaperones, their mechanism of action, and new findings in models of various diseases are discussed below.

1.1
Osmolytes

The most common chemical chaperones are usually osmolytes, such as glycerol, or trimethylamines, e.g., trimethylamine N-oxide (TMAO), and amino acid derivatives, such as proline. Osmolytes are the ancient members of stress responses. Prolin as well as glycine betaine are known osmoprotectors, defending bacterial and plant cells against osmotic and freezing stress. These osmolytes exert their beneficial activity by limiting the free movement of proteins by elevating the density of the solvent, thus preventing aggregation of unfolded proteins. Their ability to reduce the frequency of "folding-detours" to nonproductive folding pathways has been proved for several mutant proteins involved in conformational diseases (Table 1).

Chemical Chaperones

Table 1 Osmolytes used in folding problem-related diseases

Disease	Protein	Agent used	References
Alzheimer's disease	Beta-amyloid	Glycerol, TMAO	Yang et al. 1999
Cancer	Ubiquitin-activating enzyme E1	Glycerol, TMAO, D^2O	Brown et al. 1996
	Glucocorticoid receptor	Glycerol, TMAO, D^2O	Baskakov et al. 1999
	p53	Glycerol, TMAO, D^2O	Brown et al. 1996
	pp60	Glycerol, TMAO, D^2O	Brown et al. 1996
Cystic fibrosis	CFTR	Glycerol, TMAO, DMSO	Sato et al. 1996
Emphysema and liver disease	Alpha-1-antitrypsin	Glycerol, TMAO	Burrows et al. 2000
Machado-Joseph disease	Ataxin-3	Glycerol, TMAO, DMSO	Yoshida et al. 2002
Maple syrup urine disease	BCKD complex	TMAO	Song et al. 2001
Menkes disease	MNK	Glycerol	Kim et al. 2002
Nephrogenic diabetes insipidus	Aquaporin-2	Glycerol, TMAO, DMSO	Tamarappoo et al. 1999
	V2R	Glycerol	Tan et al. 2003

The most frequently examined target among the conformational diseases is the cAMP-activated chloride ion channel protein, CFTR and its most common mutation, the D508F CFTR. Many cell lines used in laboratory experiments express this mutant protein. In most cases, the mutation does not lead to the aggregation of the protein, but the mutant CFTR is degraded rapidly. In a set of experiments carried out by Sato et al. (1996), it has been proved that glycerol and TMAO could increase the maturation of the mutant CFTR protein and rescue the cAMP-activated chloride conductance of cells expressing DF508 CFTR. DMSO, the well-known cryoprotectant, was also shown to have chemical chaperone activity, since it helped the transport of mutant CFTR in a cell culture model (Bebök et al. 1998).

The mutation of the water channel, aquaporin-2, is responsible for the development of nephrogenic diabetes insipidus. Its folding deficiency was correctable with chemical chaperones (Tamarappoo and Verkman 1998). The onset of diabetes insipidus is triggered by the mutation of a vasopressin receptor, V2R. The mutated V2R protein escaped from the quality control machinery, integrated to the membrane, and functioned as a normal vasopressin receptor with the help of the osmolyte glycerol (Tan et al. 2002). The sequestration of mutant alpha-1 antitrypsin Z can be enhanced by chemical chaperones, as was proven by Burrows et al. (2000).

Other types of diseases were also involved in the investigation of chemical chaperone effects. The central nervous system is sensitive to folding diseases due to its poor regenerating ability, making neurodegeneration diseases "hot spots" in chemical chaperone research. In Alzheimer's disease, where the beta-amyloid plaque formation causes the death of neuronal cells, which leads to mental deterioration, the effect of glycerol as well as TMAO was investigated. Both molecules successfully inhibited the formation of beta amyloid plaques. In Creutzfeldt-Jacob disease the prion protein, besides the specific action of doxycycline, quinacrine, and chlorpromazine, DMSO and glycerol was also found to revert the mutated form of PrP (Sc), the protein responsible for the onset of prion disease (Gu and Singh 2004), giving new hope of curing these folding disorders.

An entirely different group of diseases is cancer. In tumors, not one but many genes have to be mutated to exert oncogenic features. However, in more than 50% of tumor cells, the tumor suppressor protein p53 was found to be mutated. Mutation of the viral oncogene protein, pp60src, is also of key importance in tumors. The active osmolytes, glycerol, TMAO, and DMSO, corrected mislocalization of many of the oncogenic mutations of these proteins, such as p53, pp60src, or the ubiquitin-activating enzyme E1 (Brown et al. 1996).

The immune system can also improve its efficiency with the help of chemical chaperones. Glycerol, TMAO, and DMSO were also found to enhance antigen-presentation by promoting the folding of MHC molecules (Ghumman et al. 1998).

An interesting member of the often used osmolytes is prolin, which is a known agent protecting against high saline and freezing in yeast and also in *Escherichia coli* (Csonka 1989). However, prolin has controversial effects on protein folding. In *E. coli* prolin was found to increase thermotolerance and was able to substitute the chaperone DnaK in deficient strains (Chattopadhyay 2004). Low physiological concentrations of prolin had an activator effect on prokaryotic chaperons, while higher concentrations had a rather inhibitory effect on chaperone–protein interactions (Diamant et al. 2001). On the other hand, prolin strongly inhibited the refolding of denatured porcine lactate dehydrogenase, probably by limiting the interactions of the side chains of the protein amino acid backbone (Chilson and Chilson 2003). It is noteworthy that in this model system TMAO was also found to have anti-chaperone activity.

As an extraordinary chemical chaperon, deuterated water is also used as an osmolyte, since it can increase the viscosity of the fluids. Deuterated water was shown to stabilize the native conformation of the mutant CFTR and many proteins having oncogenic properties (Sato et al. 1996, Brown et al. 1996).

Osmolytes, is spite of the overwhelming data on their efficiency in many models, have relatively minor significance in clinical practice, due to their nonspecific means of action.

1.2
Hydrophobic Compounds

In addition to the osmolyte effect detailed above, a new mechanism for chemical chaperones has been discovered. Different compounds with a shorter or longer hydrophobic part can be solved in different fluids and can bind to proteins. Accordingly, as lysophosphatidic acids or butyrate derivatives were found to mask mutations of proteins and stabilize their structure in the native conformation. The suggested mechanism of action is that these hydrophobic molecules have the ability to bind to the hydrophobic segments, which remain surface-exposed in unfolded proteins, and protect them from aggregation or degradation in this manner.

1.2.1
PBA

The most prominent member of this group is sodium 4-phenylbutyrate (PBA). Sodium 4-phenylbutyrate is an orally bioavailable short-chain fatty acid, which was originally used as an ammonia-scavenging agent in urea metabolism disorders. In recent years, PBA has also been shown to help the mutant CFTR protein to get to the membrane (Zeitlin et al. 2002). PBA appears to help in correcting the transport of mutant alpha-1 antitrypsin (AAT) in AAT-deficiency models. PBA enhances the secretion of mutant AAT in cell culture and also in transgenic mouse model (Burrows et al. 2000). Since mutant AAT has a significant residual elastase activity and since PBA can be used safely in human patients, this opens a promising, new therapeutic avenue for AAT deficiency.

Its ability to bind stretched hydrophobic surfaces of the protein, thus protecting it from aggregation and avoiding the check of the quality control system, was proposed as the mechanism of action, thus supporting its transport and integration into the plasma membrane. But the chaperone-like activity of PBA turned out to be much more complicated. In newer experiments, PBA was found to influence many levels of regulation. For example, PBA downregulated the general protein synthesis, but induced the synthesis of cellular chaperones in the case of mutant CFTR expressing the IBS-3 cell line (Wright et al. 2004). It was also shown that PBA activated the transcription of beta- and gamma-globin, which makes PBA a promising candidate in the treatment of thalassemias (Collins et al. 1995). In a recent study, PBA protected against cerebral ischemia through inhibition of ER stress-mediated apoptosis and inflammation (Qi et al. 2004).

Sodium phenylbutyrate (PBA) treatment seems to have no severe side effects, and it can be utilized with a good efficiency by oral supplementation. These features make PBA a promising chemical chaperone with hope of clinical use in the future.

1.2.2
Lipids and Detergents

In bacterial models, many short-chain fatty acids were found to have chaperone properties. Kern et al. (2001) found that lysophosphatidic acid can prevent *E. coli* strains from heat denaturation as well as facilitate the refolding of heat-denatured citrate-synthase. The nonionic detergent, Brij 58P was tested in many in vitro models (Krause et al. 2002). The group found that Brij 58P has a favorable effect on the refolding of denatured alpha-glucosidase, rhodanese, and citrate synthase in the presence of the aggregation-prone DnaJ molecule.

A series of cationic, zwitterionic, and nonionic detergents were tested on the course of the refolding of three different model proteins (Daugherty 1998). In this study, all types of detergents promoted the refolding of citrate synthase, unlike carbonic anhydrase B and lysozyme, which required zwitterionic detergent for the augmentation of successful folding.

2
Pharmacological Chaperones

The discovery that compounds selectively binding to intracellularly retained proteins can promote their proper folding and targeting opened the way to developing a new class of chemical compounds having chaperone activity. These specific molecules are called pharmacological chaperones. Table 2 shows the most important pharmacological chaperones used in different diseases. Pharmacological chaperones are similar to the chemical chaperones in their effect: they can promote the folding and transport of mutant proteins alleviating many folding disease (Bernier et al. 2004). The difference distinguishing these molecules from a separate group from chemical chaperones lies in their specificity. These molecules can bind to one definite protein, thus aiding its folding and transport. Pharmacological chaperones can be, for example, ligands for a receptor promoting its proper folding, or, more specifically, a molecule designed especially to bind the native conformation of the target protein, stabilizing its conformation and pushing the balance toward the native state.

2.1
Enzyme Antagonists

A study conducted on an energy-dependent transporter known as P-glycoprotein or multidrug-resistance gene-1 product (MDR1) showed that while synthetic mutations resulted in ER retention and rapid degradation of the protein (Loo and Clarke 1994), treatments with substrates (vinblastine and capsaicin) or inhibitors (cyclosporin and verapamil) of the transporter led to the appearance of functional MDR1 at the cell surface (Loo and Clarke 1995).

Table 2 Pharmacological chaperones used in different diseases

Disease	Protein	Agent used	References
Misfolding/aggregation			
Gaucher disease	β-Glucosidase	N-(n-nonyl) deoxynojirmycin	Sawkar et al. 2002
β-Galactosidosis	β-Galactosidase	Galactonojirmycin derivatives	Matsuda et al. 2003
Long QT syndrome	HERG Kp channel	Cisapride, E-4031, astemizole	Curran et al. 1995
Prion disease	Prion	IPrP13 quinacrine chlorpromazine	Soto et al. 2000; Korth et al. 2001
Misfolding/degradation			
Cancer	Smo	Cyclopamine	Chen et al. 2002
Cystic fibrosis	CFTR	Benzo(c)quinolizinium compounds	Galietta et al. 2001
Fabry disease	Alpha-Gal A	DGJ Galactose	Fan et al. 1999; Frustaci et al. 2001
Hyperinsulinemic hypoglycemia	SUR1	Sulfonylurea	Yan et al. 2004
Hypogonadotropic hypogonadism	GnRHR	GnRH peptidomimetic antagonist	Janovick et al. 2002
Drug resistance	P-glycoprotein	Cyclosporin, capsaicin, vinblastine, verapamil	Loo et al. 1995
Immunoglobulin secretion	Anti-phenyl-phosphocholine	Hapten p-nitrophenyl-phosphocholine	Wiens et al. 2001
Pain	dOR	Naltrexone	Petaja-Repo et al. 2002
Menkes disease	MNK	Copper	Kim et al. 2002
Nephrogenic diabetes insipidus	V2R	SR121463, VPA-985	Morello et al. 2000

This led the authors to propose that the drug-binding site forms early during MDR1 biosynthesis and that occupation of this site could stabilize a folding intermediate in a near-native conformation that can escape the quality control system (Loo and Clarke 1999). Similarly, an antagonist of the vasopressin receptor can increase the activity of mutant vasopressin receptor associated with nephrogenic diabetes insipidus (Morello and Bichet 2001). Potassium channel mutation associated with the long QT syndrome was also correctable with the help of selective inhibitors of the channel (Zhou et al. 1999).

A new concept of enhancing protein folding and secretion of immunoglobulins was to use hapten ligands as chemical chaperones. Secretion of antiphenylphosphocholine antibody was enhanced by phosphocholine treatment, suggesting that hapten binding can promote antibody maturation by stabilizing heavy- and light-chain assembly (Wiens et al. 2001). As an additional example of pharmacological chaperones, alpha-galactosidase A deficiency related to Fabry disease was correctable by administering galactose in a cell culture model (Okumiya et al. 1995).

2.2
Folding Agonists

The temperature-sensitive mutants of tumor suppressor p53 protein can be stabilized by a set of small molecule compounds, which can bind specifically to the p53 protein and help the folding into its active conformation (Foster et al. 1999). High throughput chemical screening led to new compounds that could stabilize p53 in its active conformation, opening a new chapter in tumor suppression strategies (Wang et al. 2003; Issaeva et al. 2003). As a result, a peptide designed to bind specifically to the native conformation of p53 could restore the activity of the R249S mutation of p53, which is the most frequent cause of cancer, especially in hepatocellular carcinomas (Friedler et al. 2004).

In Menkes disease, which is a congenital copper deficiency, results from the mutation of copper-ATPase. In this case, copper itself was found to be advantageous for the proper folding of the ATPase (Kaler et al. 1998; Kim et al. 2002).

A mutation in the sulfonylurea receptor, Sur is responsible for the onset of familial persistent hyperinsulinemic hypoglycemia (Thomas et al. 1995). Correction of the loss of function of the receptor was carried out by sulfonylurea (Yan et al. 2004) and also with diazoxide; however, the results are inconsequent at this point (Partridge et al. 2001; Yan et al. 2004). As an additional example, Dormer et al. found (2001) that benzoquinolizine derivatives can facilitate the folding of mutant CFTR in cystic fibrosis.

An interesting target of pharmacological intervention is pain. Opioid receptors were shown to fold in the endoplasmic reticulum. Still, the majority of the proteins is transported directly to the protein degradation system. Membrane permeable opioid ligands can stabilize the structure of the receptor and augment its insertion into the plasma membrane (Petaja-Repo et al. 2002)

Pharmacological chaperones, although their approach to the disease is similar to chemical chaperones, have the advantage of higher specificity. In this treatment, only the folding of the selectively targeted protein will be influenced. The specific mechanism of action of the pharmacological chaperones may be a significant advantage considering their potential use for disease-related folding deficiencies with clinical importance.

The growing pool of chemical and pharmacological chaperones opens a new field for applied research aimed to help in folding diseases. Supplementation of mutated proteins, as well as the emerging role of gene therapy, are also powerful tools for correcting dangerous mutations, but artificial chaperones constitute a real alternative in the course of folding disorders.

References

Baskakov IV, Kumar R, Srinivasan G, Ji YS, Bolen DW, Thompson EB (1999) Trimethylamine N-oxide-induced cooperative folding of an intrinsically unfolded transcription-activating fragment of human glucocorticoid receptor. J Biol Chem 274:10693–10696

Bebok Z, Venglarik CJ, Panczel Z, Jilling T, Kirk KL, Sorscher EJ (1998) Activation of DeltaF508 CFTR in an epithelial monolayer. Am J Physiol 275:599–607

Bernier V, Lagace M, Bichet DG, Bouvier M (2004) Pharmacological chaperones: potential treatment for conformational diseases. Trends Endocrinol Metab 15:222–228

Brown CR, Hong-Brown LQ, Biwersi J, Verkman AS, Welch WJ (1996) Chemical chaperones correct the mutant phenotype of the DF508 cystic fibrosis transmembrane conductance regulator protein. Cell Stress Chaperones 1:117–125

Brown CR, Hong-Brown LQ, Welch WJ (1997) Correcting temperature-sensitive protein folding defects. J Clin Invest 99:1432–1444

Burrows JA, Willis LK, Perlmutter DH (2000) Chemical chaperones mediate increased secretion of mutant a1-antitrypsin (a1-AT) Z: a potential pharmacological strategy for prevention of liver injury and emphysema in a1-AT deficiency. Proc Natl Acad Sci U S A 97:1796–1801

Chen JK, Taipale J, Cooper MK, Beachy PA (2002) Inhibition of Hedgehog signaling by direct binding of cyclopamine to smoothened. Genes Dev 16:2743–2748

Chilson OP, Chilson AE (2003) Perturbation of folding and reassociation of lactate dehydrogenase by proline and trimethylamine oxide. Appl Environm Microbiol 69:6527–6532

Cohen FE, Kelly JW (2003) Therapeutic approaches to protein misfolding diseases. Nature 426:905–909

Collins AF, Pearson HA, Giardina P, McDonagh KT, Brusilow SW, Dover GJ (1995) Oral sodium phenylbutyrate therapy in homozygous beta thalassemia: a clinical trial. Blood 85:43–49

Conn PM, Leanos-Miranda A, Janovick JA (2002) Protein origami: therapeutic rescue of misfolded gene products. Mol Intervent 2:308–316

Csonka LN (1989) Physical and genetic responses of bacteria to osmotic stress. Microbiol Rev 53:121–147

Curran ME, Splawski I, Timothy KW, Vincent GM, Green ED, Keating MT (1995) A molecular basis for cardiac arrhythmia: HERG mutations cause long QT syndrome. Cell 80:795–803

De Fost M, Aerts JM, Hollak CE (2003) Gaucher disease: from fundamental research to effective therapeutic interventions. Neth J Med 61:3–8

Denning GM, Anderson MP, Amara JF, Marshall J, Smith AE, Welsh MJ (1992) Processing of mutant cystic fibrosis transmembrane conductance regulator is temperature-sensitive. Nature 358:761–764

Diamant S, Eliahu N, Rosenthal D, Goloubinoff P (2001) Chemical chaperones regulate molecular chaperones in vitro and in cells under combined salt and heat stresses. J Biol Chem 276:39586–39591

Dormer RL, Derand R, McNeilly CM, Mettey Y, Bulteau-Pignoux L, Metaye T, Vierfond JM, Gray MA, Galietta LJ, Morris MR, Pereira MM, Doull IJ, Becq F, McPherson MA (2001) Correction of delF508-CFTR activity with benzo(c)quinolizinium compounds through facilitation of its processing in cystic fibrosis airway cells. J Cell Sci 114:4073–4081

Daugherty DL, Rozema D, Hanson PE, Gellman SH (1998) Artificial chaperone-assisted refolding of citrate synthase. J Biol Chem 273:33961–33971

Fan JQ, Ishii S, Asano N, Suzuki Y (1999) Accelerated transport and maturation of lysosomal a-galactosidase A in Fabry lymphoblasts by an enzyme inhibitor. Nat Med 5:112–115

Friedler A, DeDecker BS, Freund SM, Blair C, Rudiger S, Fersht AR (2004) Structural distortion of p53 by the mutation R249S and its rescue by a designed peptide: implications for "mutant conformation". J Mol Biol 336:187–196

Frustaci A, Chimenti C, Ricci R, Natale L, Russo MA, Pieroni M, Eng CM, Desnick RJ (2001) Improvement in cardiac function in the cardiac variant of Fabry's disease with galactose-infusion therapy. N Engl J Med 345:25–32

Foster BA, Coffey HA, Morin MJ, Rastinejad F (1999) Pharmacological rescue of mutant p53 conformation and function. Science 286:2507–2510

Gu Y, Singh N (2004) Doxycycline and protein folding agents rescue the abnormal phenotype of familial CJD H187R in a cell model. Brain Res Mol Brain Res 123:37–44

Ghumman B, Bertram EM, Watts TH (1998) Chemical chaperones enhance superantigen and conventional antigen presentation by HLA-DM-deficient as well as HLA-DM-sufficient antigen-presenting cells and enhance IgG2a production in vivo. J Immunol 161:3262–3270

Galietta LJ, Springsteel MF, Eda M, Niedzinski EJ, By K, Haddadin MJ, Kurth MJ, Nantz MH, Verkman AS (2001) Novel CFTR chloride channel activators identified by screening of combinatorial libraries based on flavone and benzoquinolizinium lead compounds. J Biol Chem 276:19723–19728

Issaeva N, Friedler A, Bozko P, Wiman KG, Fersht AR, Selivanova G (2003) Rescue of mutants of the tumor suppressor p53 in cancer cells by a designed peptide. Proc Natl Acad Sci U S A 100:13303–13307

Janovick JA, Maya-Nunez G, Conn PM (2002) Rescue of hypogonadotropic hypogonadism causing and manufactured GnRH receptor mutants by a specific protein-folding template: misrouted proteins as a novel disease etiology and therapeutic target. J Clin Endocrinol Metab 87:3255–3262

Kaler SG (1998) Diagnosis and therapy of Menkes syndrome, a genetic form of copper deficiency. Am J Clin Nutr 67:1029–1034

Kern R, Joseleau-Petit D, Chattopadhyay MK, Richarme G (2001) Chaperone-like properties of lysophospholipids. Biochem Biophys Res Commun 289:1268–1274

Kim BE, Smith K, Meagher CK, Petris MJ (2002) A conditional mutation affecting localization of the Menkes disease copper ATPase. Suppression by copper supplementation. J Biol Chem 277:44079–44084

Kopito RR (1999) Biosynthesis and degradation of CFTR. Physiol Rev 79:167–173

Kopito RR, Ron D (2000) Conformational disease. Nat Cell Biol 2:207–209

Korth C, May BC, Cohen FE, Prusiner SB (2001) Acridine and phenothiazine derivatives as pharmacotherapeutics for prion disease. Proc Natl Acad Sci U S A 98:9836–9841

Krause M, Rudolph R, Schwarz E (2002) The non-ionic detergent Brij 58P mimics chaperone effects. FEBS Lett 532:253–255

Loo TW, Clarke DM (1994) Prolonged association of temperature sensitive mutants of human P-glycoprotein with calnexin during biogenesis. J Biol Chem 269:28683–28689

Loo TW, Clarke DM (1995) P-glycoprotein. Associations between domains and between domains and molecular chaperones. J Biol Chem 270:21839–21844

Loo TW, Clarke DM (1999) Determining the structure and mechanism of the human multidrug resistance P-glycoprotein using cysteine-scanning mutagenesis and thiol-modification techniques. Biochim Biophys Acta 1461:315–325

Matsuda J, Suzuki O, Oshima A, Yamamoto Y, Noguchi A, Takimoto K, Itoh M, Matsuzaki Y, Yasuda Y, Ogawa S, Sakata Y, Nanba E, Higaki K, Ogawa Y, Tominaga L, Ohno K, Iwasaki H, Watanabe H, Brady RO, Suzuki Y (2003) Chemical chaperone therapy for brain pathology in G(M1)-gangliosidosis. Proc Natl Acad Sci U S A 100:15912–15917

Morello JP, Bichet DG (2001) Nephrogenic diabetes insipidus. Annu Rev Physiol 63:607–630

Morello JP, Petaja-Repo UE, Bichet DG, Bouvier M (2000a) Pharmacological chaperones: a new twist on receptor folding. Trends Pharmacol Sci 21:466–469

Morello JP, Salahpour A, Laperriere A, Bernier V, Arthus MF, Lonergan M, Petaja-Repo U, Angers S, Morin D, Bichet DG, Bouvier M (2000b) Pharmacological chaperones rescue cell surface expression and function of misfolded V2 vasopressin receptor mutants. J Clin Invest 105:887–895

Needham M, Stockley RA (2004) Alpha 1-antitrypsin deficiency. 3: Clinical manifestations and natural history. Thorax 59:441–445

Okumiya T, Ishii S, Takenaka T, Kase R, Kamei S, Sakuraba H, Suzuki Y (1995) Galactose stabilizes various missense mutants of alpha-galactosidase in Fabry disease. Biochem Biophys Res Commun 214:1219–1224

Partridge CJ, Beech DJ, Sivaprasadarao A (2001) Identification and pharmacological correction of a membrane trafficking defect associated with a mutation in the sulfonylurea receptor causing familial hyperinsulinism. J Biol Chem 276:35947–35952

Petaja-Repo UE, Hogue M, Bhalla S, Laperriere A, Morello JP, Bouvier M (2002) Ligands act as pharmacological chaperones and increase the efficiency of d opioid receptor maturation. EMBO J 21:1628–1637

Qi X, Hosoi T, Okuma Y, Kaneko M, Nomura Y (2004) Sodium 4-phenylbutyrate protects against cerebral ischemic injury. Mol Pharmacol 66:899–908

Sato S, Ward CL, Krouse ME, Wine JJ, Kopito RR (1996) Glycerol reverses the misfolding phenotype of the most common cystic fibrosis mutation. J Biol Chem 271:635–638

Sawkar AR, Cheng WC, Beutler E, Wong CH, Balch WE, Kelly JW (2002) Chemical chaperones increase the cellular activity of N370S b-glucosidase: a therapeutic strategy for Gaucher disease. Proc Natl Acad Sci U S A 99:15428–15433

Selkoe DJ (2003) Folding proteins in fatal ways. Nature 426:900–904

Sitia R, Braakman I (2003) Quality control in the endoplasmic reticulum protein factory. Nature 426:891–894

Smith DF, Whitesell L, Katsanis E (1998) Molecular chaperones: biology and prospects for pharmacological intervention. Pharmacol Rev 50:493–514

Song JL, Chuang DT (2001) Natural osmolyte trimethylamine N-oxide corrects assembly defects of mutant branched-chain a-ketoacid decarboxylase in maple syrup urine disease. J Biol Chem 276:40241–40246

Soto C (2003) Unfolding the role of protein misfolding in neurodegenerative diseases. Nat Rev Neurosci 4:49–60

Soto C, Kascsak RJ, Saborio GP, Aucouturier P, Wisniewski T, Prelli F, Kascsak R, Mendez E, Harris DA, Ironside J, Tagliavini F, Carp RI, Frangione B (2000) Reversion of prion protein conformational changes by synthetic β-sheet breaker peptides. Lancet 355:192–197

Tamarappoo BK, Verkman AS (1998) Defective aquaporin-2 trafficking in nephrogenic diabetes insipidus and correction by chemical chaperones. J Clin Invest 101:2257–2267

Tamarappoo BK, Yang B, Verkman AS (1999) Misfolding of mutant aquaporin-2 water channels in nephrogenic diabetes insipidus. J Biol Chem 274:34825–34831

Tan CM, Nickols HH, Limbird LE (2003) Appropriate polarization following pharmacological rescue of v2 vasopressin receptors encoded by X-linked nephrogenic diabetes insipidus alleles involves a conformation of the receptor that also attains mature glycosylation. J Biol Chem 278:35678–35686

Wang W, Rastinejad F, El-Deiry WS (2003) Restoring p53-dependent tumor suppression. Cancer Biol Ther 2:55–63

Welch WJ, Brown CR (1996) Influence of molecular and chemical chaperones on protein folding. Cell Stress Chaperones 1:109–115

Wiens GD, O'Hare T, Rittenberg MB (2001) Recovering antibody secretion using a hapten ligand as a chemical chaperone. J Biol Chem 276:40933–40939

Wright JM, Zeitlin PL, Cebotaru L, Guggino SE, Guggino WB (2004) Gene expression profile analysis of 4-phenylbutyrate treatment of IB3-1 bronchial epithelial cell line demonstrates a major influence on heat-shock proteins. Physiol Genomics 16:204–211

Zeitlin PL, Diener-West M, Rubenstein RC, Boyle MP, Lee CK, Brass-Ernst L (2002) Evidence of CFTR function in cystic fibrosis after systemic administration of 4-phenylbutyrate. Mol Ther 6:119–126

Zhou Z, Gong Q, January CT (1999) Correction of defective protein trafficking of a mutant HERG potassium channel in human long QT syndrome. Pharmacological and temperature effects. J Biol Chem 274:31123–31126

Yan F, Lin CW, Weisiger E, Cartier EA, Taschenberger G, Shyng SL (2004) Sulfonylureas correct trafficking defects of ATP-sensitive potassium channels caused by mutations in the sulfonylurea receptor. J Biol Chem 279:11096–11105

Yang DS, Yip CM, Huang THJ, Chakrabartty A, Fraser PE (1999) Manipulating the amyloid beta aggregation pathway with chemical chaperones. J Biol Chem 274:32970–32974

Yoshida H, Yoshizawa T, Shibasaki F, Shoji S, Kanazawa I (2002) Chemical chaperones reduce aggregate formation and cell death caused by the truncated Machado-Joseph disease gene product with an expanded polyglutamine stretch. Neurobiol Dis 10:88–99

Pharmacological Modulation of the Heat Shock Response

C. Sőti · P. Csermely (✉)

Department of Medical Chemistry, Semmelweis University, P.O. Box 260, 1444 Budapest, Hungary
csermely@puskin.sote.hu

1	Stress and the Heat Shock Response	418
2	Chaperone-Mediated Cytoprotection	418
3	Possible Therapeutic Use of Chaperone Induction	419
3.1	Ischemia Reperfusion	420
3.2	Inflammation and Sepsis	421
3.3	Aging and Chaperone Overload	422
4	Heat Shock Response Modulators	422
4.1	Aspirin as a Chaperone Co-inducer	423
4.2	Glutamine Is a Remedy for the Critically Ill	423
4.3	Zinc Supplementation Is a Prerequisite of Proper Chaperone Induction	423
4.4	Hsp90 Inhibitors Are Useful Cytoprotective Agents	424
4.5	Proteasome Inhibition Activates Cytoprotection	424
4.6	Anti-ulcer Drugs Induce Chaperones	425
4.7	Prescription Drugs May Induce a Heat Shock Response	425
4.8	Herbal Medicines Contain Potent Cytoprotective Compounds	426
4.9	Cyclopentenone Prostaglandins: Antiviral Drug Candidates	426
4.10	Nutrition State Influences Hsp Response	427
4.11	Low-Frequency Electromagnetic Fields: Beneficial Potential and Health Hazard	427
4.12	Chaperone Co-inducers: A Safer Opportunity to Induce the Heat Shock Response	428
5	Conclusions and Perspectives	429
References		430

Abstract Life presents a continuous series of stresses. Increasing the adaptation capacity of the organism is a long-term survival factor of various organisms and has become an attractive field of intensive therapeutic research. Induction of the heat shock response promotes survival after a wide variety of environmental stresses. Preclinical studies have proven that physiological and pharmacological chaperone inducers and co-inducers are an efficient therapeutic approach in different acute and chronic diseases. In this chapter, we summarize current knowledge of the current state of chaperone modulation and give a comprehensive list of the main drug candidates.

Keywords Chaperone inducers · Chaperones · Heat shock proteins · Heat shock response · Stress proteins

1
Stress and the Heat Shock Response

Life is stress, a continuous challenge from molecules to mind. Sophisticated and robust protective mechanisms have been and are still evolving to promote survival under constantly changing environmental conditions. One of the most ancient and highly conserved adaptive mechanisms in the cellular setting is the heat shock or stress response. Molecular chaperones or heat shock proteins maintain the conformational homeostasis of the proteome (Thirumalai and Lorimer 2001; Young et al. 2004). Most chaperones are essential proteins present at high concentrations even under nonstress conditions, emphasizing their vital housekeeping role. However, many types of toxic insults (e.g., heat shock, ethanol, oxidative stress) lead to a sudden rise in chaperone levels. Chaperone induction is mediated at the transcriptional level by an autoregulatory feedback loop. An increase in misfolded proteins results in the release of heat shock transcription factor 1 (HSF-1) from the repressing Hsp90/Hsp70/Hsp40 complex and a subsequent activation of heat shock gene transcription (see the chapter by R. Voellmy, this volume).

2
Chaperone-Mediated Cytoprotection

A sublethal stress exposure protects the cell from the deleterious effect of a subsequent otherwise lethal stress. The stress tolerance (thermotolerance in case of the prototype, heat shock) is mediated by the elevated levels of chaperones, especially Hsp70, Hsp27, and Hsp90 (Welch 1992). Chaperones play a crucial role in such vital processes as signal transduction, transport processes, cell division, migration, and differentiation, and they are indispensable for proper immune function. Moreover, from a general point of view, chaperones are stabilizing hubs of the cellular protein–protein/lipid networks (reviewed in Csermely 2004, 2005; Sőti et al. 2005).

The cytoprotective role of chaperones involves a direct stabilization of macromolecular structure of proteins and lipids (Török et al. 2001; Tsvetkova et al. 2002). Moreover, Hsp70 and Hsp110 stabilize mRNA structure (Henics et al. 1999), and cytokine mRNAs are stabilized through Hsp70 induction or proteasome downregulation (Laroia et al. 1999). Intriguingly, chaperones are critical factors in apoptosis. They play both a direct role by supporting key molecules in (anti)apoptotic signaling (APAF-1, Bcl-2), and an indirect role by being involved in antioxidative defense and scaffolding macromolecular assemblies (Sreedhar and Csermely 2004). Various studies have demonstrated that chaperone-mediated cytoprotection can be largely attributed to the suppression of apoptosis. Cells failing to mount a stress response are sensitive to apoptosis (Sreedhar et al. 1999). ATP depletion is hallmark of many disease states, including a variety of ischemic conditions. It results in a rapid metabolic incompetence accompanied by profound cell death. Both stress preconditioning and transient Hsp70 overexpression in rat cardiac myoblasts were shown to exert a marked reduction in cell death as well as in the amount of total denatured protein (Kabakov et al. 2002). In separate studies, hypoxia/reoxygenation-induced cell death was rescued by a preceding heat shock in rat cardiac myocytes (Tanonaka et al. 2003). Heat shock attenuated poly(ADP-ribose) polymerase (PARP) activation, and pharmacological PARP inhibition prevented the cell death evoked by the treatment. Interestingly, Hsp70 expression and nuclear translocation were observed upon heat shock, raising the possibility of an inhibitory Hsp70–PARP interaction during heat shock/hypoxia. Moreover, Hsp70 contributes to the increased antioxidative defense during ischemia/reperfusion (Chong et al. 1998).

Accumulation of mutant, misfolded proteins is a threat for postmitotic cells, especially for the nervous system. Protein aggregation is a complex process, harboring a wide variety of cytotoxic events, inducing cell death (Stefani and Dobson 2003; Bossy-Wetzel et al. 2004). Chaperones were shown to be associated both with oligomeric and aggregated species and protected from cell death in *Drosophila*, in mammalian cells, and in transgenic mice (Warrick et al. 1999; Carmichael et al. 2000; Cummings et al. 2001).

Intriguingly, not only cytosolic chaperones are able to confer cytoprotection. Endoplasmic reticulum chaperones, normally induced by the unfolded protein response, are also upregulated upon excitotoxic and oxidative insults and display protective effect via diminishing the production of reactive oxygen species and stabilizing calcium homeostasis (Yu et al. 1999). Similarly, upregulation of endoplasmic reticulum chaperones in epithelial cells was shown to exert a robust effect against ATP-depletion-induced cell damage (Bush et al. 1999).

3
Possible Therapeutic Use of Chaperone Induction

Since chaperones protect the cells from a wide variety of physiological and pathological of stressors, virtually any conditions associated with (a) increased cellular/organismal stress and/or (b) decreased protective potential may be therapeutic target (Table 1). The first category encompasses two subgroups: the first with increased oxidative stress/ATP depletion, the prototype is the ischemic diseases; the second is the direct toxicity of aberrant proteins, including the progressive neurodegenerative diseases. The remainder of the conditions is best represented by aging, where the stress continuously uses up the different adaptation mechanisms, including a proper mounting of the heat shock response, resulting in a vicious circle and finally exhaustion.

Table 1 Pathological states as possible therapeutic targets for chaperone upregulation[a]

Pathology	Reference
Increased stress:	Christians et al. 2002
Ischemia-reperfusion	
Cardiovascular disease	Snoeckx et al. 2001
Stroke	DeFranco et al. 2004
Inflammation, shock	van Molle et al. 2002
Toxin exposure	
Ethanol	?
Drugs	?
Oxidants	Papp et al. 2003
Trauma and regeneration	Kalmar et al. 2002, 2003
Transplantation	Perdrizet et al. 1993
Low-frequency electromagnetic field	Goodman and Blank 2002
Proteinopathies	Welch 2004
Amyloidosis and neurodegeneration	Stefani and Dobson 2003
Cystic fibrosis (and others)	Amaral 2004
Diabetes mellitus	Nánási and Jednákovits 2001
Mixed, with decreased protection	
Aging	Sőti and Csermely 2003; Hsu et al. 2003
Hyperlipidemia	Csont et al. 2002
Overnutrition	?
Recurrent (viral) infections	Rosi et al. 1996
Chronic fatigue syndrome	?

[a] Question marks indicate lack of experimental evidence

3.1
Ischemia Reperfusion

Ischemic heart disease is one of the most intensely studied diseases with respect to chaperone protection. Besides their anti-apoptotic role, chaperones contribute to proper functioning by the sustained activation of endothelial nitric oxide synthase (eNOS) provided by Hsp90 (Kupatt et al. 2004), the maintenance of redox homeostasis (Chong et al. 1998; Papp et al. 2003), and the enhancement of mitochondrial respiratory complex activity (Sammut and Harrison 2003). Female heart contains twice as much inducible Hsp70 than male heart, due to the presence of estradiol (Vos et al. 2003), which induces HSF-1 transactivation (Knowlton and Sun 2001). Both the HSF-1-inducing action and cardioprotective effect of estrogen is mediated by NFκB activation (Hamilton et al. 2004). Ovariectomy reduces and estrogen replacement re-establishes cardiac Hsp70 expression (Vos et al. 2003), supporting the cardioprotective role of estrogen replacement therapy after menopause, and raising the question of administering estrogen agonists to patients with high cardiac risks. On the other hand, exercise preferentially induced cardiac Hsp70 and exerted cardioprotection only in male as well as ovariectomized female rats, suggesting that training may be much more important for males than for females in defending against the effects of heart disease and offers a novel manner by which males may reduce the sex gap in susceptibility to adverse cardiac events (Paroo et al. 2002). However, cardiac function and postischemic adaptation is deteriorated already in middle-aged rats, despite a relatively maintained chaperone inducibility (Honma et al. 2003), arguing against considering chaperone induction as a miraculous remedy. The central importance of a healthy diet is emphasized by the finding that experimental hyperlipidemia attenuated the heat shock response in rat hearts (Csont et al. 2002). Further research will reveal the proper place of chaperone-inducing therapies in cardiovascular diseases.

There are several examples suggesting that chaperones play an important role in protecting the nervous system both in acute ischemic conditions such as stroke (DeFranco et al. 2004) and chronic degenerative states such as vascular dementia and neuronal proteinopathies such as Alzheimer's, Parkinson's, and other diseases (Stefani and Dobson 2003). Interestingly, glial Hsp70 is released and taken up by nerve cells and enhances neuronal stress tolerance (Guzhova et al. 2001). Another intriguing finding is that chaperone induction by the amino acid analog canavanine attenuated the retinopathic complication of streptozotocin-induced diabetes in rats, prompting further investigations of chaperone inducers in easing the chronic consequences of diabetes such as retinopathy (Mihály et al. 1998).

3.2
Inflammation and Sepsis

Inflammation poses a stress on immune cells. Indeed, Hsp70 is overexpressed in polymorphonuclear cells under sepsis (Hashiguchi et al. 2001), and Hsp70 conferred an augmented antioxidative response and inhibited apoptosis (Hashiguchi et al. 2001). Hsp70 overexpression also conferred cytoprotection by inhibiting NFκB activation and iNOS upregulation in rats, ameliorating cardiac shock and hypotension upon endotoxic shock (Hauser et al. 1996; Chan et al. 2004). Another protective feature of Hsp70 overexpression is that it prevented high production of LPS-induced inflammatory cytokines in human macrophages (Ding et al. 2001). Similarly, heat shock inhibited IL-6 and iNOS expression upon TNFα treatment in wild type, but not in heat-inducible hsp70.1 gene-deficient mice (van Molle et al. 2002). Heat shock reduced bowel damage and apoptosis, circumventing the severe side effects of antitumor protocols based on TNF and interferon-gamma, leading to a significant inhibition of lethality but not to a reduction of antitumoral capacity (van Molle et al. 2002).

3.3
Aging and Chaperone Overload

The positive correlation between longevity and a robust heat shock response is an experiential fact and is well documented in several studies using *Caenorhabditis elegans* as a model system, implicating the chaperone network as one of the critical adaptive mechanisms (Garigan et al. 2002; Hsu et al. 2003). Chaperone inducibility generally decreases during aging. Hsp70 overdose, under the control of its own promoter, extended the lifespan in *Drosophila*. Many experimental manipulations inducing life-span extension also upregulate chaperones and lead to a stress response of higher intensity (Sőti and Csermely 2003). Chaperones may be overburdened by the individual life in aging, as well as during the improvement of living standards over the last two centuries. This phenomenon, the so-called chaperone overload, may be a causative factor in a number of degenerative diseases and aging (Csermely 2001; Sőti and Csermely 2003). Whether strengthening the heat shock response would further aggravate or solve this problem may be a key question in the coming years of research.

The above-mentioned examples show the power of a properly mounted stress response in different pathological conditions.

4
Heat Shock Response Modulators

The heat shock or stress response is a multistep process involving cross-talk among different processes such as growth and stress-activated signaling pathways, protein misfolding, aggregation and degradation, heat shock, and other transcription factor activation, with a subsequent production of heat shock genes. For the sake of simplicity and specificity, we will focus on the inhibitors of misfolded protein degradation and activators of Hsp synthesis as candidates for therapeutic intervention.

Ancient cultures understood that sauna is a well known practice in a healthy lifestyle. Heat shock is the archetype of preconditioning, and was already extensively discussed in the literature. Mild heat stress leads to a series of beneficial effects in cells (Park et al. 2005) and hormesis induced by repeated mild heat shock is a promising means to preserve the adaptation capacity of cells (Rattan 2004). In this section, we will focus on different approaches to induce the heat shock response.

4.1
Aspirin as a Chaperone Co-inducer

The first pharmacological agents shown to affect the stress response were sodium salicylate and aspirin (Jurivich et al. 1992). Salicylate induced HSF-1 DNA binding, did not lead to hsp transcription per se, but augmented the stress response upon a challenging insult. Unfortunately, these compounds are not specific, and besides the inhibition of cyclo-oxygenase, both compounds bind to Grp78 (BiP) and inhibit its ATPase activity and may interfere with its activity (Deng et al. 2001).

4.2
Glutamine Is a Remedy for the Critically Ill

Several physiologically occurring compounds activate chaperone expression. The amino acid glutamine, an important nutrient for bowel and muscle cells, is a very potent inducer of chaperone expression in *Drosophila* cells as well as in human patients (Sanders and Kon 1991; Wischmeyer 2002, 2004). It shows all the beneficial actions of heat stress, including a marked improvement in survival during sepsis, antiapoptotic and anti-inflammatory properties, and it depends on the induction of Hsp70. Even after a single glutamine injection, Hsp70 induction can be observed in the gut, blood cells, lung and heart, among other organs, suggesting a fairly general phenomenon. For instance, glutamine administration before cardiopulmonary bypass reduced the inflammation and improved clinical outcome in rats (Hayashi et al. 2002). Though the mechanism of action of glutamine is not known, it is devoid of any toxicity and is already widely used in enteral and parenteral nutrition, by heavy exercise athletes, and by naturopathic doctors for bowel problems. It must be noted that during prolonged exercise and sepsis or major trauma, the blood glutamine level is decreased, and such glutamine depletion impairs the stress response, probably contributing to susceptibility to exhaustion, and worsening (Oehler et al. 2002). There is ample opportunity for clinical use in critically ill patients with major trauma, surgery, or sepsis.

4.3
Zinc Supplementation Is a Prerequisite of Proper Chaperone Induction

Zinc is an important trace element that supports the function of several enzymes, including antioxidant enzymes and transcription factors with zinc finger motifs, such as steroid receptors. Zinc is critical for proper immune function, and it was shown that a 3-week dietary zinc depletion reduces Hsc70, Hsp40, and Hsp60 expression to approximately 50% in murine thymus (Moore et al. 2003). The same study showed that zinc overdose led to a reduction in heat shock mRNA levels, highlighting the importance of proper zinc consumption and supplementation. Both zinc bioavailability and immune function is decreased in the aged population. On the contrary, both are better preserved in the successfully aged centenarians. While in vitro zinc supplementation somewhat augmented the heat-induced Hsp70 expression in lymphoblasts from aged human donors, it diminished it both in young and centenarian samples (Ambra et al. 2004). This finding may be explained by the hypothesis that chaperones may regulate metallothionein metabolism (Mocchegiani et al. 2000). Zinc induces a heat shock response in cell culture (Hatayama et al. 1993), as well as in gastric mucosal and hepatic cells in vivo (Odashima et al. 2002; Cheng et al. 2002). Besides the general protective effect of zinc, it can be used before major surgery, or even before transplantation for organ preservation (Cheng et al. 2002), since it is well known that chaperones are helpful in graft preservation (Perdrizet et al. 1993).

4.4
Hsp90 Inhibitors Are Useful Cytoprotective Agents

Hsp90 inhibitors are multitarget antitumor drugs: by targeting the ATP-binding site of Hsp90 they compromise several growth and survival pathways in parallel (Neckers 2003; see also the chapter by L. Neckers, this volume). Clinical implications of geldanamycin, radicicol, and other Hsp90 inhibitors are summarized in the chapter by S. Pacey et al. in this volume. It seems to be reasonable to assume that heat shock protein inhibitors may lead a compensatory stress response. Indeed, at a very low dose where cytotoxic effects cannot be observed, geldanamycin releases HSF-1 from Hsp90 and leads to heat shock protein expression. This phenomenon was already successfully used in different experimental models to induce cytoprotection. Geldanamycin treatment improved Parkinsonism in *Drosophila* (Auluck and Bonini 2002), and both geldanamycin and radicicol were effective in cellular and mouse models of polyglutamine diseases, actually better than overexpression of Hsp70 (Sittler et al. 2001; Hay et al. 2004). Moreover, geldanamycin also binds to Grp94, the ER-resident Hsp90, and induces cytoprotective ER chaperones via the unfolded protein response (Lawson et al. 1998), which widens the therapeutic potential of these drugs. Radicicol, representing another class of Hsp90 inhibitors, exerted a cardioprotective effect in ischemia/reperfusion injury (Griffin et al. 2004). Herbimycin A, a geldanamycin-related compound and a Tyr-phosphatase inhibitor, was also able to reduce intimal hyperplasia (restenosis) by upregulating Hsp27 after balloon angioplasty (Conolly et al. 2003).

4.5
Proteasome Inhibition Activates Cytoprotection

As HSF-1 is activated by an increased flux of misfolded proteins, proteasome inhibitors are also potent inducers of both the cytosolic and the ER chaperones and thermotolerance (Bush et al. 1997). MG-132 and lactacystin increased IL-6 production by human intestinal epithelial cells (Pritts et al. 2002), while MG-132 and MG-262 treatment resulted in better contractile function and faster recovery of heart papillary muscle after ischemia (Stangl et al. 2002). It must be mentioned, though, that the protective response evoked by proteasome inhibitors is not merely based on chaperone induction. First, activation of heat shock elements is not sufficient to induce αB-crystallin expression upon proteasome inhibition (Aki et al. 2003). Second, in some model systems these compounds inhibit HSF-1 dephosphorylation and transactivation if they are present after heat shock (Kim and Li 1999). Third, a low dose of MG-132 and lactacystin protects neural cells from excitotoxic and oxidative stress without chaperone induction (Lee et al. 2004). Thus, further studies are needed to clarify the cytoprotective mechanisms of proteasome inhibitors.

4.6
Anti-ulcer Drugs Induce Chaperones

Other heat shock protein inducers are emerging as promising tools in a wide variety of pathophysiological states. Geranylgeranylacetone is nontoxic anti-ulcer drug, and recently has been shown to induce the heat shock response. This property makes it a useful intervention in dysfunctions of the stomach and probably the digestive tract (Rokutan 2000). Its use is also proven in hepatocellular damage originating either from ethanol- or oxidant-related pathology (Ikeyama et al. 2001), or from hepatectomy (Oda et al. 2002). Systemic geranylgeranylacetone treatment protected retinal ganglion cells in a rat glaucoma model (Ishii et al. 2003) and hippocampal neurons against ischemia (Fujiki et al. 2004); however, in addition to chaperones, other protective mechanisms may be involved. Carbenoxolone is another anti-ulcer drug that possesses Hsp70-inducing activity (Nagayama et al. 2001). However, we lack further studies to prove its therapeutic benefit.

4.7
Prescription Drugs May Induce a Heat Shock Response

Several medicines were studied with respect to cytoprotective action. Estrogen was found to affect the inducibility of chaperones. Other steroids such as the glucocorticoids, hydrocortisone, or the agonists methylprednisolone and dexamethasone were shown to elicit a cardioprotective effect by inducing Hsp70, transcriptionally and/or post-transcriptionally (Sun et al. 2000; Valen et al. 2000). Whether other hormones behave the same way is an open question.

Cyclosporine A (CsA) is a toxic immunosuppressive drug. On the one hand, both heat shock and CsA preconditioning produce tolerance against the CsA-toxicity (Yuan et al. 1996). On the other hand, CsA is an activator of HSF-1 and -2, probably an inhibitor of the proteasome, but leads only to Hsp27 upregulation (Paslaru et al. 2000). These findings raise the possibility that toxic side effects of CsA may be circumvented by a low-dose administration preceding the therapeutic dose.

Surprisingly, the commonly used antibiotics ampicillin and ceftriaxone—previously thought to be specifically damaging to bacteria—upregulate Hsp27 and Hsp60 in human lymphocytes and protect them from staurosporine-induced apoptosis (Romano et al. 2004). The generality of this phenomenon with respect to cell types, antibiotics, and other harmful insults demands a systematic study. On the other hand, it should be noted that many cell culture facilities routinely use antibiotics, especially ampicillin, and this may produce misleading results in apoptosis and chaperone induction assays.

4.8
Herbal Medicines Contain Potent Cytoprotective Compounds

Traditional medicine has used herbal compounds for thousands of years for pain relief, fever reduction, infection and inflammation, stimulation of physical production, and for treating tumors. Among others, celastrols have recently been documented as cytoprotective agents and heat shock response inducers (Westerheide et al. 2004). The naturally occurring antioxidant ergothioneine augmented Hsp70 induction and conferred protection against liver ischemia/reperfusion in the rat (Bedirli et al. 2004).

Curcumin is dietary pigment of turmeric, a favorite spice in the culinary arts. It is also a remedy in Indian medicine. Curcumin increased Hsp70 levels and had a cytoprotective effect (Sood et al. 2001). It was harmless in normal cells; however, it induced a pronounced apoptosis in different tumor cell lines (Khar et al. 2001). Intriguingly, resistant lines had a robust heat shock response upon curcumin treatment (Khar et al. 2001), suggesting that among the pleiotropic effects of curcumin, the accentuated heat shock response may be critical for tumor survival. This finding raises two ideas. First, curcumin may be effectively combined with heat shock protein inhibitors in antitumor protocols to elicit maximal beneficial potential. Second, curcumin is a unique drug that kills tumor cells, while strengthening normal cells, which may make it an ideal panacea for cancer patients.

4.9
Cyclopentenone Prostaglandins: Antiviral Drug Candidates

Naturally occurring anti-inflammatory cyclopentenone prostaglandins also induce Hsp70. These molecules are characterized by the presence of a reactive α,β-unsaturated carbonyl group in the cyclopentane ring (cyclopentenone), which is the key structure for triggering HSF-1 activation. 2-Cyclopenten-1-one selectively induces Hsp70 in human cells, and this is associated with antiviral activity (Rossi et al. 1996). These prostaglandin derivatives also display an HSF-1/Hsp70-dependent anti-inflammatory action (Ianaro et al. 2003), which makes them promising antiviral drug candidates. However, many of the effects are mediated via the NFκB pathway, given that these compounds are direct inhibitors of the IκB kinase (Rosi et al. 2000). They may be very attractive compounds in reducing viral infection and overwhelming inflammatory reactions.

4.10
Nutrition State Influences Hsp Response

Though not pharmacological compounds, some effects from the environment also have the potential to regulate the heat shock response. Calorie restriction retards aging and oxidative stress, and induces heat shock proteins (Sőti and Csermely 2003). Both calorie restriction and its mimetic, 2-deoxyglucose, reduce focal ischemic brain damage, and induce Hsp70 (Yu and Mattson 1999), which implies that a modest diet may protect from several age-related pathologies by strengthening the natural adaptation mechanisms, including the stress response.

4.11
Low-Frequency Electromagnetic Fields: Beneficial Potential and Health Hazard

Our environment is not merely material. There is a continuous spectrum of electromagnetic radiation surrounding and penetrating us. While higher energy UV and ionizing radiations are recognized as life-threatening and stress response-inducing stimuli, we have only just begun to explore the biological effects of lower-frequency fields. Two segments of low-frequency fields are especially interesting. The first is the radio-frequency (athermal) microwave field (in the MHz range) widely used in wireless communication, e.g., in cell phones, while the other is the extremely low-frequency field (ELF, below 300 Hz) present everywhere. While proper ELF exposure leads to a variety of regenerative processes, improper exposure to both fields are proven to pose a health hazard to humans. Interestingly, acute exposure to both athermal microwave radiation (915 MHz) and to ELF (60 Hz) induced a heat shock response and cytoprotection against hypoxia-reperfusion in chick embryos (Shallom et al. 2002). Indeed, it is well documented that there is an ELF responsive element in several genes, including Hsp70, which may directly sense the electromagnetic radiation (Goodman and Blank 2002).

There are several advantageous properties to ELF radiation and preconditioning:

- It is noninvasive, safe, comfortable, and simple to administer, even repeatedly before, during, and after cardiac surgery to mount a higher stress response.

- It is more effective in inducing Hsp70 than heat: the energy density required is 14 orders of magnitude lower for ELF than for heat (which may also mean a harmful effect at this low energy input!).
- ELF easily penetrates the body, allowing systemic administration, but even can be directed to the target region.
- The same ELF exposure induces Hsp70 in all systems from bacteria to human, emphasizing the presently unknown importance of this range (Goodman and Blank 2002).

Promising biomedical applications of ELF and chaperones are emerging. One opportunity is that ELF responsive elements can be used as novel, noninvasive techniques in transgene expression in a highly regulated fashion, a possible application in gene therapy with a sensory/ELF-generating circuit (e.g., a glucose sensor, an ELF generator, and cells containing the insulin gene with ELF element in its promoter; Goodman and Blank 2002). The second application is the preconditioning driven by ELF treatment, as mentioned above. The third is that chaperones may be much better markers of ELF exposure than the traditionally used heat absorption of tissues.

From a public health standpoint, it is important that more studies be performed to determine if repeated exposures, a condition likely to be found in the everyday setting (e.g., in cell phone use), are still beneficial. It is also important how our body is saturated with chronic ELF and whether a therapeutic ELF exposure would result in chaperone expression. One study showed that similarly to heat shock and other stresses, chronic ELF (only for 4 days) stimulation led to a distress: the exhaustion of the heat shock response and downregulation of Hsp70 with a consequent decline in cytoprotection (Di Carlo et al. 2002). This may explain epidemiologic correlations between chronic ELF exposure and cancer and may question the use of chaperone expression to determine the extent of chronic ELF exposure.

4.12
Chaperone Co-inducers: A Safer Opportunity to Induce the Heat Shock Response

The previous section highlighted that instead of the expected hormesis an overdose of chaperone inducers may lead to distress over the long term. On the contrary, a multicopy of inducible Hsp70 with its own promoter extended life in *Drosophila* (Tatar et al. 1997). Therefore drugs only augmenting the naturally occurring stress response (called chaperone co-inducers; Vígh et al. 1997) may be more beneficial without the side effects of chronic stress. Chaperone co-inducers are representatives of multitarget, low-affinity drugs, which may have a much better efficiency than single-target high-affinity drugs developed by rational drug design (Csermely et al. 2005). The lead compound, a nontoxic hydroxylamine derivative, Bimoclomol, helps the induction of Hsp synthesis by both perturbing various membrane structures and helping the release of putative lipid-signaling molecules as well as by the prolongation of the binding of HSF-1 to the heat shock elements on the DNA (Hargitai et al. 2003; Török et al. 2003; Vígh et al. 1997). Bimoclomol binds to HSF-1 with a low affinity, which may contribute to its effect to prolong HSF-1 binding to DNA. Chaperone co-inducers also stabilize membranes, which may be of special importance to prevent apoptotic events related to the decomposition of cardiolipin and the consecutive destabilization of the mitochondrial membrane (Török et al. 2003).

Chaperone co-inducers lead to heat shock protein expression after stress, display strong cytoprotective effect (Vígh et al. 1997), and have great beneficial potential in a wide variety of pathological conditions. Bimoclomol was successfully used in cardiovascular and diabetic complications (Nánási and Jednákovits 2001). BRX-220, a potent analog of Bimoclomol, protected against acute pancreatitis, by diminishing serum markers and enhancing antioxidative capacity (Rakonczay et al. 2002a). However, it should be noted that this effect may not be solely due to Hsp induction, since arsenite treatment effectively inducing Hsp70 did sufficiently protect in this experimental pancreatitis model (Rakonczay et al. 2002b). Kürthy et al. (2002) showed a beneficial effect of BRX-220 against insulin resistance and peripheral neuropathy in diabetic rats. BRX-220 was a useful remedy in peripheral nerve injury (Kalmar et al. 2002, 2003). Recently arimoclomol was successfully applied in an inherited neurodegenerative disease, amyotrophic lateral sclerosis (ALS) (Kieran et al. 2004). ALS is characterized by a mutation in Cu/Zn superoxide dismutase-1, with loss of function (increased oxidative stress) and gain of function (protein aggregation) in spinal motoneurons. Arimoclomol delayed the onset of and led to a 22% life extension in amyotrophic lateral sclerosis in a mouse model (Kieran et al. 2004), suggesting that this class of drugs may prove to be effective in other neurodegenerative diseases.

5
Conclusions and Perspectives

Chaperone inducers are physiological compounds or potent drugs with pleiotropic beneficial actions. Instead of the old paradigm of meticulously targeting single molecules or eliminating disease-causing reasons, they focus on enhancing the natural protective capacity of our own body (Csermely et al. 2005). This protection is brought about by a higher level of heat shock or stress proteins and may provide an important novel therapeutic approach in a number of acute and chronic diseases and aging, even with a combination of traditional medications. Chaperone co-inducers may circumvent the distress-exhaustion cycle by overstimulation, since they only augment the naturally occurring Hsp induction. Further research will clarify the effect of long-term treatment and clinical applications, especially the relationship with chaperone overload.

Acknowledgements Work in the authors' laboratory was supported by research grants from the EU (FP6-506850, FP6-518230), Hungarian Science Foundation (OTKA-T37357 and F47281), Hungarian Ministry of Social Welfare (ETT-32/03), and by the Hungarian National Research Initiative (NKFP-1A/056/2004 and KKK-0015/3.0).

References

Aki T, Yoshida K, Mizukami Y (2003) The mechanism of αB-crystallin gene expression by proteasome inhibition. Biochem Biophys Res Commun 311:162–167

Amaral MD (2004) CFTR and chaperones: processing and degradation. J Mol Neurosci 23:41–48

Ambra R, Mocchegiani E, Giacconi R, Canali R, Rinna A, Malavolta M, Virgili F (2004) Characterization of the hsp70 response in lymphoblasts from aged and centenarian subjects and differential effects of in vitro zinc supplementation. Exp Gerontol 39:1475–1484

Auluck PK, Bonini NM (2002) Pharmacological prevention of Parkinson disease in Drosophila. Nat Med 8:1185–1186

Bedirli A, Sakrak O, Muhtaroglu S, Soyuer I, Guler I, Riza Erdogan A, Sozuer EM (2004) Ergothioneine pretreatment protects the liver from ischemia-reperfusion injury caused by increasing hepatic heat shock protein 70. J Surg Res 122:96–102

Bossy-Wetzel E, Schwarzenbacher R, Lipton SA (2004) Molecular pathways to neurodegeneration. Nat Med 10 [Suppl]:S2–S9

Bush, KT, Goldberg AL, Nigam SK (1997) Proteasome inhibition leads to a heat-shock response, induction of endoplasmic reticulum chaperones, and thermotolerance. J Biol Chem 14:9086–9092

Bush KT, George SK, Zhang PL, Nigam SK (1999) Pretreatment with inducers of ER molecular chaperones protects epithelial cells subjected to ATP depletion. Am J Physiol 277:F211–F218

Carmichael J, Chatellier J, Woolfson A, Milstein C, Fersht AR, Rubinsztein DC (2000). Bacterial and yeast chaperones reduce both aggregate formation and cell death in mammalian cell models of Huntington's disease. Proc Natl Acad Sci U S A 97:9701–9705

Chan JY, Ou CC, Wang LL, Chan SH (2004) Heat shock protein 70 confers cardiovascular protection during endotoxemia via inhibition of nuclear factor-kappaB activation and inducible nitric oxide synthase expression in the rostral ventrolateral medulla. Circulation 110:3560–3566

Cheng Y, Liu YF, Liang J (2002) Protective effect of zinc: a potent heat shock protein inducer in cold preservation of rat liver. Hepatobiliary Pancreat Dis Int 1:258–261

Chong KY, Lai CC, Lille S, Chang C, Su CY (1998) Stable overexpression of the constitutive form of heat shock protein 70 confers oxidative protection. J Mol Cell Cardiol 30:599–608

Connolly EM, Kelly CJ, Chen G, O'Grady T, Kay E, Leahy A, Bouchier-Hayes DJ (2003) Pharmacological induction of HSP27 attenuates intimal hyperplasia in vivo. Eur J Vasc Endovasc Surg 25:40–47

Christians ES, Yan LJ, Benjamin IJ (2002) Heat shock factor 1 and heat shock proteins: critical partners in protection against acute cell injury. Crit Care Med 30:S43–S50

Csermely P (2001) Chaperone overload as a possible contributor to civilization diseases. Trends Genet 17:701–704

Csermely P (2004) Strong links are important – but weak links stabilize them. Trends Biochem Sci 29:331–334

Csermely P, Ágoston V, Pongor S (2005) The efficiency of multi-target drugs: the network approach might help drug design. www.arxiv.org/q-bio.MN/0412045 Trends Pharmacol Sci 26:178–182

Csermely P (2005) Weak links: stabilizers of complex systems from proteins to social networks. Springer Verlag, Berlin Heidelberg New York

Csont T, Balogh G, Csonka C, Boros I, Horváth I, Vígh L, Ferdinándy P (2002) Hyperlipidemia induced by high cholesterol diet inhibits heat shock response in rat hearts. Biochem Biophys Res Commun 290:1535–1538

Cummings CJ, Sun Y, Opal P, Antalffy B, Mestril R, Orr HT, Dillmann WH, Zoghbi HY (2001) Over-expression of inducible HSP70 chaperone suppresses neuropathology and improves motor function in SCA1 mice. Hum Mol Genet 10:1511–1518

DeFranco DB, Ho L, Falke E, Callaway CW (2004) Small molecule activators of the heat shock response and neuroprotection from stroke. Curr Atheroscler Rep 6:295–300

Deng WG, Ruan KH, Du M, Saunders MA, Wu KK (2001) Aspirin and salicylate bind to immunoglobulin heavy chain binding protein (BiP) and inhibit its ATPase activity in human fibroblasts. FASEB J 15:2463–2470

Ding XZ, Fernandez-Prada CM, Bhattacharjee AK, Hoover DL (2001) Over-expression of hsp-70 inhibits bacterial lipopolysaccharide-induced production of cytokines in human monocyte-derived macrophages. Cytokine 16:210–219

Fujiki M, Kobayashi H, Inoue R, Tatsuya A, Ishii K (2004) Single oral dose of geranylgeranylacetone for protection against delayed neuronal death induced by transient ischemia. Brain Res 1020:210–213

Garigan D, Hsu AL, Fraser AG, Kamath RS, Ahringer J, Kenyon C (2002) Genetic analysis of tissue aging in Caenorhabditis elegans: a role for heat-shock factor and bacterial proliferation. Genetics 161:1101–1112

Griffin TM, Valdez TV, Mestril R (2004) Radicicol activates heat shock protein expression and cardioprotection in neonatal rat cardiomyocytes. Am J Physiol Heart Circ Physiol 287:H1081–H1088

Guzhova I, Kislyakova K, Moskaliova O, Fridlanskaya I, Tytell M, Cheetham M, Margulis B (2001) In vitro studies show that Hsp70 can be released by glia and that exogenous Hsp70 can enhance neuronal stress tolerance. Brain Res 914:66–73

Hamilton KL, Gupta S, Knowlton AA (2004) Estrogen and regulation of heat shock protein expression in female cardiomyocytes: cross-talk with NF kappa B signaling. J Mol Cell Cardiol 36:577–584

Hargitai J, Lewis H, Boros I, Rácz T, Fiser A, Kurucz I, Benjamin I, Pénzes Z, Vígh L, Csermely P, Latchman DS (2003) Bimoclomol, a heat shock protein co-inducer acts by the prolonged activation of heat shock factor-1. Biochem Biophys Res Commun 307:689–695

Hashiguchi N, Ogura H, Tanaka H, Koh T, Nakamori Y, Noborio M, Shiozaki T, Nishino M, Kuwagata Y, Shimazu T, Sugimoto H (2001) Enhanced expression of heat shock proteins in activated polymorphonuclear leukocytes in patients with sepsis. J Trauma 51:1104–1109

Hatayama T, Asai Y, Wakatsuki T, Kitamura T, Imahara H (1993) Regulation of hsp70 synthesis induced by cupric sulfate and zinc sulfate in thermotolerant HeLa cells. J Biochem (Tokyo) 114:592–597

Hauser GJ, Dayao EK, Wasserloos K, Pitt BR, Wong HR (1996) HSP induction inhibits iNOS mRNA expression and attenuates hypotension in endotoxin-challenged rats. Am J Physiol 271:H2529–H2535

Hay DG, Sathasivam K, Tobaben S, Stahl B, Marber M, Mestril R, Mahal A, Smith DL, Woodman B, Bates GP (2004) Progressive decrease in chaperone protein levels in a mouse model of Huntington's disease and induction of stress proteins as a therapeutic approach. Hum Mol Genet 13:1389–1405

Hayashi Y, Sawa Y, Fukuyama N, Nakazawa H, Matsuda H (2002) Preoperative glutamine administration induces heat-shock protein 70 expression and attenuates cardiopulmonary bypass-induced inflammatory response by regulating nitric oxide synthase activity. Circulation 106:2601–2607

Henics T, Nagy E, Oh HJ, Csermely P, von Gabain A, Subjeck JR (1999) Mammalian Hsp70 and Hsp110 proteins bind to RNA motifs involved in mRNA stability. J Biol Chem 274:17318–17324

Honma Y, Tani M, Yamamura K, Takayama M, Hasegawa H (2003) Preconditioning with heat shock further improved functional recovery in young adult but not in middle-aged rat hearts. Exp Gerontol 38:299–306

Hsu AL, Murphy CT, Kenyon C (2003) Regulation of aging and age-related disease by DAF-16 and heat-shock factor. Science 300:1142–1145

Ianaro A, Ialenti A, Maffia P, Di Meglio P, Di Rosa M, Santoro MG (2003) Anti-inflammatory activity of 15-deoxy-delta12,14-PGJ2 and 2-cyclopenten-1-one: role of the heat shock response. Mol Pharmacol 64:85–93

Ikeyama S, Kusumoto K, Miyake H, Rokutan K, Tashiro S (2001) A non-toxic heat shock protein 70 inducer, geranylgeranylacetone, suppresses apoptosis of cultured rat hepatocytes caused by hydrogen peroxide and ethanol. J Hepatol 35:53–61

Ishii Y, Kwong JM, Caprioli J (2003) Retinal ganglion cell protection with geranylgeranylacetone, a heat shock protein inducer, in a rat glaucoma model. Invest Ophthalmol Vis Sci 44:1982–1992

Jurivich DA, Sistonen L, Kroes RA, Morimoto RI (1992) Effect of sodium salicylate on the human heat shock response. Science 255:1243–1245

Kabakov AE, Budagova KR, Latchman DS, Kampinga HH (2002) Stressful preconditioning and HSP70 overexpression attenuate proteotoxicity of cellular ATP depletion. Am J Physiol Cell Physiol 283:C521–C534

Kalmar B, Burnstock G, Vrbova G, Urbanics R, Csermely P, Greensmith L (2002) Upregulation of heat shock proteins rescues motoneurones from axotomy-induced cell death in neonatal rats. Exp Neurol 176:87–97

Kalmar B, Greensmith L, Malcangio M, Macmahon SB, Csermely P, Burnstock G (2003) The effect of treatment with BRX-220, a co-inducer of heat shock proteins, on sensory fibres of the rat following peripheral nerve injury. Exp Neurol 184:636–647

Khar A, Ali AM, Pardhasaradhi BV, Varalakshmi CH, Anjum R, Kumari AL (2001) Induction of stress response renders human tumor cell lines resistant to curcumin-mediated apoptosis: role of reactive oxygen intermediates. Cell Stress Chaperones 6:368–376

Kieran D, Kalmar B, Dick JR, Riddoch-Contreras J, Burnstock G, Greensmith L (2004) Treatment with arimoclomol, a coinducer of heat shock proteins, delays disease progression in ALS mice. Nat Med 10:402–405

Kim D, Li GC (1999) Proteasome inhibitors lactacystin and MG132 inhibit the dephosphorylation of HSF1 after heat shock and suppress thermal induction of heat shock gene expression. Biochem Biophys Res Commun 264:352–358

Knowlton AA, Sun L (2001) Heat-shock factor-1, steroid hormones, and regulation of heat-shock protein expression in the heart. Am J Physiol Heart Circ Physiol 280:H455–H464

Kupatt C, Dessy C, Hinkel R, Raake P, Daneau G, Bouzin C, Boekstegers P, Feron O (2004) Heat shock protein 90 transfection reduces ischemia-reperfusion-induced myocardial dysfunction via reciprocal endothelial NO synthase serine 1177 phosphorylation and threonine 495 dephosphorylation. Arterioscler Thromb Vasc Biol 24:1435–1441

Kürthy M, Mogyorósi T, Nagy K, Kukorelli T, Jednákovits A, Tálosi L, Bíró K (2002) Effect of BRX-220 against peripheral neuropathy and insulin resistance in diabetic rat models. Ann N Y Acad Sci 96:482–489

Laroia G, Cuesta R, Brewer G, Schneider RJ (1999) Control of mRNA decay by heat shock-ubiquitin-proteasome pathway. Science 284:499–502

Lawson B, Brewer JW, Hendershot LM (1998) Geldanamycin, an hsp90/GRP94-binding drug, induces increased transcription of endoplasmic reticulum (ER) chaperones via the ER stress pathway. J Cell Physiol 174:170–178

Lee CS, Tee LY, Warmke T, Vinjamoori A, Cai A, Fagan AM, Snider BJ (2004) A proteasomal stress response: pre-treatment with proteasome inhibitors increases proteasome activity and reduces neuronal vulnerability to oxidative injury. J Neurochem 91:996–1006

Mocchegiani E, Muzzioli M, Giacconi R (2000) Zinc, metallothioneins, immune responses, survival and ageing. Biogerontology 1:133–143

Moore JB, Blanchard RK, Cousins RJ (2003) Dietary zinc modulates gene expression in murine thymus: results from a comprehensive differential display screening. Proc Natl Acad Sci U S A 100:3883–3888

Nagayama S, Jono H, Suzaki H, Sakai K, Tsuruya E, Yamatsu I, Isohama Y, Miyata T, Kai H (2001) Carbenoxolone, a new inducer of heat shock protein 70. Life Sci 69:2867–2873

Nánási PP, Jednákovits A (2001) Multilateral in vivo and in vitro protective effects of the novel heat shock protein co-inducer, bimoclomol: results of preclinical studies. Cardiovasc Drug Rev 19:133–151

Neckers L (2003) Development of small molecule Hsp90 inhibitors: utilizing both forward and reverse chemical genomics for drug identification. Curr Med Chem 10:733–739

Oda H, Miyake H, Iwata T, Kusumoto K, Rokutan K, Tashiro S (2002) Geranylgeranylacetone suppresses inflammatory responses and improves survival after massive hepatectomy in rats. J Gastrointest Surg 6:464–472

Odashima M, Otaka M, Jin M, Konishi N, Sato T, Kato S, Matsuhashi T, Nakamura C, Watanabe S (2002) Induction of a 72-kDa heat-shock protein in cultured rat gastric mucosal cells and rat gastric mucosa by zinc L-carnosine. Dig Dis Sci 47:2799–2804

Oehler R, Pusch E, Dungel P, Zellner M, Eliasen MM, Brabec M, Roth E (2002) Glutamine depletion impairs cellular stress response in human leucocytes. Br J Nutr 87 [Suppl 1]:S17–S21

Papp E, Nardai G, Soti C, Csermely P (2003) Molecular chaperones, stress proteins and redox homeostasis. Biofactors 17:249–257

Park HG, Han SI, Oh SY, Kang HS (2005) Cellular responses to mild heat stress. Cell Mol Life Sci 62:10–23

Paroo Z, Haist JV, Karmazyn M, Noble EG (2002) Exercise improves postischemic cardiac function in males but not females: consequences of a novel sex-specific heat shock protein 70 response. Circ Res 90:911–917

Paslaru L, Rallu M, Manuel M, Davidson S, Morange M (2000) Cyclosporin A induces an atypical heat shock response. Biochem Biophys Res Commun 269:464–469

Perdrizet GA, Kaneko H, Buckley TM, Fishman MS, Pleau M, Bow L, Schweizer RT (1993) Heat shock recovery protects renal allografts from warm ischemic injury and enhances HSP72 production. Transplant Proc 25:1670–1673

Pritts TA, Hungness ES, Hershko DD, Robb BW, Sun X, Luo GJ, Fischer JE, Wong HR, Hasselgren PO (2002) Proteasome inhibitors induce heat shock response and increase IL-6 expression in human intestinal epithelial cells. Am J Physiol Regul Integr Comp Physiol 282:R1016–R1026

Rakonczay Z Jr, Ivanyi B, Varga I, Boros I, Jednakovits A, Nemeth I, Lonovics J, Takacs T (2002a) Nontoxic heat shock protein coinducer BRX-220 protects against acute pancreatitis in rats. Free Radic Biol Med 32:1283–1292

Rakonczay Z Jr, Mandi Y, Kaszaki J, Ivanyi B, Boros I, Lonovics J, Takacs T (2002b) Induction of HSP72 by sodium arsenite fails to protect against cholecystokinin-octapeptide-induced acute pancreatitis in rats. Dig Dis Sci 47:1594–1603

Rattan SI (2004) Mechanisms of hormesis through mild heat stress on human cells. Ann N Y Acad Sci 1019:554–558

Rokutan K (2000) Role of heat shock proteins in gastric mucosal protection. J Gastroenterol Hepatol 15 [Suppl]:D12–D19

Romano CC, Benedetto N, Catania MR, Rizzo A, Galle F, Losi E, Hasty DL, Rossano F (2004) Commonly used antibiotics induce expression of Hsp 27 and Hsp 60 and protect human lymphocytes from apoptosis. Int Immunopharmacol 4:1067–1073

Rossi A, Elia G, Santoro MG (1996) 2-Cyclopenten-1-one, a new inducer of heat shock protein 70 with antiviral activity. J Biol Chem 271:32192–32196

Rossi A, Kapahi P, Natoli G, Takahashi T, Chen Y, Karin M, Santoro MG (2000) Anti-inflammatory cyclopentenone prostaglandins are direct inhibitors of IkappaB kinase. Nature 403:103–108

Sammut IA, Harrison JC (2003) Cardiac mitochondrial complex activity is enhanced by heat shock proteins. Clin Exp Pharmacol Physiol 30:110–115

Sanders MM, Kon C (1991) Glutamine is a powerful effector of heat shock protein expression in Drosophila Kc cells. J Cell Physiol 146:180–190

Shallom JM, Di Carlo AL, Ko D, Penafiel LM, Nakai A, Litovitz TA (2002) Microwave exposure induces Hsp70 and confers protection against hypoxia in chick embryos. J Cell Biochem 86:490–496

Sittler A, Lurz R, Lueder G, Priller J, Lehrach H, Hayer-Hartl MK, Hartl FU, Wanker EE (2001) Geldanamycin activates a heat shock response and inhibits huntingtin aggregation in a cell culture model of Huntington's disease. Hum Mol Genet 10:1307–1315

Snoeckx LH, Cornelussen RN, Van Nieuwenhoven FA, Reneman RS, Van der Vusse GJ (2001) Heat shock proteins and cardiovascular pathophysiology. Physiol Rev 81:1461–1497

Sood A, Mathew R, Trachtman H (2001) Cytoprotective effect of curcumin in human proximal tubule epithelial cells exposed to shiga toxin. Biochem Biophys Res Commun 283:36–41

Sőti C, Csermely P (2003) Aging and molecular chaperones. Exp Gerontol 38:1037–1040

Sőti C, Pál C, Papp B, Csermely P (2005) Molecular chaperones as regulatory elements of cellular networks. Curr Opin Cell Biol 17:210–215

Sreedhar AS, Pardhasaradhi BV, Begum Z, Khar A, Srinivas UK (1999) Lack of heat shock response triggers programmed cell death in a rat histiocytic cell line. FEBS Lett 456:339–342

Sreedhar AS, Csermely P (2004) Heat shock proteins in the regulation of apoptosis: new strategies in tumor therapy: a comprehensive review. Pharmacol Ther 101:227–257

Stangl K, Gunther C, Frank T, Lorenz M, Meiners S, Ropke T, Stelter L, Moobed M, Baumann G, Kloetzel PM, Stangl V (2002) Inhibition of the ubiquitin-proteasome pathway induces differential heat-shock protein response in cardiomyocytes and renders early cardiac protection. Biochem Biophys Res Commun 291:542–549

Stefani M, Dobson CM (2003) Protein aggregation and aggregate toxicity: new insights into protein folding, misfolding diseases and biological evolution. J Mol Med 81:678–699

Sun L, Chang J, Kirchhoff SR, Knowlton AA (2000) Activation of HSF and selective increase in heat-shock proteins by acute dexamethasone treatment. Am J Physiol Heart Circ Physiol 278:H1091–H1097

Tanonaka K, Toga W, Takahashi M, Kawana K, Miyamoto Y, Yoshida H, Takeo S (2003) Hsp70 attenuates hypoxia/reoxygenation-induced activation of poly(ADP-ribose) synthetase in the nucleus of adult rat cardiomyocytes. Mol Cell Biochem 248:149–155

Tatar M, Khazaeli AA, Curtsinger JW (1997) Chaperoning extended life. Nature 390:30

Thirumalai D, Lorimer GH (2001) Chaperonin-mediated protein folding. Annu Rev Biophys Biomol Struct 30:245–269

Török Z, Goloubinoff P, Horváth I, Tsvetkova NM, Glatz A, Balogh G, Varvasovszki V, Los DA, Vierling E, Crowe JH, Vígh L (2001) Synechocystis HSP17 is an amphitropic protein that stabilizes heat-stressed membranes and binds denatured proteins for subsequent chaperone-mediated refolding. Proc Natl Acad Sci U S A 98:3098–3103

Török Z, Tsvetkova NM, Balogh G, Horváth I, Nagy E, Pénzes Z, Hargitai J, Bensaude O, Csermely P, Crowe JH, Maresca B, Vígh L (2003) Heat shock protein co-inducers with no effect on protein denaturation specifically modulate the membrane lipid phase. Proc Natl Acad Sci U S A 100:3131–3136

Tsvetkova NM, Horváth I, Török Z, Wolkers WF, Balogi Z, Shigapova N, Crowe LM, Tablin F, Vierling E, Crowe JH, Vígh L (2002) Small heat-shock proteins regulate membrane lipid polymorphism. Proc Natl Acad Sci U S A 99:13504–13509

Valen G, Kawakami T, Tahepold P, Dumitrescu A, Lowbeer C, Vaage J (2000) Glucocorticoid pretreatment protects cardiac function and induces cardiac heat shock protein 72. Am J Physiol Heart Circ Physiol 279:H836–H843

Van Molle W, Wielockx B, Mahieu T, Takada M, Taniguchi T, Sekikawa K, Libert C (2002) HSP70 protects against TNF-induced lethal inflammatory shock. Immunity 16:685–695

Vígh L, Literáti Nagy P, Horváth I, Török Z, Balogh G, Glatz A, Kovács E, Boros I, Ferdinándy P, Farkas B, Jaszlits L, Jednákovits A, Korányi L, Maresca B (1997) Bimoclomol: a nontoxic, hydroxilamine derivative with stress protein-inducing activity and cytoprotective effects. Nat Med 3:1150–1154

Voss MR, Stallone JN, Li M, Cornelussen RN, Knuefermann P, Knowlton AA (2003) Gender differences in the expression of heat shock proteins: the effect of estrogen. Am J Physiol Heart Circ Physiol 285:H687–H692

Warrick JM, Chan HY, Gray-Board GL, Chai Y, Paulson HL, Bonini NM (1999) Suppression of polyglutamine-mediated neurodegeneration in Drosophila by the molecular chaperone HSP70. Nat Genet 23:425–428

Welch WJ (1992) Mammalian stress response: cell physiology, structure/function of stress proteins, and implications for medicine and disease. Physiol Rev 72:1063–1081

Welch WJ (2004) Role of quality control pathways in human diseases involving protein misfolding. Semin Cell Dev Biol 15:31–38

Westerheide SD, Bosman JD, Mbadugha BN, Kawahara TL, Matsumoto G, Kim S, Gu W, Devlin JP, Silverman RB, Morimoto RI (2004) Celastrols as inducers of the heat shock response and cytoprotection. J Biol Chem 279:56053–56060

Wischmeyer P (2002) Glutamine and heat shock protein expression. Nutrition 18:225–228

Wischmeyer P (2004) Glutamine, heat shock protein, and inflammation–opportunity from the midst of difficulty. Nutrition 20:583–585

Young JC, Agashe VR, Siegers K, Hartl FU (2004): Pathways of chaperone-mediated protein folding in the cytosol. Nat Rev Mol Cell Biol 5:781–791

Yu Z, Luo H, Fu W, Mattson MP (1999) The endoplasmic reticulum stress-responsive protein GRP78 protects neurons against excitotoxicity and apoptosis: suppression of oxidative stress and stabilization of calcium homeostasis. Exp Neurol 155:302–314

Yu ZF, Mattson MP (1999) Dietary restriction and 2-deoxyglucose administration reduce focal ischemic brain damage and improve behavioral outcome: evidence for a preconditioning mechanism. J Neurosci Res 57:830–839

Yuan CM, Bohen EM, Musio F, Carome MA (1996) Sublethal heat shock and cyclosporine exposure produce tolerance against subsequent cyclosporine toxicity. Am J Physiol 271:F571–F578

Subject Index

α-1-antitrypsin deficiency 200, 209
α-synuclein 206, 210, 224
αB-crystallin 97, 102, *see* HspB5
14-3-3 97
17-AAG 188, 337, 342, 345–349
17-AG 342
17-DMAG 337, 349, 350
17-allylamino, 17-demethoxy geldanamycin (17-AAG) 332
17-demethoxy, dimethylaminoethylamino geldanamycin (17-DMAG) 332

Aβ peptides 210, 212, 213, 242
activatory ligand 294
ADD70 188
aggresome 199, 207, 208, 210
aging 419, 422
AIDS 367
AIF 172, 174, 179, 184, 185, 188
allograft 364
Alzheimer's disease (AD) 83, 84, 200, 201, 206, 209, 210, 212, 213, 242, 333, 421
amyloid 199–201, 203, 206, 210–213
amyloid precursor protein (APP) 206
amyloidosis 200, 201, 211, 212
amyotrophic lateral sclerosis (ALS) 429
androgen insensitivity syndrome 381
angiogenesis 382
antigen 312, 314
– DNA 314
– peptide 312
– proteins 312
Apaf-1 172, 178, 181, 184, 185
APIases 363
apoptosis 77, 79, 102, 104, 171–175, 177–186, 188, 189, 366, 418

– anti-apoptotic 171, 175, 178, 180–182, 184, 185
– pro-apoptotic 171, 172, 175, 181, 182, 188
arrhythmias 385
arteriosclerosis 372
arthritis 333, 372
– rheumatic 372
asthma 372
ataxin 242
ATP 178, 180, 184
autoantibody 376

BAG-1 98, 100
BAG5 230
barrier 362
– torsion 362
Bax 172, 182, 183
Bcl-2 172–174, 180, 181, 183, 184
BiP 95
BiP/Grp78 242
bond 360
– disulfide bond 360
– prolyl 361
bortezomib 347
bovine spongiform encephalopathy (BSE) 232
breast cancer 346, 381

calcineurin 365
calmodulin 373
cancer 174, 178, 187–189, 262, 387
candida 334, 339
capsid protein 367
carcinogenesis 171
cardioprotective 421
cardiopulmonary bypass 423
caspase 172–174, 176–179, 181, 182, 185, 189

CCT018159   338
cdc37   100
cell cycle   386
CF1010   338
chaperone   2, 94, 96–104, 171, 174–177, 180–186, 204, 308, 309, 314
– co-chaperone   179, 180, 184
– co-inducers   428, 429
– cross-priming   309
– innate stimulatory   310
– molecular chaperone   199, 202–205, 207, 208, 261
– overload   422, 429
– peptide-binding   308
– protein holding   314
chemical chaperone   238, 405
– lysophosphatidic acid   410
– osmolyte   405
– PBA   409
chemotaxis   372
chemotherapy   334
CHIP   95, 128, 129, 230, 243
chromatin   162
chromatin remodeling   102
chronic wasting disease (CWD)   232
client proteins   335
clinical trial   318
– colorectal cancer   318
– infectious disease   319
– melanoma   318
ClpB   208
conformation   360
Creutzfeldt-Jakob disease (CJD)   201, 210, 232
crystallin   97
cyclinE-Cdk2   366
cyclophilin   361
cyclosporin A   363
CYP3A4   344
Cystic fibrosis   333
cytochrome c   172, 176–178, 181, 182, 185
cytokine   156, 365
cytostatic   347

DAXX   95
death   171, 172, 176, 178, 179, 183, 185
– apoptotic   176, 178
dementia   389
deposits   199, 200, 203, 210, 211

dermatitis   376
development   81
dexamethasone   176, 189
diabetes   82, 429
differentiation   76, 80, 81
disaggregation   199, 208
DISC   179
disease   199–202, 204–209, 211, 212
– autoimmune   364
– graft-versus-host   364
– inflammatory   372
DJ-1   231
domain   370, 373
– catalytic   373
– tetratricopeptide repeat   370
dose limiting toxicity   344
drug development   342, 343
– phase I   342, 343, 345, 347, 348
– phase II   342, 343, 345, 347, 348
– phase III   342, 343
dynactin   125
dynamitin   125
dynein   123, 125, 381

endocytosis   366
endothelial nitric oxide synthase (eNOS)   420
endotoxic shock   421
epidermal growth factor receptor   382
ER   69–71, 73, 75–80, 82–85
– ER chaperone   70
– ER stress   72, 74–80, 82–84
– ER-associated degradation (ERAD)   71, 78, 79, 83, 237
– ERSE   76
– rough ER   70, 83
estrogen   421

feedback regulation   43
fgf   163
fibril   200, 212, 213
FK506   363
FK506-binding proteins   361
folding disease   406
– alpha-1 antitrypsin deficiency   406
– Alzheimer's disease   406
– cystic fibrosis   406
– Fabry disease   412
– Menkes disease   412

Subject Index

gain-of-function   365
geldanamycin   95, 99–101, 112, 125, 128, 181, 188, 240, 259
genetic plasticity   260
Gerstmann-Sträussler-Scheinker syndrome (GSS)   232
glucocorticoid receptor   111–138
glucocorticoids   369
glucose   81, 82
glycosylphosphatidylinositol (GPI) anchor   235
GroEL   237
GroES   237
growth factor   99
– receptor   99
Grp94   240, 424

hall mark traits (of cancer)   336
Hdj1   94
heat shock   94
– heat shock response   94, 96, 97
heat shock response modulators (allreviated as HSRM!)   422
– antibiotics   426
– arimoclomol   429
– aspirin   423
– bimoclomol   429
– calorie restriction   427
– curcumin   426
– cyclopentenone prostaglandins   426
– cyclosporine A   425
– estrogen   425
– geldanamycin   424
– geranylgeranylacetone   425
– glutamine   423
– herbal medicines   426
– herbimycin A   424
– lactacystin   424
– low-frequency electromagnetic fields   427
– MG-132   424
– radicicol   424
– sodium salicylate   423
– Zinc   423
hepatitis C   368
histamine   373
histone   102
Hop   97, 117, 119
hormesis   422
hormone   369
– steroid   369
hormone receptor   97, 98
– glucocorticoid receptor   98
– nuclear   97
– steroid   97
– vitamin D_3 receptor   98
Hsc70   98, 99, 143
HSF1   94–97, 104, 139, 142–146, 148, 149, 186, 226, 240, 418, 420, 424, 429
HSF1 phosphorylation   53
– DAXX   56
– negative controlled by phosphorylation   54
– phosphorylation of Ser^{230} 54
– phosphorylation of Ser^{326} 54
– promyelocytic leukemia oncogenic domains (PODs)   57
– role of phosphorylation in the HSF1 activation process   54
HSF2   139, 142, 143, 145, 146, 148, 149
HSF4   139, 142, 143, 145, 146
Hsp10   175
Hsp100   199, 208
Hsp104   208
Hsp105   143
Hsp110   174, 183, 186
Hsp25/27   96, 143, 144
Hsp27   101, 102, 104, 171, 174–178, 184–187, 189, 226, 242
Hsp28   95, 240
Hsp33   96
Hsp40   97, 231, 242–244
Hsp60   143, 174, 175, 180, 182–184, 186
Hsp70   3, 17, 94, 95, 97–101, 104, 111–138, 143–147, 171, 174, 175, 177–180, 183–189, 204, 205, 208, 225, 226, 230, 231, 242–244, 363
– Hsp70.1   146, 147
– Hsp70.3   146, 147
Hsp72   95, 96, 240
Hsp90   94, 95, 97–101, 111–138, 143, 145, 174, 175, 180–182, 186–188, 243, 259
HspB2   101, 141, 148
HspB5   141, 148
Hsps   94, 171, 174, 175, 184–189, 204
human immunodeficiency virus type-1   366
huntingtin   242

Huntington's disease (HD)   201, 206, 209–211, 213, 241, 333
Hymenolepis microstoma   369
hyperlipidemia   421
hypoxia   372

IκBα   177, 181
IAPs   172, 174
IL-2   365
immunity   310
- CTLs   310
- DCs   310
- NK   310
immunoglobulins   211
immunophilin   98, 99, 112, 363
- CyP-40   113
- FKBP52   113, 116, 123
- PP5   113
inclusion bodies   199, 207, 210
infections   369
infectivity   366
- protein   366
- viral   366
inhibition   371
insulin resistance   429
intermediates   200, 202–205, 213
ischaemic   334
ischemia reperfusion   420
isomerase   361
- peptidyl prolyl *cis/trans*   361
- secondary amide peptide bond *cis/trans*   361
isomerization   361

KF58333   338
kinase   176, 177, 179–181, 183
knockout   146–149, 155
KOS-953   338, 345

Lebers congenital amarosis   378
Legionnaires disease   378
Lens   162
Lewis bodies (LBs)   199, 207, 208
Lewy bodies   222
lipid networks   418
lupus erythematosus   376
lymphocyte   365

malaria, leishmaniasis, trypanosomiasis, schistosomiasis, and filariasis   369

mastocytosis   346
maximum tolerated dose   340, 342, 344, 345
MEFs   143, 144, 147
melanoma   344, 346, 347
metallothionein   424
microarray   146, 148
Mip protein   379
mitochondria   172, 173, 175, 176, 178, 179, 182, 183, 188
MKBP   *see* HspB2
molecular therapeutic   335
- imatinib trastuzumab   335
molecularly targeted therapeutics   260
multidrug-resistant   370
multitarget drugs   428
mycograb   339
myeloma   345, 347

National Cancer Institute   345
necrosis   366
nematodes   164
nephrotoxicity   365
nervous system   159
- brain   160
neurodegenerative disease   210
neurodegenerative disorder   206
neurofibrillary tangles (NFT)   243
neuroimmunophilins   377
neuropathy   429
neuroprotection   377
neuroregeneration   377
new variant CJD   233
NF-κB   177, 181, 182, 189
NK cells   279, 283, 286, 287, 292–295
- CTL   283, 285, 286
NQO1   344

oligomerization   175, 177, 181, 184
oncogenes   259
oogenesis   145
ovarian cancer   346

P-glycoprotein   370
p23   95, 98–100, 119, 126
p50cdc37   99, 100
p53   388
pancreatitis   429
parkin   227
Parkinson's disease (PD)   83, 84, 200, 206, 208–211, 222, 333, 377, 421

- α-synuclein   200
penile   377
permeability transition pore   366
pharmacodynamic (PD)   340, 342, 344, 349
pharmacokinetic (PK)   340, 342, 344, 349
pharmacological chaperone   410
- enzyme antagonist   410
- folding agonist   412
- hapten ligands   412
phosphorylation   96, 97, 99–102, 175–177, 179, 184, 186
pimecrolimus   376
Pin1   101, 386
plaque   200, 212, 213
plasmodium   334
pneumonia   378
poly(ADP-ribose) polymerase (PARP)   419
polyglutamine   213, 241, 244
polymorphism   387
- conformational   387
post-transcriptional aspects of feedback regulation   58
- disposal of stress-unfolded proteins by refolding or proteasome-mediated degradation   58
- Hsp mRNA stability and translatability   59
PPIases   361
prion   209, 232
Prion disease   210, 232, 333
- Creutzfeldt Jacob disease   333
Prion protein (PrP)   201, 206
proline   361
prostate   377
prostate cancer   345, 346
proteasome   128, 205, 207, 213, 225, 226, 239
protein
- capsid   366
- viral   366
protein kinase   96, 97, 99, 100, 102, 176, 177, 179–181, 183
- AKT   177, 180, 181
- B-RAF   347
- calcium/calmodulin-dependent kinase   97
- cascades   96, 97

- casein kinase 2   97
- CK2, casein   99, 100
- cyclin-dependent kinase 11   100
- ErbB-2   100
- ERK   96, 97, 179, 180
- focal adhesion kinase   99
- GSK3   97
- I-kappa-B kinase   100, 101
- IRAK1-1   100
- JNK   96, 97, 179–181
- MAPK   96, 97
- MK2   96
- mTOR kinase   375
- myotonic dystrophy protein kinase (DMPK)   101
- nucleophosmin-anaplastic lymphoma kinase   101
- p38   96, 101
- PKR   99, 100
- polo-like kinase   97, 100
- raf-1   99, 100, 104
- SAPK   96, 97
- Ste11   99, 100
- v-src   99
protein phosphatase   98, 101, 374, 386
protein phosphatase 2a   386
protofibril   224, 225
psoriasis   376
PU24FCl   338

quality control   199, 202, 206, 209

rapamycin   370
raptor   383
receptor   99, 310
- endocytic   310
- ErbB-2   99
- glucocorticoid receptor   101
- interleukin-1   100
- signaling   310
- toll-like   100
- VEGF   99
redundancy   164
renal cancer   346
replication   368
repression of HSF1 transcriptional activity   49
- Hsp70 and Hsp40 (Hdj1)   52
- Hsp90-containing multichaperone complex   49

repression of trimerization of HSF1   45
– Hsp90-containing multichaperone complexes   46
– repressor of HSF1 homotrimerization   46
restriction factor   367
retinopathy   421
RNA   100, 103, 104
ROS   96
Russell bodies   199, 207, 208, 210
ryanodine receptors   384

sanglifehrin   366
sauna   422
sclerosis   371
– familial amyotrophic lateral   371
scrapie   232
sepsis   372, 421
signals   96
– extracellular   96
siRNA   367
sirolimus   375
spermatogenesis   146, 157, 378
spinocerebellar ataxias   241
stress   171, 174, 176, 180, 183, 185, 365
– heat   174, 176
– oxidative   174, 176, 365
stress granules   57
stress protein   280, 282
– extracellular   282, 290
– membrane-bound   282
stress proteins   306
– glucose-regulated proteins   306
– heat shock protein   306
stress response   240
stroke   371
superoxide dismutase-1   429
Syndecan   382
synucleinopathies   223

T cells   279, 281, 289, 365
tau   243
tau-protein   388
tauopathies   243
TGFβ receptor   382
thyroid cancer   346
Tid1   99
TNF   140, 144, 147, 172, 176, 179–182
torsion   363
toxicity   202, 212, 213
– toxic   199, 212, 213
TRAIL   179
transcription   75–77, 81
transcription factor   98, 99
translation   73, 77, 81, 82
transmissible spongiform encephalopathies   232
transmissible spongiform encephalopathies (TSEs)   201, 206, 209, 210, 213
transplantation   368
Trypanosoma cruzi   369
tumor   172, 183, 185–187, 189
tumorigenicity   174, 178, 185

ubiquitin-proteasome pathway (UPP)   205, 206, 221
ubiquitin-proteasome system (UPS)   227, 241
unfolded protein response (UPR)   70–72, 75, 76, 78, 79, 81–85, 239
– ERSE   76
– UPRE   72

VER49009   338

Western blot   344
Williams Beuren syndrome   378
WW domain   386

Printing: Strauss GmbH, Mörlenbach
Binding: Schäffer, Grünstadt